Springer Texts in
Electrical Engineering

Consulting Editor: John B. Thomas

Springer Texts in Electrical Engineering

Multivariable Feedback Systems
F.M. Callier/C.A. Desoer

Linear Programming
M. Sakarovitch

Introduction to Random Processes
E. Wong

Stochastic Processes in Engineering Systems
E. Wong/B. Hajek

Introduction to Probability
J.B. Thomas

Elements of Detection and Signal Design
C.L. Weber

An Introduction to Communication Theory and Systems
J.B. Thomas

Signal Detection in Non-Gaussian Noise
S.A. Kassam

An Introduction to Signal Detection and Estimation
H.V. Poor

H. Vincent Poor

An Introduction to Signal Detection and Estimation

With 47 Illustrations

Springer-Verlag
New York Berlin Heidelberg
London Paris Tokyo

A Dowden &
Culver Book

H. Vincent Poor
Department of Electrical
and Computer Engineering
University of Illinois at
Urbana-Champaign
Urbana, IL 61801
USA

Library of Congress Cataloging-in-Publication Data
Poor, H. Vincent.
An introduction to signal detection and estimation / H. Vincent Poor.
p. cm.—(Springer texts in electrical engineering)
Adapted from a one semester graduate course taught at the University of Illinois.
Bibliography: p.
Includes index.
ISBN 0-387-96667-6
1. Signal theory (Telecommunication) 2. Estimation theory.
I. Title. II. Series.
TK5102.5.P654 1988
621.38′043—dc19 87-36920

Camera-ready copy provided by the author.
Printed and bound by R.R. Donnelley & Sons, Harrisonburg, Virginia.
Printed in the United States of America.

9 8 7 6 5 4 3 2 1

ISBN 0-387-96667-6 Springer-Verlag New York Berlin Heidelberg
ISBN 3-540-96667-6 Springer-Verlag Berlin Heidelberg New York

To my family

PREFACE

The purpose of this book is to introduce the reader to the basic theory of signal detection and estimation. It is assumed that the reader has a working knowledge of applied probability and random processes such as that taught in a typical first-semester graduate engineering course on these subjects. This material is covered, for example, in the book by Wong (1983) in this series. More advanced concepts in these areas are introduced where needed, primarily in Chapters VI and VII, where continuous-time problems are treated.

This book is adapted from a one-semester, second-tier graduate course taught at the University of Illinois. However, this material can also be used for a shorter or first-tier course by restricting coverage to Chapters I through V, which for the most part can be read with a background of only the basics of applied probability, including random vectors and conditional expectations. Sufficient background for the latter option is given for example in the book by Thomas (1986), also in this series.

This treatment is also suitable for use as a text in other modes. For example, two smaller courses, one on signal detection (Chapters II, III, and VI) and one on estimation (Chapters IV, V, and VII), can be taught from the material as organized here. Similarly, an introductory level course (Chapters I through IV) followed by a more advanced course (Chapters V through VII) is another possibility.

TABLE OF CONTENTS

Page

I INTRODUCTION

Generally speaking, signal detection and estimation is the area of study that deals with the processing of information-bearing signals for the purpose of extracting information from them. Applications of the theory of signal detection and estimation are found in many areas, such as communications and automatic control. For example, in communications applications such as data transmission or radar, detection and estimation provides the theoretical and analytical basis for the design of effective communication receivers. Alternatively, in automatic control applications, detection and estimation theory leads to techniques for making accurate inferences concerning the conditions present in a process or system to be controlled.

An example of an application in which detection and estimation techniques are useful is radar, in which one transmits a pulsed electromagnetic signal and then waits for a possible return signal to be reflected from a target. Due to electrical receiver noises, atmospheric disturbances, spurious reflections from the ground and other objects, and other signal distortions, it is usually not possible to determine with absolute certainty whether or not a target is present. Thus one must infer from the (imperfect) observation of the receiving antenna output whether or not a target is present, and detection theory provides a means for choosing a good technique for doing so. Furthermore, having determined with some degree of certainty that a target is present, one may then wish to estimate some characteristics of the target such as its position and velocity, a task that falls within the general context of estimation theory. Such estimates may then be useful in controlling the antenna to track the target or in remotely controlling the target itself to maintain a desired trajectory. Other specific applications in which detection and estimation techniques are useful include seismology, radio astronomy, sonar, speech and image processing, medical signal processing, and optical communications.

In general, detection and estimation applications involve making inferences from observations that are distorted or corrupted in some unknown manner. Moreover, the information that one wishes to extract from such observations is *a fortiori* unknown to the observer. Thus it is very useful to cast detection and estimation problems in a probabilistic framework in which unknown behavior is assumed to be random. In this light, detection and estimation theory fits properly within the province of statistical inference, and this is the interpretation to be used throughout this treatment.

Basic to the study of signal detection and estimation theory is the concept of a random observation Y taking values in an observation set Γ, which may be a set of vectors, waveforms, real numbers, or any other set. From our observation of Y we wish to extract information about some phenomenon related to Y. There are two types of such problems in which we are interested: *detection* problems, in which we wish to decide among some finite number of possible situations or "states of nature," and *estimation* problems, in which we wish to estimate the value(s) of some quantity or quantities that are not observed directly. In either case the connection between the observation and the desired information is probabilistic rather than direct, in the sense that the statistical behavior of Y is influenced by the state of nature or the values of quantities to be estimated. Thus a model for this situation must involve a family of probability distributions on Γ, the members of which correspond to statistical conditions present under the various states of nature or under the various values of the quantities to be estimated. Given such a model, the detection and estimation problem is to find an optimum way of processing the observation Y in order to extract the desired information. The basic features that distinguish such problems from one another are the nature of the desired information (discrete or continuous), the amount of *a priori* knowledge that we have about the desired quantities or state of nature, and the performance criteria by which we compare various detection and estimation procedures.

It is the purpose of this book to introduce the reader to the fundamentals of detection and estimation theory. Chapters II, III, and VI deal with problems of signal detection.

Chapter II presents the basics of binary hypothesis testing, which provide the basis for most signal detection problems. In Chapter III these basics are applied to derive optimum procedures for models specific to signal detection problems, and the performance analysis of these procedures is also considered in this chapter. Chapter III also deals with several special signal detection methods that can be applied to problems of detecting signals in nonstandard situations. Chapters II through V deal primarily with situations in which the observations are vectors. This case corresponds to signal detection with discrete-time (i.e., sampled) observations. The problem of signal detection with continuous-time observations is treated in Chapter VI. This problem, although philosophically the same as the discrete-time case, is treated separately because of the more advanced analytical methods required in its analysis. Chapters IV, V, and VII deal with problems of estimation. In Chapter IV the elements and structure of parameter estimation problems are explored, while Chapters V and VII treat the problem of signal estimation. As in the signal detection case, discrete-time estimation (Chapters IV and V) and continuous-time estimation (Chapter VII) are treated separately because of the disparity in analytical difficulty between the two problems.

A Note on Notation

The specification of a probability distribution on the observation set Γ requires the assignment of probabilities to subsets of Γ. For some observation spaces of interest it is not possible to assign consistent probabilities to all subsets of Γ; thus we will always associate with Γ a class G of subsets of Γ to which we wish to assign probabilities. The sets in G are called *observation events*, and the pair (Γ, G) is termed the *observation space*. For analytical reasons we will always assume that the collection G is a σ-algebra; that is, we assume that G contains all complements (relative to Γ) and countable (i.e., denumerable) unions of its members.†

Throughout most of this book we will be interested in two cases for (Γ, G). The first is the case in which $\Gamma = \mathbb{R}^n$, the set of n-dimensional vectors with real components, and the

---

† In other words, G has the property that $A \in G$ implies $A^c \in G$ (here, and throughout, a superscript c denotes complementation) and that $A_1, A_2, \ldots \in G$ implies $\cup_{i=1}^{\infty} A_i \in G$.

second is the case in which Γ is a discrete (i.e., countable) set, $\Gamma=\{\gamma_1, \gamma_2, ...\}$. In the first of these cases, it is natural to wish to assign probabilities to sets of the form $\{y=(y_1, ..., y_n)^T \in \mathbb{R}^n | a_1 \leqslant y_1 \leqslant b_1, ..., a_n \leqslant y_n \leqslant b_n\}$, where the a_i's and b_i's are arbitrary real numbers. Thus, for $\Gamma=\mathbb{R}^n$, we will take G be the smallest σ-algebra containing all of these sets with the a_i's and b_i's ranging throughout the reals. This σ-algebra is usually denoted by B^n and is termed the class of *Borel sets* in $\mathbb{R}^n$. For the second of these two observation sets it is possible to take G be the set of *all* subsets of Γ. This σ-algebra is usually denoted by 2^Γ and is called the *power set* of Γ. These two observation spaces will be sufficient for considering most of the problems of discrete-time detection and estimation treated in Chapters II through V. Until otherwise stated, we will always assume that (Γ, G) is one of these two cases. More abstract observation spaces are required for treatment of continuous-time problems, and these will be introduced as needed in Chapters VI and VII.

For the discrete observation space $(\Gamma, 2^\Gamma)$, probabilities can be assigned to subsets of Γ in terms of a *probability mass function*, $p : \Gamma \rightarrow [0, 1]$, by way of

$$P(A) = \sum_{\gamma_i \in A} p(\gamma_i), \; A \in 2^\Gamma, \tag{I.1}$$

where $P(A)$ denotes the probability that the observation Y lies in the set A. Any function mapping Γ to $[0, 1]$ is a legitimate probability mass function provided that it satisfies the condition $\sum_{i=1}^{\infty} p(\gamma_i)=1$. For the observation space $(\mathbb{R}^n, B^n)$, we will be interested primarily in so-called *continuous random vectors* for which probabilities can be assigned in terms of a *probability density function*, $p : \mathbb{R}^n \rightarrow [0, \infty)$, by way of

$$P(A) = \int_A p(y)dy, \; A \in B^n. \tag{I.2}$$

(Note that the integral in (I.2) is n-fold.) Any integrable function mapping $\mathbb{R}^n$ to $[0, \infty)$ is a legitimate probability density function provided that it satisfies the condition $\int_{\mathbb{R}^n} p(y)dy=1$. For compactness of terminology and notation we will use the term *density* for both the probability mass function and the probability density function, and we will use the notation

$$P(A) = \int_A p(y)\mu(dy) \tag{I.3}$$

to denote both the sum of (I.1) and the n-fold integral of (I.2). Where the variable of integration is understood, we will sometimes use the alternative notation

$$P(A) = \int_A p \, d\mu. \tag{I.4}$$

For a real-valued function g of the random observation Y, we are often interested in the *expected value* of $g(Y)$, denoted by $E\{g(Y)\}$. In the case of a discrete observation space $(\Gamma, 2^\Gamma)$ this quantity is given by

$$E\{g(Y)\} = \sum_{i=1}^{\infty} g(\gamma_i)p(\gamma_i), \tag{I.5}$$

and in the case of a continuous random vector in $(\mathbb{R}^n, B^n)$ we have

$$E\{g(Y)\} = \int_{\mathbb{R}^n} g(y)p(y)dy, \tag{I.6}$$

where in each case we have assumed the existence of the required sum or integral. Again, for compactness of notation, we will use the following notations for both (I.5) and (I.6):

$$E\{g(Y)\} = \int_\Gamma g(y)p(y)\mu(dy) = \int_\Gamma gp \, d\mu. \tag{I.7}$$

Further meaning will be given to this notation in Chapter VI. Note that (I.3) and (I.4) are special cases of (I.7) with g given by

$$g(y) = \begin{cases} 1, \text{ if } y \in A \\ \\ 0, \text{ if } y \in A^c. \end{cases} \tag{I.8}$$

Throughout this treatment we will use uppercase letters to denote random quantities and lowercase letters to denote specific values of those quantities. Thus the random observation Y may take on the value y.

II ELEMENTS OF HYPOTHESIS TESTING

II.A Introduction

Most signal detection problems can be cast in the framework of *M-ary hypothesis testing*, in which we have an observation (possibly a vector or function) on the basis of which we wish to decide among M possible statistical situations describing the observations. For example, in an M-ary communications receiver we observe an electrical waveform that consists of one of M possible signals corrupted by random channel or receiver noise, and we wish to decide which of the M possible signals is present. Obviously, for any given decision problem, there are a number of possible decision strategies or rules that could be applied; however, we would like to choose a decision rule that is optimum in some sense. There are several useful definitions of optimality for such problems, and in this chapter we consider the three most common formulations - Bayes, minimax, and Neyman-Pearson - and derive the corresponding optimum solutions. In general, we consider the particular problem of binary ($M=2$) hypothesis testing, although the extension of many of the results of this chapter to the general M-ary case is straightforward and will be developed in the exercises. The application of this theory to those models specific to signal detection is considered in detail in Chapters III and VI.

II.B Bayesian Hypothesis Testing

The primary problem that we consider in this chapter is the simple hypothesis-testing problem in which we assume that there are two possible *hypotheses* or "states or nature," H_0 and H_1, corresponding to two possible probability distributions P_0 and P_1, respectively, on the observation space (Γ, G). We may write this problem as

$$H_0: Y \sim P_0$$
$$\text{versus} \qquad \text{(II.B.1)}$$
$$H_1: Y \sim P_1,$$

where the notation "$Y \sim P$" denotes the condition "Y has distribution P." The hypotheses H_0 and H_1 are sometimes referred to as the *null* and *alternative* hypotheses, respectively. A *decision rule* (or *hypothesis test*) δ for H_0 versus H_1 is any partition of the observation set Γ into sets $\Gamma_1 \in G$ and $\Gamma_0 = \Gamma_1^c$ such that we choose H_j when $y \in \Gamma_j$ for $j=0$ or 1. The set Γ_1 is sometimes known as the *rejection region* (or *critical region*) and Γ_0 as the *acceptance region*. We can also think of the decision rule δ as a function on Γ given by

$$\delta(y) = \begin{cases} 1 \text{ if } y \in \Gamma_1 \\ \\ 0 \text{ if } y \in \Gamma_1^c, \end{cases} \qquad \text{(II.B.2)}$$

so that the value of δ for a given $y \in \Gamma$ is the index of the hypothesis accepted by δ.

We would like to choose Γ_1 in some optimum way and, with this in mind, we would like to assign costs to our decisions; in particular, we will assume for now that we have positive numbers C_{ij} for $i=0$, 1 and $j=0$, 1, where C_{ij} is the cost incurred by choosing hypothesis H_i when hypothesis H_j is true. We can then define the *conditional risk* for each hypothesis as the average or expected cost incurred by decision rule δ when that hypothesis is true; i.e.,

$$R_j(\delta) = C_{1j} P_j(\Gamma_1) + C_{0j} P_j(\Gamma_0),\ j = 0, 1. \qquad \text{(II.B.3)}$$

Note that $R_j(\delta)$ is the cost of choosing H_1 when H_j is true times the probability of doing so, plus the cost of choosing H_0 when H_j is true times the probability of doing this.

Now assume further that we can also assign probabilities π_0 and $\pi_1 = (1-\pi_0)$ to the occurrences of hypotheses H_0 and H_1, respectively. That is, π_j is the probability that hypothesis H_j is true unconditioned on the value of Y. These probabilities π_0 and π_1 are known as the *prior* or *a priori* probabilities of the two hypotheses. For given priors we can define an average or *Bayes risk* as the overall average cost incurred by decision rule δ. This quantity is given by

$$r(\delta) = \pi_0 R_0(\delta) + \pi_1 R_1(\delta). \tag{II.B.4}$$

We may now define an optimal decision rule for H_0 versus H_1 as one that minimizes, over all decision rules, the Bayes risk. Such a decision rule is known as a *Bayes rule* for H_0 versus H_1.

Note that (II.B.3) and (II.B.4) can be combined to give

$$\begin{aligned} r(\delta) &= \sum_{j=0}^{1} \pi_j [C_{0j}(1-P_j(\Gamma_1)) + C_{1j}P_j(\Gamma_1)] \\ &= \sum_{j=0}^{1} \pi_j C_{0j} + \sum_{j=0}^{1} \pi_j (C_{1j} - C_{0j}) P_j(\Gamma_1), \end{aligned} \tag{II.B.5}$$

where we have used the fact that $P_j(\Gamma_1^c) = 1 - P_j(\Gamma_1)$. Assuming that P_j has density p_j for $j=0, 1$, and using the notation introduced in Chapter I, (II.B.5) implies that

$$\begin{aligned} r(\delta) &= \sum_{j=0}^{1} \pi_j C_{0j} \\ &+ \int_{\Gamma_1} \left[\sum_{j=0}^{1} \pi_j (C_{1j} - C_{0j}) p_j(y) \right] \mu(dy), \end{aligned} \tag{II.B.6}$$

and thus we see that $r(\delta)$ is a minimum over all Γ_1 if we choose

$$\Gamma_1 = \{y \in \Gamma | \sum_{j=0}^{1} \pi_j (C_{1j} - C_{0j}) p_j(y) \leqslant 0\}$$

$$= \{y \in \Gamma | \pi_1(C_{11} - C_{01}) p_1(y) \leqslant \pi_0(C_{00} - C_{10}) p_0(y)\}. \qquad \text{(II.B.7)}$$

Assuming that $C_{11} < C_{01}$ (i.e., that the cost of correctly choosing H_1 is less than the cost of incorrectly rejecting H_1), (II.B.7) can be rewritten as

$$\Gamma_1 = \{y \in \Gamma | p_1(y) \geqslant \tau p_0(y)\} \qquad \text{(II.B.8)}$$

where

$$\tau \triangleq \frac{\pi_0(C_{10} - C_{00})}{\pi_1(C_{01} - C_{11})}. \qquad \text{(II.B.9)}$$

Note that the region $\{y \in \Gamma | p_1(y) = \tau p_0(y)\}$ does not contribute to the average error and thus can be omitted in whole or in part from Γ_1 if desired without affecting the risk incurred.

The decision rule described by the rejection region of (II.B.8) is known as a *likelihood-ratio test* (or *probability-ratio test*); this test plays a central role in the theory of hypothesis testing. Note that Γ_1 of (II.B.8) can be rewritten as

$$\Gamma_1 = \{y \in \Gamma | [p_1(y)/p_0(y)] \geqslant \tau\}, \qquad \text{(II.B.10)}$$

where we interpret $(k/0)$ as ∞ for any $k \geqslant 0$. The quantity

$$L(y) = \frac{p_1(y)}{p_0(y)}, \; y \in \Gamma, \qquad \text{(II.B.11)}$$

is known as the *likelihood ratio* (or the *likelihood-ratio statistic*) between H_0 and H_1. Thus the Bayes decision rule corresponding to (II.B.8) computes the likelihood ratio for the observed value of Y and then makes its decision by comparing this ratio to the threshold τ; i.e., a Bayes rule for (II.B.1) is

$$\delta_B(y) = \begin{cases} 1 \text{ if } L(y) \geqslant \tau \\ 0 \text{ if } L(y) < \tau. \end{cases} \qquad \text{(II.B.12)}$$

A commonly used cost assignment is the *uniform* cost assignment given by

$$C_{ij} = \begin{cases} 0 \text{ if } i = j \\ 1 \text{ if } i \neq j. \end{cases} \qquad \text{(II.B.13)}$$

The Bayes risk for a decision rule δ with critical region Γ_1 is given in this case by

$$r(\delta) = \pi_0 P_0(\Gamma_1) + P_1(\Gamma_0). \qquad \text{(II.B.14)}$$

Note that $P_i(\Gamma_j)$ is the probability of choosing H_j when H_i is true. Thus $P_i(\Gamma_j)$ for $i \neq j$ is the conditional probability of making an error given that H_i is true, and so in this case $r(\delta)$ is the *average probability of error* incurred by the decision rule δ. Since the likelihood-ratio test with $\tau = (\pi_0 / \pi_1)$ minimizes $r(\delta)$ for the cost structure of (II.B.13), it is thus a *minimum-probability-of-error* decision scheme.

Bayes' formula [see, e.g., Thomas (1986)] implies that the conditional probability that hypothesis H_j is true given that the random observation Y takes on value y is given by

$$\pi_j(y) = P(H_j \text{ } true \,|Y = y) = \frac{p_j(y)\pi_j}{p(y)}, \qquad \text{(II.B.15)}$$

where $p(y)$ is the average or overall density of Y given by $p(y) = \pi_0 p_0(y) + \pi_1 p_1(y)$. The probabilities $\pi_0(y)$ and $\pi_1(y)$ are called the *posterior* or *a posteriori* probabilities of the two hypotheses. By using (II.B.15), the critical region of the Bayes rule (II.B.7) can be rewritten as

$$\Gamma_1 = \{y \in \Gamma | C_{10}\pi_0(y) + C_{11}\pi_1(y) \leqslant C_{00}\pi_0(y) + C_{01}\pi_1(y)\}. \tag{II.B.16}$$

Note that the quantity

$$C_{i0}\pi_0(y) + C_{i1}\pi_1(y) \tag{II.B.17}$$

is the average cost incurred by choosing hypothesis H_i given that Y equals y. This quantity is called the *posterior cost* of choosing H_i when the observation is y, and thus the Bayes rule makes its decision by choosing the hypothesis that yields the minimum posterior cost. For example, for the uniform cost criterion (II.B.13), the Bayes rule can be thus written as

$$\delta_B(y) = \begin{cases} 1 \text{ if } \pi_1(y) \geqslant \pi_0(y) \\ 0 \text{ if } \pi_1(y) < \pi_0(y). \end{cases} \tag{II.B.18}$$

Thus the minimum-probability-of-error decision rule chooses the hypothesis that has the maximum *a posteriori* probability of having occurred give that $Y=y$. This decision rule is sometimes known as the *MAP* decision rule for the binary hypothesis test (II.B.1).

The following simple examples will serve to illustrate the Bayes decision rule.

Example II.B.1: The Binary Channel

Suppose that a binary digit (i.e., a "zero" or a "one") is to be transmitted over a communication channel. Our observation Y is the output of the channel, which can be either 0 or 1. Because of channel noises and imperfect modulation or demodulation, a transmitted "zero" is received as a 1 with probability λ_0 and as a 0 with probability $(1-\lambda_0)$, where $0<\lambda_0<1$. Similarly, a transmitted "one" is received as a 0 with probability λ_1 and as a 1 with probability $(1-\lambda_1)$. (These relationships are depicted in Fig. II.B.1.) Thus, observing Y does not tell us exactly whether the transmitted digit was a "zero" or a

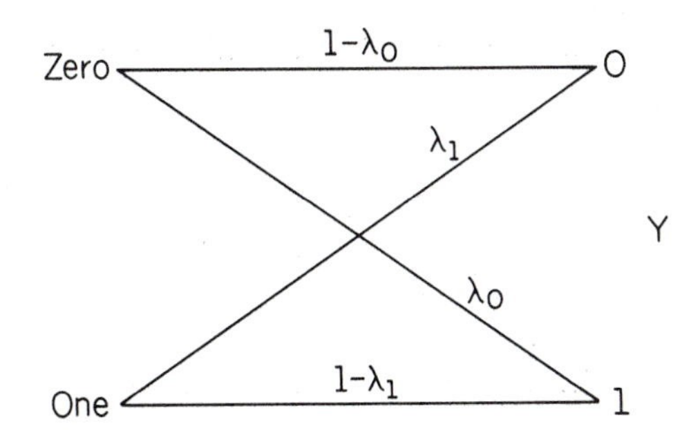

Fig. II.B.1: The binary channel.

"one," and we would like to find an optimum way to decide what was transmitted.

This situation can be modeled as a binary hypothesis testing problem in which the hypothesis H_j is that a "j" was transmitted ($j=0, 1$), the observation set Γ is $\{0, 1\}$, and the observation Y has densities (i.e., probability mass functions)

$$p_j(y) = \begin{cases} \lambda_j \text{ if } y \neq j \\ (1-\lambda_j) \text{ if } y = j \end{cases} \tag{II.B.19}$$

for $j=0$ and 1. The likelihood ratio is thus given by

$$L(y) = \frac{p_1(y)}{p_0(y)} = \begin{cases} \dfrac{\lambda_1}{1-\lambda_0} \text{ if } y = 0 \\ \\ \dfrac{1-\lambda_1}{\lambda_0} \text{ if } y = 1. \end{cases} \tag{II.B.20}$$

For a Bayes test, the test threshold τ is determined by the costs and prior probabilities from (II.B.9). If λ_1, λ_0, and τ are such that $\lambda_1 \geq \tau(1-\lambda_0)$, the likelihood-ratio test of (II.B.12) interprets a received 0 as a transmitted "one"; otherwise, a received 0 is interpreted as a transmitted "zero." Similarly, if $(1-\lambda_1) \geq \tau\lambda_0$, the likelihood-ratio test interprets a received 1 as a transmitted "one," and if $(1-\lambda_1) < \tau\lambda_0$, a received 1 is interpreted as a transmitted "zero." The minimum Bayes risk $r(\delta_B)$ for this example can be computed straightforwardly from (II.B.5) (see Exercise 1).

For example, with uniform costs and equal priors ($\pi_0 = \pi_1 = ½$), we have $\tau = 1$ and the Bayes rule of (II.B.12) becomes

$$\delta_B(0) = \begin{cases} 1 \text{ if } \lambda_1 \geq (1-\lambda_0) \\ \\ 0 \text{ if } \lambda_1 < (1-\lambda_0). \end{cases} \tag{II.B.21a}$$

$$\delta_B(1) = \begin{cases} 1 \text{ if } (1-\lambda_1) \geq \lambda_0 \\ \\ 0 \text{ if } (1-\lambda_1) < \lambda_0. \end{cases} \tag{II.B.21b}$$

As noted above, boundary points where $L(y) = \tau$ can be assigned to either Γ_1 or Γ_0. Thus an equivalent (in terms of risk) Bayes test for this case is given by

$$\delta_B(y) = \begin{cases} y \text{ if } (1-\lambda_1) \geq \lambda_0 \\ \\ 1-y \text{ if } (1-\lambda_1) < \lambda_0. \end{cases} \tag{II.B.22}$$

If we further assume a *symmetric* channel $\lambda_1 = \lambda_0 = \lambda$, then (II.B.22) becomes

$$\delta_B(y) = \begin{cases} y \text{ if } \lambda \leq ½ \\ \\ 1-y \text{ if } \lambda > ½. \end{cases} \tag{II.B.23}$$

The interpretation of (II.B.23) is straightforward--if the channel is more likely than not to invert bits (i.e., $\lambda > \frac{1}{2}$), we make our decision by inverting the received bit. Otherwise, we accept the received bit as being correct. For the latter situation, the minimum Bayes risk turns out to be

$$r(\delta_B) = \min\{\lambda, 1-\lambda\}. \tag{II.B.24}$$

Thus the performance improves as the channel becomes more reliable in either transmitting the bit directly or in inverting the transmitted bit. Note that because of uniform costs and equal priors, simply guessing the transmitted bit without observing y yields a risk of ½. So if $\lambda = \frac{1}{2}$, the observation is worthless.

Example II.B.2: Location Testing with Gaussian Error

Consider the following two hypotheses concerning a real-valued observation Y:

$$\begin{aligned} &H_0: Y = \epsilon + \mu_0 \\ \text{versus} \\ &H_1: Y = \epsilon + \mu_1 \end{aligned} \tag{II.B.25}$$

where ϵ is a Gaussian random variable with zero mean and variance σ^2, and where μ_0 and μ_1 are two fixed numbers with $\mu_1 > \mu_0$. Note that the addition of μ_0 or μ_1 to ϵ changes only the mean value of the observation, so that we are testing about which of two possible values or "locations" the observation is distributed. Applications of a more general form of this simple model will be discussed later. In terms of distributions on the observation space, the hypothesis pair of (II.B.25) can be rewritten as

$$\begin{aligned} &H_0: Y \sim N(\mu_0, \sigma^2) \\ \text{versus} \\ &H_1: Y \sim N(\mu_1, \sigma^2), \end{aligned} \tag{II.B.26}$$

where $N(\mu, \sigma^2)$ denotes the Gaussian (or normal) distribution with mean μ and variance σ^2. [Recall that a $N(\mu, \sigma^2)$

random variable is one with probability density function

$$(1/\sqrt{2\pi}\sigma)\exp\{(x-\mu)^2/2\sigma^2\},\ x\in\mathbb{R}.]$$

The likelihood ratio for (II.B.26) is given by

$$L(y)=\frac{p_1(y)}{p_0(y)}=\frac{\frac{1}{\sqrt{2\pi}\sigma}e^{-(y-\mu_1)^2/2\sigma^2}}{\frac{1}{\sqrt{2\pi}\sigma}e^{-(y-\mu_0)^2/2\sigma^2}}$$

$$=\exp\left\{\frac{\mu_1-\mu_0}{\sigma^2}\left(y-\frac{\mu_0+\mu_1}{2}\right)\right\}. \tag{II.B.27}$$

Thus a Bayes test for (II.B.26) is

$$\delta_B(y)=\begin{cases}1 \text{ if } \exp\left\{\frac{\mu_1-\mu_0}{\sigma^2}\left(y-\frac{\mu_0+\mu_1}{2}\right)\right\}\geqslant\tau \\ 0 \ \text{ otherwise}\end{cases} \tag{II.B.28}$$

where τ is the appropriate threshold. Since $\mu_1>\mu_0$, the likelihood ratio of (II.B.27) is a strictly increasing function of the observation y (i.e., $dL(y)/dy=(\mu_1-\mu_0)L(y)/\sigma^2>0$). So comparing $L(y)$ to the threshold τ is equivalent to comparing y itself to another threshold $\tau'=L^{-1}(\tau)$, where L^{-1} is the inverse function of L. In particular, taking logarithms in the inequality of (II.B.28) and rearranging terms yields

$$\delta_B(y) = \begin{cases} 1 \text{ if } y \geqslant \tau' \\ \\ 0 \text{ if } y < \tau', \end{cases} \tag{II.B.29}$$

where

$$\tau' = \frac{\sigma^2}{\mu_1 - \mu_0} \log(\tau) + \frac{\mu_0 + \mu_1}{2}. \tag{II.B.30}$$

For example, with uniform costs and equal priors we have $\tau=1$ and $\tau'=(\mu_0+\mu_1)/2$. Thus, in this particular case, the Bayes rule compares the observation to the average of μ_0 and μ_1. If y is greater than or equal to the average, we choose H_1; if y is less than this average, we choose H_0. This test is illustrated in Fig. II.B.2.

The minimum Bayes risk, $r(\delta_B)$, for this problem can be computed from (II.B.5) if we have $P_j(\Gamma_1)$ for $j=0, 1$. Since $\Gamma_1=\{y \in \mathbb{R} | y \geqslant \tau'\}$, we have that

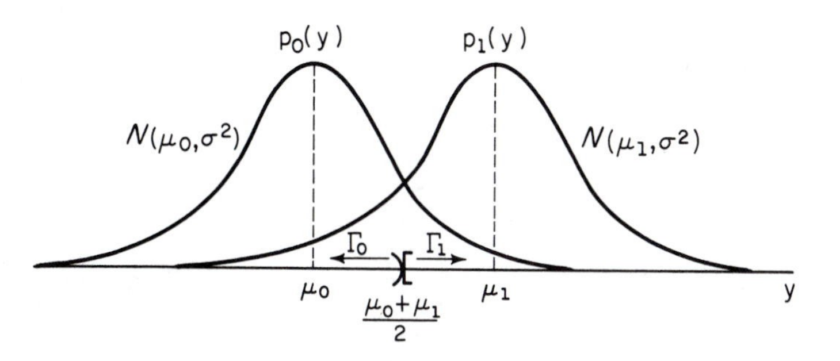

Fig. II.B.2: Illustration of location testing with Gaussian errors, uniform costs, and equal priors.

$$P_j(\Gamma_1) = \int_{\tau'}^{\infty} p_j(y)dy = 1-\Phi\left(\frac{\tau'-\mu_j}{\sigma}\right)$$

$$= \begin{cases} 1-\Phi\left(\dfrac{\log\tau}{d}+\dfrac{d}{2}\right), & j=0 \\ 1-\Phi\left(\dfrac{\log\tau}{d}-\dfrac{d}{2}\right), & j=1, \end{cases} \qquad \text{(II.B.31)}$$

where Φ denotes the cumulative probability distribution function (cdf) Of a $N(0,1)$ random variable, and where $d=(\mu_1-\mu_0)/\sigma$. If we again consider the particular case of uniform costs and equal priors, we have straightforwardly that

$$r(\delta_B) = 1-\Phi(d/2), \qquad \text{(II.B.32)}$$

which is graphed in Fig. II.B.3 as a function of d. Note that the performance improves monotonically as the separation in means, $(\mu_1-\mu_0)$, increases relative to the standard deviation of

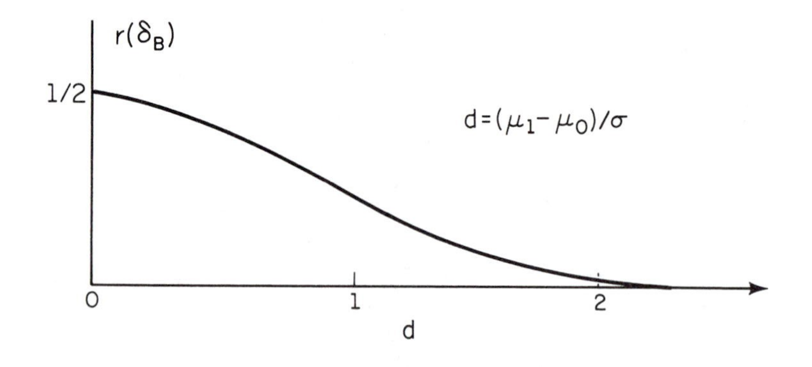

Fig. II.B.3: Bayes risk in location testing with Gaussian error.

the error, σ. This quantity d is a simple version of a signal-to-noise ratio, and will be given further meaning in subsequent chapters.

Examples II.B.1 and II.B.2, although quite simple, illustrate the basic principles of Bayesian hypothesis testing. Further examples will be discussed in Chapters III and VI, and a number of related exercises are included at the end of this chapter.

The primary result of this section is that the Bayes definition of optimality yields the likelihood ratio test (II.B.12) as optimal when the costs C_{ij} and priors π_i are specified. In the following sections we consider other definitions of optimality for situations in which the priors and/or the costs are unknown.

II.C Minimax Hypothesis Testing

Suppose that in the formulation of Section II.B, the prior probabilities π_0 and π_1 are unknown to the designer. Such situations can arise frequently in practice since the designer of a decision rule may not have control over or access to the mechanism generating the state of nature. In such cases the average or Bayes risk is not an acceptable design criterion since it is unlikely that a single decision rule would minimize the average risk for every possible prior distribution. Thus in this case it is necessary to seek an alternative design criterion. One such criterion is to seek a decision rule that minimizes, over all δ, the maximum of the conditional risks, $R_0(\delta)$ and $R_1(\delta)$; i.e., a possible design criterion is

$$\max\{R_0(\delta), R_1(\delta)\}. \tag{II.C.1}$$

A decision rule minimizing the quantity in (II.C.1) is known as a *minimax* rule, and in this subsection we discuss the structure of such rules.

To seek a decision rule minimizing the quantity in (II.C.1), it is useful to consider the function $r(\pi_0, \delta)$, defined for a given prior $\pi_0 \epsilon [0, 1]$ and decision rule δ as the average risk,

$$r(\pi_0, \delta) = \pi_0 R_0(\delta) + (1-\pi_0) R_1(\delta). \qquad \text{(II.C.2)}$$

Note that as a function of π_0, $r(\pi_0, \delta)$ is a straight line from $r(0, \delta) = R_1(\delta)$ to $r(1, \delta) = R_0(\delta)$, as depicted in Fig. II.C.1. Thus, for fixed δ, the maximum value of $r(\pi_0, \delta)$ as π_0 ranges from 0 to 1 occurs at either $\pi_0 = 0$ or $\pi_0 = 1$, and the maximum value is $\max\{R_0(\delta), R_1(\delta)\}$. So the problem of minimizing (II.C.1) over δ is the same as that of minimizing the quantity

$$\max_{0 \leqslant \pi_0 \leqslant 1} r(\pi_0, \delta) \qquad \text{(II.C.3)}$$

over δ. The latter quantity is more convenient to consider.

For each prior $\pi_0 \in [0, 1]$, let δ_{π_0} denote a Bayes rule corresponding to that prior, and let $V(\pi_0) = r(\pi_0, \delta_{\pi_0})$; that is, $V(\pi_0)$ is the minimum possible Bayes risk for the prior π_0. It is straightforward to show that $V(\pi_0)$ is a continuous concave

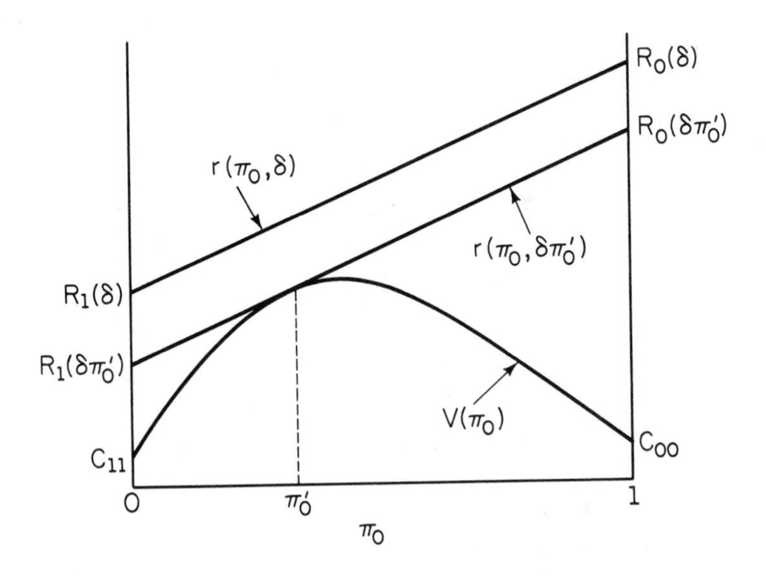

Fig. II.C.1: Illustration of the functions $r(\pi_0, \delta)$ and $V(\pi_0)$.

function of π_0 for $\pi_0 \epsilon [0, 1]$ with $V(0)=C_{11}$ and $V(1)=C_{00}$ (see Exercise 8). A typical $V(\pi_0)$ curve is sketched in Fig. II.C.1.

For the purposes of discussion suppose that $V(\pi_0)$ and $r(\pi_0, \delta)$ are as depicted in Fig. II.C.1. Also shown in Fig. II.C.1 is the line, labeled $r(\pi_0, \delta_{\pi_0'})$, that is both parallel to $r(\pi_0, \delta)$ and tangent to $V(\pi_0)$. Note that, for this case, δ cannot be a minimax rule because the risk line shown as $r(\pi_0, \delta_{\pi_0'})$ lies completely below $r(\pi_0, \delta)$ and thus has a smaller maximum value. Since $r(\pi_0, \delta_{\pi_0'})$ touches $V(\pi_0)$ at $\pi_0 = \pi_0'$, $\delta_{\pi_0'}$ is a Bayes rule for the prior π_0'. Since a similar tangent line (i.e., one that lowers both conditional risks) can be drawn for any decision rule δ, it is easily seen that only Bayes rules can possibly be minimax rules for this figure. Moreover, by examination of Fig. II.C.2, we see that the minimax rule for this case is a Bayes rule for the prior value π_L that maximizes $V(\pi_0)$ over $\pi_0 \epsilon [0, 1]$. Note that for this

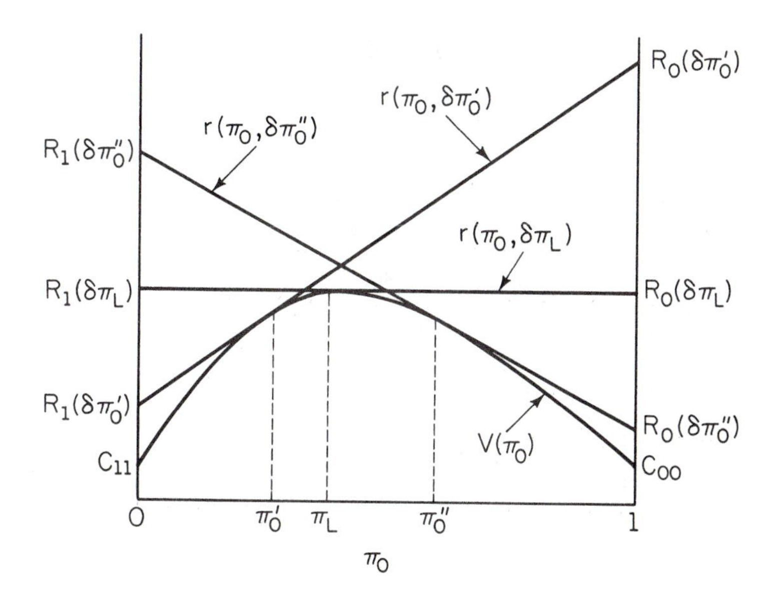

Fig. II.C.2: Illustration of the minimax rule when V has an interior maximum.

prior we have that $r(\pi_0, \delta_{\pi_L})$ is constant over π_0, so $\max\{R_0(\delta_{\pi_L}), R_1(\delta_{\pi_L})\}=R_0(\delta_{\pi_L})=R_1(\delta_{\pi_L})$ (a decision rule with equal conditional risks is called an *equalizer rule*). The fact that δ_{π_L} is minimax follows from the figure, since if $\pi_0'<\pi_L$, we have $\max\{R_0(\delta_{\pi_0'}), R_1(\delta_{\pi_0'})\}=R_0(\delta_{\pi_0'})>R_0(\delta_{\pi_L})$, and if $\pi_0''>\pi_L$, we have that $\max\{R_0(\delta_{\pi_0''}), R_1(\delta_{\pi_0''})\}= R_1(\delta_{\pi_0''})>R_1(\delta_{\pi_L})$, as depicted.

Because π_L in Fig. II.C.2 maximizes the minimum Bayes risk, it is called the *least-favorable* prior. Thus for this case a minimax decision rule is the Bayes rule for the least-favorable prior. In arguing above, we have not considered the possibility that $\max_{0\leqslant\pi_0\leqslant 1}V(\pi_0)$ may occur at $\pi_0=0$ or $\pi_0=1$, or that $V(\pi_0)$ may not have a tangent line at every point (i.e., it may not be differentiable everywhere). However, even in these cases it is always true that the minimax rule is a Bayes rule for the least-favorable prior. The following results develop formally the general solution to the minimax hypothesis-testing problem. We begin with the following proposition, which essentially summarizes the cases in which $V(\pi_0)$ is as depicted in Figs. II.C.1 and II.C.2, or in which $\pi_L=0$ or $\pi_L=1$.

Proposition II.C.1: The Minimax Test

Suppose that π_L is a solution to $V(\pi_L)=\max_{0\leqslant\pi_0\leqslant 1}V(\pi_0)$. Suppose further that either $\pi_L=0$, $\pi_L=1$, or $R_1(\delta_{\pi_L})=R_0(\delta_{\pi_L})$. Then δ_{π_L} is a minimax rule.

Proof: First, consider the case $R_1(\delta_{\pi_L})=R_0(\delta_{\pi_L})$. Then, for any prior π_0, we have

$$\max_{0\leqslant\pi_0\leqslant 1}\min_{\delta} r(\pi_0, \delta) = r(\pi_L, \delta_{\pi_L}) = r(\pi_0, \delta_{\pi_L}), \qquad \text{(II.C.4)}$$

where the first equality is by definition of V and π_L, and where the second equality follows from the fact that $r(\pi_0, \delta_{\pi_L})$ is constant in π_0. Thus we have that

$$\max_{0\leqslant\pi_0\leqslant 1}\ \min_{\delta} r(\pi_0, \delta) = \max_{0\leqslant\pi_0\leqslant 1} r(\pi_0, \delta_{\pi_L})$$

$$\geqslant \min_{\delta}\ \max_{0\leqslant\pi_0\leqslant 1} r(\pi_0, \delta). \tag{II.C.5}$$

But for each δ, we always have

$$\max_{0\leqslant\pi_0\leqslant 1} r(\pi_0, \delta) \geqslant \max_{0\leqslant\pi_0\leqslant 1}\ \min_{\delta} r(\pi_0, \delta), \tag{II.C.6}$$

which implies that

$$\min_{\delta}\ \max_{0\leqslant\pi_0\leqslant 1} r(\pi_0, \delta) \geqslant \max_{0\leqslant\pi_0\leqslant 1}\ \min_{\delta} r(\pi_0, \delta). \tag{II.C.7}$$

On combining (II.C.5) and (II.C.7), we have

$$\min_{\delta}\ \max_{0\leqslant\pi_0\leqslant 1} r(\pi_0, \delta) = \max_{0\leqslant\pi_0\leqslant 1}\ \min_{\delta} r(\pi_0, \delta), \tag{II.C.8}$$

and the left-hand equality of (II.C.5) implies

$$r(\pi_L, \delta_{\pi_L}) = \min_{\delta}\ \max_{0\leqslant\pi_0\leqslant 1} r(\pi_0, \delta), \tag{II.C.9}$$

which was to be shown.

Now suppose that $\pi_L=0$. In this case it is straightforward to show that $\max_{0\leqslant\pi_0\leqslant 1} r(\pi_0, \delta_{\pi_L}) = R_1(\delta_{\pi_L}) = r(\pi_L, \delta_{\pi_L})$. Figure II.C.3 depicts a typical such case. This fact and the argument of Eqs. (II.C.6) through (II.C.9) imply that δ_{π_L} is a minimax rule. A similar argument holds for the case $\pi_L=1$. This completes the proof.

□

Note that for any $\pi_0' \epsilon [0, 1]$, we always have $r(\pi_0, \delta_{\pi_0'}) \geqslant V(\pi_0')$ for all $\pi_0 \epsilon [0, 1]$, and $r(\pi_0', \delta_{\pi_0'}) = V(\pi_0')$. Thus since $r(\pi_0, \delta_{\pi_0'})$ as a function of π_0 describes a straight line, it must be a tangent to V at $\pi_0=\pi_0'$, as illustrated, for example, in Fig. II.C.2. Thus if V is differentiable at π_0', we

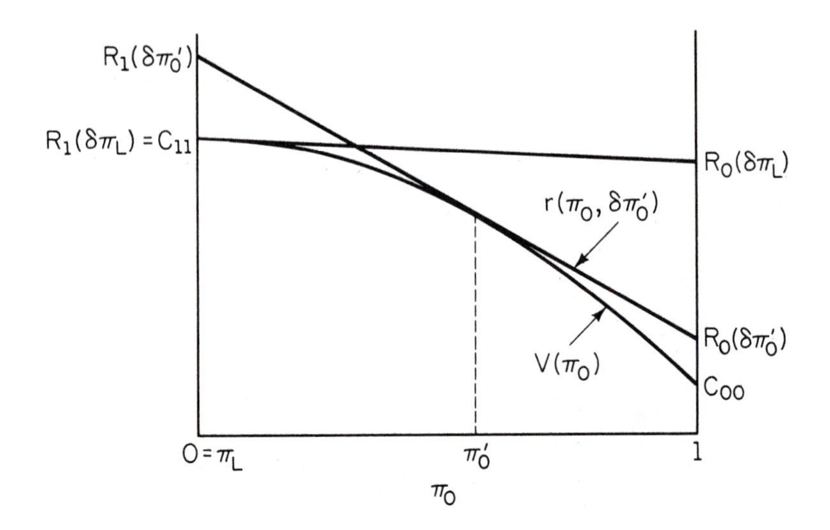

Fig. II.C.3: Depiction of the minimax rule when V has an endpoint maximum.

must have

$$V'(\pi_0') = dr(\pi_0, \delta_{\pi_0'})/d\,\pi_0 = [R_0(\delta_{\pi_0'}) - R_1(\delta_{\pi_0'})],$$

where V' denotes the derivative of V. If V has an interior maximum [i.e., $\pi_L \in (0, 1)$], then $V'(\pi_L)$ equals zero if V is differentiable at π_L. (Note that the concavity of V implies that $V'(\pi_0)$ can only equal zero at a maximum.) So the condition $R_1(\delta_{\pi_L}) = R_0(\delta_{\pi_L})$ holds whenever $\pi_L \in (0, 1)$ and $V'(\pi_L)$ exists, and thus the hypothesis of proposition II.C.1 is not very restrictive.

We now have characterized minimax rule for all cases except those in which V has an interior maximum at which it is not differentiable. For such cases, define two decision rules by $\delta^-_{\pi_L} = \lim_{\pi_0 \uparrow \pi_L} \delta_{\pi_0}$ and $\delta^+_{\pi_L} = \lim_{\pi_0 \downarrow \pi_L} \delta_{\pi_0}$. Note that $\delta^-_{\pi_L}$ necessarily has critical region

$$\Gamma_1^- = \{y \in \Gamma | (1-\pi_L)(C_{11}-C_{01})p_1(y) \\ \leqslant \pi_L (C_{00}-C_{10})p_0(y)\}, \tag{II.C.10}$$

and $\delta_{\pi_L}^+$ has critical region

$$\Gamma_1^+ = \{y \in \Gamma | (1-\pi_L)(C_{11}-C_{01})p_1(y) \\ < \pi_L (C_{00}-C_{10})p_0(y)\}, \tag{II.C.11}$$

regardless of which particular Bayes rules δ_{π_0} are used to define them. Consider the decision rule $\tilde{\delta}_{\pi_L}$, which uses Γ_1^- with probability q and uses Γ_1^+ with probability $(1-q)$; i.e., $\tilde{\delta}_{\pi_L}$ chooses H_1 if $y \in \Gamma_1^+$, chooses H_0 if $y \in (\Gamma_1^-)^c$, and chooses H_1 with probability q if y is on the boundary of Γ_1^-. Note from (II.B.6) that the boundary decision is irrelevant to the Bayes risk, so that $r(\pi_L, \tilde{\delta}_{\pi_L}) = V(\pi_L)$ and thus $\tilde{\delta}_{\pi_L}$ is a Bayes rule for π_L. However, the *conditional* risks do depend on the boundary, and they become

$$R_j(\tilde{\delta}_{\pi_L}) = qR_j(\delta_{\pi_L}^-) + (1-q)R_j(\delta_{\pi_L}^+), \tag{II.C.12}$$

so that the condition $R_0(\tilde{\delta}_{\pi_L}) = R_1(\tilde{\delta}_{\pi_L})$ is achieved by choosing

$$q = \frac{R_0(\delta_{\pi_L}^+) - R_1(\delta_{\pi_L}^+)}{R_0(\delta_{\pi_L}^+) - R_1(\delta_{\pi_L}^+) + R_1(\delta_{\pi_L}^-) - R_0(\delta_{\pi_L}^-)}. \tag{II.C.13}$$

Thus, as in Proposition II.C.1, $\tilde{\delta}_{\pi_L}$ with q chosen by (II.C.13) is a minimax rule.

Note that, because V is concave, it must have left- and right-hand derivatives at π_L, which we denote by $V'(\pi_L^-)$ and $V'(\pi_L^+)$, respectively. It is straightforward to see that $V'(\pi_L^+) = [R_0(\delta_{\pi_L}^+) - R_1(\delta_{\pi_L}^+)]$ and $V'(\pi_L^-) = [R_0(\delta_{\pi_L}^-) - R_1(\delta_{\pi_L}^-)]$, so that (II.C.13) becomes

$$q = \frac{V'(\pi_L^+)}{V'(\pi_L^+) - V'(\pi_L^-)}. \tag{II.C.14}$$

Note further that if the region $\{y \in \Gamma | (1-\pi_L)(C_{11}-C_{01})p_1(y) = \pi_L(C_{00}-C_{10})p_0(y)\}$ occurs with zero probability under H_0 and H_1, then the need for using $\tilde{\delta}_{\pi_L}$ rather than any other version of δ_{π_L} will not occur. The decision rule $\tilde{\delta}_{\pi_L}$ is an example of a *randomized decision rule*, a concept that will be discussed further Section II.D.

The action of the decision rule $\tilde{\delta}_{\pi_L}$ can be seen from the example depicted in Fig. II.C.4. Note that the lines $r(\pi_0, \delta_{\pi_L}^-)$ and $r(\pi_0, \delta_{\pi_L}^+)$ cross at $\pi_0 = \pi_L$ and have slopes equal to $V'(\pi_0^-)$ and $V'(\pi_0^+)$, respectively. By varying the probability q from 0 to 1, any line between these two lines can be obtained. The particular choice of q given by (II.C.14) yields the horizontal line that lies between these two. The minimaxity of the corresponding decision rule is obvious from the

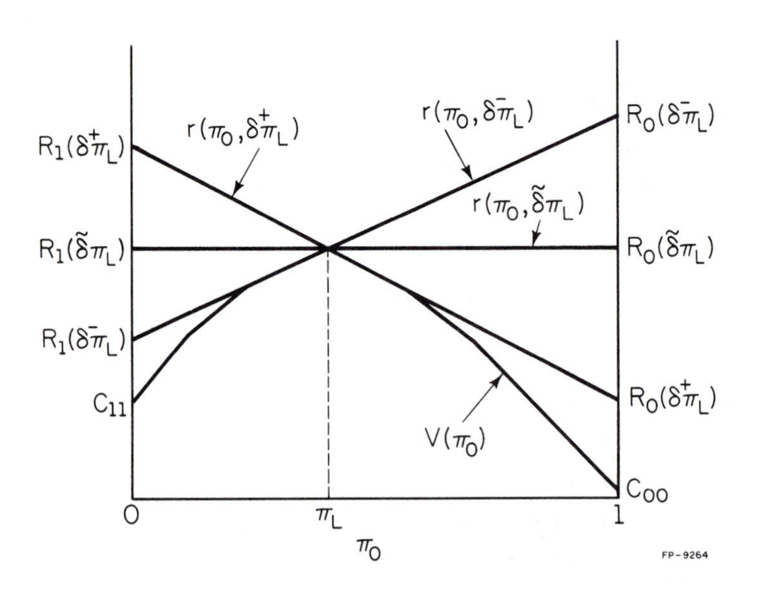

Fig. II.C.4: Depiction of a randomized decision rule.

figure as was discussed in connection with Fig. II.C.2

To illustrate the design of minimax decision rules, we consider the two examples presented in Section II.B.

Example II.C.1: Location Testing With Gaussian Error

Consider the location testing problem of Example II.B.2 with Gaussian error and with uniform costs. The function V follows from (II.B.4) and (II.B.3) and is given by

$$V(\pi_0) = \pi_0 \left(1 - \Phi\left(\frac{\tau' - \mu_0}{\sigma}\right)\right) + (1-\pi_0)\Phi\left(\frac{\tau' - \mu_1}{\sigma}\right) \tag{II.C.15}$$

with

$$\tau' = \frac{\sigma^2}{\mu_1 - \mu_0} \log\left(\pi_0/(1-\pi_0)\right) + \frac{\mu_1 + \mu_0}{2}.$$

Since $V(0) = C_{11} = 0 = C_{00} = V(1)$, the least favorable prior π_L is in the interior (0, 1) in this case. Moreover, since (II.C.15) is a differentiable function of π_0 randomization is unnecessary, and π_L can be found by setting $R_0(\delta_{\pi_L}) = R_1(\delta_{\pi_L})$. (That randomization is unnecessary also follows by noting that $P_0(L(Y)=\tau) = P_1(L(Y)=\tau) = 0$ for any τ since $L(Y)$ is a continuous random variable.) The prior π_0 enters $R_0(\delta_{\pi_0})$ and $R_1(\delta_{\pi_0})$ only through τ', so an equalizer rule is found by solving

$$1 - \Phi\left(\frac{\tau' - \mu_0}{\sigma}\right) = \Phi\left(\frac{\tau' - \mu_1}{\sigma}\right) \tag{II.C.16}$$

for τ'. By inspection of Fig. II.C.5, we see that the unique solution is

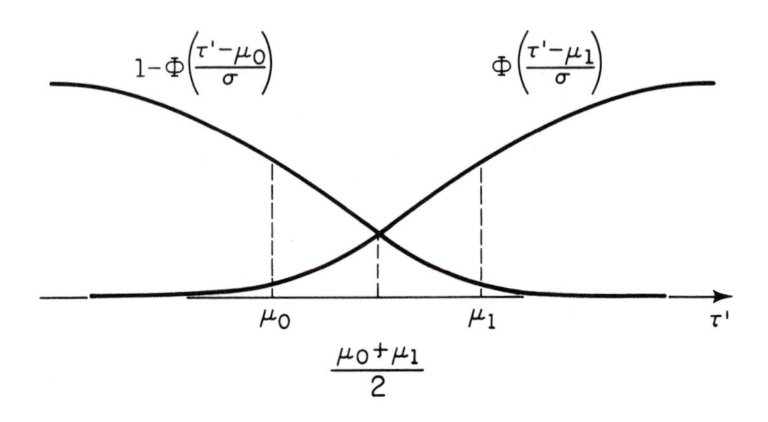

Fig. II.C.5: Conditional risks for location testing with Gaussian error and uniform costs.

$$\tau_L' = \frac{\mu_0+\mu_1}{2}, \tag{II.C.17}$$

so the minimax decision rule is

$$\delta_{\pi_L}(y) = \begin{cases} 1 \text{ if } y \geq (\mu_0+\mu_1)/2 \\ \\ 0 \text{ if } y < (\mu_0+\mu_1)/2. \end{cases} \tag{II.C.18}$$

From (II.C.17) it follows that the least-favorable prior is $\pi_L = \frac{1}{2}$, so the minimax risk is

$$V(\tfrac{1}{2}) = 1-\Phi\left(\frac{\mu_1-\mu_0}{2\sigma}\right).$$

Example II.C.2: The Binary Channel

As a second example, consider the binary channel of Example II.B.1 with uniform costs. The minimum Bayes risk function for this case is given by (see Exercise 1)

$$V(\pi_0) = \min\{(1-\pi_0)\lambda_1, \pi_0(1-\lambda_0)\} + \min\{(1-\pi_0)(1-\lambda_1), \pi_0\lambda_0\}. \tag{II.C.19}$$

This function can be rewritten as

$$V(\pi_0) = \begin{cases} \pi_0 \text{ if } 0 \leqslant \pi_0 \leqslant \underline{\pi} \\ \underline{\pi} + C\,\pi_0 \text{ if } \underline{\pi} < \pi_0 < \overline{\pi} \\ (1-\pi_0) \text{ if } \overline{\pi} \leqslant \pi_0 \leqslant 1, \end{cases} \tag{II.C.20}$$

where

$$\underline{\pi} = \min\left\{\frac{\lambda_1}{1-\lambda_0+\lambda_1}, \frac{1-\lambda_1}{1-\lambda_1+\lambda_0}\right\},$$

$$\overline{\pi} = \max\left\{\frac{\lambda_1}{1-\lambda_0+\lambda_1}, \frac{1-\lambda_1}{1-\lambda_1+\lambda_0}\right\},$$

and $C=(1-\overline{\pi}-\underline{\pi})/(\overline{\pi}-\underline{\pi})$. Note that $V(\pi_0)$ is piecewise linear with changes in slope at $\underline{\pi}$ and $\overline{\pi}$. Thus since $V(0)=V(1)=0$, the sign of the slope between $\underline{\pi}$ and $\overline{\pi}$ will determine π_L. In particular, $\pi_L=\underline{\pi}$ if $C<0$, $\pi_L=\overline{\pi}$ if $C>0$, and π_L is any prior in $[\underline{\pi}, \overline{\pi}]$ if $C=0$. In either case the minimax risk is $V(\pi_L)=\max\{\underline{\pi}, 1-\overline{\pi}\}$.

Since $V(\pi_0)$ is not differentiable at either $\underline{\pi}$ or $\overline{\pi}$ it is necessary here to consider a randomized test. For the purposes of illustration we consider the case $C>0$ as depicted in Fig. II.C.6. We then have $\pi_L=\overline{\pi}$. By inspection we have

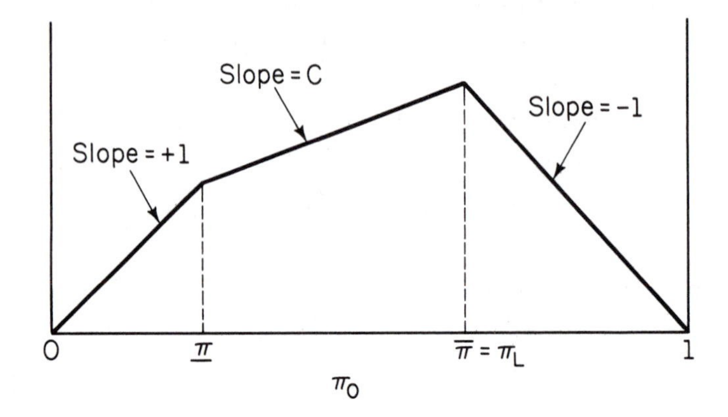

Fig. II.C.6: $V(\pi_0)$ *for the binary channel.*

$V'(\pi_L^+)=-1$ and $V'(\pi_L^-)=C$, so that the randomization constant q is given by

$$q = \frac{1}{1+C} = \overline{\pi}-\underline{\pi}. \tag{II.C.21}$$

It is straightforward from Example II.B.1 to show that if $\pi_0>\overline{\pi}$, then the Bayes rule is $\delta_{\pi_0}(0)=\delta_{\pi_0}(1)=0$. Thus $\delta_{\pi_L}^+(0)=\delta_{\pi_L}^+(1)=0$ and Γ_1^+ is the empty set. Similarly, for $\underline{\pi}<\pi_0<\overline{\pi}$, we have $\delta_{\pi_0}(y)=y$ if $\lambda_1<\lambda_0$ and $\delta_{\pi_0}(y)=1-y$ if $\lambda_1>1-\lambda_0$. Thus, for example, if $\lambda_1<1-\lambda_0$, we have $\delta_{\pi_L}^-(y)=y$, and the minimax rule $\tilde{\delta}_{\pi_L}$ chooses 0 if $y=0$ and it chooses 1 with probability q (and 0 with probability $(1-q)$) if $y=1$. In this case the minimax risk is

$$V(\pi_L) = 1-\overline{\pi} = \frac{\lambda_0}{1-\lambda_1+\lambda_0}. \tag{II.C.22}$$

If we have further that $\lambda_0=\lambda_1=\lambda$ (in which case $\lambda_1<1-\lambda_0$ if $\lambda<½$), then $q=1$ and the minimax risk is

$$V(\pi_L) = \lambda,$$

which is the same as the Bayes risk for any $\pi_0 \epsilon [\lambda, 1-\lambda]$. Thus, for the latter case, uncertainty is the prior does not cost any performance as long as $\lambda \leqslant \pi_0 \leqslant 1-\lambda$.

Summarizing this section, we have seen that optimum tests can be designed without the assumption of known priors by adopting a minimax design criterion. The solution is a Bayes test for the least-favorable prior, with randomization on the boundary of the decision region being necessary to given minimaxity in some problems. Again, we see that a likelihood ratio test emerges as the optimum decision rule. In the following section, we consider yet another formulation of the binary hypothesis testing problem.

II.D Neyman-Pearson Hypothesis Testing

In the Bayesian formulation of Section II.B optimality in testing (II.B.1) was defined in terms of minimizing the overall expected cost, defined as the average risk. Similarly, in the minimax formulation of Section II.C, priors were not assumed known and optimality was defined in terms of minimizing the maximum of the conditional expected costs under the two hypotheses. In many problems of practical interest, the imposition of a specific cost structure on the decisions made in testing (II.B.1) is not possible or desirable. In such cases an alternative design criterion, known as the *Neyman-Pearson criterion*, is often imposed. In this section we describe this alternative formulation.

In testing H_0 versus H_1 in (II.B.1) there are two types of errors that can be made: H_0 can be falsely rejected or H_1 can be falsely rejected. The first of these two error types is called a *Type I error* or a *false alarm*. The second type is called a *Type II error* or a *miss*. The terms "false alarm" and "miss" come from radar problems in which H_0 and H_1 usually represent the absence and presence of a target, respectively. Correct acceptance of H_1 is similarly called a *detection*. For a decision rule δ, the probability of a Type I error is known as the *size* or *false-alarm probability* (or *false-alarm rate*) of δ, and we will denote it by $P_F(\delta)$. Similarly the probability of a Type II error is called the *miss probability*, $P_M(\delta)$. However, in discussing the latter quantity we usually talk about the *detection probability*, $P_D(\delta)=1-P_M(\delta)$, which is also called the *power* of δ.

Obviously, the design of a test for H_0 versus H_1 involves a trade-off between the probabilities of the two types of errors, since one can always be made arbitrarily small at the expense of the other. (The Bayes and minimax criteria are two ways of trading these off.) The Neyman-Pearson criterion for making this trade-off is to place a bound on the false-alarm probability and then to minimize the miss probability within this constraint; i.e., the Neyman-Pearson design criterion is

$$\max_{\delta} P_D(\delta) \text{ subject to } P_F(\delta)\leqslant\alpha, \tag{II.D.1}$$

where α is the above-mentioned bound, which is known as the *level* or *significance level* of the test. Thus the Neyman-Pearson design goal is to find the most powerful α-level test of H_0 versus H_1. Note that as opposed to the Bayes and minimax criteria, the Neyman-Pearson criterion recognizes a basic asymmetry in importance of the two hypotheses.

In order to give the general solution to the Neyman-Pearson problem (II.D.1), it is necessary to consider randomized tests similar to the test $\tilde{\delta}_{\pi_L}$ defined in the solution to the minimax problem in Section II.C. For our purposes it is convenient to define a *randomized decision rule* $\tilde{\delta}$ for H_0 versus H_1 as a function mapping Γ to the interval [0, 1] with the interpretation that for $y \in \Gamma$, $\tilde{\delta}(y)$ is the conditional probability with which we accept H_1 given that we observe $Y=y$.

For example, the randomized minimax rule $\tilde{\delta}_{\pi_L}$ introduced in Section II.C can be written as

$$\tilde{\delta}_{\pi_L}(y) = \begin{cases} 1 \text{ if } L(y)>\tau_L \\ q \text{ if } L(y)=\tau_L \\ 0 \text{ if } L(y)<\tau_L \end{cases} \tag{II.D.2}$$

where τ_L is the threshold corresponding to the least-favorable prior π_L. With this definition we see that a nonrandomized decision rule is a special case of a randomized decision rule. In particular, a nonrandomized rule δ corresponds to the randomized rule $\tilde{\delta}(y)=\delta(y)$. The difference between the two is that the value of δ is the index of the accepted hypothesis and the value of $\tilde{\delta}$ is the probability with which we accept H_1. These coincide as long as $\tilde{\delta}$ takes on only the two values 0 and 1.

The false-alarm probability of a decision rule is the probability with which it accepts H_1 given that H_0 is true. For a randomized rule $\tilde{\delta}$ this quantity is given by

$$P_F(\tilde{\delta}) = E_0\{\tilde{\delta}(Y)\} = \int_\Gamma \tilde{\delta}(y) p_0(y) \mu(dy), \qquad \text{(II.D.3)}$$

where $E_0\{\cdot\}$ denotes expectation under hypothesis H_0. Equation (II.D.3) follows because the probability of accepting H_1 given that H_0 is true is just the probability of accepting H_1 given Y [that is $\tilde{\delta}(Y)$], averaged over the distribution of Y under H_0. Similarly, the detection probability of a randomized rule $\tilde{\delta}$ is given by

$$P_D(\tilde{\delta}) = E_1\{\tilde{\delta}(Y)\} = \int_\Gamma \tilde{\delta}(y) p_1(y) \mu(dy). \qquad \text{(II.D.4)}$$

The general solution to the Neyman-Pearson design problem can be summarized in the following result.

Proposition II.D.1: The Neyman-Pearson Lemma

Consider the hypothesis pair of (II.B.1) in which P_j has density p_j for $j=0$ and $j=1$, and suppose that $\alpha>0$. Then the following statements are true.

(i) (Optimality) Let $\tilde{\delta}$ be any decision rule satisfying $P_F(\tilde{\delta}) \leqslant \alpha$, and let $\tilde{\delta}'$ be any decision rule of the form

$$\tilde{\delta}'(y) = \begin{cases} 1 \text{ if } p_1(y) > \eta p_0(y) \\ \gamma(y) \text{ if } p_1(y) = \eta p_0(y) \\ 0 \text{ if } p_1(y) < \eta p_0(y), \end{cases} \tag{II.D.5}$$

where $\eta \geqslant 0$ and $0 \leqslant \gamma(y) \leqslant 1$ are such that $P_F(\tilde{\delta}') = \alpha$. Then $P_D(\tilde{\delta}') \geqslant P_D(\tilde{\delta})$. That is, any size-$\alpha$ decision rule of the form (II.D.5) is a Neyman-Pearson rule.

(ii) (Existence) For every $\alpha \in (0, 1)$ there is a decision rule, $\tilde{\delta}_{NP}$, of the form of (II.D.5) with $\gamma(y) = \gamma_0$ (a constant), for which $P_F(\tilde{\delta}_{NP}) = \alpha$.

(iii) (Uniqueness) Suppose that $\tilde{\delta}''$ is any α-level Neyman-Pearson decision rule for H_0 versus H_1. Then $\tilde{\delta}''$ must be of the form of (II.D.5) except possibly on a subset of Γ having zero probability under H_0 and H_1.

Proof: (i) Assume that $\tilde{\delta}$ and $\tilde{\delta}'$ are as defined above. Note that because of the way $\tilde{\delta}'$ is defined, we always have $[\tilde{\delta}'(y) - \tilde{\delta}(y)][p_1(y) - \eta p_0(y)] \geqslant 0$ for every $y \in \Gamma$. Thus

$$\int_\Gamma [\tilde{\delta}'(y) - \tilde{\delta}(y)][p_1(y) - \eta p_0(y)]\mu(dy) \geqslant 0. \tag{II.D.6}$$

Expanding the terms in (II.D.6) and rearranging we have

$$\int_\Gamma \tilde{\delta}' p_1 d\mu - \int_\Gamma \tilde{\delta} p_1 d\mu \geqslant \int_\Gamma \tilde{\delta}' p_0 d\mu - \int_\Gamma \tilde{\delta} p_0 d\mu. \tag{II.D.7}$$

Applying (II.D.3) and (II.D.4), (II.D.7) becomes

$$P_D(\tilde{\delta}') - P_D(\tilde{\delta}) \geqslant P_F(\tilde{\delta}') - P_F(\tilde{\delta}) = \alpha - P_F(\tilde{\delta}) \geqslant 0, \tag{II.D.8}$$

where we have used the fact that $P_F(\tilde{\delta}) \leqslant \alpha$. Thus $P_D(\tilde{\delta}') \geqslant P_D(\tilde{\delta})$, which was to be shown.

(ii) Let η_0 be the smallest number such that

$$P_0(p_1(Y) > \eta_0 p_0(Y)) \leqslant \alpha. \tag{II.D.9}$$

Then if $P_0(p_1(Y) > \eta_0 p_0(Y)) < \alpha$, choose

$$\gamma_0 = \frac{\alpha - P_0(p_1(Y) > \eta_0 p_0(Y))}{P_0(p_1(Y) = \eta_0 p_0(Y))}; \tag{II.D.10}$$

otherwise, choose γ_0 arbitrarily. These relationships are illustrated in Fig. II.D.1. Then, on defining $\tilde{\delta}_{NP}$ to be the decision rule of (II.D.5) with $\eta = \eta_0$ and $\gamma(y) = \gamma_0$, we have

$$\begin{aligned} P_F(\tilde{\delta}_{NP}) = E_0\{\tilde{\delta}_{NP}(Y)\} &= P_0(p_1(Y) > \eta_0 p_0(Y)) \\ &+ \gamma_0 P_0(p_1(Y) = \eta_0 p_0(Y)) = \alpha. \end{aligned} \tag{II.D.11}$$

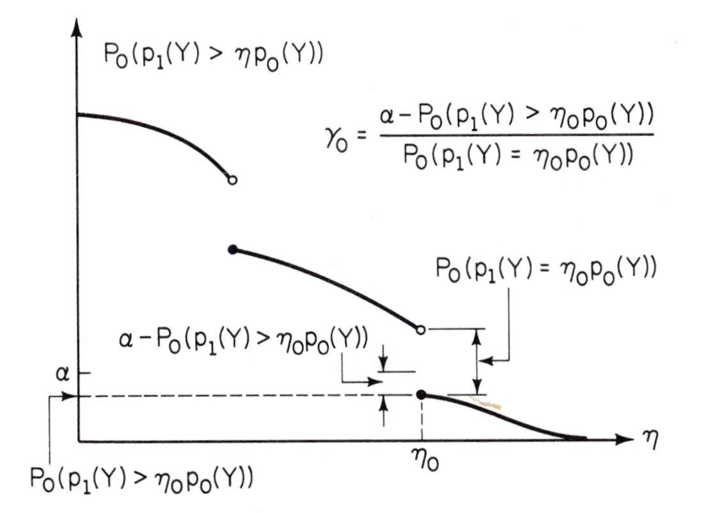

Fig. II.D.1: Threshold and randomization for an α-level Neyman-Pearson test.

Thus we have chosen a decision rule of the form of (II.D.5) with $\gamma(y)$ constant and false-alarm probability α.

(iii) Suppose that $\tilde{\delta}'$ is an α-level Neyman-Pearson rule of the form (II.D.5) and let $\tilde{\delta}''$ be any other α-level Neyman-Pearson rule. Then $P_D(\tilde{\delta}'')$ must equal $P_D(\tilde{\delta}')$, so from (II.D.8) we have $0 \geqslant \alpha - P_F(\tilde{\delta}') \geqslant 0$, or $P_F(\tilde{\delta}') = \alpha$. Thus $\tilde{\delta}''$ must be of size α. Using the facts that $P_D(\tilde{\delta}'') = P_D(\tilde{\delta}')$ and $P_F(\tilde{\delta}'') = P_F(\tilde{\delta}')$, and working backward from (II.D.8) to (II.D.6), we conclude that

$$\int_{\Gamma} [\tilde{\delta}'(y) - \tilde{\delta}''(y)][p_1(y) - \eta p_0(y)]\mu(dy) = 0. \quad \text{(II.D.12)}$$

Since the integrand is nonnegative (as discussed above), (II.D.12) implies that it is zero except possibly on a set of zero probability under H_0 and H_1. Thus $\tilde{\delta}'$ and $\tilde{\delta}''$ differ only on the set $\{y \in \Gamma | p_1(y) = \eta p_0(y)\}$, which implies that $\tilde{\delta}''$ is also of the form (II.D.5), possibly differing from $\tilde{\delta}'$ only in the function $\gamma(y)$.

This completes the proof of the proposition.

□

The result above again indicates the optimality of the likelihood ratio test. The Neyman-Pearson test for a given hypothesis pair differs from the Bayes and minimax tests only in the choice of threshold and randomization. (Note that, for $\alpha=0$, the Neyman-Pearson test is given straightforwardly by the nonrandomized test with critical region $\Gamma_1 = \{y \in \Gamma | p_0(y) = 0\}$.) The design of Neyman-Pearson tests is illustrated by the following two examples.

Example II.D.1: Location Testing with Gaussian Error

Consider first the location testing problem with Gaussian errors as introduced in Example II.B.2. Here we have

$$\begin{aligned} P_0(p_1(Y) > \eta p_0(Y)) &= P_0(L(Y) > \eta) = P_0(Y > \eta') \\ &= 1 - \Phi\left(\frac{\eta' - \mu_0}{\sigma}\right), \end{aligned} \quad \text{(II.D.13)}$$

where $\eta'=\sigma^2 \log(\eta)/(\mu_1-\mu_0)+(\mu_0+\mu_1)/2$. This curve is illustrated in Fig. II.D.2 as a function of η'. Note that any value of α can be achieved exactly by choosing

$$\eta_0' = \sigma\Phi^{-1}(1-\alpha)+\mu_0, \qquad \text{(II.D.14)}$$

where Φ^{-1} is the inverse function of Φ. Since $P(Y=\eta_0')=0$, the randomization can be chosen arbitrarily, say $\gamma_0=1$. An α-level Neyman-Pearson test for this case is then given by

$$\tilde{\delta}_{NP}(y) = \begin{cases} 1 \text{ if } y \geqslant \eta_0' \\ 0 \text{ if } y < \eta_0', \end{cases} \qquad \text{(II.D.15)}$$

where η_0' is from (II.D.14).

The detection probability of $\tilde{\delta}_{NP}$ is given by

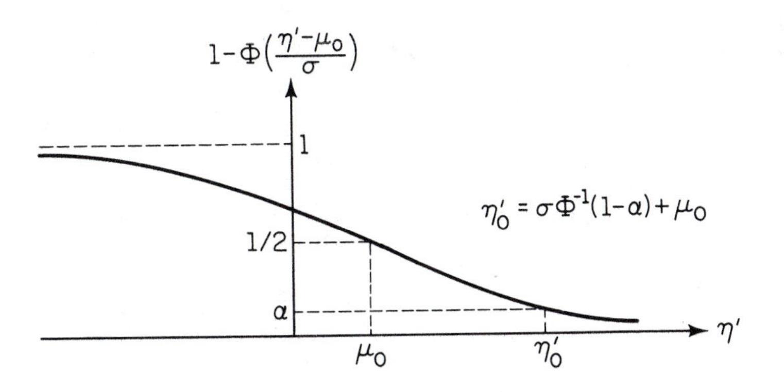

Fig. II.D.2: Illustration of threshold (η_0') for Neyman-Pearson testing of location with Gaussian error.

$$P_D(\tilde{\delta}_{NP}) = E_1\{\tilde{\delta}_{NP}(Y)\} = P_1(Y \geq \eta_0')$$

$$= 1-\Phi\left(\frac{\eta_0'-\mu_1}{\sigma}\right) \qquad \text{(II.D.16)}$$

$$= 1-\Phi(\Phi^{-1}(1-\alpha)-d\,),$$

where $d = (\mu_1-\mu_0)/\sigma$ is the signal-to-noise ratio defined in Example II.B.2. For fixed α, (II.D.16) gives the detection probability as a function of d for the test of (II.D.15). This relationship is sometimes known as the *power function* of the test. A plot of this relationship is shown in Fig. II.D.3. Equation (II.D.16) also gives the detection probability as a function of the false-alarm probability for fixed d. Again borrowing from radar terminology, a parametric plot of this relationship is called the *receiver operating characteristics* (ROCs). The

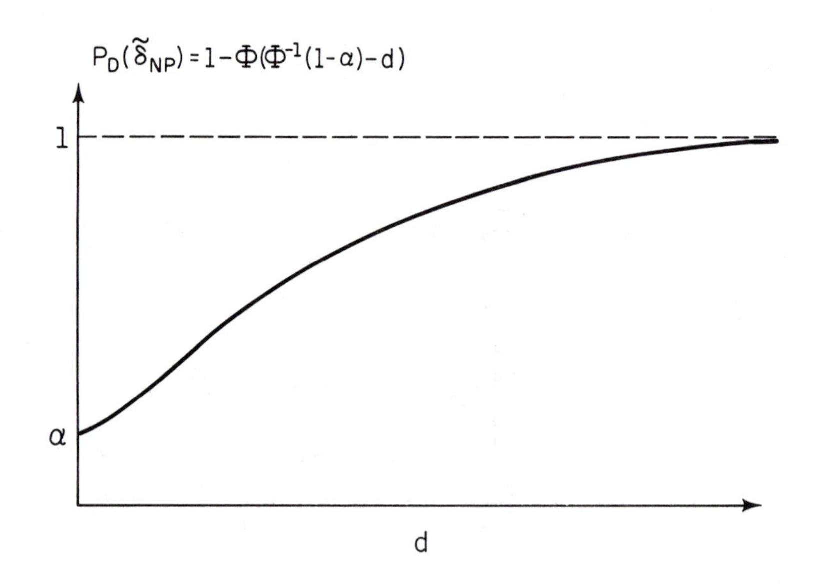

Fig. II.D.3: Power function for Neyman-Pearson testing of location with Gaussian error [$d = (\mu_1-\mu_0)/\sigma$].

ROCs for the test of (II.D.15) are shown in Fig. II.D.4. Figures II.D.2 through II.D.4 actually apply to a large class of problems involving signal detection in Gaussian noise, and these will be discussed in subsequent chapters.

Example II.D.2: The Binary Channel

To illustrate a Neyman-Pearson problem in which randomization is necessary, consider again the binary channel of Example II.B.1. The likelihood ratio for this problem is given by (II.B.20). To find the threshold for achieving an α-level Neyman-Pearson test, we must consider $P_0(L(Y)>\eta)$. For the sake of simplicity we assume that $\lambda_0+\lambda_1<1$, in which case $\lambda_1/(1-\lambda_0)<(1-\lambda_1)/\lambda_0$. We then have

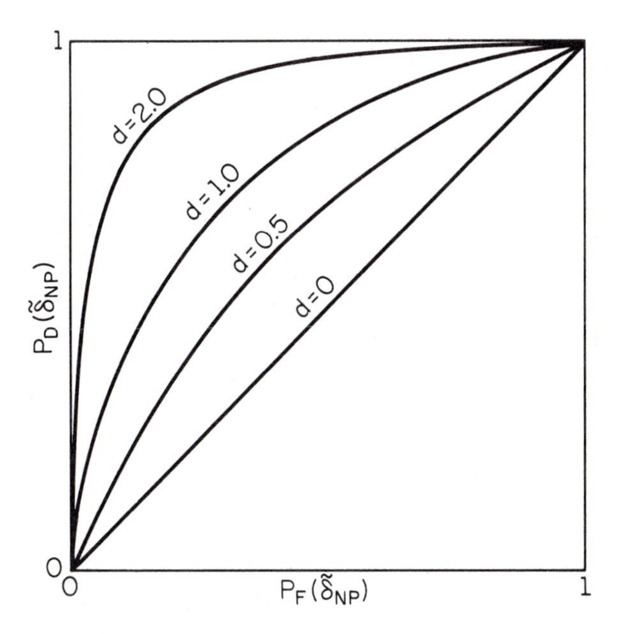

Fig. II.D.4: Receiver operating characteristics (ROCs) for Neyman-Pearson location testing with Gaussian error $[d=(\mu_1-\mu_0)/\sigma]$.

$$P_0(L(Y) > \eta) = \begin{cases} 1 \text{ if } \eta < \dfrac{\lambda_1}{(1-\lambda_0)} \\ \lambda_0 \text{ if } \dfrac{\lambda_1}{(1-\lambda_0)} \leq \eta < \dfrac{(1-\lambda_1)}{\lambda_0} \\ 0 \text{ if } \eta \geq \dfrac{(1-\lambda_1)}{\lambda_0}. \end{cases} \tag{II.D.17}$$

This function is depicted in Fig. II.D.5. By inspection we see that the desired threshold for α-level Neyman-Pearson testing is given by

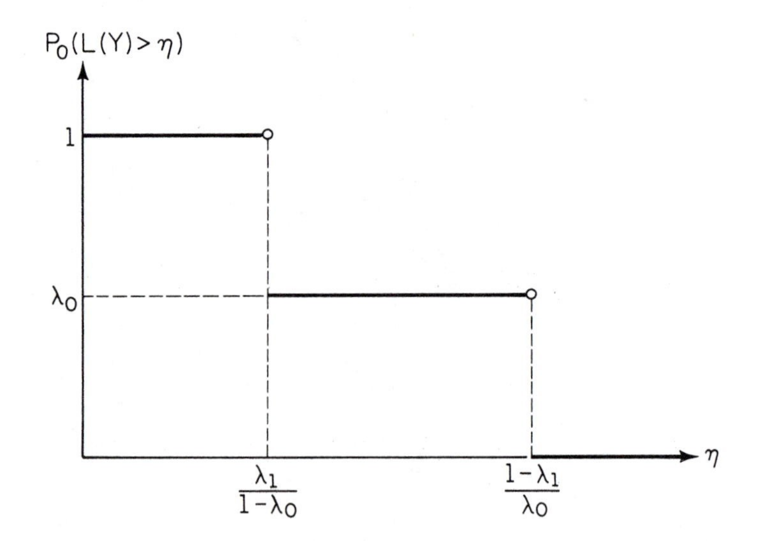

Fig. II.D.5: Curve for threshold and randomization selection for a binary channel.

$$\eta_0 = \begin{cases} \dfrac{(1-\lambda_1)}{\lambda_0} \text{ if } 0 \leqslant \alpha < \lambda_0 \\ \\ \dfrac{\lambda_1}{(1-\lambda_0)} \text{ if } \lambda_0 \leqslant \alpha < 1 \\ \\ 0 \text{ if } \alpha = 1. \end{cases} \tag{II.D.18}$$

Also, by inspection we see that the randomization constant must be given by

$$\gamma_0 = \begin{cases} \alpha - \lambda_0 \text{ if } 0 \leqslant \alpha < \lambda_0 \\ \\ \dfrac{(\alpha-\lambda_0)}{(1-\lambda_0)} \text{ if } \lambda_0 \leqslant \alpha < 1 \\ \\ \textit{arbitrary} \text{ if } \alpha = 1. \end{cases} \tag{II.D.19}$$

By using (II.B.20), (II.D.18), and (II.D.19), the resulting Neyman-Pearson test is seen to be

$$\delta_{NP}(y) = \begin{cases} \dfrac{\alpha}{\lambda_0} \text{ if } y = 1 \\ \\ 0 \text{ if } y = 0, \end{cases} \tag{II.D.20}$$

for $0 \leqslant \alpha < \lambda_0$, and

$$\tilde{\delta}_{NP}(y) = \begin{cases} 1 \text{ if } y = 1 \\ \\ \dfrac{(\alpha-\lambda_0)}{(1-\lambda_0)} \text{ if } y = 0, \end{cases} \tag{II.D.21}$$

for $\lambda_0 \leq \alpha \leq 1$.

The detection probability of the Neyman-Pearson test is given by $P_D(\tilde{\delta}_{NP}) = P_1(L(Y) > \eta_0) + \gamma_0 P_1(L(Y) = \eta_0)$, which is straightforwardly seen here to be

$$P_D(\tilde{\delta}_{NP}) = \begin{cases} \alpha \dfrac{(1-\lambda_1)}{\lambda_0} & \text{if } 0 \leq \alpha < \lambda_0 \\ \\ (1-\lambda_1) + \lambda_1 \dfrac{(\alpha-\lambda_0)}{(1-\lambda_0)} & \text{if } \lambda_0 \leq \alpha \leq 1. \end{cases} \tag{II.D.22}$$

Note that the ROCs (P_D versus P_F) are piecewise linear with a change in slope at $P_F = \lambda_0$. This behavior is depicted in Fig. II.D.6 for the symmetric-channel case, $\lambda_0 = \lambda_1 = \lambda \leq 1/2$.

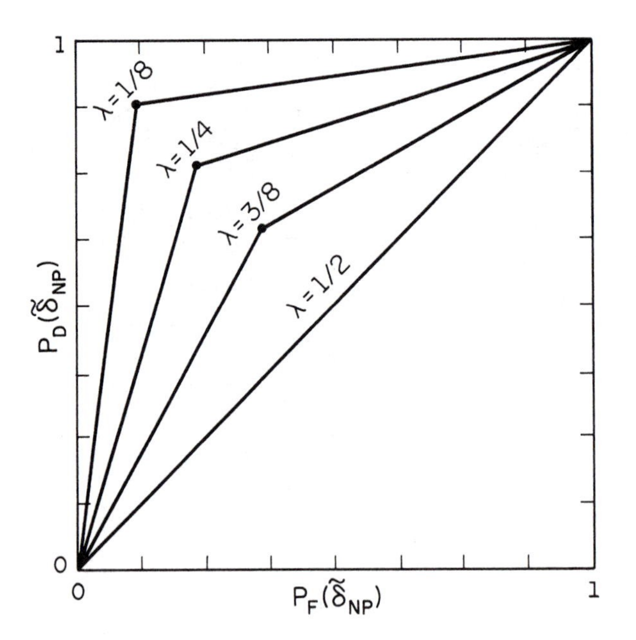

Fig. II.D.6: ROCs for a binary symmetric channel.

II.E Composite Hypothesis Testing

The hypothesis testing problems discussed in the preceding sections are sometimes known as *simple* hypothesis-testing problems because each of the two hypotheses in (II.D.1) corresponds to only a single distribution for the observation. In many hypothesis testing problems, however, there are many possible distributions that can occur under each of the-hypotheses. Hypotheses of this type are known as *composite* hypotheses. An example of where this type of problem might arise is in radar detection, in which the returned signal, if present, has unknown parameters, such as its exact time of arrival (related to position) and its Doppler shift (related to velocity); thus the "target present" hypothesis is composite. In this section we discuss briefly the design of hypothesis tests for composite problems.

To model the most general type of composite hypothesis-testing problem, we can consider a family of probability distributions on Γ indexed by a parameter θ taking values in a parameter set Λ. That is, we have a family $\{P_\theta; \theta \epsilon \Lambda\}$, where P_θ is the probability distribution of the observation given that Λ is the true parameter value. The parameter values in Λ represent the set of all possible states of nature. For the simple hypothesis pair of (II.B.1), we have the special case in which $\Lambda=\{0, 1\}$. More generally, we might have a parameter space that is the union of two disjoint parameters sets Λ_0 and Λ_1 representing the ranges of the parameter under the two hypotheses.

In a Bayesian formulation of the composite hypothesis-testing problem, we assume that the parameter is a random quantity, Θ, taking on the values in Λ. In this case P_θ is interpreted as the conditional distribution of Y given that $\Theta=\theta$. We wish to make a binary decision about Θ and, for the sake of simplicity, we will consider only nonrandomized decision rules. To choose an optimum decision rule we must first assign costs to our decisions through a cost function $C[i, \theta]$ where $C[i, \theta]$ is the cost of choosing decision i when $Y \sim P_\theta$, for $i \epsilon [0, 1]$ and $\theta \epsilon \Lambda$. For example, in the radar detection problem in which the parameter θ may be related to target position and velocity, the cost structure might assign higher costs to missing close or fast-moving targets than to missing slower, more distant targets. For simplicity we assume that C is nonnegative and bounded.

For a decision rule δ, we can define conditional risks analogous to those for the simple case via

$$R_\theta(\delta) = E_\theta\{C[\delta(Y), \theta]\}, \ \theta \in \Lambda, \tag{II.E.1}$$

where E_θ denotes expectation assuming that $Y \sim P_\theta$. Also an average or Bayes risk can be defined as

$$r(\delta) = E\{R_\Theta(\delta)\}, \tag{II.E.2}$$

and a Bayes rule is defined as one that minimizes $r(\delta)$.

Since $E_\theta\{C[\delta(Y), \theta]\} \triangleq E\{C[\delta(Y), \Theta] | \Theta = \theta\}$, $r(\delta)$ can be written as

$$r(\delta) = E\{E\{C[\delta(Y), \Theta] | \Theta\}\} = E\{C[\delta(Y), \Theta]\}, \tag{II.E.3}$$

where the second equality follows from the use of iterated expectations, $E\{X\} = E\{E\{X|Y\}\}$. Thus $r(\delta)$ is simply the cost of using δ averaged over Θ and Y. Again using iterated expectations we can write

$$r(\delta) = E\{E\{C[\delta(Y), \Theta] | Y\}\}. \tag{II.E.4}$$

Note from (II.E.4) that $r(\delta)$ is minimized over δ if for each $y \in \Gamma$, we choose $\delta(y)$ to be the decision that minimizes the posterior cost,

$$E\{C[\delta(y), \Theta] | Y = y\}. \tag{II.E.5}$$

Since $\delta(y)$ can only be 0 or 1, we thus see that a Bayes rule for this problem is given by

$$\delta_B(y) = \begin{cases} 1 & \text{if } E\{C[1,\Theta]|Y=y\} < E\{C[0,\Theta]|Y=y\} \\ 0 \text{ or } 1 & \text{if } E\{C[1,\Theta]|Y=y\} = E\{C[0,\Theta]|Y=y\} \\ 0 & \text{if } E\{C[1,\Theta]|Y=y\} > E\{C[0,\Theta]|Y=y\}. \end{cases} \quad \text{(II.E.6)}$$

The interpretation of (II.E.6) is simple; δ_B chooses the hypothesis that is least costly, on the average, given our observation. In the case where $\Lambda=\{0, 1\}$, (II.E.6) reduces of course to the Bayes rule for simple hypothesis-testing, which as discussed in Section II.B, also has the interpretation of minimizing the posterior cost.

For many problems of interest the parameter space can be decomposed into two disjoint sets Λ_0 and Λ_1, representing hypothesis H_0 and H_1, respectively, with the costs being uniform over these sets; i.e.,

$$C[i,\theta] = C_{ij}, \; \theta \epsilon \Lambda_j. \quad \text{(II.E.7)}$$

In this case it is easily seen that under the assumption $C_{11} < C_{01}$, (II.E.6) reduces to

$$\delta_B(y) = \begin{cases} 1 & > \\ 0 \text{ or } 1 \text{ if } \dfrac{P(\Theta \epsilon \Lambda_1 | Y=y)}{P(\Theta \epsilon \Lambda_0 | Y=y)} & = \dfrac{C_{10}-C_{00}}{C_{01}-C_{11}}, \\ 0 & < \end{cases} \quad \text{(II.E.8)}$$

where $P(\Theta \epsilon \Lambda_j | Y=y)$ denotes the conditional probability that Θ lies in Λ_j given that $Y=y$. Assuming that Y has conditional densities $p(y | \Theta \epsilon \Lambda_j)$ for $j=0, 1$, Bayes' formula implies that

$$P(\Theta\epsilon\Lambda_j|Y=y)=\frac{p(y|\Theta\epsilon\Lambda_j)P(\Theta\epsilon\Lambda_j)}{p(y)} \quad \text{(II.E.9)}$$

for $j=0,1$ with $p(y)=\sum_{j=0}^{1}p(y|\Theta\epsilon\Lambda_j)P(\Theta\epsilon\Lambda_j)$. Thus (II.E.8) reduces to

$$\delta_B(y)=\begin{cases}1 & > \\ 0 \text{ or } 1 \text{ if } L(y) & = \frac{\pi_0(C_{10}-C_{00})}{\pi_1(C_{01}-C_{11})} \\ 0 & < \end{cases} \quad \text{(II.E.10)}$$

with $\pi_j=P(\Theta\epsilon\Lambda_j)$ and $L(y)=p(y|\Theta\epsilon\Lambda_1)/p(y|\Theta\epsilon\Lambda_0)$. From (II.E.10), we see that this problem is equivalent to the simple Bayesian hypothesis-testing problem with $p_j(y)=p(y|\Theta\epsilon\Lambda_j)$, a fact that should be more or less obvious in retrospect.

Assuming that Θ has density $w(\theta)$ and P_θ has density p_θ for each $\theta\epsilon\Lambda$, then, using the notation introduced in Chapter I, the densities $p(y|\Theta\epsilon\Lambda_j)$ are given by

$$p(y|\Theta\epsilon\Lambda_j)=\int_\Lambda p_\theta(y)w_j(\theta)\mu(d\theta), \quad \text{(II.E.11)}$$

where $w_j(\theta)$ is the conditional density of Θ given that $\Theta\epsilon\Lambda_j$; i.e.,

$$w_j(\theta)=\begin{cases}0 & \text{if } \theta\epsilon\Lambda_j \\ w(\theta)/\pi_j & \text{if } \theta\epsilon\Lambda_j,\end{cases} \quad \text{(II.E.12)}$$

with $\pi_j=\int_{\Lambda_j}w(\theta)\mu(d\theta)$.

Note that the hypothesis pair in this problem is defined by the observation densities of (II.E.11), which depend only on the conditional densities w_j. Thus one can also pose composite minimax and Neyman-Pearson problems in which the

w_j's are known but not the π_j's. Of course, aside from possible differing physical interpretations of the parametric model, these problems are no different from the simple hypothesis-testing problems of Section II.B-D, as was noted above. The following example illustrates this type of problem.

Example II.E.1: Testing on the Radius of a Point in the Plane

Suppose that $\Gamma=\mathbb{R}^2$ [i.e., $Y=(Y_1, Y_2)^T$] and our hypotheses are as follows:

$$\begin{aligned} H_0: \quad & Y_1 = \epsilon_1 \\ & Y_2 = \epsilon_2 \\ \text{versus} \quad & \\ H_1: \quad & Y_1 = A\cos\Psi + \epsilon_1 \\ & Y_2 = A\sin\Psi + \epsilon_2, \end{aligned} \tag{II.E.13}$$

where A is a positive constant, Ψ is a random variable distributed uniformly in $[0, 2\pi]$, and ϵ_1 and ϵ_2 are $N(0, \sigma^2)$ random variables that are independent of one another and of Ψ. The observation here can be thought of as a noisy measurement of the coordinates of a point in the plane which is either at the origin or is uniformly distributed on a circle of radius A. (Applications of this model will arise in Chapter III.)

The parameter in this case can be taken to be $\Theta=(\Theta_1, \Theta_2)$ with $\Theta_1 \epsilon \{0, A\}$ and $\Theta_2 \epsilon [0, 2\pi]$. The parameter set Λ is thus $\{0, A\}\times[0, 2\pi]$ with $\Lambda_0=\{\theta \epsilon \Lambda | \theta_1=0\}$ and $\Lambda_1=\{\theta \epsilon \Lambda | \theta_1=A\}$. The density of Y given $\Theta=\theta$ is the joint density of two independent $N(0, \sigma^2)$ random variables shifted in mean by $\theta_1 \cos\theta_2$ and $\theta_1 \sin\theta_2$, respectively; i.e.,

$$\begin{aligned} p_\theta(y) &= \frac{1}{2\pi\sigma^2}\exp\{q(y,\theta)/2\sigma^2\}, \; y \,\epsilon\, \mathbb{R}^2, \\ q(y, \theta) &\triangleq [(y - \theta_1\cos\theta_2)^2 + (y - \theta_1\sin\theta_2)^2]. \end{aligned} \tag{II.E.14}$$

It then follows straightforwardly that

$$p(y|\Theta \in \Lambda_0) = p_\theta(y)\Big|_{\theta_1=0}$$

(II.E.15)

$$= \frac{1}{2\pi\sigma^2} \exp\{-(y_1^2+y_2^2)/2\sigma^2\},$$

and

$$p(y|\Theta \in \Lambda_1) = \frac{1}{2\pi}\int_0^{2\pi} p_\theta(y)\Big|_{\theta_1=A} d\,\theta_2$$

(II.E.16)

$$= \frac{1}{4\pi\sigma^2}\int_0^{2\pi} \exp\{q(y,\theta)\Big|_{\theta_1=A}/2\sigma^2\} d\,\theta_2.$$

From (II.E.15) and (II.E.16) the likelihood ratio is given by

$$L(y) = \frac{p(y|\Theta \in \Lambda_1)}{p(y|\Theta \in \Lambda_0)}$$

(II.E.17)

$$= \frac{e^{-A^2/2\sigma^2}}{2\pi}\int_0^{2\pi} \exp\left\{\frac{A}{\sigma^2}(y_1 \cos\theta_2 + y_2 \sin\theta_2)\right\} d\,\theta_2,$$

where we have used the identity $\sin^2(\theta_2)+\cos^2(\theta_2)=1$. To simplify (II.E.17) we introduce variables $r=[y_1^2+y_2^2]^{1/2}$ and $\phi = Tan^{-1}(y_2/y_1)$ so that $y_1 = r\cos\phi$ and $y_2 = r\sin\phi$. Then, using the identity $\cos\phi\cos\theta_2+\sin\phi\sin\theta_2=\cos(\theta_2-\phi)$, we have

$$L(y) = \frac{e^{-A^2/2\sigma^2}}{2\pi}\int_0^{2\pi} \exp\left\{\frac{Ar}{\sigma^2}\cos(\theta_2-\theta)\right\} d\,\theta_2$$

(II.E.18)

$$= e^{-A^2/2\sigma^2} I_0\left(\frac{Ar}{\sigma^2}\right),$$

where I_0 is the zeroth-order modified Bessel function of the first kind, defined by the integral in (II.E.18).

The function $I_0(x)$ is monotone increasing in its argument. So a test that compares $L(y)$ from (II.E.18) to a threshold τ is equivalent to one that compares r to another threshold τ' given by $\tau' = \sigma^2 / A I_0^{-1}(\tau e^{A^2/2\sigma^2})$. The Bayes, minimax, and Neyman-Pearson tests for (II.E.13) thus are of the form

$$\tilde{\delta}_0(y) = \begin{cases} 1 \text{ if } r > \tau' \\ \gamma \text{ if } r = \tau' \\ 0 \text{ if } r < \tau'. \end{cases} \tag{II.E.19}$$

Thus optimum tests for (II.E.13) operate by comparing r, the distance of the point (y_1, y_2) from the origin, to a threshold. This decision region is illustrated in Fig. II.E.1. (Note that the randomization is irrelevant here.) Further aspects of this and related models are discussed in Chapter III.

For composite hypothesis-testing problems in which we do not have a prior distribution (or conditional priors) for the parameter, the development of hypothesis tests which satisfy precise analytical definitions of optimality is very often an illusive task. One way of defining optimality in such problems is a generalization of the Neyman-Pearson criterion of Section II.D. Suppose that the parameter space is decomposed into two disjoint sets Λ_0 and Λ_1 as before. For a randomized decision rule $\tilde{\delta}$, we can define false-alarm and detection probabilities as follows

$$P_F(\tilde{\delta}; \theta) = E_\theta\{\tilde{\delta}(Y)\}, \; \theta \in \Lambda_0 \tag{II.E.20}$$

and

$$P_D(\tilde{\delta}; \theta) = E_\theta\{\tilde{\delta}(Y)\}, \; \theta \in \Lambda_1.$$

Suppose, as in the Neyman-Pearson formulation, we wish to be assured that the false-alarm probability does not exceed a

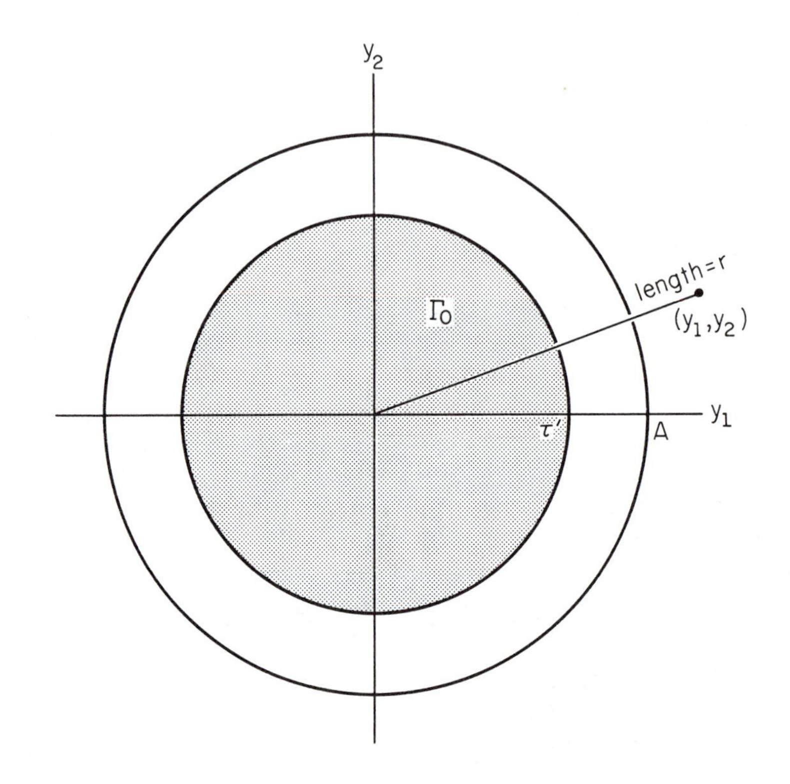

Fig. II.E.1: Decision regions for Example II.E.1 $(\Gamma_1 = \Gamma_0^c)$.

given value, say α. Then an ideal test would be one that maximizes $P_D(\tilde{\delta};\theta)$ for every $\theta \in \Lambda_1$ subject to this constraint $(P_F(\tilde{\delta};\theta) \leqslant \alpha, \theta \in \Lambda_0)$. Such a test is known as a *uniformly most powerful* (UMP) test of level α.

Unfortunately, although UMP tests are very desirable, they exist only under special circumstances. To see this consider the situation in which the null hypothesis (H_0) is simple, so that Λ_0 consists of a single element θ_0. Assuming that P_θ has density p_θ for each $\theta \in \Lambda$, the most powerful α-level test for H_0 versus the hypothesis that $Y \sim P_\theta$ has critical region $\Gamma_\theta = \{y \in \Gamma | p_\theta(y) > \tau p_{\theta_0}(y)\}$ with τ (and possibly a randomization) chosen to give size α. Also, from the Neyman-

Pearson lemma we know that this test is essentially unique so that any other α-level test will have smaller power. For example, if we choose two elements θ' and θ'' of Λ_1, the test with critical region $\Gamma_{\theta'}$ will have smaller power in testing H_0 versus $Y \sim P_{\theta''}$ than does the test with critical region $\Gamma_{\theta''}$ (and vice versa) unless these two critical regions are essentially identical. Thus, it follows that a UMP test exists for H_0 versus the composite hypothesis $H_1: Y \sim P_\theta, \theta \epsilon \Lambda_1$, if and only if the critical region Γ_θ is the same for all $\theta \epsilon \Lambda_1$. We illustrate this with the following example.

Example II.E.2: UMP Testing of Location

Consider the parametric family of distributions $\{P_\theta; \theta \epsilon \Lambda\}$, where P_θ is the $N(\theta, \sigma^2)$ distribution and Λ is a subset of $\mathbb{R}$. Suppose that we have the hypothesis pair

$$\begin{array}{c} H_0: \theta = \mu_0 \\ \text{versus} \\ H_1: \theta > \mu_0 \end{array} \tag{II.E.21}$$

where μ_0 is a fixed number. This is a problem with a simple null hypothesis $\Lambda_0 = \{\mu_0\}$ and a composite alternative $\Lambda_1 = (\mu_0, \infty)$. From Example II.D.1 we know that for each $\theta \epsilon \Lambda$, the most powerful α-level test of H_0 versus $Y \sim N(\theta, \sigma^2)$ has critical region $\Gamma_\theta = \{y \epsilon \Gamma | y > \sigma \Phi^{-1}(1-\alpha) + \mu_0\}$. Since this region does not depend on θ it thus gives a UMP test for (II.E.21) which we will denote by $\tilde{\delta}_1$. Note that (II.D.16) implies that

$$P_D(\tilde{\delta}_1; \theta) = 1 - \Phi\left(\Phi^{-1}(1-\alpha) - \frac{\theta - \mu_0}{\sigma}\right). \tag{II.E.22}$$

Alternatively, for the same family of distributions, suppose that we consider the hypothesis pair

$$\begin{array}{c} H_0: \theta = \mu_0 \\ \text{versus} \\ H_1: \theta \neq \mu_0 \end{array} \tag{II.E.23}$$

We now have the composite alternative $\Lambda_1=(-\infty, \mu_0)\cup(\mu_0, \infty)$. For $\theta>\mu_0$, the most powerful critical region is as given in the preceding paragraph. However, for $\theta<\mu_1$ it is straightforward to see that the most powerful α-level test has critical region

$$\Gamma_\theta=\{y\in\Gamma \mid y<\sigma\Phi^{-1}(\alpha)+\mu_0\}. \tag{II.E.24}$$

Although this region is independent of θ, it is quite different from Γ_θ for $\theta>\mu_0$. Thus no UMP test exists for (II.E.23).

If we denote by $\tilde{\delta}_2$ the test with critical region (II.E.24), then we have straightforwardly that

$$P_D(\tilde{\delta}_2;\theta)=\Phi\left(\Phi^{-1}(\alpha)+\frac{\theta-\mu_0}{\sigma}\right). \tag{II.E.25}$$

This quantity, together with $P_D(\tilde{\delta}_1;\theta)$ from (II.E.22), is plotted versus θ in Fig. II.E.2. Note that neither test performs well when θ is outside of its region of optimality. [A more reasonable test for (II.E.23) than either $\tilde{\delta}_1$ or $\tilde{\delta}_2$ is one that compares $|y-\mu_0|$ to a threshold; however, this test cannot be

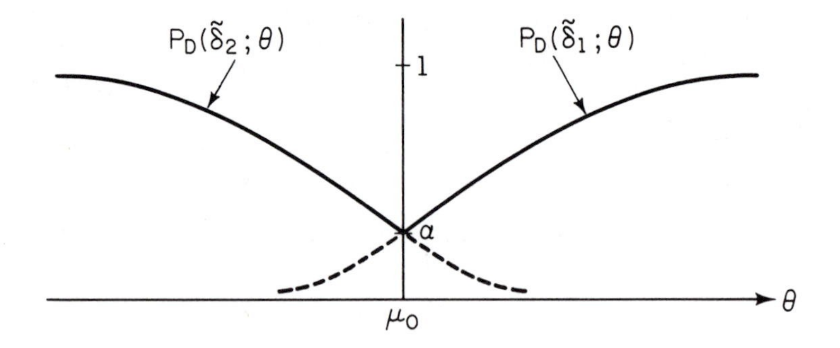

Fig. II.E.2: Power curves for test of $\theta=\mu_0$ versus $\theta>\mu_0$ and $\theta=\mu_0$ versus $\theta<\mu_0$, for location testing with Gaussian error.

UMP for (II.E.23).]

Example II.E.2 illustrates that the UMP criterion is too strong for many situations. Sometimes this can be overcome by applying other constraints to eliminate unreasonable tests from consideration. One such condition is *unbiasedness*, which requires that $P_D(\tilde{\delta};\theta) \geqslant \alpha$ for all $\theta \in \Lambda_1$ in addition to the constraint $P_F(\tilde{\delta};\theta) \leqslant \alpha$. Note that this requirement would eliminate both $\tilde{\delta}_1$ and $\tilde{\delta}_2$ in the example from consideration as tests for (II.E.23). Further discussion of this and related issues can be found, for example, in the book by Lehmann (1986).

In many situations of interest, the parameter set Λ is of the form $[\theta_0, \infty)$ with $\Lambda_0 = \{\theta_0\}$ and $\Lambda_1 = (\theta_0, \infty)$; so that we have the hypothesis pair

$$\begin{aligned} &H_0: \theta = \theta_0 \\ \text{versus} \quad & \\ &H_1: \theta > \theta_0. \end{aligned} \tag{II.E.26}$$

This type of situation arises, for example, in many signal detection problems in which $\theta_0 = 0$ and θ is a signal amplitude parameter. In many situations of this type, we are interested primarily in the case in which, under H_1, θ is near θ_0. If, for example, θ is a signal amplitude parameter, the latter case represents the situation in which the signal strength is small.

Consider a decision rule $\tilde{\delta}$. Within regularity we can expand $P_D(\tilde{\delta};\theta)$ in a Taylor series about θ_0; i.e.,

$$\begin{aligned} P_D(\tilde{\delta};\theta) = P_D(\tilde{\delta};\theta_0) &+ (\theta-\theta_0)P_D'(\tilde{\delta};\theta_0) \\ &+ O((\theta-\theta_0)^2), \end{aligned} \tag{II.E.27}$$

where $P_D'(\tilde{\delta};\theta) = \partial P_D(\tilde{\delta};\theta)/\partial\theta$. Note that $P_D(\tilde{\delta};\theta_0) = P_F(\tilde{\delta})$; so for all size-$\alpha$ tests, $P_D(\tilde{\delta};\theta)$ is given for θ near θ_0 by

$$P_D(\tilde{\delta};\theta) \cong \alpha + (\theta-\theta_0)P_D'(\tilde{\delta};\theta_0). \tag{II.E.28}$$

Thus for θ near θ_0 we can achieve approximate maximum power with size α by choosing $\tilde{\delta}$ to maximize $P_D'(\tilde{\delta};\theta_0)$. A test which maximizes $P_D'(\tilde{\delta};\theta_0)$ subject to false-alarm constraint $P_F(\tilde{\delta})\leq\alpha$, is called an α-level *locally most powerful* (LMP) test, or simply a *locally optimum* test.

To see the general structure of LMP tests we note that, assuming that P_θ has density p_θ for each $\theta\in\Lambda_1$, we can write

$$\begin{aligned} P_D(\tilde{\delta};\theta) &= E_\theta\{\tilde{\delta}(Y)\} \\ &= \int_\Gamma \tilde{\delta}(y)p_\theta(y)\mu(dy). \end{aligned} \tag{II.E.29}$$

Assuming sufficient regularity on $\{p_\theta(y);\theta\in\Lambda_1\}$ that we can interchange order of integration and differentiation, we have

$$P_D'(\tilde{\delta};\theta_0) = \int_\Gamma \tilde{\delta}(y)\frac{\partial}{\partial\theta}p_\theta(y)\Big|_{\theta=\theta_0}\mu(dy). \tag{II.E.30}$$

Comparison of (II.E.30) with (II.D.4) indicates that the α-level LMP design problem is the same as the α-level Neyman-Pearson design problem, in which we replace $p_1(y)$ with $\partial p_\theta(y)/\partial\theta|_{\theta=\theta_0}$. Using this analogy, it is straightforward to show that, within regularity, an α-level LMP test for (II.E.26) is given by

$$\tilde{\delta}_{lo}(y) = \begin{cases} 1 & > \\ \gamma \text{ if } \frac{\partial}{\partial\theta}p_\theta(y)\Big|_{\theta=\theta_0} & = \eta p_{\theta_0}(y), \\ 0 & < \end{cases} \tag{II.E.31}$$

where η and γ are chosen so that $P_F(\tilde{\delta}_{lo})=\alpha$. Details of this development can be found in the book by Ferguson (1968). LMP tests are discussed further in Chapter III.

In the absence of applicability of any of the above mentioned optimality criteria, a test that is often used for composite problems in which θ is the union of disjoint Λ_0 and Λ_1 is that based on comparing the quantity

$$\frac{\max_{\theta \in \Lambda_1} p_\theta(y)}{\max_{\theta \in \Lambda_0} p_\theta(y)} \tag{II.E.32}$$

to a threshold. This test is sometimes known as the *generalized likelihood-ratio test* or a *maximum-likelihood test*, and further motivation for tests of this type is found in Chapter IV.

II.F Exercises

1. Find the minimum Bayes risk for the binary channel of Example II.B.1.

2. Suppose Y is a random variable which, under hypothesis H_o, has pdf

$$p_0(y) = \begin{cases} (2/3)(y+1), & 0 \leqslant y \leqslant 1 \\ 0, & \text{otherwise} \end{cases}$$

and, under hypothesis H_1, has pdf

$$p_1(y) = \begin{cases} 1, & 0 \leqslant y \leqslant 1 \\ 0, & \text{otherwise.} \end{cases}$$

(a) Find the Bayes rule and minimum Bayes risk for testing H_0 versus H_1 with uniform costs and equal priors.
(b) Find the minimax rule and minimax risk for uniform costs.
(c) Find the Neyman-Pearson rule and the corresponding detection probability for false-alarm probability $\alpha \in (0, 1)$.

3. Repeat Exercise 2 for the situation in which p_j is given instead by

$$p_j(y) = \frac{(j+1)}{2} e^{-(j+1)|y|}, \ y \in \mathbb{R}, \ j = 0, 1.$$

For parts (a) and (b) assume costs

$$C_{ij} = \begin{cases} 0 & , \text{ if } i=j \\ 1 & , \text{ if } i=1 \text{ and } j=0 \\ 3/4 & , \text{ if } i=0 \text{ and } j=1, \end{cases}$$

and for part (a) assume priors $\pi_0 = 1/4$.

4. Repeat Exercise 2 for the situation in which p_0 and p_1 are given instead by

$$p_0(y) = \begin{cases} e^{-y} & , y \geq 0 \\ 0 & , y < 0 \end{cases}$$

and

$$p_1(y) = \begin{cases} \sqrt{2/\pi}\, e^{-y^2/2} & , y \geq 0 \\ 0 & , y < 0. \end{cases}$$

For part (a) consider arbitrary priors.

5. Repeat Exercise 2 for the hypothesis pair

$$H_0: Y \text{ has density } p_0(y) = \frac{1}{\sqrt{2\pi}} e^{-y^2/2}, \; y \in \mathbb{R}$$

versus

$$H_1: Y \text{ has density } p_1(y) = \begin{cases} 1/5, & \text{if } y \in [0, 5] \\ 0, & \text{if } y \notin [0, 5] \end{cases}$$

For part (a) assume priors $\pi_0 = 3/4$ and $\pi_1 = 1/4$.

6. Repeat Exercise 2 for the hypothesis pair

$$H_0: Y = N - s$$

versus

$$H_1: Y = N + s$$

where $s > 0$ is a fixed real number and N is a continuous random variable with density

$$p_N(n) = \frac{1}{\pi(1 + n^2)}, \; n \in \mathbb{R}.$$

7. (a) Consider the hypothesis pair

$$H_0: Y = N$$

versus

$$H_1: Y = N + S$$

where N and S are independent random variables each having pdf

$$p(x) = \begin{cases} e^{-x}, & x \geq 0 \\ 0, & x < 0. \end{cases}$$

Find the likelihood ratio between H_0 and H_1.
(b) Find the threshold and detection probability for α-level Neyman-Pearson testing in (a).

(c) Consider the hypothesis pair

$$H_0: Y_k = N_k, \quad k = 1, \ldots, n$$

versus

$$H_1: Y_k = N_k + S, \quad k = 1, \ldots, n$$

where $n > 1$ and $N_1, \ldots, N_n$, and S are independent random variables each having the pdf given in (a). Find the likelihood ratio.

(d) Find the threshold for α-level Neyman-Pearson testing in (c).

8. Show that the minimum-Bayes-risk function V (defined in Section C) is concave and continuous in $[0, 1]$. (After showing that V is concave you may use the fact that any concave function on $[0, 1]$ is continuous on $(0, 1)$.)

9. Suppose we have a real observation Y and binary hypotheses described by the following pair of pdf's:

$$p_0(y) = \begin{cases} (1-|y|), & \text{if } |y| \leq 1 \\ 0, & \text{if } |y| > 1 \end{cases}$$

and

$$p_1(y) = \begin{cases} (2-|y|)/4, & \text{if } |y| \leq 2 \\ 0, & \text{if } |y| > 2. \end{cases}$$

(a) Assume that the costs are given by

$$C_{01} = 2C_{10} > 0$$

$$C_{00} = C_{11} = 0.$$

Find the minimax test of H_0 versus H_1 and the corresponding minimax risk.

(b) Find the Neyman-Pearson test of H_0 versus H_1 with false-alarm probability α. Find the corresponding power of the test.

10. Suppose we observe a random variable Y given by

$$Y = N + \theta\lambda$$

where θ is either 0 or 1, λ is a fixed number between 0 and 2, and where N is a random variable which has a uniform density on the interval (-1, 1). We wish to decide between the hypotheses

$$H_0: \theta = 0$$

versus

$$H_1: \theta = 1.$$

(a) Find the Neyman-Pearson decision rule for false-alarm probability ranging from 0 to 1.
(b) Find the power of the Neyman-Pearson decision rule as a function of the false-alarm probability and the parameter λ. Sketch the receiver operating characteristics.

11. Consider the simple hypothesis testing problem for the real-valued observation Y:

$$H_0: p_0(y) = \exp(-y^2/2)/\sqrt{2\pi}, \quad y \in \mathbb{R}$$

$$H_1: p_1(y) = \exp(-(y-1)^2/2)/\sqrt{2\pi}, \quad y \in \mathbb{R}.$$

Suppose the cost assignment is given by $C_{00}=C_{11}=0$, $C_{10}=1$, and $C_{01}=N$. Investigate the behavior of the Bayes rule and risk for equally likely hypotheses and the minimax rule and risk when N is very large.

12. Consider the following Bayes decision problem:
The conditional density of the real observation Y given the real parameter $\Theta=\theta$ is given by

$$p_\theta(y) = \begin{cases} \theta e^{-\theta y}, & y \geqslant 0 \\ 0, & y < 0. \end{cases}$$

Θ is random variable with density

$$w(\theta) = \begin{cases} \alpha e^{-\alpha\theta}, & \theta \geqslant 0 \\ 0, & \theta < 0 \end{cases}$$

where $\alpha > 0$. Find the Bayes rule and minimum Bayes risk for the hypotheses

$$H_0: \Theta \in (0, \beta) \triangleq \Lambda_0$$

versus

$$H_1: \Theta \in [\beta, \infty) \triangleq \Lambda_1$$

where $\beta > 0$ is fixed. Assume the cost structure

$$C[i, \theta] = \begin{cases} 1, & \text{if } \theta \notin \Lambda_i \\ 0, & \text{if } \theta \in \Lambda_i. \end{cases}$$

13. Repeat Exercise 12 for the case in which Y consists of n independent (conditioned on Θ) and identically distributed observations $Y = Y_1, \ldots, Y_n$ each with the conditional density given in 12. You need not find the Bayes risk in closed form.

14. Consider the composite hypothesis testing problem:

$$H_0: Y \text{ has density } p_0(y) = \tfrac{1}{2} e^{-|y|}, \quad y \in \mathbb{R}$$

versus

$$H_1: Y \text{ has density } p_\theta(y) = \tfrac{1}{2} e^{-|y-\theta|}, \quad y \in \mathbb{R}, \ \theta > 0.$$

(a) Describe the locally most powerful α-level test and derive its power function.
(b) Does a uniformly most powerful test exist? If so, find it and derive its power function. If not, find the generalized likelihood ratio test for H_0 versus H_1.

15. In Section B, we formulated and solved the binary Bayesian hypothesis-testing problem. Generalize this formulation and solution to M hypotheses for $M>2$.

16. Formulate the M-ary minimax hypothesis testing problem. Show that a Bayes equalizer rule (if one exists) is minimax.

17. How would you formulate a criterion analogous to the Neyman-Pearson criterion for M hypotheses? Conjecture a solution.

18. Consider the following pair of hypotheses concerning a sequence $Y_1, Y_2, ..., Y_n$ of random variables

$$H_0: Y_k \sim N(\mu_0, \sigma_0^2), \; k = 1, 2, ..., n$$

versus

$$H_1: Y_k \sim N(\mu_1, \sigma_1^2), \; k = 1, 2, ..., n$$

where μ_0, μ_1, σ_1^2, and σ_2^2 are known constants.
(a) Show that the likelihood ratio can be expressed as a function of the parameters μ_0, μ_1, σ_0^2, and σ_1^2, and the quantities $\sum_{k=1}^n Y_k^2$ and $\sum_{k=1}^n Y_k$.
(b) Describe the Neyman-Pearson test for the two cases $(\mu_0=\mu_1, \sigma_1^2>\sigma_0^2)$ and $(\sigma_0^2=\sigma_1^2, \mu_1>\mu_0)$.
(c) Find the threshold and ROC's for the case $\mu_0=\mu_1$, $\sigma_1^2>\sigma_0^2$ with $n=1$.

19. Consider the hypotheses of Exercise 18 with $\mu \triangleq \mu_1>\mu_0=0$ and $\sigma^2 \triangleq \sigma_0^2=\sigma_1^2>0$. Does there exist a uniformly most powerful test of these hypotheses under the assumption that μ is known and σ^2 is not? If so, find it and show that it is UMP. If not, show why and find the generalized likelihood ratio test.

III SIGNAL DETECTION IN DISCRETE TIME

III.A Introduction

In Chapter II we discussed several basic optimality criteria and design methods for binary hypothesis-testing problems. In this chapter we apply these and related methods to derive optimum procedures for detecting signals embedded in noise. To avoid analytical complications, we consider exclusively the case of discrete-time detection, leaving the continuous-time case for Chapter VI. The discrete-time case is of considerable practical interest due to the trend of increased digitization of signal processing functions.

In Section III.B we discuss various models for signal detection problems and derive the resulting optimum detector structures corresponding to the criteria set forth in Chapter II. Section III.C deals with some methods of analyzing performance of these structures for situations in which the closed-form computation of relevant error probabilities is not tractable. There are several useful design methods for detection procedures other than those discussed in Chapter II, and in Sections III.D and III.E we introduce three such methods, namely sequential, robust, and nonparametric detection.

III.B Models and Detector Structures

The basic physical observation model that we wish to consider is that of an observed continuous-time waveform that consists of one of two possible signals corrupted by additive noise. Our objective is to decide which of the two possible signals is present, and we wish to do so by processing a finite number (say n) of samples taken from the observed waveform.

This problem can be modeled statistically by the following hypothesis pair for the observation space $(\Gamma, G)=(\mathbb{R}^n, B^n)$:

$$H_0: Y_k = N_k + S_{0k}, \quad k = 1, 2, ..., n$$

versus (III.B.1)

$$H_1: Y_k = N_k + S_{1k}, \quad k = 1, 2, ..., n,$$

where $\underline{Y}=(Y_1, ..., Y_n)^T$ is an observation vector consisting of the samples from the observed waveform, $\underline{N}=(N_1, ..., N_n)^T$ is a vector of noise samples, and $\underline{S}_0=(S_{01}, ..., S_{0n})^T$ and $\underline{S}_1=(S_{11}, ..., S_{1n})^T$ are vectors of samples from the two possible signals.† Note that the interpretation of $\underline{Y}$ as a vector of time samples is not the only possibility for (III.B.1) since the same model also arises if, for example, we take simultaneous (in time) samples from n spatially separated signal sensors or from the outputs of a bank of n parallel filters. In any case, we will refer to this subscript as a time parameter, although the results of course apply equally well to other situations modeled by (III.B.1).

Optimum procedures for deciding between H_0 and H_1 can be derived using the results of Chapter II if we have models for the statistical behavior of the signals and noise. For practical purposes the signals $\underline{S}_0$ and $\underline{S}_1$ can usually be classified as one of three basic types. They can be completely known (i.e., deterministic), they can be known except for a set of unknown (possibly random) parameters, or they can be completely random and thus specified only by their probability distributions. Sometimes (e.g., in radar/sonar problems) one of the signals, usually $\underline{S}_0$, is identically zero, so that we are actually trying to *detect* a signal embedded in noise. For the purposes of this treatment we will assume that the noise is independent of the signals under each hypothesis and that its probability distribution does not depend on which hypothesis is true. This assumption is valid for most applications, although in some applications the noise can depend on the signal (an example of this is the radar/sonar problem, in which the noise is sometimes partially composed of spurious signal reflections from the ground or from objects other than

† Here, and elsewhere in this book, vectors are taken to be columnar and superscript T denotes transposition.

the intended target.) We will also assume throughout that this noise distribution is determined by a (continuous or discrete) density p_N on $\mathbb{R}^n$.

Under the foregoing assumptions, the likelihood ratio for (III.B.1) can be computed if we know the statistics of $\underline{S}_j$ for $j=0, 1$. In particular, given $\underline{S}_j = \underline{s}_j \in \mathbb{R}^n$, the observation $\underline{Y}$ has conditional density (under H_j)

$$p_{\underline{N}}(\underline{y} - \underline{s}_j),\ \underline{y} \in \mathbb{R}^n. \qquad \text{(III.B.2)}$$

From (III.B.2) we see that the density of $\underline{Y}$ under H_j is given by

$$p_j(\underline{y}) = E\{p_{\underline{N}}(\underline{y} - \underline{S}_j)\},\ \underline{y} \in \mathbb{R}^n, \qquad \text{(III.B.3)}$$

where it should be noted that the expectation is with respect to the signal $\underline{S}_j$. The likelihood ratio then becomes

$$L(\underline{y}) = \frac{E\{p_{\underline{N}}(\underline{y} - \underline{S}_1)\}}{E\{p_{\underline{N}}(\underline{y} - \underline{S}_0)\}},\ \underline{y} \in \mathbb{R}^n. \qquad \text{(III.B.4)}$$

Thus optimum procedures for (III.B.1) are derived by computing (III.B.4) and are analyzed by finding the appropriate probabilities of the resulting critical regions. We now consider a number of important particular cases of this problem.

Case III.B.1: Detection of Deterministic Signals in Independent Noise

In many problems of interest, the two signals $\underline{S}_0$ and $\underline{S}_1$ are completely deterministic. In particular, we have $\underline{S}_j = \underline{s}_j$, with $\underline{s}_j \in \mathbb{R}^n$ being known to the designer. This is sometimes knows as the *coherent* detection problem. In this case $L(\underline{y})$ of (III.B.4) becomes

$$L(\underline{y}) = \frac{p_{\underline{N}}(\underline{y}-\underline{s}_1)}{p_{\underline{N}}(\underline{y}-\underline{s}_0)}$$

$$= \frac{p_{\underline{N}}(y_1-s_{11}, \ldots, y_n-s_{1n})}{p_{\underline{N}}(y_1-s_{01}, \ldots, y_n-s_{0n})}, \qquad \text{(III.B.5)}$$

and thus the optimum detector structure is easily determined if $p_{\underline{N}}$ is known. Not much can be said about tests based on (III.B.5) without making further simplifying assumptions. In general, (III.B.5) could be a complicated function of the observations which may be very difficult to implement and to analyze, particularly if the number of samples (n) is large. For example, the setting of thresholds in the minimax or Neyman-Pearson problems and the computation of performance require the evaluation of n-fold integrals of the form $\int_{\{L(\underline{y})>\tau\}} p_j(\underline{y})\mu(d\underline{y})$, a task that is often formidable.

An important special case of (III.B.1) with known signals is that in which the noise samples $N_1, \ldots, N_n$ are statistically independent. In this case we have

$$p_{\underline{N}}(\underline{y}) = \prod_{k=1}^{n} p_{N_k}(y_k), \qquad \text{(III.B.6)}$$

where p_{N_k} is the marginal density of N_k, so that $L(\underline{y})$ becomes

$$L(\underline{y}) = \prod_{k=1}^{n} L_k(y_k) \qquad \text{(III.B.7)}$$

with $L_k(y_k) = p_{N_k}(y_k - s_{1k})/p_{N_k}(y_k - s_{0k})$. Since $\log(x)$ is a strictly increasing function of x, comparison of $L(\underline{y})$ to a threshold τ is equivalent to the comparison of $\log L(\underline{y})$ to the threshold $\log(\tau)$. Thus the optimum tests for this case can be written as

$$\tilde{\delta}_o(\underline{y}) = \begin{cases} 1 & > \\ \gamma \text{ if } \sum_{k=1}^{n} \log L_k(y_k) & = \log \tau. \\ 0 & < \end{cases} \qquad \text{(III.B.8)}$$

As illustrated in Fig. III.B.1, this structure consists of a time-varying instantaneous nonlinearity, $\log L_k$, followed by an accumulator, which is in turn followed by a threshold comparator. The following two examples illustrate two particular cases of (III.B.8) arising in practice.

Example III.B.1: Coherent Detection in I.i.d. Gaussian Noise

Suppose that the noise samples $N_1, ..., N_n$ are independent and identically distributed (i.i.d.) with marginal distribution $N(0, \sigma^2)$. Such a noise model arises, for example, in communication receivers when the principal source of noise is the so-called *thermal noise* generated by the motion of the electrons in the receiver electronics. Also suppose for simplicity that $\underline{s}_0=\underline{0}$ where $\underline{0}$ denotes the n-vector of all zeros, and denote $\underline{s}_1$ by $\underline{s}$. (Note that this assumption does not result in any loss in generality since we could always redefine our observations as $\underline{y}'=\underline{y}-\underline{s}_0$ so that the signal would be $\underline{0}$ under H_0 and $\underline{s}=\underline{s}_1-\underline{s}_0$ under H_1.) We then have (as in Example II.B.2) that $\log L_k(y_k)=s_k(y_k-s_k/2)/\sigma^2$, so that the optimum test becomes

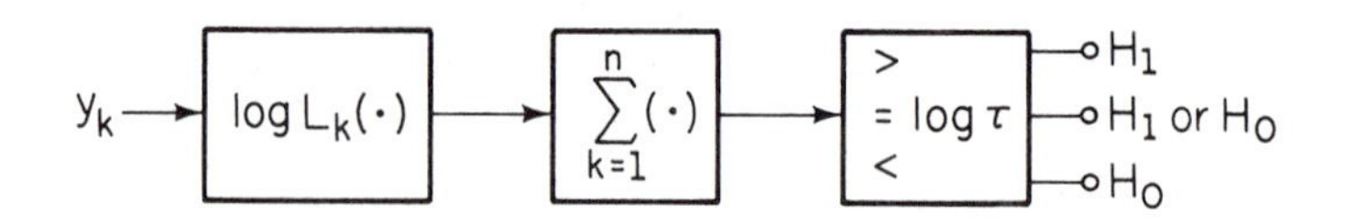

Fig. III.B.1: Detector structure for coherent signals in independent noise.

$$\tilde{\delta}_o(\underline{y}) = \begin{cases} 1 & > \\ \gamma \text{ if } \sum_{k=1}^{n} s_k(y_k - s_k/2) & = \tau' \\ 0 & <. \end{cases} \quad \text{(III.B.9)}$$

with $\tau' \triangleq \sigma^2 \log\tau$. This structure is depicted in Fig. III.B.2(a). Note that the term $-\frac{1}{2}\sum_{k=1}^{n} s_k^2$ can be incorporated into the threshold so that a test equivalent to (III.B.9) is one comparing $\sum_{k=1}^{n} s_k y_k$ to the threshold $\tau'' = \tau' + \frac{1}{2}\sum_{k=1}^{n} s_k^2$. This structure is depicted in Fig. III.B.2(b) and is known as a *correlation detector* or simply as a *correlator*.

An important feature of this optimum detector for Gaussian noise is that is operates by comparing the output of a *linear* system to a threshold. In particular, we can write $\sum_{k=1}^{n} s_k y_k = \sum_{k=-\infty}^{n} h_{n-k} y_k$, where

$$h_k = \begin{cases} s_{n-k} \text{ for } 0 \leqslant k \leqslant n-1 \\ 0 \text{ otherwise.} \end{cases} \quad \text{(III.B.10)}$$

Thus this detector can be viewed as a system that puts the observation sequence $y_1, ..., y_n$ into a digital linear filter and then samples the output at time n for comparison to a threshold. This structure is known as a *matched filter*.

Example III.B.2: Coherent Detection in I.i.d. Laplacian Noise

Suppose, as in Example III.B.1, that the noise samples $N_1, ..., N_n$ are i.i.d., but with the Laplacian marginal probability density

$$p_{N_k}(y_k) = \frac{\alpha}{2} e^{-\alpha|y_k|}, \quad y_k \in \mathbb{R}, \quad \text{(III.B.11)}$$

where $\alpha > 0$ is a scale parameter of the density. This model is sometimes used to represent the behavior of impulsive noises

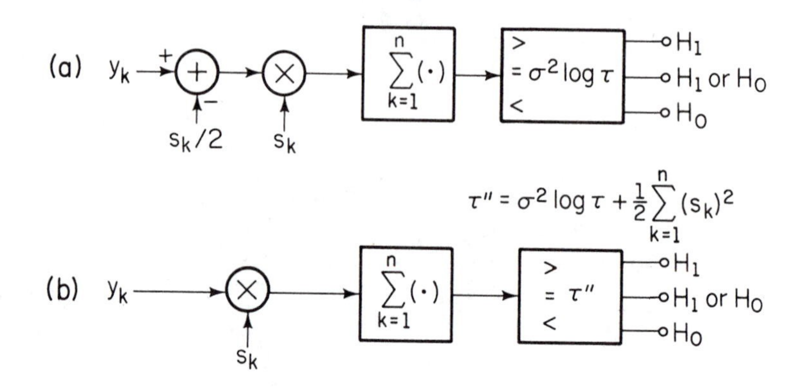

Fig. III.B.2: Optimum detector for coherent signals i.i.d. Gaussian noise.

in communications receivers. In comparison to the Gaussian model it is characterized by longer "tails," representing higher probabilities of large observations.

The function $\log L_k(y_k)$ for (III.B.11) is given by $\log L_k(y_k) = \alpha(|y_k| - |y_k - s_k|)$, which can be written as

$$\log L_k(y_k) = \begin{cases} -\alpha|s_k| & \text{if } sgn(s_k)y_k \leqslant 0 \\ \alpha|2y_k - s_k| & \text{if } 0 < sgn(s_k)y_k < |s_k| \\ +\alpha|s_k| & \text{if } sgn(s_k)y_k \geqslant |s_k|, \end{cases} \tag{III.B.12}$$

where sgn denotes the signum function

$$sgn(x) = \begin{cases} +1 \text{ if } x > 0 \\ 0 \text{ if } x = 0 \\ -1 \text{ if } x < 0. \end{cases} \qquad \text{(III.B.13)}$$

This function $\log L_k(y_k)$ is depicted in Fig. III.B.3 for both cases: $s_k < 0$ and $s_k > 0$. By inspection of these figures it can be seen that the optimum detectors based on $\sum_{k=1}^{n} \log L_k(y_k)$ can be implemented as

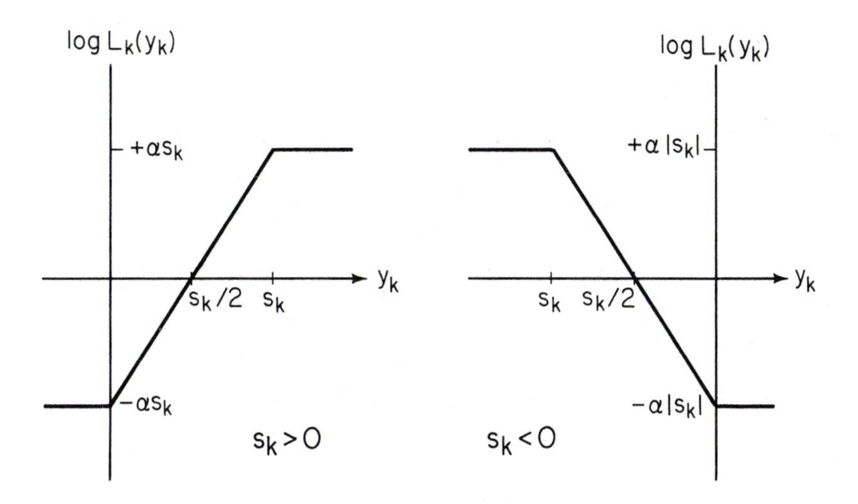

Fig. III.B.3: Per-sample log-likelihood ratio for coherent detection in Laplacian noise.

$$\tilde{\delta}_o(\underline{y}) = \begin{cases} 1 & > \\ \gamma \text{ if } \sum_{k=1}^{n} sgn(s_k) l_k(y_k) = \log \tau/2\alpha \triangleq \tau', \\ 0 & < \end{cases} \quad \text{(III.B.14)}$$

where $\tau' \triangleq \log\tau/2d$, and where the functions l_k are given by

$$l_k(x) = \begin{cases} -|s_k|/2 & \text{if } x \leqslant -|s_k|/2 \\ x & \text{if } -|s_k|/2 < x < |s_k|/2 \\ +|s_k|/2 & \text{if } x \geqslant +|s_k|/2. \end{cases} \quad \text{(III.B.15)}$$

Note that $l_k(x)$ is linear in x if $x \in (-|s_k|/2, +|s_k|/2)$, and otherwise $l_k(x)$ equals $|s_k| sgn(x)/2$. This function is known sometimes as a *soft limiter* or *amplifier limiter*. The system of (III.B.14) is illustrated in Fig. III.B.4.

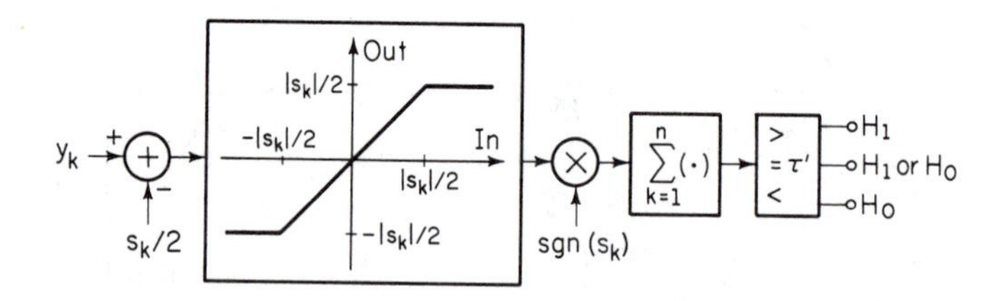

Fig. III.B.4: Optimum detector for coherent signals in Laplacian noise.

It is interesting to contrast the detectors of Figs. III.B.2(a) and III.B.4. Both systems "center" the observations by subtracting $s_k/2$ from each y_k. The system of

Fig. III.B.2(a) then correlates the centered data with the known signal and compares the output of this correlation with a threshold. Alternatively, the detector of Fig. III.B.4 soft-limits the centered data and then correlates these soft-limited observations with the sequence of signal signs. The effect of this soft limiting is to reduce the effect of large observations on the sum, thus making this system more tolerant to large noise values than the linear system of Fig. III.B.2(a). This process is consistent with the behavior of the Laplacian random variables as compared to the Gaussian as discussed above.

Example III.B.3: Locally Optimum Detection of Coherent Signals in I.i.d. Noise

In many detection problems the form of the received signal is known but not its amplitude. To model this problem we consider the composite hypothesis testing problem described by

$$\begin{aligned} &H_0: Y_k = N_k, \quad k = 1, 2, ..., n \\ \text{versus} \\ &H_1: Y_k = N_k + \theta s_k, \quad k = 1, 2, ..., n, \theta > 0, \end{aligned} \tag{III.B.16}$$

where $\underline{s} = (s_1, ..., s_n)^T$ is a known signal, $\underline{N} = (N_1, ..., N_n)^T$ is a continuous random noise vector with i.i.d. components and marginal probability density functions p_{N_k}, and θ is a signal-strength parameter. Given θ, the likelihood ratio between H_0 and H_1 is given by

$$L_\theta(\underline{y}) = \prod_{k=1}^{n} \frac{p_{N_k}(y_k - \theta s_k)}{p_{N_k}(y_k)}. \tag{III.B.17}$$

Note that the critical region $\Gamma_\theta = \{\underline{y} \in \mathbb{R}^n \,|\, L_\theta(\underline{y}) > \tau\}$ will generally depend on θ in this situation except for some special cases (the Gaussian noise case of Example III.B.1 is one such special case). Thus UMP tests for (III.B.16) exist only for particular noise models. However, LMP tests for (III.B.16) have a particularly simple and intuitively reasonable structure, and thus it is of interest to consider locally optimum detection for this case.

Assuming sufficient regularity on p_{N_k}, the locally optimum test for H_0 versus H_1 is given by [see (II.E.31)]

$$\tilde{\delta}_{lo}(\underline{y}) = \begin{cases} 1 & < \\ \gamma \text{ if } \frac{\partial}{\partial\theta} L_\theta(\underline{y})\Big|_{\theta=0} & = \tau. \\ 0 & < \end{cases} \qquad \text{(III.B.18)}$$

Upon differentiation of (III.B.17) we have

$$\frac{\partial}{\partial\theta} L_\theta(\underline{y})\Big|_{\theta=0} = \sum_{k=1}^{n} s_k \, g_{lo}(y_k), \qquad \text{(III.B.19)}$$

where $g_{lo}(x) \triangleq -p'_{N_1}(x)/p_{N_1}(x)$, and where $p'_{N_1}(x)=dp_{N_1}(x)/dx$. This structure is depicted in Fig. III.B.5. It consists of the memoryless nonlinearity g_{lo} followed by a correlator, a combination known as a *nonlinear correlator*.

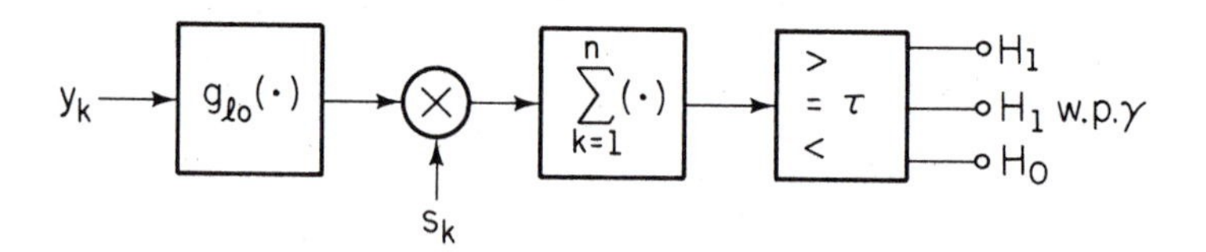

Fig. III.B.5: Locally optimum detector structure for coherent signals in i.i.d. noise.

Like the likelihood ratio, the locally optimum nonlinearity g_{lo} shapes the observations to reduce the detrimental effects of the noise as much as is possible. For example, with $N(0, \sigma^2)$ noise, we have $g_{lo}(x)=x/\sigma^2$, so that Fig. III.B.5 is simply the correlation detector of Fig. III.B.2(b). (This must be so; since this detector is UMP it is also LMP.) For Laplacian noise with density (III.B.11) we have $g_{lo}(x)=\alpha sgn(x)$, so that the locally optimum detector correlates the signal with the sequence of signs of the observations, as depicted in Fig. III.B.6. The function $g_{lo}(x)$ in this case is known as a *hard limiter*. An even heavier-tailed noise model than the Laplacian is described by the Cauchy density given by

$$p_{N_1}(x) = \frac{1}{\pi(1+x^2)}, \quad -\infty < x < \infty. \qquad \text{(III.B.20)}$$

For this case $g_{lo}(x)=2x/(1+x^2)$, which is approximately linear near $x=0$ and then redescends (asymptotically) to zero. Thus this detector ignores observations with very large magnitudes. This behavior is depicted in Fig. III.B.7. (An approximation to this nonlinearity is the so-called *noise blanker*,

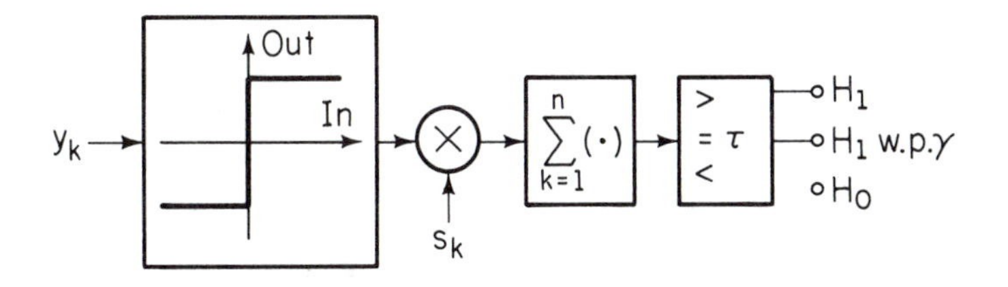

Fig. III.B.6: Locally optimum detector for Laplacian noise.

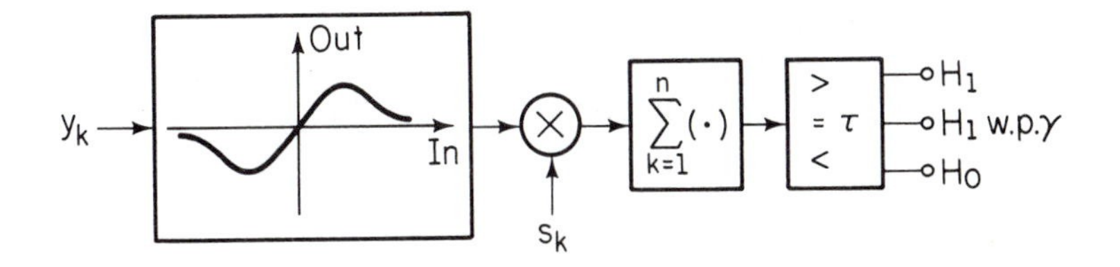

Fig. III.B.7: Locally optimum detector for Cauchy noise.

$$g(x) = \begin{cases} x \text{ if } |x| \leqslant K \\ \\ 0 \text{ if } |x| > K, \end{cases} \qquad \text{(III.B.21)}$$

where $K > 0$ is a constant, which is sometimes used in practice to combat extremely impulsive noise.)

Case III.B.2: Detection of Deterministic Signals in Gaussian Noise

If the noise samples N_k in (III.B.1) are not independent of one another, then the optimum solutions of (III.B.4) do not exhibit any particular structure [beyond, say, (III.B.5)] even for the case of deterministic signals. An important exception to this lack of general structure is the situation in which the signals are deterministic ($\underline{S}_j = \underline{s}_j$) and the noise vector $\underline{N}$ has a multivariate Gaussian distribution. In this case, the optimum detectors have simple, easily implemented structures and the performance of the optimum systems can be analyzed thoroughly. Moreover, the assumption of Gaussian noise is often justifiable in practice, and the systems derived to be optimum under this assumption are intuitively reasonable systems to use even when the noise is not Gaussian. Thus it is of interest to consider this particular case in some detail.

Let us assume, then, that the signals $\underline{S}_j$ take known values $\underline{s}_j \in \mathbb{R}^n$ and that the noise vector $\underline{N}$ is a Gaussian random vector with mean vector $\underline{0}$ and covariance matrix Σ_N. (The assumption of zero-mean noise does not reduce the generality of these results since we can always subtract a nonzero noise mean from $\underline{y}$ to produce a new observation with zero-mean noise.)

Recall that a *Gaussian random vector* in $\mathbb{R}^n$ with mean vector $\underline{\mu} \triangleq E\{\underline{X}\} \in \mathbb{R}^n$ and $n \times n$ covariance matrix $\Sigma \triangleq E\{(\underline{X}-\underline{\mu})(\underline{X}-\underline{\mu})^T\}$ is one with probability density function

$$p_{\underline{X}}(\underline{x}) = \frac{1}{(2\pi)^{n/2}|\Sigma|^{1/2}} \times \exp\{-\tfrac{1}{2}(\underline{x}-\underline{\mu})^T \Sigma^{-1}(\underline{x}-\underline{\mu})\}, \ \underline{x} \in \mathbb{R}^n , \tag{III.B.22}$$

where $|\Sigma|$ denotes the determinant of Σ and where Σ^{-1} denotes the inverse of Σ. We denote such a random variable as a $N(\underline{\mu}, \Sigma)$ random variable. Note that a covariance matrix is always nonnegative definite (i.e., $\underline{x}^T \Sigma \underline{x} \geqslant 0$ for all $\underline{x} \in \mathbb{R}^n$ and, in writing (III.B.22), we have assumed that Σ is actually positive definite ($\underline{x}^T \Sigma \underline{x} > 0$ for all $\underline{x} \in \mathbb{R}^n$ except $\underline{x}=0$). Positive definiteness of Σ implies that $|\Sigma|>0$ and that Σ^{-1} exists. If Σ is not positive definite, this implies that at least one of the components of $\underline{X}$ can be written as a linear combination of the others and is thus redundant. In dealing with Gaussian random vectors here we will assume, unless otherwise noted, that the covariance matrix is positive definite. For a discussion of Gaussian random vectors, the reader is referred to Thomas (1986).

Given the foregoing assumptions on $\underline{S}_j$ and $\underline{N}$, the likelihood ratio can be derived straightforwardly. Since the signals are known, their effect on the distribution of $\underline{Y}$ is merely to shift the mean from that of $\underline{N}$, and thus $\underline{Y} \sim N(\underline{s}_j, \Sigma_N)$ under H_j for $j=0$ and 1. We see that the likelihood ratio is given by

$$L(\underline{y}) = \frac{p_1(\underline{y})}{p_0(\underline{y})}$$

$$= \frac{\dfrac{1}{(2\pi)^{n/2}|\Sigma_N|^{1/2}} \exp\left\{-\tfrac{1}{2}(\underline{y}-\underline{s}_1)^T \Sigma_N^{-1}(\underline{y}-s_1)\right\}}{\dfrac{1}{(2\pi)^{n/2}|\Sigma_N|^{1/2}} \exp\left\{-\tfrac{1}{2}(\underline{y}-\underline{s}_0)^T \Sigma_N^{-1}(\underline{y}-s_0)\right\}}$$

$$= \exp\left\{\underline{s}_1^T \boldsymbol{\Sigma}_N^{-1}\underline{y} - \underline{s}_0^T \boldsymbol{\Sigma}_N^{-1}\underline{y} - ½\underline{s}_1\boldsymbol{\Sigma}_N^{-1}\underline{s}_1 + ½\underline{s}_0^T \boldsymbol{\Sigma}_N^{-1}\underline{s}_0\right\}$$

$$= \exp\left\{(\underline{s}_1 - \underline{s}_0)^T \boldsymbol{\Sigma}_N^{-1}\left(\underline{y} - \frac{\underline{s}_0 + \underline{s}_1}{2}\right)\right\}, \ \underline{y} \in \mathbb{R}^n, \tag{III.B.23}$$

where in the next to last equality we have used the fact that $\boldsymbol{\Sigma}_N$ (and hence $\boldsymbol{\Sigma}_N^{-1}$) is a symmetric matrix to write $\underline{s}_j^T \boldsymbol{\Sigma}_N^{-1}\underline{y} = \underline{y}^T \boldsymbol{\Sigma}_N^{-1}\underline{s}_j$.

It is interesting to compare (III.B.23) with its scalar counterpart (II.B.27). In doing so we see that the problem under consideration here is the vector version of the simple location testing problem of Example II.B.2, where the locations μ_0 and μ_1 are replaced with location vectors $\underline{s}_0$ and $\underline{s}_1$ and the noise variance σ^2 is replaced with noise covariance matrix $\boldsymbol{\Sigma}_N$.

The optimum tests based on (III.B.23) are more simply written by taking logarithms, and since the term $½(\underline{s}_1 - \underline{s}_0)^T \boldsymbol{\Sigma}_N^{-1}(\underline{s}_0 + \underline{s}_1)$ does not depend on $\underline{y}$ it can be incorporated into the decision threshold. The optimum tests become

$$\tilde{\delta}_o(\underline{y}) = \begin{cases} 1 & > \\ \gamma \text{ if } (\underline{s}_1 - \underline{s}_0)^T \boldsymbol{\Sigma}_N^{-1}\underline{y} & = \tau' \\ 0 & < \end{cases} \tag{III.B.24}$$

with $\tau' = \log\tau + ½(\underline{s}_1 - \underline{s}_0)^T \boldsymbol{\Sigma}_N^{-1}(\underline{s}_0 + \underline{s}_1)$. Note that we can write

$$(\underline{s}_1 - \underline{s}_0)^T \boldsymbol{\Sigma}_N^{-1}\underline{y} = \underline{\tilde{s}}^T \underline{y} = \sum_{k=1}^{n} \tilde{s}_k y_k, \tag{III.B.25}$$

with $\underline{\tilde{s}} = \boldsymbol{\Sigma}_N^{-1}(\underline{s}_1 - \underline{s}_0)$, and thus the detector structure here is identical to the correlation detector of Fig. III.B.2(b) with the actual signal $\underline{s}$ replaced by the "pseudosignal" $\underline{\tilde{s}}$. So for this

Gaussian case, detector implementation is no more difficult for dependent noise than for independent noise.

Further aspects of the structure of (III.B.24) will be discussed below. However, it is of interest first to consider the performance of (III.B.24). Note that the quantity $T(\underline{Y}) \triangleq (\underline{s}_1-\underline{s}_0)^T \Sigma_N^{-1}\underline{Y}$ is a linear transformation of the Gaussian random vector $\underline{Y}$. A basic property of the multivariate Gaussian distribution is that linear transformations of Gaussian vectors are also Gaussian. In this case, the transformation is to $\mathbb{R}$, so that $T(\underline{Y})$ is a Gaussian random variable, and thus we can characterize its distributions under H_0 and H_1 completely by finding its means and variances under the two hypotheses. Under H_j the mean of $T(\underline{Y})$ is given by [with $\underline{\tilde{s}}=\Sigma_N^{-1}(\underline{s}_1-\underline{s}_0)$]

$$\begin{aligned} E\{T(\underline{Y})|H_j\} &= E\{\underline{\tilde{s}}^T \underline{Y}|H_j\} = \underline{\tilde{s}}^T E\{\underline{Y}|H_j\} \\ &= \underline{\tilde{s}}^T E\{\underline{N}|H_j\}+\underline{\tilde{s}}^T \underline{s}_j = \underline{\tilde{s}}^T \underline{s}_j \triangleq \tilde{\mu}_j . \end{aligned} \tag{III.B.26}$$

Similarly, the variance of $T(\underline{Y})$ under H_j is

$$\begin{aligned} Var(T(\underline{Y})|H_j) &= E\{(\underline{\tilde{s}}^T \underline{Y}-\underline{\tilde{s}}^T \underline{s}_j)^2|H_j\} = E\{(\underline{\tilde{s}}^T \underline{N})^2\} \\ &= E\{\underline{\tilde{s}}^T \underline{N}\underline{N}^T \underline{\tilde{s}}\} = \underline{\tilde{s}}^T E\{\underline{N}\underline{N}^T\}\underline{\tilde{s}} \\ &= \underline{\tilde{s}}^T \Sigma_N \underline{\tilde{s}} = (\underline{s}_1-\underline{s}_0)^T \Sigma_N^{-1}(\underline{s}_1-\underline{s}_0) \triangleq d^2. \end{aligned} \tag{III.B.27}$$

Note that the variance of $T(\underline{Y})$ is independent of the hypothesis. Also note that the positive definiteness of Σ_N implies that Σ_N^{-1} is also positive definite and thus that $d^2>0$ unless $\underline{s}_1=\underline{s}_0$.

From the analysis above we see that $T(\underline{Y})\sim N(\tilde{\mu}_j, d^2)$ under H_j for $j=0, 1$. This implies, among other things, that the randomization γ in (III.B.24) is irrelevant. The probability of choosing H_1 under H_j is thus given by

$$P_j(\Gamma_1) = \frac{1}{\sqrt{2\pi}d} \int_{\tau'}^{\infty} e^{-(x-\tilde{\mu}_j)^2/2d^2} dx$$
$$= 1 - \Phi\left(\frac{\tau' - \tilde{\mu}_j}{d}\right) \tag{III.B.28}$$

with τ' from (III.B.24) and where d is the positive square root of d^2. For the Bayesian problem it is convenient to write (III.B.28) in terms of the original threshold τ, in which case we have

$$P_j(\Gamma_1) = \begin{cases} 1 - \Phi\left(\frac{\log \tau}{d} + \frac{d}{2}\right) & \text{for } j = 0 \\ \\ 1 - \Phi\left(\frac{\log \tau}{d} - \frac{d}{2}\right) & \text{for } j = 1. \end{cases} \tag{III.B.29}$$

Comparing (III.B.29) and (II.B.31), we see that Bayesian performance in the problem under consideration here is identical with that in the simple location testing problem of Example II.B.2 with d defined in (II.B.31) being the $n=1$ case of d defined in (III.B.27). Similarly, performance and threshold selection for the minimax and Neyman-Pearson problems here are the same as the scalar cases of Examples II.C.1 and II.D.1, respectively, with the identification of $\tilde{\mu}_j$ and d of (III.B.28) with μ_j and σ of (II.B.30). For example, for α-level Neyman-Pearson testing we set $P_F(\tilde{\delta}_0) = P_0(\Gamma_1) = \alpha$ to yield a threshold of

$$\tau' = d\,\Phi^{-1}(1-\alpha) + \tilde{\mu}_0, \tag{III.B.30}$$

and the corresponding detection probability becomes

$$P_D(\tilde{\delta}_{NP}) = P_1(\Gamma_1) = 1-\Phi\left(\frac{\tau'-\tilde{\mu}_1}{d}\right)$$

(III.B.31)

$$= 1-\Phi(\Phi^{-1}(1-\alpha)-d).$$

Thus comparing (III.B.31) and (II.D.16), we note that the power curves and ROCs for this general problem are those of Figs. II.D.3 and II.D.4.

In the following remarks we discuss several interesting features of this general problem of detecting coherent signals in Gaussian noise.

Remark III.B.1: Interpretation of $\mathbf{d}^2$

In view of the discussion above we see that the performance of optimum detection of deterministic signals in Gaussian noise generally improves monotonically with increasing d. As we indicated in Example II.D.2 this quantity (or more properly its square) can be interpreted as a measure of signal-to-noise ratio. To see this, consider first the case treated in Example III.B.1, in which the signals are $\underline{s}_0=\underline{0}$ and $\underline{s}_1=\underline{s}$ and the noise is i.i.d. $N(0, \sigma^2)$, which corresponds to the multivariate Gaussian case with $\mathbf{\Sigma}_N=\sigma^2\mathbf{I}$, where $\mathbf{I}$ denotes the $n \times n$ identity matrix; i.e.,

$$\mathbf{\Sigma}_N = \begin{pmatrix} \sigma^2 & & & & \\ & \sigma^2 & 0 & & \\ & & \cdot & & \\ & 0 & & \cdot & \\ & & & & \cdot \\ & & & & & \sigma^2 \end{pmatrix}. \qquad \text{(III.B.32)}$$

In this case $\mathbf{\Sigma}_N^{-1}=\sigma^{-2}\mathbf{I}$, so d^2 becomes

$$d^2 = (\underline{s}_1 - \underline{s}_0)^T \Sigma_N^{-1} (\underline{s}_1 - \underline{s}_0) = \frac{\underline{s}^T \mathbf{I} \underline{s}}{\sigma^2} = \frac{\underline{s}^T \underline{s}}{\sigma^2}$$

$$= \frac{1}{\sigma^2} \sum_{k=1}^{n} s_k^2 = n \frac{\overline{s^2}}{\sigma^2}, \tag{III.B.33}$$

where $\overline{s^2} \triangleq (1/n) \sum_{k=1}^{n} s_k^2$ is the average signal power. Note that $\sigma^2 = (1/n) \sum_{k=1}^{n} E\{N_k^2\}$ is the average noise power, so that d^2 here is given by the signal-to-noise average power ratio times the number of samples. Thus performance is enhanced by increasing either of these quantities, and as either of the two increases without bound perfect performance can result.

A similar interpretation can be given to d^2 in the non-i.i.d. case with $\underline{s}_0 = \underline{0}$ and $\underline{s}_1 = \underline{s}$. In particular, as in (III.B.10), we can write the quantity $\sum_{k=1}^{n} \tilde{s}_k y_k$ as the input at time n of a linear time-invariant filter with impulse response

$$\tilde{h}_k = \begin{cases} \tilde{s}_{n-k}, & 0 \leqslant k \leqslant n-1 \\ 0, & \text{otherwise.} \end{cases} \tag{III.B.34}$$

If the input to this filter consisted of the signal $s_1, \ldots, s_n$ only, then the output at time n would be

$$\sum_{k=1}^{n} \tilde{s}_k s_k = \underline{\tilde{s}}^T \Sigma_N^{-1} \underline{s} = d^2.$$

Thus the output *power* at the sampling time due to *signal only* is d^4. On the other hand, if the noise only were put into this filter, the output at time n would be $\sum_{k=1}^{n} \tilde{s}_k N_k$, a random quantity with a power

$$E\left\{\left(\sum_{k=1}^{n} \tilde{s}_k N_k\right)^2\right\} = E\{(\underline{\tilde{s}}^T \underline{N})^2\} = d^2, \tag{III.B.35}$$

where we have used (III.B.27). So the ratio of the power output of $\tilde{h}_k$ due to signal only to that due to noise only is

$$\frac{\left(\sum_{k=1}^{n} \tilde{s}_k s_k\right)^2}{E\left(\sum_{k=1}^{n} \tilde{s}_k N_k\right)^2} = \frac{d^4}{d^2} = d^2. \qquad \text{(III.B.36)}$$

Thus the quantity d^2 in the general case is the signal-to-noise power ratio at the *output* of the filter used for optimum detection at the sampling time n. It is intuitively reasonable that the higher this output SNR is, the better the signal can be detected by comparing the sampled output to a threshold, and this intuition is borne out by the monotonicity of detection performance as a function of d^2 shown above.

It is interesting to note that the filter $\tilde{h}_k$ of (III.B.34) has maximum output signal-to-noise power ratio at time n among all linear filters with impulse response of length n (see Exercise 1). This result relies only on the fact that $\underline{N}$ is zero-mean with covariance Σ_N, and thus is true even for non-Gaussian noise. However, the optimality of (III.B.34) for Bayes, minimax, and Neyman-Pearson testing depends heavily on the assumption of Gaussian noise.

The quantity d^2 also has another interpretation for the i.i.d. case with general signals. In this case we can write

$$d^2 = \frac{1}{\sigma^2}\|\underline{s}_1 - \underline{s}_0\|^2, \qquad \text{(III.B.37)}$$

where $\|\underline{s}_1 - \underline{s}_0\|$ denotes the Euclidean distance between the signal vectors $\underline{s}_0$ and $\underline{s}_1$ given by

$$\|\underline{s}_1 - \underline{s}_0\| = [\textstyle\sum_{k=1}^{n}(s_{1k} - s_{0k})^2]^{1/2}.$$

Thus the farther apart the signal vectors are, the better performance can be achieved. A similar interpretation can be made in the non-i.i.d. noise case, as will be discussed below.

Remark III.B.2: Reduction to the I.i.d. Noise Case

Since Σ_N is an $n \times n$ symmetric positive-definite matrix, it has several structural properties that can be examined to give some insight into the structure of the optimum detection system. The eigenvalues $\lambda_1, ..., \lambda_n$ and corresponding eigenvectors $\underline{v}_1, ..., \underline{v}_n$ of an $n \times n$ matrix Σ_N are the solutions to the equation $\Sigma_N \underline{v}_k = \lambda_k \underline{v}_k$. (The set of eigenvalues of a matrix is unique but the set of eigenvectors is not.) Since Σ_N in our case is symmetric and positive definite, all of its eigenvalues are real and positive and its eigenvectors can be chosen to be orthonormal (i.e., $\underline{v}_k^T \underline{v}_l = 0$ if $k \neq l$ and $\underline{v}_k^T \underline{v}_k = 1$, for all $l, k = 1, ..., n$). With this choice of eigenvectors we can write Σ_N as

$$\Sigma_N = \sum_{k=1}^{n} \lambda_k \underline{v}_k \underline{v}_k^T. \qquad \text{(III.B.38)}$$

Equation (III.B.38) is called the *spectral decomposition* of Σ_N and its validity follows easily from the fact that the orthonormal set $\underline{v}_1, ..., \underline{v}_n$ forms a basis for $\mathbb{R}^n$ (as does any set of n linearly independent vectors in $\mathbb{R}^n$)†. Note that the matrix $\underline{v}_k \underline{v}_k^T$, when multiplied by a vector $\underline{x}$, gives the *projection* of $\underline{x}$ onto $\underline{v}_k$.

Using (II.B.38) it follows straightforwardly that $\Sigma_N^{-1} = \sum_{k=1}^{n} \lambda_k^{-1} \underline{v}_k \underline{v}_k^T$, from which the optimum detection statistic $T(\underline{y})$ is given by

$$T(\underline{y}) = (\underline{s}_1 - \underline{s}_0)^T \Sigma_N^{-1} \underline{y} = \sum_{k=1}^{n} (\hat{s}_{1k} - \hat{s}_{0k}) \hat{y}_k, \qquad \text{(III.B.39)}$$

where

$$\hat{y}_k = \underline{v}_k^T \underline{y} / \sqrt{\lambda_k}, \quad k = 1, ..., n \qquad \text{(IIII.B.40)}$$

---

† Indeed, for any $\underline{x} \in \mathbb{R}^n$, we can write $\underline{x} = \sum_{k=1}^{n} c_k \underline{v}_k$ with $c_k = \underline{v}_k^T \underline{x}$, so we have

$$\Sigma_N \underline{x} = \sum_{k=1}^{n} c_k \Sigma_N \underline{v}_k = \sum_{k=1}^{n} \lambda_k \underline{v}_k c_k = \sum_{k=1}^{n} \lambda_k \underline{v}_k \underline{v}_k^T \underline{x} = \left(\sum_{k=1}^{n} \lambda_k \underline{v}_k \underline{v}_k^T \right) \underline{x}.$$

and

$$\hat{s}_{jk} = \underline{v}_k^T \underline{s}_j / \sqrt{\lambda_k}, \quad k = 1, ..., n \text{ and } j = 0, 1.$$

Note that $\underline{y}$ can be obtained from $\hat{\underline{y}}$ by $\underline{y} = \sum_{k=1}^n \sqrt{\lambda_k}\, \hat{y}_k \underline{v}_k$, so that $\hat{\underline{y}}$ is an equivalent observation to $\underline{y}$. In terms of $\hat{\underline{Y}}$ (the random vector corresponding to $\hat{\underline{y}}$), the hypothesis pair II.B.1 becomes

$$\begin{array}{c} H_0: \hat{\underline{Y}} = \hat{\underline{N}} + \hat{\underline{s}}_0 \\ \text{versus} \\ H_1: \hat{\underline{Y}} = \hat{\underline{N}} + \hat{\underline{s}}_1, \end{array} \tag{III.B.41}$$

where $\hat{N}_k = \underline{v}_k^T \underline{N} / \sqrt{\lambda_k}$. Note that $\hat{\underline{N}}$ is a Gaussian random vector since it is a linear transformation of $\underline{N}$. Also,

$$\begin{aligned} E\{\hat{N}_k \hat{N}_l\} &= E\{\underline{v}_k^T \underline{N} \underline{v}_l^T \underline{N}\} / \sqrt{\lambda_k \lambda_l} \\ &= E\{\underline{v}_k^T \underline{N}\underline{N}^T \underline{v}_l\} / \sqrt{\lambda_k \lambda_l} \\ &= \underline{v}_k^T E\{\underline{N}\underline{N}^T\} \underline{v}_l / \sqrt{\lambda_k \lambda_l} \\ &= \underline{v}_k^T \Sigma_N \underline{v}_l / \sqrt{\lambda_k \lambda_l} = \underline{v}_k^T \underline{v}_l \sqrt{\lambda_l / \lambda_k}, \end{aligned} \tag{III.B.42}$$

where we have used the fact that $\underline{v}_k$ is an eigenvector of Σ_N. By the orthonormality of $\underline{v}_1, ..., \underline{v}_n$, (III.B.42) implies that

$$E\{\hat{N}_k \hat{N}_l\} = \begin{cases} 1 \text{ if } k = l \\ 0 \text{ if } k \neq l, \end{cases} \tag{III.B.43}$$

so $\hat{N}_1, ..., \hat{N}_n$ are i.i.d. $N(0, 1)$ random variables. Thus by the appropriate linear transformation of $\underline{Y}$, we have transformed a problem with dependent Gaussian noise into an equivalent problem with i.i.d. Gaussian noise. Of course, (III.B.39) gives the optimum detection statistic for this transformed problem.

Essentially what we have done above is to change from the original standard coordinate system in $\mathbb{R}^n$ to a different coordinate system in which the usual axes are aligned with the vectors $\underline{v}_1, \ldots, \underline{v}_n$. In the latter system the noise coordinates $\hat{N}_1, \ldots, \hat{N}_n$ are independent. Another way of looking at this change of coordinates is to write $\mathbf{\Sigma}_N = \mathbf{B}^2$, where $\mathbf{B}$ is the matrix $\sum_{k=1}^n \lambda_k^{1/2} \underline{v}_k \underline{v}_k^T$. ($\mathbf{B}$ is called the *square root* of $\mathbf{\Sigma}_N$.) This matrix $\mathbf{B}$ has inverse $\mathbf{B}^{-1} = \sum_{k=1}^n \lambda_k^{-1/2} \underline{v}_k \underline{v}_k^T$, and $\mathbf{\Sigma}_N^{-1} = \mathbf{B}^{-2} \triangleq (\mathbf{B}^{-1})^2$. If we define $\underline{s}_j^* = \mathbf{B}^{-1} \underline{s}_j$ and $\underline{y}^* = \mathbf{B}^{-1} \underline{y}$, we have that

$$(\underline{s}_1 - \underline{s}_0)^T \mathbf{\Sigma}_N^{-1} \underline{y} = (\underline{s}_1^* - \underline{s}_0^*)^T \underline{y}^*. \qquad \text{(III.B.44)}$$

Moreover, under H_j, we can write $\underline{Y}^* = \underline{N}^* + \underline{s}_j^*$ with $\underline{N}^* = \mathbf{B}^{-1} \underline{N}$, and

$$E\{\underline{N}^*(\underline{N}^*)^T\} = E\{\mathbf{B}^{-1} \underline{N}\underline{N}^T \mathbf{B}^{-1}\} = \mathbf{B}^{-1} E\{\underline{N}\underline{N}^T\} \mathbf{B}^{-1}$$

$$= \mathbf{B}^{-1} \mathbf{\Sigma}_N \mathbf{B}^{-1} = \mathbf{B}^{-1} \mathbf{B}\,\mathbf{B}\,\mathbf{B}^{-1} = \mathbf{I}.$$

Thus $N_1^*, \ldots, N_n^*$ are i.i.d. $N(0, 1)$ variables and we have a situation similar to that with the observables $\hat{\underline{Y}}$ above. In fact, $\underline{Y}^*$ and $\hat{\underline{Y}}$ are the same random vectors in two different coordinate systems since $\underline{Y}^* = \sum_{k=1}^n \hat{Y}_k \underline{v}_k$ and $\hat{\underline{Y}} = \sum_{k=1}^n \hat{Y}_k \underline{e}_k$, where $\underline{e}_1, \ldots, \underline{e}_n$ are the standard basis vectors for $\mathbb{R}^n$, i.e., $\underline{e}_k$ is all 0's except for a 1 in the its $k^{\underline{th}}$ component.

The observation vector $\underline{Y}$ can be transformed in another interesting way to give an equivalent observation with i.i.d. noise. In particular because $\mathbf{\Sigma}_N$ is positive definite it can be written as

$$\mathbf{\Sigma}_N = \mathbf{C}\mathbf{C}^T \qquad \text{(III.B.45)}$$

where $\mathbf{C}$ is an $n \times n$ invertible lower triangular matrix (i.e., all above-diagonal elements of $\mathbf{C}$ are zero). Equation (III.B.45) is called the *Cholesky decomposition* of $\mathbf{\Sigma}_N$, and these are several standard algorithms for find $\mathbf{C}$ from $\mathbf{\Sigma}_N$ (see, e.g., Bierman (1977)). We can then write $\mathbf{\Sigma}_N^{-1} = (\mathbf{C}^T)^{-1} \mathbf{C}^{-1} = (\mathbf{C}^{-1})^T \mathbf{C}^{-1}$. On defining new observables

$\underline{\tilde{Y}}=\mathbf{C}^{-1}\underline{Y}=\mathbf{C}^{-1}\underline{N}+\mathbf{C}^{-1}\underline{s}_j \triangleq \underline{\tilde{N}}+\underline{\tilde{s}}_j$, we have straightforwardly that $\underline{\tilde{N}} \sim N(\underline{0}, \mathbf{I})$. So we again have an i.i.d. noise situation and the optimum detection statistic is $(\underline{\tilde{s}}_1-\underline{\tilde{s}}_0)^T \underline{\tilde{Y}}$.

The interesting thing about this particular transformation is that the lower triangularity of $\mathbf{C}$ implies that $\mathbf{C}^{-1}$ is also lower triangular. This in turn implies that we can write

$$\tilde{y}_k = \sum_{l=1}^{k} h_{k,l}\, y_l, \qquad \text{(III.B.46)}$$

where $h_{k,l}$ is the $k-l^{\underline{th}}$ element of $\mathbf{C}^{-1}$. Note that (III.B.46) is a *causal* operation, and in fact (III.B.46) shows that $\tilde{y}_1, ..., \tilde{y}_n$ can be produced by a causal, but possibly time-varying, linear filtration of $y_1, ..., y_n$. Since the noise in the output of this filter is white (i.e., i.i.d.), this filter is sometimes known as a *whitening filter*. So the optimum detector structure of (III.B.24) can be represented as the causal linear filter with impulse response $\{h_{k,l}\}$ driven by $y_1, ..., y_n$ and followed by a correlator in which the filter output is correlated with $(\tilde{s}_{11}-\tilde{s}_{01}), ..., (\tilde{s}_{1n}-\tilde{s}_{0n})$, the output of the same filter driven by the difference signal $(s_{11}-s_{01}), ..., (s_{1n}-s_{0n})$. This structure is depicted in Fig. III.B.8.

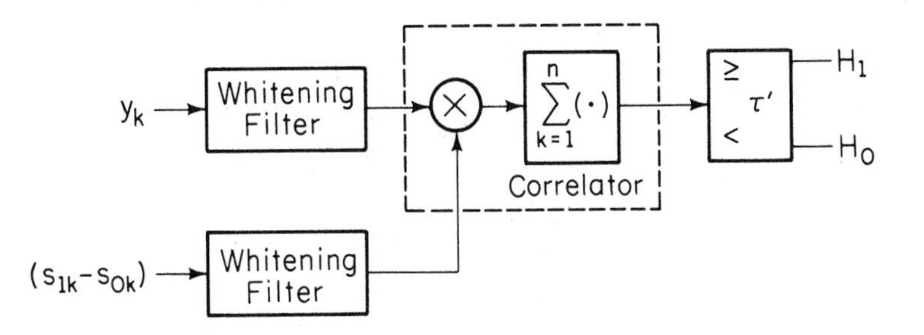

Fig. III.B.8: Optimum detector for coherent signals in dependent Gaussian noise.

As a final comment we note that the signal-to-noise ratio $d^2 = (\underline{s}_1 - \underline{s}_0)^T \Sigma_N^{-1} (\underline{s}_1 - \underline{s}_0)$ can be written in terms of any of the transformed signal pairs as

$$d^2 = \|\hat{\underline{s}}_1 - \hat{\underline{s}}_0\|^2 = \|\underline{s}_1^* - \underline{s}_0^*\|^2 = \|\tilde{\underline{s}}_1 - \tilde{\underline{s}}_0\|^2. \quad \text{(III.B.47)}$$

Thus the performance of coherent detection in dependent noise depends on how far apart the signal are when transformed to a coordinate system in which the noise components are i.i.d. [Compare with (III.B.37)]. It should be noted that all three signal pairs in (III.B.47) are the same distance apart because they are all representations of the same pair of vectors in different coordinate systems that are simple rotations of one another.

Remark III.B.3: Signal Selection

The performance of optimum coherent detection in Gaussian noise is improved by increasing the quantity $d^2 \triangleq (\underline{s}_1 - \underline{s}_0)^T \Sigma_N^{-1} (\underline{s}_1 - \underline{s}_0)$. In many of the applications in which coherent detection arises, there is often some flexibility in the choice of the signals $\underline{s}_0$ and $\underline{s}_1$. In such situations it is reasonable to choose these signals to maximize d^2.

As noted in the preceding discussion we can write $\Sigma_N^{-1} = \sum_{k=1}^n \lambda_k^{-1} \underline{v}_k \underline{v}_k^T$ where $\lambda_1, \ldots, \lambda_n$ and $\underline{v}_1, \ldots, \underline{v}_n$ are the eigenvalues and corresponding orthonormal eigenvectors of Σ_N. So for any vector $\underline{x} \in \mathbb{R}^n$, we have

$$\begin{aligned} \underline{x}^T \Sigma_N^{-1} \underline{x} &= \sum_{k=1}^n \lambda_k^{-1} \underline{x}^T \underline{v}_k \underline{v}_k^T \underline{x} \\ &\leqslant \lambda_{\min}^{-1} \sum_{k=1}^n \underline{x}^T \underline{v}_k \underline{v}_k^T \underline{x}, \end{aligned} \quad \text{(III.B.48)}$$

where $\lambda_{\min} = \min\{\lambda_1, \ldots, \lambda_n\}$. Since

$$\sum_{k=1}^n \underline{x}^T \underline{v}_k \underline{v}_k^T \underline{x} = \underline{x}^T \left(\sum_{k=1}^n \underline{v}_k \underline{v}_k^T \underline{x}\right) = \underline{x}^T \underline{x} = \|\underline{x}\|^2,$$

we have

$$\underline{x}^T \Sigma_N^{-1} \underline{x} \leqslant \lambda_{\min}^{-1} \|\underline{x}\|^2. \qquad \text{(III.B.49)}$$

Note that we can have equality in (III.B.49) if and only if $\underline{x}$ is proportional to an eigenvector corresponding to the eigenvalue $\lambda_{\min}$. [If there is more than one eigenvector corresponding to $\lambda_{\min}$, $\underline{x}$ can be any linear combination of these and still achieve equality in (III.B.49). Any such linear combination is still an eigenvector corresponding to $\lambda_{\min}$.]

From the above we see that, for fixed $\|\underline{s}_1 - \underline{s}_0\|$, the best way to choose the difference signal $\underline{s}_1 - \underline{s}_0$ is to be along an eigenvector corresponding to the minimum eigenvalue of Σ_N. The eigenvalues of Σ_N are measures of the noise power in the directions of their corresponding eigenvectors. Thus putting the signal difference along the minimum-eigenvalue eigenvector is equivalent to signaling in the least noisy direction. By doing so we get a value of d^2 given by

$$d^2 = \frac{1}{\lambda_{\min}} \|\underline{s}_1 - \underline{s}_0\|^2. \qquad \text{(III.B.50)}$$

Once we have chosen the direction of the signal difference $\underline{s}_1 - \underline{s}_0$, we can further optimize performance by maximizing $\|\underline{s}_1 - \underline{s}_0\|^2$. Obviously, this quantity can be arbitrarily large if we put no constraints on the signals. However, signals are usually constrained by their total power, and thus it is of interest to maximize (III.B.50) within such a constraint. In particular, suppose that we constrain $\|\underline{s}_1\|^2 \leqslant P$ and $\|\underline{s}_0\|^2 \leqslant P$, where $0 < P < \infty$. We have

$$\begin{aligned} d^2 &= \frac{1}{\lambda_{\min}} (\underline{s}_1 - \underline{s}_0)^T (\underline{s}_1 - \underline{s}_0) \\ &= \frac{1}{\lambda_{\min}} (\|\underline{s}_1\|^2 - 2\underline{s}_1^T \underline{s}_0 + \|\underline{s}_0\|^2). \end{aligned} \qquad \text{(III.B.51)}$$

Note that $\underline{s}_1^T \underline{s}_0$ is the dot (or inner) product between $\underline{s}_0$ and $\underline{s}_1$. With fixed $\|\underline{s}_1\|$ and $\|\underline{s}_0\|$ this is minimized (and hence d^2 is maximized) if $\underline{s}_1$ and $\underline{s}_0$ are in opposite directions; i.e., if $\underline{s}_0 = \alpha \underline{s}_1$ with $\alpha < 0$. In this case we have

$$d^2 = \frac{1}{\lambda_{\min}}(\|\underline{s}_1\|^2 - 2\alpha\|\underline{s}_1\|^2 + \alpha^2\|\underline{s}_1\|^2)$$
$$= \frac{\|\underline{s}_1\|^2(1-\alpha)^2}{\lambda_{\min}}, \tag{III.B.52}$$

and α must be given by $\alpha = -\|\underline{s}_0\| / \|\underline{s}_1\|$. So, for fixed $\|\underline{s}_0\|$ and $\|\underline{s}_1\|$, the maximum value of d^2 is

$$d^2 = \frac{(\|\underline{s}_1\| + \|\underline{s}_0\|)^2}{\lambda_{\min}}.$$

We see that d^2 is further maximized by choosing $\|\underline{s}_0\|^2 = \|\underline{s}_1\|^2 = P$, in which case we have $\alpha = -1$ and

$$\max_{\|\underline{s}_j\|^2 \leq P} d^2 = \frac{4P}{\lambda_{\min}}. \tag{III.B.53}$$

Since we have chosen $\underline{s}_1 - \underline{s}_0$ to be along a minimum-eigenvalue eigenvector, $\underline{v}_{\min}$, we can achieve (III.B.53) by choosing $\underline{s}_1 = c\underline{v}_{\min}$ and $\underline{s}_0 = -\underline{s}_1$, where c is chosen so that $\|\underline{s}_1\|^2 = \|\underline{s}_0\|^2 = P$. The correct c is thus $\sqrt{P} / \|v_{\min}\|$, so optimum signals are given by

$$\underline{s}_1 = \sqrt{P}\underline{v}_{\min} / \|\underline{v}_{\min}\| \text{ and } \underline{s}_0 = -\underline{s}_1. \tag{III.B.54}$$

The foregoing concepts are illustrated by the following simple example.

Example III.B.4: Optimum Signals for Two-sample Detection

Consider the case $n = 2$ with

$$\Sigma_N = \sigma^2 \begin{bmatrix} 1 & \rho \\ \rho & 1 \end{bmatrix}, \tag{III.B.55}$$

where $|\rho|<1$. For this Σ_N the eigenvalues and corresponding orthonormal eigenvectors are easily shown to be given by

$$\lambda_1 = \sigma^2(1-\rho) \text{ and } \lambda_2 = \sigma^2(1+\rho)$$

$$\underline{v}_1 = \frac{1}{\sqrt{2}}\begin{pmatrix} 1 \\ -1 \end{pmatrix} \text{ and } \underline{v}_2 = \frac{1}{\sqrt{2}}\begin{pmatrix} 1 \\ 1 \end{pmatrix}. \qquad \text{(III.B.56)}$$

Thus if $\rho<0$, $\lambda_{min}=\lambda_1$ and optimum signals are given by

$$\underline{s}_1 = \sqrt{P/2}\begin{pmatrix} 1 \\ -1 \end{pmatrix} \text{ and } \underline{s}_0 = \sqrt{P/2}\begin{pmatrix} -1 \\ 1 \end{pmatrix}, \qquad \text{(III.B.57)}$$

and if $\rho<0$, $\lambda_{min}=\lambda_2$ and optimum signals are

$$\underline{s}_1 = \sqrt{P/2}\begin{pmatrix} 1 \\ 1 \end{pmatrix} \text{ and } \underline{s}_0 = \sqrt{P/2}\begin{pmatrix} -1 \\ -1 \end{pmatrix}. \qquad \text{(III.B.58)}$$

In either case the maximum value of d^2 is

$$d^2 = \frac{4P}{\sigma^2(1-|\rho|)}. \qquad \text{(III.B.59)}$$

The optimality of the signal sets above for these two cases is easily seen from Fig. III.B.9 on which equal-density contours have been drawn for the two cases. Note that for either case the signal vectors are in the directions in which the noise density falls off the fastest, thus giving a maximum signal-to-noise ratio for fixed P, σ^2, and ρ. It is interesting to note that one only needs to know the algebraic sign of ρ, not its actual value, to choose the optimum signals in this case.

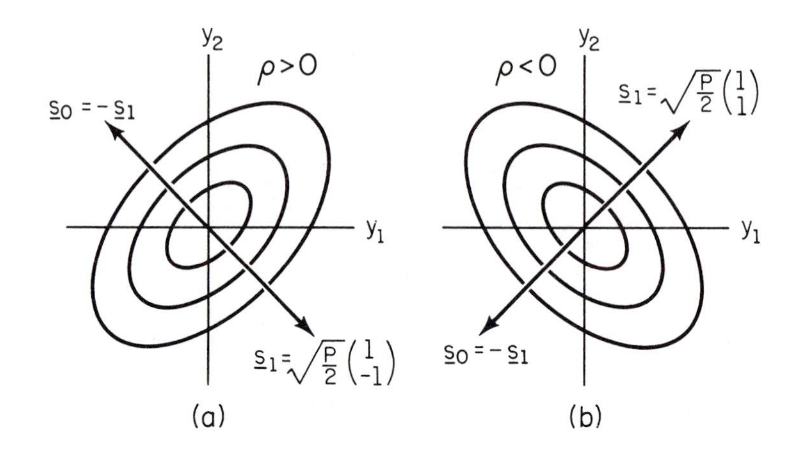

Fig. III.B.9: Illustration of optimum signals for Gaussian noise with $\mathbf{\Sigma}_N = \sigma^2 \begin{pmatrix} 1 & \rho \\ \rho & 1 \end{pmatrix}$.

Case III.B.3: Detection of Signals with Random Parameters

In Cases III.B.1 and III.B.2 we have discussed the problem of deciding between two signals that are completely known to the observer. In many applications we are often faced with the related problem of deciding between signals that are known except for a set of unknown parameters. This situation arises, for example, in digital communication systems in which one of two waveforms (representing "zero" and "one," respectively) is modulated onto a sinusoidal carrier at the transmitter and the receiver must decide which was sent. Even though the two signaling waveforms and the carrier frequency are known at the receiver, the amplitude and phase of the carrier may not be; and thus these quantities represent unknown parameters that must be considered in the detection process. Similar situations arise in radar, sonar, and other applications in which signals of unknown frequency, time of arrival, and amplitude must be detected.

For this situation it is convenient to write (III.B.1) as

$$H_0: Y_k = N_k + s_{0k}(\theta), \quad k = 1, \ldots, n$$

versus (III.B.60)

$$H_1: Y_k = N_k + s_{1k}(\theta), \quad k = 1, \ldots, n$$

where $\underline{s}_0(\theta)$ and $\underline{s}_1(\theta)$ are known vector-valued functions of θ, which is an unknown parameter taking values in a parameter set Λ. Assuming that θ is random (in which case we write it as Θ) with density w_j under hypothesis H_j, the likelihood ratio for (III.B.60) is

$$L(\underline{y}) = \frac{E_1\{p_N(\underline{y} - \underline{s}_1(\Theta))\}}{E_0\{p_N(\underline{y} - \underline{s}_0(\Theta))\}}$$

$$= \frac{\int_\Lambda p_N(\underline{y} - \underline{s}_1(\theta)) w_1(\theta) \mu(d\theta)}{\int_\Lambda p_N(\underline{y} - \underline{s}_0(\theta)) w_0(\theta) \mu(d\theta)}. \quad \text{(III.B.61)}$$

For the purposes of discussion we will assume that $\underline{s}_0(\theta) \equiv \underline{0}$ and $\underline{s}_1(\theta) \triangleq \underline{s}(\theta)$, since other cases can be handled similarly. In this case we have

$$L(\underline{y}) = \int_\Lambda \frac{p_N(\underline{y} - \underline{s}(\theta))}{p_N(\underline{y})} w(\theta) \mu(d\theta)$$

(III.B.62)

$$= \int_\Lambda L_\theta(\underline{y}) w(\theta) \mu(d\theta),$$

where $L_\theta(\underline{y})$ is the likelihood ratio conditioned on $\Theta = \theta$ and where we have dropped the subscript from w_1. From (III.B.62) we see that $L(\underline{y})$ in this case is simply the averaged (over θ) likelihood ratio for known θ. With θ known (III.B.60) is a deterministic-signal problem and so $L_\theta(\underline{y})$ is found directly as in Cases III.B.1 and III.B.2. For example, with i.i.d. $N(0, \sigma^2)$ noise samples $L(\underline{y})$ becomes

$$L(\underline{y}) = \int_{\Lambda} \exp\{(\underline{s}^T(\theta)\underline{y} - \tfrac{1}{2}\|\underline{s}(\theta)\|^2)/\sigma^2\} w(\theta)\mu(d\theta). \quad \text{(III.B.63)}$$

Similarly, for non-i.i.d. Gaussian noise (III.B.63) is valid with $\sigma^2=1$ and with $\underline{y}$ and $\underline{s}(\theta)$ being with quantities transformed to yield an i.i.d. noise problem.

This type of problem is illustrated by the following example, which arises in a number of applications.

Example III.B.5: Noncoherent Detection of a Modulated Sinusoidal Carrier

Consider the signal pair $\underline{s}_0(\theta)=\underline{0}$ and $\underline{s}_1(\theta)=\underline{s}(\theta)$ with

$$s_k(\theta) = a_k \sin[(k-1)\omega_c T_s + \theta], \quad k = 1, \ldots, n \qquad \text{(III.B.64)}$$

where $a_1, a_2, \ldots, a_n$ is a known amplitude sequence, Θ is a random phase angle independent of the noise and uniformly distributed on $[0, 2\pi]$, and where ω_c and T_s are a known carrier frequency and sampling interval with the relationship $n\omega_c T_s = m2\pi$ for some integer m (i.e., there are an integral number of periods of the sinusoid in the time interval $[0, nT_s]$). We also assume that the number of samples taken per cycle of the sinusoid (i.e., n/m) is an integer larger than 1. These signals provide a model, for example, for a digital signaling scheme in which a "zero" is transmitted by sending nothing during the interval $[0, nT_s]$ and a "one" is transmitted by sending a signal $a(t)$ modulated onto a sinusoidal carrier of frequency ω_c. This signaling scheme is known as *on-off keying* (OOK). In this case the sequence $a_1, \ldots, a_n$ is the sampled waveform $a(t)$ [i.e., $a_k = a((k-1)T_s)$] and θ represents the phase angle of the carrier, which is assumed here to be unknown at the receiver. Detection of a modulated carrier in which the carrier phase is unknown at the receiver is called *noncoherent* detection. The assumption that the phase angle is uniform on $[0, 2\pi]$ represents a belief that all phases are equally likely to occur, which is a reasonable assumption in the absence of any information to the contrary.

Assuming i.i.d. $N(0, \sigma^2)$ noise, the likelihood ratio for this problem is given from (III.B.63) to be

$$L(\underline{y}) =$$

$$\frac{1}{2\pi}\int_0^{2\pi} \exp\left\{\frac{1}{\sigma^2}\left[\sum_{k=1}^{n} y_k\, s_k(\theta) - \tfrac{1}{2}\sum_{k=1}^{n} s_k^2(\theta)\right]\right\} d\theta. \qquad \text{(III.B.65)}$$

Using the identity $\sin(a+b)=\cos a \sin b + \sin a \cos b$, the first term in parentheses in the exponent in (III.B.65) can be written as $\sum_{k=1}^{n} y_k\, s_k(\theta) = y_c \sin\theta + y_s \cos\theta$ with

$$y_c \triangleq \sum_{k=1}^{n} a_k y_k \cos[(k-1)\omega_c T_S] \qquad \text{(III.B.66)}$$

and

$$y_s \triangleq \sum_{k=1}^{n} a_k y_k \sin[(k-1)\omega_c T_S].$$

Similarly, with the identity $\sin^2 a = \tfrac{1}{2} - \tfrac{1}{2}\cos 2a$, the second term in parentheses in the exponent becomes

$$-\tfrac{1}{2}\sum_{k=1}^{n} s_k^2(\theta) =$$

$$-\tfrac{1}{4}\sum_{k=1}^{n} a_k^2 + \tfrac{1}{4}\sum_{k=1}^{n} a_k^2 \cos(2(k-1)\omega_c T_S + 2\theta). \qquad \text{(III.B.67)}$$

For most situations arising in practice, the second term on the right-hand side of (III.B.67) is zero or approximately zero for all values of θ. For example, if the signal sequence $a_1, \ldots, a_n$ is a constant times a sequence or +1's and -1's, or if $a_1, \ldots, a_n$ has a raised-cosine shape of the form $a_k = A(1-\cos((k-1)2\pi/(n-1)), k=1, \ldots, n$, then this second term is identically zero. In other cases of interest in practice $a_1^2, \ldots, a_n^2$ is usually slowly varying as compared to twice the carrier frequency. So this second term amounts to a low-pass filtering of a high frequency signal, an operation that results in a negligible output. In any case we will assume that $a_1, \ldots, a_n$ is such that this second term is zero for all θ, and thus $L(\underline{y})$ becomes

$$L(\underline{y}) = e^{-n\overline{a^2}/4\sigma^2} \times \frac{1}{2\pi}\int_0^{2\pi} \exp\left\{\frac{1}{\sigma^2}(y_c \sin\theta + y_s \cos\theta)\right\} d\theta \tag{III.B.68}$$

with $\overline{a^2} = (1/n)\sum_{k=1}^{n} a_k^2$.

Expression (III.B.68) is similar to that for the likelihood ratio in the Example II.E.1. In particular, comparing (III.B.68) and (II.E.17), we see that

$$L(\underline{y}) = e^{-n\overline{a^2}/4\sigma^2} I_0(r/\sigma^2), \tag{III.B.69}$$

where $r = +[y_c^2 + y_s^2]^{1/2}$ and I_0 is the zeroth-order modified Bessel function of the first kind. In view of the monotonicity of $I_0(\cdot)$, the optimum tests in this case are thus given by

$$\tilde{\delta}_o(\underline{y}) = \begin{cases} 1 & > \\ \gamma & \text{if } r = \tau' \triangleq \sigma^2 I_0^{-1}(e^{n\overline{a^2}/4\sigma^2}). \\ 0 & < \end{cases} \tag{III.B.70}$$

The structure of this detector is shown in Fig. III.B.10. Note that the observed signal $y_1, \ldots, y_n$ is split into two channels, one of which multiplies each y_k by $\cos[(k-1)\omega_c T_s]$ and the other of which multiplies each y_k by $\sin[(k-1)\omega_c T_s]$. (These are sometimes called *in-phase* and *quadrature* channels, respectively.) Each channel correlates the resulting product with the amplitude sequence $a_1, \ldots, a_n$. The channel outputs are then combined to give r, which is compared to a threshold. Intuitively, the principle of operation of this detector is that when a signal is present, each channel picks up an amount of

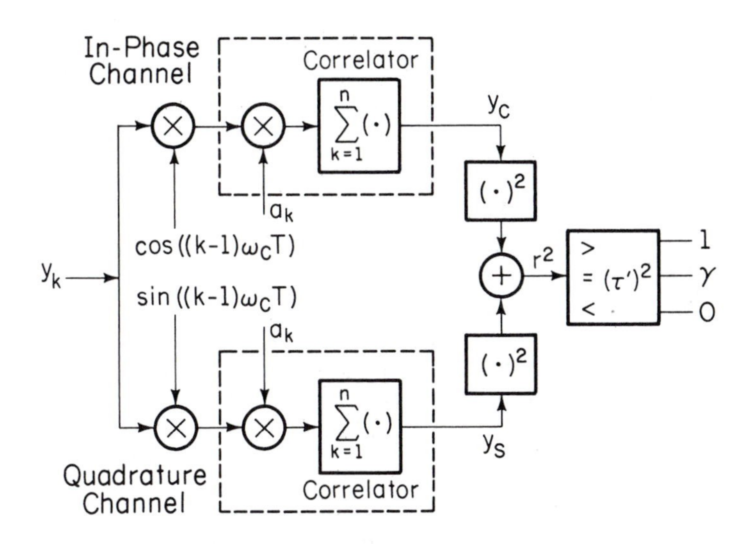

Fig. III.B.10: Optimum system for noncoherent detection of a modulated sinusoid in i.i.d. Gaussian noise.

the signal energy depending on the actual phase angle of the carrier. However, regardless of the carrier phase, the combination of the two channels picks up all the signal energy.

To analyze the performance of the detector of (III.B.70), we must find $P_j(R>\tau')=P_j(R^2>(\tau')^2)$ for $j=0,1$. Since $R^2=Y_c^2+Y_s^2$, where $Y_c \triangleq \sum_{k=1}^{n} a_k Y_k \cos[(k-1)\omega_c T_S]$ and $Y_s \triangleq \sum_{k=1}^{n} a_k Y_k \sin[(k-1)\omega_c T_S]$, the desired probabilities can be found from the joint probability density function of Y_c and Y_s under the two hypotheses. Under H_0 $\underline{Y}$ is $N(0, \sigma^2 I)$, and since Y_c and Y_s are linear in $\underline{Y}$ they are jointly Gaussian under H_0. Thus we can specify the joint density of (Y_c, Y_s) under H_0 by finding the means and variances of Y_c and Y_s and the correlation coefficient between Y_c and Y_s. We have straightforwardly that

$$E\{Y_c|H_0\} = \sum_{k=1}^{n} a_k E\{N_k\} \cos[(k-1)\omega_c T_S] = 0$$

and

$$Var\ [Y_c|H_0] = E\{Y_c^2|H_0\} = \sum_{k=1}^{n}\sum_{l=1}^{n} a_k a_l E\{N_k N_l\}$$

$$\times \cos[(k-1)\omega_c T_S]\cos[(l-1)\omega_c T_S]$$

$$= \sigma^2 \sum_{k=1}^{n} a_k^2 \cos^2[(k-1)\omega_c T_S] = \frac{n\,\sigma^2\overline{a^2}}{2},$$

where we have again used the assumption that the second term in (III.B.67) is zero for all θ. Similarly, we have $E\{Y_s|H_0\}=0$ and $Var\ (Y_s|H_0)=Var\ (Y_c|H_0)$. The correlation coefficient between Y_c and Y_s under H_0 is given by $Cov\ (Y_c, Y_s|H_0)/[Var\ (Y_c|H_0)Var\ (Y_s|H_0)]^{1/2}$. Since Y_c and Y_s have zero means under H_0, we have

$$Cov\,(Y_c, Y_s|H_0)$$

$$= E\{Y_c Y_s|H_0\}$$

$$= \sum_{k=1}^{n}\sum_{l=1}^{n} a_k a_l E\{N_k N_l\}\cos[(k-1)\omega_c T_S]\sin[(l-1)\omega_c T_S] \tag{III.B.71}$$

$$= \sum_{k=1}^{n} a_k^2 \cos[(k-1)\omega_c T_S]\sin[(k-1)\omega_c T_S]$$

$$= \tfrac{1}{2}\sum_{k=1}^{n} a_k^2 \sin[2(k-1)\omega_c T_S] = 0.$$

From (III.B.71) we see that under H_0 Y_c and Y_s are uncorrelated, and since they are jointly Gaussian they are thus independent.

We conclude that under H_0, Y_c and Y_s are independent $N(0, n\,\sigma^2\overline{a^2}/2)$ random variables. Noting that the randomization is irrelevant here, the false-alarm probability of (III.B.70) thus becomes

$$P_0(\Gamma_1) = \int\limits_{\{y_c^2+y_c^2 \geq (\tau')^2\}} \int \frac{1}{\pi n\, \sigma^2 \overline{a^2}} e^{-(y_c^2+y_s^2)/n\,\sigma^2\overline{a^2}} dy_c\, dy_s$$

$$= \frac{1}{\pi n\, \sigma^2 \overline{a^2}} \int_0^{2\pi} \int_{\tau'}^{\infty} r e^{-r^2/n\,\sigma^2\overline{a^2}} dr d\psi \qquad \text{(III.B.72)}$$

$$= e^{-(\tau')^2/n\,\sigma^2\overline{a^2}},$$

where to get the second equality we have introduced polar coordinates r and ψ.

To determine the detection probability of (III.B.70) we need to find the joint density of Y_c and Y_s under H_1. Note that given $\Theta=\theta$, $\underline{Y}$ has a conditional $N(\underline{s}(\theta), \sigma^2\mathbf{I})$ distribution under H_1. Thus given $\Theta=\theta$, Y_c and Y_s are conditionally jointly Gaussian. We have straightforwardly that

$$E\{Y_c | H_1, \Theta = \theta\} = \sum_{k=1}^{n} a_k E\{Y_k | H_1, \Theta = \theta\} \cos[(k-1)\omega_c T_S]$$

$$= \sum_{k=1}^{n} a_k^2 \sin[(k-1)\omega_c T_S + \theta] \cos[(k-1)\omega_c T_S]$$

$$= \frac{n\overline{a^2}}{2} \sin\theta,$$

and similarly, $E\{Y_s | H_1, \Theta=\theta\} = (n\overline{a^2}/2)\cos\theta$. With θ fixed, the variances and covariance under H_1 of Y_c and Y_s are unchanged from their H_0 values since the only change in $\underline{Y}$ is a shift in mean.

The unconditioned density of Y_c, Y_s under H_1 is found by averaging the conditional density over θ. We have

$$p_{Y_c,Y_s}(y_c, y_s | H_1)$$

$$= \frac{1}{2\pi}\int_0^{2\pi} \frac{1}{\pi n \sigma^2 \overline{a^2}}$$

$$\times \exp\left\{-\frac{1}{n\sigma^2\overline{a^2}} q\left(y_c, y_s ; \frac{n\overline{a^2}}{2}, \theta\right)\right\} d\theta \tag{III.B.73}$$

$$= p_{Y_c,Y_s}(y_c, y_s | H_0) e^{-n\overline{a^2}/4\sigma^2} I_0\left(\frac{[y_c^2+y_s^2]^{1/2}}{\sigma^2}\right),$$

where $p_{Y_c,Y_s}(y_c, y_s | H_0)$ is the joint density of Y_c and Y_s under H_0, and where q is defined in (II.E.14). The detection probability of (III.B.70) thus becomes

$$P_D(\tilde{\delta}_o) = P_1(\Gamma_1) = \iint_{\{y_c^2+y_2^2>(\tau')^2\}} p_{Y_c,Y_s}(y_c, y_s | H_1) dy_c \, dy_s$$

$$= \frac{e^{-n\overline{a^2}/4\sigma^2}}{\pi n \sigma^2 \overline{a^2}} \int_0^{2\pi}\int_{\tau'}^{\infty} r e^{-r^2/n\sigma^2\overline{a^2}} I_0\left(\frac{r}{\sigma^2}\right) dr \, d\psi \tag{III.B.74}$$

$$= \int_{\tau_0}^{\infty} x e^{(-x^2+b^2)/2} I_0(bx) dx \triangleq Q(b, \tau_o),$$

where in the third equality we have defined $b^2 = n\overline{a^2}/2\sigma^2$ and $\tau_0 = \tau'/\sigma^2 b$ and made the substitution $x = r/\sigma^2 b$. The function Q defined in (III.B.74) is sometimes known as *Marcum's Q-function*. Note from (III.B.72) that $Q(0, \tau_o) = P_F(\tilde{\delta}_o)$.

From (III.B.72) we can easily set the threshold for α-level Neyman-Pearson detection in this problem. We have

$$\tau' = [n\sigma^2\overline{a^2} \log(1/\alpha)]^{1/2}, \tag{III.B.75}$$

from which (III.B.74) gives the ROCs as

$$P_D(\tilde{\delta}_o) = Q[b, [2 \log(1/\alpha)]^{1/2}]. \qquad \text{(III.B.76)}$$

These receiver operating characteristics look very similar to those for the coherent problem (i.e., Fig. II.D.4), and are plotted, for example, in Fig. V.3 of Helstrom (1968).

We see from (III.B.76) that the performance of Neyman-Pearson detection here depends only on the parameter b. Note that the average signal energy is

$$E\left\{\sum_{k=1}^{n} s_k^2(\Theta)\right\} = \frac{1}{n}\frac{1}{2\pi}\int_0^{2\pi}\sum_{k=1}^{n} a_k^2 \cos^2[(k-1)\omega_c T_S + \theta]d\theta = \frac{\overline{a^2}}{2}.$$

Thus $b^2 = n\overline{a^2}/2\sigma^2$ has a signal-to-noise ratio interpretation analogous to d^2 in the coherent detection problem. If we were to detect the same signal coherently (i.e., with θ known), the corresponding value of d^2 would be

$$d^2 = \frac{1}{\sigma^2}\sum_{k=1}^{n} s_k^2(\theta) = \frac{n\overline{a^2}}{2\sigma^2} = b^2. \qquad \text{(III.B.77)}$$

Thus these signal-to-noise ratios are actually the same. However, the performance for fixed α is different for the two systems with $b = d$. For the ranges of SNR and α occurring in most practical problems we have

$$Q[b, [2\log(1/\alpha)]^{1/2}] = 1 - \Phi[\Phi^{-1}(1-\alpha) - d] \qquad \text{(III.B.78)}$$

when $b \cong d + 0.4$ [see, Helstrom (1968)]. So if we wish to use a noncoherent technique, slightly higher SNR is required to achieve the same performance as the corresponding coherent technique. The disadvantage of a coherent system is that some means for deriving the carrier phase must be provided.

In concluding this example, it should be noted that the likelihood ratio derivation above can be applied to problems in which we have signals of the form of (III.B.64) but with differing amplitude sequences present under both hypotheses, i.e., in which we have

$$s_{jk}(\theta) = a_{jk} \sin[(k-1)\omega_c T_S + \theta], \quad k = 1, \ldots, n, \qquad \text{(III.B.79)}$$

for $j = 0, 1$. Here the likelihood ratio for i.i.d. noise can be found from (III.B.69) by using a third hypothesis of noise only as a "catalyst." In particular,

$$L(\underline{y}) = \frac{E_1\{p_N(\underline{y} - \underline{s}_1(\Theta))\}}{E_0\{p_N(\underline{y} - \underline{s}_0(\Theta))\}}$$

$$= \frac{E_1\{p_N(\underline{y} - \underline{s}_1(\Theta))/p_N(\underline{y})\}}{E_0\{p_N(\underline{y} - \underline{s}_0(\Theta))/p_N(\underline{y})\}} = \frac{L_1(\underline{y})}{L_0(\underline{y})},$$

where $L_j(\underline{y})$ is the likelihood ratio for this third hypotheses versus H_j. From (III.B.69), we then conclude that

$$L(y) = \frac{e^{-n\overline{a_1^2}/4\sigma^2} I_0\left(\frac{r_1}{\sigma^2}\right)}{e^{-n\overline{a_0^2}/4\sigma^2} I_0\left(\frac{r_0}{\sigma^2}\right)}, \qquad \text{(III.B.80)}$$

where $\overline{a_j^2} = (1/n)\sum_{k=1}^n a_{jk}^2$ and $r_j = [y_{cj}^2 + y_{sj}^2]^{1/2}$ with

$$y_{cj} = \sum_{k=1}^n a_{jk} y_k \cos[(k-1)\omega_c T_S]$$

and

$$y_{sj} = \sum_{k=1}^n a_{jk} y_k \sin[(k-1)\omega_c T_S].$$

Optimum detection thus involves combining the outputs of two systems like the one in Fig. III.B.10, one "matched" to each of the amplitude sequences. For example, if the signals have balanced energies ($\overline{a_0^2} = \overline{a_1^2}$) and we assume uniform costs and equal priors ($\tau = 1$), then the Bayes test in this situation becomes

$$\delta_B(\underline{y}) = \begin{cases} 1 \text{ if } & r_1 > r_0 \\ 0 \text{ or } 1 \text{ if } & r_1 = r_0 \\ 0 \text{ if } & r_1 < r_0, \end{cases} \qquad \text{(III.B.81)}$$

since I_0 is monotone increasing. For the latter situation, if we assume that the amplitude sequences are orthogonal, i.e.,

$$\sum_{k=1}^{n} a_{jk}\, a_{lk} = 0 \text{ if } j \neq l$$

and that

$$\sum_{k=1}^{n} a_{jk}\, a_{lk} \sin[(k-1)\omega_c T_s + \theta] = 0$$

for all θ (this assumption holds for several common signal sets used in practice), then the error probability can be shown straightforwardly to be (see Exercise 8)

$$P_e = \tfrac{1}{2} e^{-b^2/4}, \qquad \text{(III.B.82)}$$

where $b^2 = n\overline{a_1^2}/2\sigma^2 (\equiv n\overline{a_0^2}/2\sigma^2)$.

Other aspects of this problem, including detection with unknown amplitude and phase, are developed in the exercises.

Case III.B.4: Detection of Stochastic Signals

In some applications, signals arise that are best modeled as being purely random or stochastic. This type of model is useful, for example, in radio astronomy, sonar, and other applications in which signals are perturbed by propagation through turbulent media or along multiple paths. For this case we have the general model of (III.B.1) and the simplest general likelihood-ratio formula is that of (III.B.4).

An important special case of this problem is that in which both signals and noise are Gaussian random vectors. To study this case we first note that for the following hypothesis testing problem in $\mathbb{R}^n$:

$$\begin{gathered} H_0: \underline{Y} \sim N(\underline{\mu}_0, \Sigma_0) \\ \text{versus} \\ H_1: \underline{Y} \sim N(\underline{\mu}_1, \Sigma_1) \end{gathered} \tag{III.B.83}$$

the logarithm of the likelihood ratio is given by

$$\begin{aligned} \log L(\underline{y}) &= \tfrac{1}{2} \log [|\Sigma_0|/|\Sigma_1|] + \tfrac{1}{2}(\underline{y}-\underline{\mu}_0)^T \Sigma_0^{-1}(\underline{y}-\underline{\mu}_0) \\ &\quad - \tfrac{1}{2}(\underline{y}-\underline{\mu}_1)^T \Sigma_1^{-1}(\underline{y}-\underline{\mu}_1) \\ &= \tfrac{1}{2}\underline{y}^T [\Sigma_0^{-1}-\Sigma_1^{-1}]\underline{y} + [\underline{\mu}_1^T\Sigma_1^{-1}-\underline{\mu}_0^T\Sigma_0^{-1}]\underline{y} + C, \end{aligned} \tag{III.B.84}$$

where $C = \tfrac{1}{2}(\log(|\Sigma_0|/|\Sigma_1|) + \underline{\mu}_0^T\Sigma_0^{-1}\underline{\mu}_0 - \underline{\mu}_1^T\Sigma_1^{-1}\underline{\mu}_1)$. Note that $\log L(\underline{y})$ here consists of a quadratic term in $\underline{y}$, a linear term in $\underline{y}$, and a constant. If the two covariances are the same, say $\Sigma_0 = \Sigma_1 = \Sigma$, then the quadratic term disappears, and we essentially have a linear test statistic, $(\underline{\mu}_1 - \underline{\mu}_0)^T \Sigma^{-1}\underline{y}$, since C can be incorporated into the threshold. This is the case of coherent detection in Gaussian noise treated in Case III.B.2. If, on the other hand the mean vectors are the same under both hypotheses ($\underline{\mu}_0 = \underline{\mu}_1$), we can (without loss of generality) take them to be $\underline{0}$ and the structure of $\log L(\underline{y})$ is thus quadratic.

The latter case is applicable to the problem of detecting zero-mean stochastic signals in Gaussian noise. In particular, consider the hypothesis pair

$$\begin{gathered} H_0: \underline{Y} = \underline{N} \\ \text{versus} \\ H_1: \underline{Y} = \underline{N} + \underline{S}, \end{gathered} \tag{III.B.85}$$

where $\underline{N} \sim N(\underline{0}, \sigma^2 I)$ and $\underline{S} \sim N(0, \Sigma_S)$. Cases in which $\underline{N}$ is Gaussian but not i.i.d. can be fit within this model by prewhitening since a linear transformation of the Gaussian signal $\underline{S}$ will still be Gaussian. Also, cases in which there are signals present under each hypothesis can be handled by using (III.B.85) as a "catalyst" as in the preceding example.

The hypothesis pair of (III.B.85) is a particular case of (III.B.83) with $\Sigma_0=\sigma^2\mathbf{I}$ and $\Sigma_1=\sigma^2\mathbf{I}+\Sigma_S$. (We assume, as always, that signal and noise are independent.) Thus from (III.B.84) we see that optimum tests for (III.B.85) are of the form

$$\tilde{\delta}_o(\underline{y}) = \begin{cases} 1 & > \\ \gamma \text{ if } \underline{y}^T \mathbf{Q} \underline{y} = & \tau' \\ 0 & < \end{cases} \qquad \text{(III.B.86)}$$

with $\tau' \triangleq \log\tau - C$ and $\mathbf{Q} \triangleq \sigma^{-2}\mathbf{I} - (\sigma^2\mathbf{I}+\Sigma_S)^{-1} \equiv \sigma^{-2}\Sigma_s(\sigma^2\mathbf{I}+\Sigma_s)^{-1}$. From (III.B.86) we see that the optimum detector computes the quadratic form $\underline{y}^T\mathbf{Q}\underline{y}$ and compares it to a threshold. This structure is known as a *quadratic detector*.

For example, if the signal samples are i.i.d. $N(0, \sigma_S^2)$ random variables, then $\Sigma_S=\sigma_S^2\mathbf{I}$ and

$$\underline{y}^T\mathbf{Q}\underline{y} = \frac{\sigma_S^2}{\sigma^2(\sigma^2+\sigma_S^2)} \sum_{k=1}^{n} y_k^2. \qquad \text{(III.B.87)}$$

Thus in this particular case, the optimum detector compares the quantity $\sum_{k=1}^n y_k^2$ to a threshold. Since $(1/n)\sum_{k=1}^n y_k^2$ is the average energy in the observed waveform, the resulting detector structure is sometimes known as an *energy detector*. (This is also known as a *radiometer*.) This is an intuitively reasonable way of detecting the signal in this case since we have to choose here between two situations in which the random observations $Y_1, \ldots, Y_n$ differ only in terms of the (statistical) average energy they contain. This can be contrasted with the problem of choosing between two constant signals, $\underline{\mu}_0=s_0\underline{1}$ and $\underline{\mu}_1=s_1\underline{1}$, where $\underline{1}=(1, \ldots, 1)^T$ and $s_1>s_0$, in white noise. In this case the optimum detector compares $\sum_{k=1}^n y_k$ to a threshold; since the random observations differ here under the two hypotheses only in terms of their (statistical) average amplitudes, it is intuitively reasonable to decide between them by comparing the average observation, $(1/n)\sum_{k=1}^n y_k$, to a threshold. The latter structure is sometimes known as a *linear detector.*

In order to analyze the performance of the detector in (III.B.86), we must compute the probabilities $P_j(\underline{Y}^T \mathbf{Q} \underline{Y} > \tau')$ for $j=0, 1$. This problem can be discussed more easily if we first transform the observations in a way similar to that discussed in the coherent detection problem. In particular, suppose that $\lambda_1, \ldots, \lambda_n$ and $\underline{v}_1, \ldots, \underline{v}_n$ are the eigenvalues and corresponding orthonormal eigenvectors of the *signal* covariance matrix $\mathbf{\Sigma}_S$. Then we can write $\mathbf{\Sigma}_S = \sum_{k=1}^n \lambda_k \underline{v}_k \underline{v}_k^T$, and it is easily seen that

$$\mathbf{I} = \sum_{k=1}^n \underline{v}_k \underline{v}_k^T.$$

Thus

$$(\sigma^2 \mathbf{I} + \mathbf{\Sigma}_S)^{-1} = \sum_{k=1}^n (\sigma^2 + \lambda_k)^{-1} \underline{v}_k \underline{v}_k^T,$$

and

$$\mathbf{Q} = \sum_{k=1}^n \frac{1}{\sigma^2} \underline{v}_k \underline{v}_k^T - \sum_{k=1}^n \frac{1}{\sigma^2 + \lambda_k} \underline{v}_k \underline{v}_k^T$$
$$= \sum_{k=1}^n \frac{\lambda_k}{\sigma^2(\sigma^2 + \lambda_k)} \underline{v}_k \underline{v}_k^T. \tag{III.B.88}$$

We see that the detection statistic can be written as

$$\underline{y}^T \mathbf{Q} \underline{y} = \sum_{k=1}^n (\tilde{y}_k)^2 \tag{III.B.89}$$

with $\tilde{y}_k \triangleq [\lambda_k / \sigma^2(\sigma^2 + \lambda_k)]^{1/2} \underline{v}_k^T \underline{y}$.

Since $\underline{v}_1, \ldots, \underline{v}_n$ is a set of orthonormal eigenvectors for both $\sigma^2 \mathbf{I}$ and for $(\sigma^2 \mathbf{I} + \mathbf{\Sigma}_S)$, it is straightforward to show [similarly to (III.B.42)] that $\tilde{Y}_1, \ldots, \tilde{Y}_n$ are independent zero-mean Gaussian random-variables under *both* hypotheses with variances

$$\sigma_{jk}^2 \triangleq Var\,(\bar{Y}_k | H_j) = \begin{cases} \dfrac{\lambda_k}{\sigma^2+\lambda_k} & \text{if } j = 0 \\ \\ \dfrac{\lambda_k}{\sigma^2} & \text{if } j = 1. \end{cases} \tag{III.B.90}$$

This implies that under H_j, $\underline{Y}^T \mathbf{Q} \underline{Y}$ is the sum of independent random variables in which the $k^{\underline{th}}$ term $\bar{Y}_k^2$ has the distribution of a $N(0, \sigma_{jk}^2)$ random variable squared. The probability density function of $T_k \triangleq \bar{Y}_k^2$ under H_j can be shown to be [see, e.g., Papoulis (1984)]

$$p_{T_k}(t | H_j) = \begin{cases} \dfrac{1}{\sqrt{2\pi t}\ \sigma_k} e^{-t/2\sigma_{jk}^2}, & t>0 \\ \\ 0, & t \leqslant 0 \end{cases} \tag{III.B.91}$$

which is a *gamma* (½, $1/2\sigma_{jk}^2$) density. The probability density, p_T, of $T \triangleq \sum_{k=1}^n \bar{y}_k^2$ is the n-fold convolution $p_{T_1} * p_{T_2} * \cdots * p_{T_n}$, which is more easily expressed using Fourier transforms as

$$p_T = F^{-1}\left\{ \prod_{k=1}^n \phi_{T_k} \right\}, \tag{III.B.92}$$

where $\phi_{T_k}(u) = F\{f_{T_k}\}(u) = E\{e^{iuT_k}\}$ is the *characteristic function* of T_k (here $i = \sqrt{-1}$). The characteristic function of the gamma (½, $1/2\sigma_{jk}^2$) density is given by (Lukacs, 1960)

$$\phi_{T_k}(u) = \frac{1}{[1-2iu\,\sigma_{jk}^2]^{1/2}},\ u \in \mathbb{R}. \tag{III.B.93}$$

Thus

$$p_T(t \mid H_j) = \int_{-\infty}^{\infty} e^{-iut} \prod_{k=1}^{n} [1 - 2iu\,\sigma_{jk}^2]^{-1/2} du. \quad \text{(III.B.94)}$$

No general closed form is known for (III.B.94). However, in the particular case in which $\sigma_{j1}^2 = \ldots = \sigma_{jn}^2 \triangleq \sigma_j^2$, (III.B.94) can be inverted to give

$$p_T(t \mid H_j) = \begin{cases} \dfrac{1}{(2\sigma_j^2)^{n/2}} \Gamma(n/2) t^{(n/2-1)} e^{-t/2\sigma_j^2}, & t > 0 \\ 0, & t \leqslant 0, \end{cases} \quad \text{(III.B.95)}$$

where $\Gamma(x) = \int_0^\infty e^{-y} y^{x-1} dy$ is the gamma function. This case corresponds to the situation in which $\lambda_1 = \lambda_2 = \ldots = \lambda_n \triangleq \sigma_S^2$, from which

$$\mathbf{\Sigma}_S = \sigma_S^2 \sum_{k=1}^{n} \underline{v}_k \underline{v}_k^T = \sigma_S^2 \mathbf{I}.$$

That is, in this case the signal samples are i.i.d. $N(0, \sigma_S^2)$, and we have $\sigma_0^2 = \sigma_S^2/(\sigma^2 + \sigma_S^2)$ and $\sigma_1^2 = \sigma_S^2/\sigma^2$.

Equation (III.B.95) is the gamma $(n/2, 1/2\sigma_j^2)$ density, and from this we have

$$P_j(\underline{Y}^T \mathbf{Q} \underline{Y} > \tau') = 1 - \Gamma\left(\frac{n}{2}; \frac{\tau'}{2\sigma_j^2}\right), \quad \text{(III.B.96)}$$

where

$$\Gamma(x; t) \triangleq \int_0^t e^{-y} y^{x-1} dy / \Gamma(x)$$

is the *incomplete gamma function*†. For Neyman-Pearson detection with false-alarm probability α, we thus choose

$$\tau' = 2\sigma_0^2\Gamma^{-1}\left(\frac{n}{2}\,;\,1-\alpha\right)$$

where $\Gamma^{-1}(x\,;\cdot)$ is the inverse function of $\Gamma(x\,;\cdot)$ in its second variable. The ROCs are given by

$$P_D(\tilde{\delta}_{NP}) = 1-\Gamma\left[\frac{n}{2}\,;\,\frac{\sigma_0^2}{\sigma_1^2}\Gamma^{-1}\left(\frac{n}{2};\,1-\alpha\right)\right]. \qquad \text{(III.B.97)}$$

Thus the performance is parameterized here by the two parameters n and $\sigma_0^2/\sigma_1^2=1/(1+\sigma_S^2/\sigma^2)$. Note that σ_S^2/σ^2 is the ratio of the average signal power to the average noise power in this case, and the performance of (III.B.84) improves as this quantity and/or n increases.

For the case in which the signal eigenvalues are not identical (i.e., for a non-i.i.d. signal), (III.B.94) cannot be found in closed from. For this case approximation or bounds can be used to analyze the detection performance. Techniques for doing this are discussed in Section III.C.

Remark III.B.4: A Relationship between the Dependent and Independent Signal Cases

Consider the problem of (III.B.85) in which $\underline{N} \sim N(\underline{0}, \sigma^2 \mathbf{I})$ and $\underline{S} \sim N(\underline{\mu}, \Sigma_S)$ with $\Sigma_S = diag\{\sigma_{S_1}^2, ..., \sigma_{S_n}^2\}$. This is the case in which the noise samples are i.i.d. $N(0, \sigma^2)$ and the signal samples are independent $N(\mu_k, \sigma_{S_k}^2)$. The log-likelihood ratio for this case is given by

† Note that for n even, (III.B.96) can be integrated by parts to yield

$$P_j(\underline{Y}^T \mathbf{Q} \underline{Y} > \tau') = e^{-\tau'/2\sigma_j^2} \sum_{k=0}^{n/2-1} \frac{(\tau'/2\sigma_j^2)^k}{k}.$$

$$\log L(\underline{y}) = \tfrac{1}{2}\sum_{k=1}^{n} y_k^2/\sigma^2 - \tfrac{1}{2}\sum_{k=1}^{n}(y_k-\mu_k)^2/(\sigma^2+\sigma_{S_k}^2)$$
$$+ \tfrac{1}{2}\sum_{k=1}^{n}\log[\sigma^2/(\sigma^2+\sigma_{S_k}^2)]. \tag{III.B.98}$$

Now consider the same problem in which $\mathbf{\Sigma}_S$ is not diagonal. With $p_j(y_1, ..., y_l)$ denoting the density of $Y_1, ..., Y_l$ under H_j, we can write

$$p_j(\underline{y}) = p_j(y_1)\prod_{k=2}^{n} p_j(y_k|y_1, ..., y_{k-1}), \tag{III.B.99}$$

where $p_j(y_k|y_1, ..., y_{k-1})$ is the conditional density of Y_k given $Y_1=y_1, ..., Y_{k-1}=y_{k-1}$. Equation (III.B.99) holds for any density on $\mathbb{R}^n$ and easily follows from the fact that $p_j(y_k|y_1, ..., y_{k-1})=p_j(y_1, ..., y_k)/p_j(y_1, ..., y_{k-1})$. Under H_0, Y_k is independent of $Y_1, ..., Y_{k-1}$ since $\underline{N}$ is i.i.d., so $p_0(\underline{y})=\Pi_{k=1}^n p_0(y_k)$. Under H_1, Y_k is not independent of $Y_1, ..., Y_{k-1}$; however, since $\underline{Y}$ is a Gaussian random vector under H_1, Y_k is conditionally Gaussian given $Y_1=y_1, ..., Y_{k-1}=y_{k-1}$. The mean of this conditional density is given by

$$E_1\{Y_k|Y_k = y_1, ..., Y_{k-1} = y_{k-1}\}$$
$$= E_1\{S_k|Y_1 = y_1, ..., Y_{k-1} = y_{k-1}\}$$
$$+ E_1\{N_k|Y_1 = y_1, ..., Y_{k-1} = y_{k-1}\} \tag{III.B.100}$$
$$= E_1\{S_k|Y_1 = y_1, ..., Y_{k-1} = y_{k-1}\} \triangleq \hat{S}_k,$$

where we have used the fact that N_k is independent of $Y_1, ..., Y_{k-1}$ and has zero mean. Similarly, the variance of the conditional density is

$$Var_1(Y_k | Y_1 = y_1, \ldots, Y_{k-1} = y_{k-1})$$

$$= Var_1(S_k | Y_1 = y_1, \ldots, Y_{k-1} = y_{k-1})$$

$$+ Var_1(N_k | Y_1 = y_1, \ldots, Y_{k-1} = y_{k-1}) \quad \text{(III.B.101)}$$

$$= \hat{\sigma}_{S_k}^2 + \sigma^2,$$

where $\hat{\sigma}_{S_k}^2 \triangleq Var_1(S_k | Y_1 = y_1, \ldots, Y_{k-1} = y_{k-1})$.

A property of the multivariate Gaussian distribution is that $\hat{\sigma}_{S_k}^2$ does not depend on the values of $y_1, \ldots, y_{k-1}$. (This and related properties of the multivariate Gaussian distribution are developed in subsequent chapters.) Thus with $\hat{S}_1 = E\{S_1\}$ and $\hat{\sigma}_{S_1}^2 = Var(S_1)$, $p_1(\underline{y})$ is the product of $N(\hat{S}_k, \hat{\sigma}_{S_k}^2 + \sigma^2)$ densities, and the log-likelihood ratio becomes

$$\log L(\underline{y}) = \tfrac{1}{2} \sum_{k=1}^{n} y_k^2 / \sigma^2 - \tfrac{1}{2} \sum_{k=1}^{n} (y_k - \hat{S}_k)^2 / (\hat{\sigma}_{S_k}^2)$$

$$+ \tfrac{1}{2} \sum_{k=1}^{n} \log[\sigma^2 / (\sigma^2 + \hat{\sigma}_{S_k}^2)]. \quad \text{(III.B.102)}$$

Comparing (III.B.102) with (III.B.98) we see that detecting a dependent stochastic signal is analogous to detecting an independent stochastic signal with mean $\hat{\underline{S}}$ and covariance $diag\{\hat{\sigma}_{S_1}^2, \ldots, \hat{\sigma}_{S_k}^2\}$. Of course the difference is that $\hat{\underline{S}}_k$ in (III.B.102) depends on $y_1, \ldots, y_{k-1}$, whereas μ_k in (III.B.98) does not. Another way to view this is to write, under H_1,

$$Y_k = N_k + s_k = N_k + \epsilon_k + \hat{S}_k, \quad \text{(III.B.103)}$$

where we interpret $\hat{S}_k$ as the random quantity $E_1\{S_k | Y_1, \ldots, Y_{k-1}\}$ and $\epsilon_k = (S_k - \hat{S}_k)$. As we will see in subsequent chapters the quantity $\hat{S}_k$ is an optimum predictor (under H_1) of S_k from the past observations $Y_1, \ldots, Y_{k-1}$. So ϵ_k can be interpreted as the error in this prediction or, equivalently, as the part of S_k that cannot be predicted from the past observations. So as we take each observation we can think of the signal as consisting of a part, $\hat{S}_k$, known from the past, and of a new part ϵ_k which cannot be predicted from the past.† It can be shown that under H_1, ϵ_k is statistically independent of $Y_1, \ldots, Y_{k-1}$, and that it is a $N(0, \hat{\sigma}_{S_k}^2)$ random variable. By comparison, in the case in which $S_1, \ldots, S_n$ is an independent $N(\mu_k, \sigma_{S_k}^2)$ sequence, $\hat{S}_k$ equals μ_k and $\epsilon_k = s_k - \mu_k$, which is $N(0, \sigma_{S_k}^2)$.

Remark III.B.5: Estimator-Correlator Interpretation of the Optimum Detector for Stochastic Signals in I.i.d. Gaussian Noise

As a further comment on the structure of (III.B.102), note that we can write $L(\underline{y})$ as

$$\log L(\underline{y}) = \frac{1}{2\sigma^2}\left[\sum_{k=1}^{n} y_k^2 - \sum_{k=1}^{n}(y_k - \hat{S}_k)^2/(1+\hat{\sigma}_{S_k}^2/\sigma^2)\right] \tag{III.B.104}$$

$$-\tfrac{1}{2}\sum_{k=1}^{n}\log(1+\hat{\sigma}_{S_k}^2/\sigma^2).$$

Suppose that the noise variance σ^2 is large relative to the maximum prediction error variance $\max_{1 \leqslant k \leqslant n} \hat{\sigma}_{S_k}^2$, i.e., suppose that $\hat{\sigma}_{S_k}^2 << \sigma^2$ for all k. In this case, $(1+\hat{\sigma}_{S_k}^2/\sigma^2) \cong 1$ and we can write

$$\log L(\underline{y}) \cong \frac{1}{\sigma^2}\left[\sum_{k=1}^{n} y_k \hat{S}_k - \tfrac{1}{2}\sum_{k=1}^{n}(\hat{S}_k)^2\right]. \tag{III.B.105}$$

Comparing this with (III.B.9), we see that (III.B.105) is the structure for detecting $\hat{S}_1, \ldots, \hat{S}_n$ as if it were a *coherent* signal.

---

† A similar interpretation can be given to $\hat{S}_k$ and $N_k + \epsilon_k$ with regard to Y_k; i.e., $\hat{S}_k$ is the part of Y_k that is known from the past and $Y_k - \hat{S}_k = N_k + \epsilon_k$ is the part of Y_k that cannot be determined from the past. That is, $Y_k - \hat{S}_k$ contains the new information in Y_k. The sequence $(Y_1 - \hat{S}_1), \ldots, (Y_k - \hat{S}_k)$ is sometimes known as the *innovations* sequence. This idea plays an important role in filtering and will be discussed further in subsequent chapters.

That is, we can view the stochastic signal detector in this case, at least approximately, as one that estimates the signal and then treats it as a known signal.

More generally, suppose that we have a stochastic signal with multivariate probability density function $p_{\underline{S}}$, not necessarily Gaussian, embedded in $N(\underline{0}, \sigma^2 \mathbf{I})$ noise. The likelihood ratio for (III.B.85) becomes

$$L(\underline{y}) = \int_{\mathbb{R}^n} \exp\left\{\frac{1}{\sigma^2}(\underline{s}^T \underline{y} - \tfrac{1}{2}\|\underline{s}\|^2)\right\} p_{\underline{S}}(\underline{s}) d\underline{s}. \qquad \text{(III.B.106)}$$

Within regularity on $p_{\underline{S}}$, the mean-value theorem [see Apostol (1974)] implies that

$$L(\underline{y}) = \exp\left\{\frac{1}{\sigma^2}(\underline{\hat{S}}^T \underline{y} - \tfrac{1}{2}\|\underline{\hat{S}}\|^2)\right\} \qquad \text{(III.B.107)}$$

for some $\underline{\hat{S}}$ in $\mathbb{R}^n$ (of course $\underline{\hat{S}}$ depends on $\underline{y}$). Thus we see that, in general, the likelihood ratio for stochastic signals in i.i.d. Gaussian noise can be interpreted as an "estimator," $\underline{\hat{S}}$, of the signal followed by the optimum detector for $\underline{\hat{S}}$ as if it were a coherent signal.† This structure is known as an *estimator-correlator*. This structure is not particularly simple since the determination of the function $\underline{\hat{S}}(\underline{y})$ can be quite difficult. Also, in contrast to (III.B.105) $\underline{\hat{S}}$ in (III.B.107) depends in general on all of $\underline{y}$ and so could not be computed in real time. However, this structure is suggestive of how one might design a suboptimal system for detecting stochastic signals in i.i.d. Gaussian noise, by first building a system that could estimate the signal well if it were present, and then to treat this estimate as a known signal. In the continuous-time analog of the problem of detecting a stochastic signal in i.i.d. Gaussian noise, the estimator correlator structure arises in a

---

† Such an interpretation is valid for any noise density p_N that is sufficiently regular since for each $\underline{y} \in \mathbb{R}^n$,

$$L(\underline{y}) = \int_{\mathbb{R}^n} \frac{p_N(\underline{y} - \underline{s})}{p_N(\underline{y})} p_{\underline{S}}(\underline{s}) d\underline{s} = \frac{p_N(\underline{y} - \underline{\hat{S}})}{p_N(\underline{y})}$$

for some $\underline{\hat{S}} \in \mathbb{R}^n$.

more direct way with an easily characterized estimator, as we shall see in Chapter VI.

Remark III.B.6: Locally Optimum Detection of Stochastic Signals

One problem with the quadratic detector of (III.B.86) is seen by considering the composite hypothesis-testing problem

$$H_0: Y_k = N_k, \quad k = 1,2, \dots, n$$

versus (III.B.108)

$$H_1: Y_k = N_k + \theta^{1/2} S_k, \quad k = 1, 2, \dots, n, \ \theta > 0,$$

where $\underline{N}$ and $\underline{S}$ are as in (III.B.85) with unit noise variance. In this case the covariance of $\underline{S}$ is $\theta \boldsymbol{\Sigma}_S$, and so the relevant quadratic detection statistic for Neyman-Pearson testing with fixed θ is

$$\theta \underline{y}^T \boldsymbol{\Sigma}_S (\mathrm{I} + \theta \boldsymbol{\Sigma}_S)^{-1} \underline{y}. \tag{III.B.109}$$

Although the leading θ coefficient can be absorbed into the decision threshold, the θ appearing in $(\mathrm{I} + \theta \boldsymbol{\Sigma}_S)^{-1}$ cannot be decoupled from the observations $\underline{y}$. Thus, no UMP test exists for (III.B.108). However, an LMP test statistic can be found by differentiating (III.B.109) with respect to θ and setting θ to zero. Doing so, we get an LMP test statistic

$$2 \underline{y}^T \boldsymbol{\Sigma}_S \underline{y}. \tag{III.B.110}$$

The statistic (III.B.110) has an interesting interpretation for the case in which the $k-l^{th}$ element of $\boldsymbol{\Sigma}_S$, say $\rho_{k,l}$, depends only on the difference $(k-l)$. In this case we can write $\rho_{k,l} = \rho_{k-l,0} \triangleq \rho_{k-l}$, where we suppress the second index for convenience of notation. A signal with this property is said to be *wide-sense stationary*, a concept that we will return to in Chapter V. Consider the scaled LMP statistic

$$T(\underline{y}) = \frac{1}{n}\underline{y}^T \Sigma_S \underline{y}, \tag{III.B.111}$$

which is equivalent to (III.B.110), since the scaling merely rescales the threshold. After some algebra (see Exercise 14), we can write $T(\underline{y})$ as

$$T(\underline{y}) = \frac{1}{n}\sum_{k=1}^{n}\sum_{l=1}^{n} y_k y_l \rho_{k-l} = \rho_o \hat{\rho}_o + 2\sum_{k=1}^{n-1} \rho_k \hat{\rho}_k, \tag{III.B.112}$$

where $\hat{\rho}_k$ is defined by

$$\hat{\rho}_k \triangleq \frac{1}{n}\sum_{l=1}^{n-k} y_l y_{l+k} \quad , k = 0, \ldots, n-1. \tag{III.B.113}$$

The representation of (III.B.112) leads to the following interpretation of the LMP statistic (III.B.111). Note that, for $n >> k$, $\hat{\rho}_k$ is an estimate of the covariance $E\{Y_l Y_{l+k}\}$ for $l = 1, \ldots, n-k$. Thus, $T(\underline{y})$ estimates the covariance structure of the observations, and then correlates this with the signal covariance sequence. Under H_0 we have

$$E\{Y_l Y_{l+k}\} = \begin{cases} 1 \text{ if } k = 0 \\ 0 \text{ if } k \neq 0, \end{cases} \tag{III.B.114}$$

and, under H_1, we have

$$E\{Y_l Y_{l+k}\} = \begin{cases} 1 + \theta\rho_o \text{ if } k = 0 \\ \theta\rho_k \text{ if } k \neq 0. \end{cases} \tag{III.B.115}$$

So, assuming the estimates $\hat{\rho}_k$ were reasonably accurate, we would have

$$T(\underline{y}) \cong \begin{cases} \rho_0 \text{ under } H_0 \\ \\ \rho_0 + 2\theta \sum_{k=1}^{n-1} \rho_k^2 \text{ under } H_1. \end{cases} \qquad \text{(III.B.116)}$$

From (III.B.116), we see that the statistic $T(\underline{y})$ is on intuitively reasonable way of detecting the signal, particularly if the signal is highly correlated (i.e., $\sum_{k=0}^{n} \rho_k^2$ is large).

The statistic $T(\underline{y})$ also has on interesting interpretation in the frequency domain. In particular, suppose we think of $S_1, ..., S_n$ as a segment of an infinite random sequence $\{S_k\}_{k=-\infty}^{\infty}$ with $E\{S_l S_{l+k}\} = \rho_k$, for all integers l and k. Then the discrete-time Fourier transform of the sequence $\{\rho_k\}_{k=-\infty}^{\infty}$, namely,

$$\phi(\omega) \triangleq \sum_{k=-\infty}^{\infty} \rho_k \, e^{i\omega k}, \qquad \text{(III.B.117)}$$

is the *power spectrum* of $\{S_k\}_{k=-\infty}^{\infty}$. Since $\rho_k = 1/2\pi \int_{-\pi}^{\pi} \phi(\omega) e^{-i\omega k} d\omega$, $T(\underline{y})$ can be rewritten as

$$T(\underline{y}) = \frac{1}{2\pi} \int_{-\pi}^{\pi} \phi(\omega)\hat{\phi}(\omega) d\omega, \qquad \text{(III.B.118)}$$

where

$$\hat{\phi}(\omega) \triangleq \frac{1}{n} \left| \sum_{k=1}^{n} y_k \, e^{i\omega k} \right|^2, \; -\pi \leqslant \omega \leqslant \pi. \qquad \text{(III.B.119)}$$

The function $\hat{\phi}$ is known as the *periodogram* of the data, and is an estimate of the spectrum of the observations. Thus, in the form (III.B.118), $T(\underline{y})$ estimates the observation spectrum and correlates this estimate (in the frequency domain) with the signal spectrum. Since the observation spectrum equals 1 for $\omega \epsilon [-\pi, \pi]$ under H_0 and equals $1+\phi(\omega)$ for $\omega \epsilon [-\pi, \pi]$ under H_1, the operation of (III.B.118) has an interpretation similar to that of (III.B.112).

III.C Performance Evaluation of Signal Detection Procedures

In Section III.B the design and analysis of optimum procedures for discrete-time signal detection were discussed. In a sense, the design of such procedures is more straightforward than is their performance analysis because of the frequent intractability of the latter problem. We were able to compute performance in most of the examples presented in Section III.B only because the particular models considered are among the tractable ones and are thus best used to illustrate the theory. Sometimes the assumptions can vary only slightly from those in these examples and this tractability disappears. For example, in the problem of detecting a Gaussian signal in i.i.d. Gaussian noise the computation of error probabilities is tractable if the signal is also i.i.d., but it is intractable if the variance of only one signal sample changes.

The basic performance measures of a binary signal detection system using a decision rule $\tilde{\delta}$ are the two conditional error probabilities P_F and P_M defined by

$$P_F(\tilde{\delta}) = P_0(\tilde{\delta} \text{ chooses } H_1)$$

and

$$P_M(\tilde{\delta}) = P_1(\tilde{\delta} \text{ chooses } H_0)$$

Likelihood-ratio tests and most other decision rules of interest are of the form

$$\tilde{\delta}_T(y) = \begin{cases} 1 \text{ if } T(y) > \tau \\ \gamma \text{ if } T(y) = \tau \\ 0 \text{ if } T(y) < \tau, \end{cases} \qquad \text{(III.C.1)}$$

where T is a mapping from (Γ, G) to $(\mathbb{R}, B)$ (e.g., the log-likelihood ratio). Thus performance evaluation for most systems involves computing the probabilities of the regions $\{T(Y) > \tau\}$ (or $\{T(Y) < \tau\}$) and $\{T(Y) = \tau\}$ under the two hypotheses. Although this problem is conceptually simple, the actual computation of the required probabilities is often analytically difficult. For example, if $Y=(Y_1, ..., Y_n)$ has joint pdf p_0 under H_0, then

$$P_F(\tilde{\delta}_T) = \underset{\{T(y)>\tau\}}{\int \cdots \int} p_0(y_1, ..., y_n) dy_1 \cdots dy_n$$

$$+ \gamma \underset{\{T(y)=\tau\}}{\int \cdots \int} p_0(y_1, ..., y_n) dy_1 \cdots dy_n ,$$

which is difficult to compute for large n without further simplification. In this section we discuss several commonly used techniques for computing, bounding, or approximating the performance of detection systems.

III.C.1 Direct Performance Computation

Note that for a system of the form (III.C.1), we have

$$P_F(\tilde{\delta}_T) = P(T(Y) > \tau | H_0) + \gamma P(T(Y) = \tau | H_0)$$

$$= [1 - F_{T,0}(\tau)] + \gamma [F_{T,0}(\tau) - \lim_{\sigma \to \tau -} F_{T,0}(\sigma)]$$

and

$$P_M(\tilde{\delta}_T) = P(T(Y) < \tau | H_1) + (1-\gamma) P(T(Y) = \tau | H_1)$$

$$= P(T(Y) \leq \tau | H_1) - \gamma P(T(Y) = \tau | H_1)$$

$$= F_{T,1}(\tau) - \gamma [F_{T,1}(\tau) - \lim_{\sigma \to \tau -} F_{T,1}(\sigma)],$$

where $F_{T,j}$ is the cumulative distribution function (cdf) of $T(Y)$ under hypothesis H_j. Thus for detection procedures of the form of (III.C.1) (such as the likelihood-ratio detector), evaluation is facilitated if the cdf's $F_{T,j}$ can be determined easily in a neighborhood of the threshold τ.

One case in which a straightforward expression for $F_{T,j}$ can be written is that in which $\underline{Y}=(Y_1, ..., Y_n)^T$ is a vector of independent (real) random variables and $T(\underline{y})$ has the structure

$$T(\underline{y}) = \sum_{k=1}^{n} g_k(y_k),$$

where $\{g_k\}_{k=1}^{n}$ is the sequence of nonlinearities (e.g., the log-likelihood ration is of this form). In this case it is often simple to compute $F_{T,j}$ by using characteristic functions. Recall that the characteristic function (ch.f.) of a random variable X is defined by

$$\phi_X(u) = E\{e^{iuX}\}, \ u \in \mathbb{R},$$

where i denotes the imaginary unit $\sqrt{-1}$. Note that a cdf F and its ch.f. ϕ form a unique pair.† Denoting by $\phi_{T,j}$ and $\phi_{g_{k,j}}$ the ch.f.'s under H_j of $T(\underline{Y})$ and $g_k(Y_k)$, respectively, we have by the independence of the Y_k's that

$$\phi_{T,j}(u) = E\left\{ \exp\left\{ i\sum_{k=1}^{n} g_k(Y_k)\right\} \Big| H_j \right\}$$

$$= \prod_{k=1}^{n} E\{\exp\{ig_k(Y_k)\}|H_j\} = \prod_{k=1}^{n} \phi_{g_{k,j}}(u).$$

The ch.f. $\phi_{T,j}$ can be inverted to give the cdf $F_{T,j}$ via the formula [see, e.g., Billingsley (1979)]

† A through treatment of the subject of characteristic functions is given in Lukacs (1960).

$$F_{T,j}(b)-F_{T,j}(a) = \lim_{U\to\infty} \int_{-U}^{U} \frac{e^{-iua}-e^{-iub}}{iu} \phi_{T,j}(u)du, \quad \text{(III.C.2)}$$

which holds for all a and b that are continuity point of $F_{T,j}$. Knowledge of $[F_{T,j}(b)-F_{T,j}(a)]$ at all continuity points is sufficient to determine $F_{T,j}$ uniquely since $F_{T,j}(-\infty)=0$ and $F_{T,j}$ must be right-continuous. If $T(\underline{Y})$ is a continuous random variable under H_j, the inversion is simplified since $F_{T,j}$ has a probability density function $p_{T,j}$ given in this case by

$$p_{T,j}(t) = \frac{1}{2\pi} \int_{-\infty}^{\infty} \phi_{T,j}(u)e^{-iut}\,du,$$

and so $p_{T,j}$ and $\phi_{T,j}$ are a Fourier transform pair.† So, for example, in the latter case we have

$$P_F(\tilde{\delta}_T) = \frac{1}{2\pi} \int_{\tau}^{\infty} \int_{-\infty}^{\infty} e^{-iut} \left[\prod_{k=1}^{n} \phi_{g_{k,0}}(u)\right] du\; dt, \quad \text{(III.C.3)}$$

and similarly for $P_M(\tilde{\delta})$.

Example III.C.1: Correlation Detection in Cauchy Noise

As an example of the foregoing approach, consider the performance of the correlation detector [defined by $g_k(y_k)=s_k y_k$, $k=1, \ldots, n$] in detecting a coherent signal in additive Cauchy noise. In particular, consider the hypothesis pair

$$H_0: Y_k = N_k, \quad k = 1, \ldots, n$$

versus

$$H_1: Y_k = N_k + s_k, \quad k = 1, \ldots, n,$$

where $N_1, \ldots, N_n$ is a sequence if i.i.d. random variables each having the Cauchy pdf

---

† A sufficient (but not necessary) condition for $T(\underline{Y})$ to be continuous under H_j is $\int_{-\infty}^{\infty}|\phi_{T,j}(u)|du < \infty$ [see Breiman (1968)].

$$p_{N_k}(x) = \frac{1}{\pi(1+x^2)}, \; x \in \mathbb{R},$$

and where $s_1, \ldots, s_n$ is a known signal sequence. In this case we have

$$\phi_{g_{k,0}}(u) = E\{e^{ius_k N_k}\} = \phi_{N_k}(us_k)$$

$$= F\{p_{N_k}\}\Big|_{us_k} = e^{-|s_k u|}, \; u \in \mathbb{R}.$$

Thus

$$\phi_{T,0}(u) = \prod_{k=1}^{n} \phi_{g_{k,0}}(u) = e^{-n\overline{|s|}|u|}$$

where $\overline{|s|} \triangleq (1/n)\sum_{k=1}^{n}|s_k|$. Since $\phi_{T,0}(u)$ is absolutely integrable, we have

$$P_F(\tilde{\delta}_T) = \frac{1}{2\pi}\int_{\tau}^{\infty}\int_{-\infty}^{\infty} e^{-iut}\, e^{-n\overline{|s|}|u|}\,du\,dt$$

$$= \frac{1}{n\overline{|s|}\pi}\int_{\tau}^{\infty}\frac{1}{1+(t/n\overline{|s|})^2}dt$$

$$= ½-(1/\pi)\mathrm{Tan}^{-1}(\tau/n\overline{|s|}).$$

Similarly, we have $\phi_{T,1}(u) = \phi_{T,0}(u)e^{iun\overline{s^2}}$ where $\overline{s^2} \triangleq (1/n)\sum_{k=1}^{n}s_k^2$, which gives straightforwardly

$$P_M(\tilde{\delta}_T) = ½ + (1/\pi)\mathrm{Tan}^{-1}(\tau/n\overline{|s|}-\sqrt{\overline{s^2}}).$$

[As an aside, it is interesting to note here that we can achieve size-α detection by choosing $\tau = n\overline{|s|}\tan(½-\alpha)$, in which case we have ROCs

$$P_D(\tilde{\delta}_T) = 1 - P_M(\tilde{\delta}_T)$$
$$= \tfrac{1}{2} - (1/\pi)\mathrm{Tan}^{-1}(\tan(\tfrac{1}{2}-\alpha) - \sqrt{s^2}). \tag{III.C.4}$$

Equation (III.C.4) implies the surprising result that performance is improved here only by increasing the *average* signal power and not be increasing the number of samples. This odd behavior is due to the fact that the correlation detector is quite different from the *optimum* detector for coherent signals in Cauchy noise, for which performance does improve with increased sample size. Essentially, the heaviness of the tails of the Cauchy distribution cancels the effect of noise reduction that the correlator achieves in the Gaussian case.]

In Example III.C.1, exact closed-form expressions for P_F and P_M were obtained by using characteristic functions. Another example in which this method yields closed-form expressions is the case of detecting an i.i.d. Gaussian signal in i.i.d. Gaussian noise, which was developed in Section III.B. However, the characteristic-function approach does not always yield tractable closed-form expressions for error probabilities, and it is more common that (III.C.3) and the corresponding expression for P_M are used as a basis for approximating the error probabilities. For moderate values of the error probabilities, direct numerical integration of (III.C.3) and the corresponding P_M integral, using for example the fast Fourier transform (FFT) algorithm, can give close approximations to the error probabilities. In many practical problems, however, one or both of the error probabilities are quite small ($\leqslant 10^{-5}$) and other numerical methods are usually more effective. In such cases there usually is a parameter (the threshold, the signal-to-noise ratio, the number of samples, etc.) that is very large, and thus performance can be approximated by using asymptotic expansions of the error probabilities in this parameter. One technique of this type that has been applied widely in the analysis of communications systems is the *saddle-point approximation* technique [see e.g., De Bruijn (1981)]. Examples of applications of this method are found in Lugannani and Rice (1981), in which approximations for the case in which the $g_k(Y_k)$ are identically distributed are found, and in Mazo and Salz (1965), in which saddle-point approximations for the quadratic detector of Section III.B with non-i.i.d. Gaussian signals and noise are developed.

III.C.2 Chernoff and Related Bounds

It is frequently impractical (or impossible) to compute exactly the error probabilities P_F and P_M of detectors of the form of (III.C.1). However, in practice it is usually sufficient to obtain good upper bounds on these quantities. One type of performance bound that is commonly used in this context is the *Chernoff* bound. The Chernoff bound is a bound on the performance of likelihood ratio detectors and is based on the following simple inequality.

Markov Inequality: Suppose that X is a random variable. If $P(X \geqslant 0)=1$, then $P(X \geqslant a) \leqslant E\{X\}/a$ for all $a>0$.

Proof: $P(X \geqslant a)=E\{I_{[a,\infty)}(X)\}$, where $I_{[a,\infty)}$ is the indicator function of the set $[a,\infty)$ defined by

$$I_{[a,\infty)}(x) = \begin{cases} 1 \text{ if } x \geqslant a \\ 0 \text{ if } x < a . \end{cases}$$

Since $X \geqslant 0$, we note that $I_{[a,\infty)}(X) \leqslant X/a$, and Markov's inequality follows.

□

To apply the Markov inequality to the system of (III.C.1), we note that since the randomization $\gamma \leqslant 1$, we can write

$$\begin{aligned} P_M(\tilde{\delta}_T) &\leqslant P_0(T(Y) \geqslant \tau) = P_0(e^{sT(Y)} \geqslant e^{s\tau}) \\ &\leqslant e^{-s\tau} E\{e^{sT(Y)}|H_0\} = \exp\{-s\tau + \mu_{T,0}(s)\} \end{aligned} \tag{III.C.5}$$

for all $s>0$, where $\mu_{T,0}$ is the *cumulant generating function* (cgf) of $T(Y)$ under H_0 defined by

$$\mu_{T,0}(s) = \log(E\{e^{sT(Y)}|H_0\}).$$

Similarly, since $\gamma \geqslant 0$, we have

$$P_M(\tilde{\delta}_T) \leqslant P_1(T(Y) \leqslant \tau) = P_1(e^{tT(Y)} \geqslant e^{t\tau})$$
$$\leqslant \exp\{-t\tau + \mu_{T,1}(t)\} \qquad \text{(III.C.6)}$$

for $t < 0$, where $\mu_{T,1}$ is the cgf of $T(Y)$ under H_1.

The bounds of (III.C.5) and (III.C.6) can be minimized over $s > 0$ and $t < 0$ to find the tightest such bounds provided the cgf's of $T(Y)$ are known. These bounds are particularly useful for the likelihood ratio detector. To investigate this case, we assume that P_j has density p_j for $j = 0$ and 1, and we choose $T(y) = \log L(y)$, where $L = p_1/p_0$. In this case we have

$$\mu_{T,0}(s) = \log(\int_\Gamma e^{s\log L} p_0 d\mu) = \log(\int_\Gamma L^s p_0 d\mu)$$

and

$$\mu_{T,1}(t) = \log(\int_\Gamma L^t p_1 d\mu) = \log(\int_\Gamma L^{t+1} p_0 d\mu)$$

$$= \mu_{T,0}(t+1).$$

Thus we can rewrite the bound of (III.C.6) as

$$P_M(\delta_T) \leqslant \exp\{(1-s)\tau + \mu_{T,0}(s)\}, \; s < 1. \qquad \text{(III.C.7)}$$

Note that both bounds (III.C.5) and (III.C.7) achieve their minima at the same value of s if

$$\arg\{\min_{s<1}[\mu_{T,0}(s) - s\tau]\} > 0$$

and

$$\arg\{\min_{s>0}[\mu_{T,0}(s)-s\tau]\}<1.$$

It can be shown straightforwardly that $[\mu_{T,0}(s)-s\tau]$ is a convex function of s on the region where $\mu_{T,0}(s)<\infty$ [which must be an interval, see Billingsley (1979)]. Therefore, the condition

$$\mu'_{T,0}(s)=\tau, \tag{III.C.8}$$

where $\mu'_{T,0}(s)=d\,\mu_{T,0}(s)/ds$ is sufficient for a minimum of $[\mu_{T,0}(s)-s\tau]$. It is also straightforward to show that if $E\{|\log L(Y)||H_j\}<\infty$ for $j=0$ or 1, then

$$\mu'_{T,0}(j)=E\{\log L(Y)|H_j\}. \tag{III.C.9}$$

The convexity of $[\mu_{T,0}(s)-s\tau]$ implies that $\mu'_{T,0}(s)$ is an increasing function of s, and thus (III.C.9) implies that (III.C.8) has a solution with $s\in(0,1)$ if (and only if)

$$\mu_0 \triangleq E\{\log L(Y)|H_0\}<\tau<E\{\log L(Y)|H_1\}\triangleq\mu_1. \tag{III.C.10}$$

Thus, on assuming that (III.C.10) is valid, (III.C.5) and (III.C.7) become

$$P_F(\tilde{\delta}_T)\leqslant\exp\{\mu_{T,0}(s_0)-s_0\mu'_{T,0}(s_0)\} \tag{III.C.11}$$

$$P_M(\tilde{\delta}_T)\leqslant\exp\{\mu_{T,0}(s_0)+(1-s_0)\mu'_{T,0}(s_0)\} \tag{III.C.12}$$

with

$$\mu_0<\mu'_{T,0}(s_0)=\tau<\mu_1. \tag{III.C.13}$$

Equations (III.C.11) and (III.C.12) are known as the *Chernoff bounds*. It follows from the convexity of $[\mu_{0,T}(s)-s\tau]$ that if $\tau \leqslant \mu_0$, then $\min_{s \geqslant 0}[\mu_{T,0}(s)-s\tau]=0$ which implies that the bound of (III.C.5) is trivial (i.e., is $\geqslant 1$), and similarly that if $\tau \geqslant \mu_1$, then the bound of (III.C.7) is trivial, although in either case the other bound is still nontrivial. Note that Jensen's inequality† implies that $\mu_0 \leqslant 0$ and $\mu_1 \geqslant 0$ with $\mu_0=0$ and/or $\mu_1=0$ if and only if $P_0=P_1$, so a threshold of $\tau=0$ always satisfies (III.C.10).

Note that if priors π_0 and π_1 are known, then (III.C.11) and (III.C.12) yield an upper bound on the *average* probability of error. In particular,

$$\begin{aligned} P_e &= \pi_0 P_F + \pi_1 P_M \\ &\leqslant [\pi_0 + \pi_1 e^{\mu'_{T,0}(s_0)}] \exp\{\mu_{T,0}(s_0) - s_0 \mu'_{T,0}(s_0)\}. \end{aligned} \tag{III.C.14}$$

However, a better bound on P_e can be obtained by noting that (see Exercise 16)

$$P_e \leqslant \pi_0 e^{-s\tau} \int_{\Gamma_1} L^s p_0 d\mu + \pi_1 e^{(1-s)\tau} \int_{\Gamma_0} L^s p_0 d\mu \tag{III.C.15}$$

for $0 \leqslant s \leqslant 1$, where $\Gamma_1 = \{L(Y) \geqslant \tau\}$ and $\Gamma_0 = \Gamma_1^c$. Equation (III.C.15) implies

$$P_e \leqslant \max\{\pi_0, \pi_1 e^{\tau}\} \exp\{\mu_{T,0}(s) - s\tau\}, \ 0 \leqslant s \leqslant 1. \tag{III.C.16}$$

Under the condition of (III.C.13), the best bound of the form of (III.C.16) is

$$\begin{aligned} P_e &\leqslant \max\{\pi_0, \pi_1 e^{\mu'_{T,0}(s_0)}\} \\ &\times \exp\{\mu_{T,0}(s_0) - s_0 \mu'_{T,0}(s_0)\}, \end{aligned} \tag{III.C.17}$$

---

† *Jensen's Inequality*: For any random variable X and convex function C we have $E\{C(X)\} \geqslant C(E\{X\})$, with equality if and only if $P(X = E\{X\}) = 1$ when C is strictly convex.

which is tighter bound than is (III.C.14) unless π_0 is 0 or 1. Note that a minimum-probability-of-error log-likelihood ratio detector uses threshold $\tau=\log(\pi_0/\pi_1)$, in which case (III.C.16) reduces to

$$P_e \leqslant \pi_0^{1-s}\pi_1^s e^{\mu_{T,0}(s)}, \quad 0\leqslant s\leqslant 1. \qquad \text{(III.C.18)}$$

An interesting case of the foregoing analysis is that in which $\Gamma=\mathbb{R}^n$ and $Y=(Y_1,\ldots,Y_n)$ is a sequence of independent and identically distributed observations with marginal density f_j under hypothesis H_j. In this case we have straightforwardly that

$$\mu_{T,0}(s)=n\ \log\left(\int_{\mathbb{R}} f_1^s f_0^{1-s}\right),$$

and it follows from Jensen's inequality that for $0<s<1$,

$$\log\left(\int_{\mathbb{R}} f_1^s f_0^{1-s}\right)<0.$$

Thus (III.C.18) implies that the probability of error in this case decreases at least at an exponential rate as the number of samples (n) increases.

Example III.C.2: The Chernoff Bound for Quadratic Detection

To illustrate the use of the Chernoff bound consider the problem of detecting a $N(\underline{0},\Sigma_S)$ signal in $N(\underline{0},\sigma^2\mathrm{I})$ noise. After transforming $Y_1,\ldots,Y_n$ into the independent sequence $\tilde{Y}_1,\ldots,\tilde{Y}_n$ as in (III.B.87), we have that

$$L(\underline{y})=\prod_{k=1}^{n}\frac{\sigma_{0k}}{\sigma_{1k}}e^{\tilde{y}_k^2}$$

with σ_{jk}^2 from (III.B.88). Thus with $T=\log L$ we have

$$\mu_{T,0}(s) = \log\left(E\left\{\prod_{k=1}^{n}\left(\frac{\sigma_{0k}}{\sigma_{1k}}\right)^{s} e^{sY_k^2}\middle| H_0\right\}\right)$$

(III.C.19)

$$= \sum_{k=1}^{n} s \log\left(\frac{\sigma_{0k}}{\sigma_{1k}}\right) + \sum_{k=1}^{n} \log\left(E\{e^{sY_k^2/2}|H_0\}\right).$$

The expectation in (III.C.19) is given by

$$E\{e^{sY_k^2/2}|H_0\} = \begin{cases} [1-s\,\sigma_{0k}^2]^{-1/2} & \text{if } s < 1/\sigma_{0k}^2 \\ \infty & \text{if } s \geqslant 1/\sigma_{0k}^2. \end{cases}$$

Since $\sigma_{0k}^2 < 1$ for all k, (III.C.19) is finite for all k when $s \leqslant 1$, and so we can readily bound, say, the minimum error probability by

$$P_e \leqslant \pi_0^{1-s}\pi_1^{s}\prod_{k=1}^{n}\frac{\sigma^s(\sigma^2+\lambda_k)^{(1-s)/2}}{[\sigma^s+(1-s)\lambda_k]^{1/2}},\ 0 \leqslant s \leqslant 1. \qquad \text{(III.C.20)}$$

The bound is minimized by the value s_0 solving

$$2\log(\pi_0/\pi_1) + \sum_{k=1}^{n}\log(1+\lambda_k/\sigma^2)$$

(III.C.21)

$$= \sum_{k=1}^{n}\frac{\lambda_k}{\sigma^2+(1-s_0)\lambda_k},$$

which can be solved directly for the i.i.d. case ($\lambda_1=\ldots=\lambda_n$) and numerically for other cases.

There are several other bounds similar to the Chernoff bound that are useful in applications. One such bound is the particular case of (III.C.18) with $s = 1/2$, for which we have

$$P_e \leqslant \sqrt{\pi_0 \pi_1} \exp\{\mu_{T,0}(1/2)\}. \qquad \text{(III.C.22)}$$

This bound is sometimes called the *Bhattacharyya bound*. The quantity

$$\rho \triangleq \exp\{\mu_{T,0}(1/2)\} = \int_\Gamma [p_0 p_1]^{1/2} d\mu$$

is known as the *Bhattacharyya coefficient* or the *Hellinger integral* (also the *affinity*), and lower bounds on P_e are also available in terms of ρ. For example, we have [see Kobayashi and Thomas (1968)]

$$\pi_0 \pi_1 \rho^2 \leqslant P_e \leqslant (\pi_0 \pi_1)^{1/2} \rho. \qquad \text{(III.C.23)}$$

A related lower bound is based on a quantity known as the *J-divergence* defined by

$$J = \int_\Gamma (L-1) \log(L) p_0 d\mu,$$

and for which we have the bound [Kobayashi and Thomas, (1968)]

$$P_e > \pi_0 \pi_1 e^{-J/2}.$$

[The quantity J is closely related to the *relative entropy* between p_0 and p_1 which plays a role in information theory; see Kullback (1959).] Because the quantities ρ and J are more tractable than is P_e, they are sometimes used as criteria for signal selection (i.e., choice of p_0 and p_1), as are a number of related criteria.

The utility of the Chernoff and other bounds described above is based on the fact that quantities such as $E\{L^s\}$ and $E\{(L-1)\log L\}$ are usually easier to compute than are error probabilities, and the validity of these bounds is based on the fact that functions of L appearing in the bounds are pointwise no smaller than the indicator functions corresponding to the error probabilities. In other words, we should like to bound $E\{h(L)\}$ where h is the indicator function of a decision region, and we do so by finding a function $g \geqslant h$ for which $E\{g(L)\}$ is easy to compute.† A related class of error-probability bounds, known as *moment-space* bounds, are based on a similar principle in which one tries to find a function g such that $E\{g(L)\}$ is easy to compute and that the curve traced out by $(h(L), g(L))$ as L varies is nearly a straight line. These bounds were introduced by Yao and Tobin (1976) and are useful in evaluating the effects on performance of several types of noises which arises in digital communication systems, such as intersymbol interference and multiple-access noise.

III.C.3 Asymptotic Relative Efficiency

In Sections III.C.1 and III.C.2 we considered techniques for bounding or computing directly the error probabilities associated with a signal detection procedure. We see that, in general, the error probabilities may be difficult quantities to obtain exactly. There are other meaningful performance measures that are often easier to compute than error probability. One such criterion, the asymptotic relative efficiency (ARE), is particularly useful in comparing discrete-time detection systems under large-sample-size, weak-signal conditions.‡ The ARE can be motivated as follows.

Suppose that we have observations $Y_1, Y_2, \ldots$ which obey one of two statistical hypotheses, H_0 and H_1. Suppose further that $\tilde{\delta}_1$ and $\tilde{\delta}_2$ are two test for H_0 versus H_1 which have identical error probabilities but which use n_1 and n_2 samples, respectively. If $n_1 < n_2$, we might say that $\tilde{\delta}_1$ is more efficient than $\tilde{\delta}_2$ because it requires less information than $\tilde{\delta}_2$ does to achieve identical performance. Similarly, if $n_1 > n_2$,

---

† A general class of upper and lower bounds of this type has been developed by Boekee and Ruitenbeck (1981).

‡ The conditions of large sample size and weak signal arise in applications (e.g., passive sonar and radio astronomy) in which the signals are deeply embedded in noise, thus requiring long integration times to detect them. Actually, there are several definitions of ARE, but the most commonly used definition is that introduced by E. J. G. Pitman [see, Noether (1955)] which is the definition treated here.

we would say that $\tilde{\delta}_1$ is less efficient than $\tilde{\delta}_2$. Thus the ratio n_2/n_1 is a good measure of the efficiency of $\tilde{\delta}_1$ relative to $\tilde{\delta}_2$. Such a measure of relative efficiency would be particularly useful for large samples sizes (large n_1 and n_2) because, although both systems would probably achieve good error-probability performance with large samples, a ratio of $n_1/n_1=2$, say, would indicate that $\tilde{\delta}_1$ is significantly more desirable than $\tilde{\delta}_2$ if n_1 is a large number and if $\tilde{\delta}_1$ and $\tilde{\delta}_2$ are of similar complexity. Pitman's ARE is an asymptotic $(n_1, n_2 \to \infty)$ measure of the efficiency of one detector relative to another in this sense of relative sample sizes required to achieve equivalent performance.

To define this notion of ARE more precisely, we consider two sequences $\{\tilde{\delta}_1^{(n)}\}_{n=1}^{\infty}$ and $\{\tilde{\delta}_2^{(n)}\}_{n=1}^{\infty}$ of tests of H_0 versus H_1, where $\tilde{\delta}_j^{(n)}$ operates with n samples. We assume that the false-alarm probability of each test in each sequence is fixed at $\alpha \epsilon (0, 1)$. The *relative efficiency* of $\{\tilde{\delta}_1^{(n)}\}_{n=1}^{\infty}$ relative to $\{\tilde{\delta}_2^{(n)}\}_{n=1}^{\infty}$ for sample size n is defined to be the ratio n_2/n, where n_2 is the smallest number of samples such the $P_D(\tilde{\delta}_2^{(n_2)}) \geq P_D(\tilde{\delta}_1^{(n)}) \triangleq \beta_n$. We now would like to define the ARE as the limit of the relative efficiency as n approaches ∞; however, for most reasonable test sequences $\lim_{n \to \infty} \beta_n = 1$,† and thus for very large n, β_n is no longer a suitable criterion for preferring one test over another. To overcome the difficulty we consider a sequence of alternative hypotheses $\{H_1^{(n)}\}_{n=1}^{\infty}$ converging in some way to H_0 such that $\lim_{n \to \infty} \beta_n \triangleq \beta \epsilon (\alpha, 1)$. We then compute the relative efficiency assuming that both $\tilde{\delta}_1^{(n)}$ and $\tilde{\delta}_2^{(n)}$ are tests of H_0 versus $H_1^{(n)}$ and define the asymptotic efficiency of $\{\tilde{\delta}_1^{(n)}\}_{n=1}^{\infty}$ by

$$ARE_{1,2} = \lim_{n \to \infty} (n_2/n).$$

Note that $H_1^{(n)}$ being "close" to H_0 (for large n) corresponds to the local testing problem discussed in Section II.E (e.g., the case of a weak signal in a signal detection model).

In general, the quantity n_2/n is a function of α and $\{\beta_n\}_{n=1}^{\infty}$ among other things. It turns out, however, that

---

† A sequence of tests with this property is said to be *consistent*.

under mild assumptions the ARE is not dependent on these quantities. In particular, assume that the tests $\tilde{\delta}_j^{(n)}$ are of the form

$$\tilde{\delta}_j^{(n)}(\underline{y}) = \begin{cases} 1 \text{ if } T_j^{(n)}(y_1, ..., y_n) > \tau_j^{(n)} \\ \gamma_j \text{ if } T_j^{(n)}(y_1, ..., y_n) = \tau_j^{(n)} \\ 0 \text{ if } T_j^{(n)}(y_1, ..., y_n) < \tau_j^{(n)}, \end{cases}$$

and that the hypotheses are of the form

$$H_0: Y \sim P_{\theta_0}$$

$$H_1^{(n)}: Y \sim P_{\theta_n},$$

where $\theta_n > \theta_0$ and where $\{P_\theta; \theta \geqslant \theta_0\}$ is a family of distributions for Y. Thus the notion of $H_1^{(n)}$ converging to H_0 can be represented in this case by $\lim_{n\to\infty}\theta_n = \theta_0$. Define for $j=0, 1$, for $n=1, 2, \ldots,$ and for $\theta \geqslant \theta_0$, the two quantities

$$\psi_j^{(n)}(\theta) = E\{T_j^{(n)}(Y_1, ..., Y_n) | Y \sim P_\theta\}$$

and

$$\sigma_j^{(n)}(\theta) = [\ Var\ (T_j^{(n)}(Y_1, ..., Y_n) | Y \sim P_\theta)]^{1/2};$$

i.e., $\psi_j(\theta)$ and $\sigma_j(\theta)$ are the mean and standard deviation of the test statistic $T_j(Y)$ when $Y \sim P_\theta$.

Consider the following regularity conditions:

1. There exists a positive integer m such that the first through $(m-1)^{\underline{th}}$ derivatives of $\psi_j^{(n)}(\theta)$ are zero at $\theta=\theta_0$, and

$$\frac{d^m}{d\,\theta^m}\psi_j^{(n)}(\theta)|_{\theta=\theta_0}>0 \text{ for } j=0,1.$$

2. There exists $\delta>0$ such that, for $j=0, 1$,

$$\lim_{n\to\infty}\left[n^{-m\delta}\frac{d^m}{d\,\theta^m}\psi_j^{(n)}(\theta)|_{\theta=\theta_0}\sigma_j^{(n)}(\theta_0)\right]\triangleq c_j>0. \quad \text{(III.C.24)}$$

3. Define $\theta_n=\theta_0+Kn^{-\delta}$ for $n=1,2,\ldots$ Then

$$\lim_{n\to\infty}\left[\frac{d^m}{d\,\theta^m}\psi_j^{(n)}(\theta)|_{\theta=\theta_n}/\frac{d^m}{d\,\theta^m}\psi_j^{(n)}(\theta)|_{\theta=\theta_0}\right]=1$$

and

$$\lim_{n\to\infty}\left[\sigma_j^{(n)}(\theta_n)/\sigma_j^{(n)}(\theta_0)\right]=1.$$

4. Define

$$W_j^{(n)}(\underline{Y})=[T_j^{(n)}(Y_1,\ldots,Y_n)-\psi_j^{(n)}(\theta)]/\sigma_j^{(n)}(\theta).$$

Then

$$\lim_{n\to\infty}P_\theta(W_j^{(n)}(Y)\leqslant w)=\frac{1}{\sqrt{2\pi}}\int_{-\infty}^{w}e^{-x^2/2}dx\triangleq\Phi(w)$$

for all $w\in\mathbb{R}$, uniformly in θ for $\theta_0\leqslant\theta\leqslant\theta_0+d$ for some $d>0$.

We may now state the following

Proposition III.C.1: The Pitman-Noether Theorem

Suppose that $\{\tilde{\delta}_1^{(n)}\}_{n=1}^{\infty}$ and $\{\tilde{\delta}_2^{(n)}\}_{n=1}^{\infty}$ satisfy conditions 1 through 4; then for the sequence of alternatives $\theta_n = \theta_0 + Kn^{-\delta}$, we have

$$ARE_{1,2} = \eta_1 / \eta_2,$$

where η_j is defined by

$$\eta_j = (c_j)^{1/m\delta}, \; j = 0, 1$$

and c_j is from (III.C.25).

Outline of Proof: A complete proof of this result can be found in Noether (1955). Here we give an outline of the basic idea of the proof.

First, condition 4 implies that $T_j^n(Y_1, ..., Y_n)$ is approximately $N(\psi_j(\theta_0), [\sigma_j^{(n)}(\theta_0)]^2)$ under H_0. Thus, for $P_F(\tilde{\delta}_j) = \alpha$,

$$\tau_j^{(n)} \sim \sigma_j^{(n)}(\theta_0)\Phi^{-1}(1-\alpha) + \psi_j^{(n)}(\theta_0).$$

Similarly, the detection probability for $\tilde{\delta}_j$ with level α is (asymptotically)

$$P_D(\tilde{\delta}_j) \sim 1 - \Phi\left(\frac{\tau_j^{(n)} - \psi_j^{(n)}(\theta_n)}{\sigma_j^{(n)}(\theta_n)}\right)$$

$$= 1 - \Phi\left(\frac{\sigma_j^{(n)}(\theta_0)}{\sigma_j^{(n)}(\theta_n)}\Phi^{-1}(1-\alpha) - \frac{\psi_j^{(n)}(\theta_n) - \psi_j^{(n)}(\theta_0)}{\sigma_j^{(n)}(\theta_n)}\right).$$

By condition 1 we have

$$\psi_j^{(n)}(\theta_0)-\psi_j^{(n)}(\theta_0)\sim\left(\frac{\theta_n-\theta_0}{m!}\right)^m \frac{d^m}{d\theta^m}\psi_j^{(n)}(\theta)\Big|_{\theta=\theta_0}$$

and by condition 3, $\sigma_j^{(n)}(\theta_0)\sim\sigma_j^{(n)}(\theta_n)$. Thus since $\theta_n=\theta_0+Kn^{-\delta}$,

$$P_D(\delta_j)\sim 1-\Phi\left(\Phi^{-1}(1-\alpha)-\frac{K^m}{m!n^{m\delta}}c_j\right). \tag{III.C.25}$$

If we equate $P_D(\tilde{\delta}_1)$ and $P_D(\tilde{\delta}_2)$, then, letting n_1 and n_2 denote the respective sample sizes, we have, from (III.C.25),

$$\frac{c_1}{(n_1)^{m\delta}}\sim\frac{c_2}{(n_2)^{m\delta}},$$

or, equivalently,

$$\frac{n_2}{n_1}\sim\left(\frac{c_2}{c_1}\right)^{1/m\delta}=\frac{\eta_2}{\eta_1}, \tag{III.C.26}$$

which is the desired result.

□

Remarks

1. The quantity η_j is known as the (limiting) *efficacy* of the test sequence $\{\tilde{\delta}_j^{(n)}\}_{n=1}^{\infty}$. Thus the Pitman-Noether theorem asserts that the test sequence with higher efficacy is the most efficient asymptotically.

2. The regularity conditions 1 through 4 are easily satisfied by many signal detection models. For example, consider the case in which the observations $Y_1, Y_2, \ldots$ are independent and identically distributed with marginal density $f_\theta(y)$. Consider detection statistics of the form

$$T_j^{(n)}(y_1,\ldots,y_n)=\sum_{k=1}^{n} g_j(y_k),\; j=0,1. \tag{III.C.27}$$

Then we have

$$\psi_j^{(n)}(\theta) = n \int g_j f_\theta d\mu$$

and

$$\sigma_j^{(n)}(\theta) = [n\{\int (g_j)^2 f_\theta d\mu - (\psi_j^{(1)}(\theta))^2\}]^{1/2}.$$

As we will see below, it is not unreasonable to assume that $\int g_j f_{\theta_0} d\mu = 0$ [the mean of $T_j^{(n)}(Y_1, ..., Y_n)$ under H_0 is irrelevant in any case because any constant added to $T_j^{(n)}$ results in the same constant being added to $\tau_j^{(n)}$] and $\partial \int g_j f_\theta d\mu / \partial\theta|_{\theta=\theta_0} > 0$, which gives values of $m=1$ and $\delta = 1/2$; and

$$\eta_j = \left[\frac{\partial}{\partial\theta} \int g_j f_\theta d\mu \Big|_{\theta=\theta_0}\right]^2 \Big/ \int g_j^2 f_{\theta_0} d\mu. \qquad \text{(III.C.28)}$$

Thus, assuming that conditions 3 and 4 hold, it is quite simple to compute AREs of detectors of the form of (III.C.27). Condition 3 is simply a smoothness condition on the density f_θ as a function of θ and on the nonlinearity g_j, and if $\int g_j^2 f_\theta d\mu < \infty$, we always have

$$P_\theta(W_j^{(n)}(Y) \leq w) \to \Phi(w),$$

from the central limit theorem [see Breiman (1968)]. Condition 4 requires the *uniformity* of this convergence in $[\theta_0, \theta_0 + d]$. However, we have for i.i.d. samples the following result [known as the *Berry-Eseen bound*; Breiman (1968)]:

$$\sup_{w \in \mathbb{R}} |P_\theta(W_j^{(n)}(Y) \leq w) - \Phi(w)|$$
$$\leq \frac{4 \int |g_j|^3 f_\theta d\mu}{\sqrt{n}\, [\int g_j^2 f_\theta d\mu]^{3/2}}. \qquad \text{(III.C.29)}$$

So if there are positive constants a, b, and d with

$$\int |g_j|^3 f_\theta d\mu \leqslant a \text{ and } \int g_j^2 f_\theta d\mu \geqslant b \qquad \text{(III.C.30)}$$

for all $\theta \in [\theta_0, \theta_0+d]$, condition 4 is satisfied. Note that (III.C.30) is a very mild condition.

3. Under the model discussed in Remark 2, we see that efficacy of $\{\tilde{\delta}_j^{(n)}\}_{n=1}^{\infty}$ depends on the quantity $\partial \int g_j f_\theta d\mu / \partial\theta|_{\theta=\theta_0}$. We can rewrite this quantity as $\partial \int g_j l_\theta f_{\theta_0} d\mu / \partial\theta|_{\theta=\theta_0}$, where $l_\theta = f_\theta / f_{\theta_0}$. Assuming sufficient smoothness of l_θ, we can interchange order of integration and differentiation to yield

$$\frac{\partial}{\partial\theta}\int g_j f_\theta d\mu\bigg|_{\theta=\theta_0} = \int g_j \left(\frac{\partial}{\partial\theta} l_\theta\right)\bigg|_{\theta=\theta_0} f_{\theta_0} d\mu \qquad \text{(III.C.31)}$$

$$= \int g_j T_{lo} f_{\theta_0} d\mu,$$

where

$$T_{lo} = \frac{\partial}{\partial\theta} l_\theta\bigg|_{\theta=\theta_0}. \qquad \text{(III.C.32)}$$

Thus, under these assumptions the efficacy of (III.C.28) becomes

$$\eta_j = \left[\int g_j T_{lo} f_{\theta_0} d\mu\right]^2 \Big/ \int g_j^2 f_{\theta_0} d\mu. \qquad \text{(III.C.33)}$$

Equation (III.C.33) and the Schwarz inequality† imply that the most efficient system of the form of (III.C.27) is achieved by using the nonlinearity $g_j = cT_{lo}$ for any positive constant c. The value of the constant c is irrelevant since a change in c only results in a change in the threshold. However, c must be

† *Schwarz Inequality*: $[\int |fg| d\mu]^2 \leqslant \int f^2 d\mu \int g^2 d\mu$ with equality if and only if $f = cg$ for some constant $c \in \mathbb{R}$.

positive since

$$\frac{\partial}{\partial\theta}\int [cT_{lo}]f_\theta d\mu\Big|_{\theta=\theta_0} = c\int (T_{lo})^2 f_{\theta_0} d\mu > 0.$$

The maximum possible value of η_j is given by substituting $g_j = T_{lo}$ into (III.C.33) to yield

$$\max \eta_j = \int (T_{lo})^2 f_{\theta_0} d\mu, \qquad \text{(III.C.34)}$$

a quantity that will arise again in Chapter IV in the context of parameter estimation.

It is interesting to note that the detector based on the nonlinearity T_{lo} is also locally optimum for $\theta=\theta_0$ versus $\theta>\theta_0$ in the sense discussed in Chapter II. We know that for fixed n, the likelihood-ratio detector has the best performance. Thus for any finite n, the efficiency of any detector relative to the likelihood-ratio detector cannot be larger than unity. However, we see from the above that the detector based on T_{lo} is most efficient asymptotically, and thus in this sense the detector based on T_{lo} is asymptotically equivalent to the likelihood-ratio detector.

III.D Sequential Detection

All of the detection procedures discussed in Section III.B are fixed-sample-size detectors; that is, in each case we were given a fixed number of observations and we wished to derive an optimum detector based on these samples. An alternative approach to this problem might be to fix the desired performance and to allow the number of samples to vary in order to achieve this performance. That is, for some realizations of the observation sequence we may be able to make a decision after only a few samples, whereas for some other realizations we may wish to continue sampling to make a better decision. A detector that uses a random number of samples depending on the observation sequence is generally known as a *sequential* detector. To describe such detectors we may use the following model.

Suppose that our observation set $\Gamma=\mathbb{R}^\infty$, the set of all (one-sided) real sequences,† and that the observations $\{Y_k\,;\, k=1,2,...\}$ are independent and identically distributed according to

$$\begin{aligned} &H_0\colon Y_k \sim P_0,\ k = 1, 2, \cdots \\ \text{versus} \\ &H_1\colon Y_k \sim P_1,\ k = 1, 2, ..., \end{aligned} \tag{III.D.1}$$

where P_0 and P_1 are two possible distributions on $(\mathbb{R}, B)$, B denoting the Borel σ-algebra on $\mathbb{R}$. A *sequential decision rule* is pair of sequences $(\underline{\phi}, \underline{\delta})$, where $\underline{\phi}=\{\phi_j\,;\, j=0,1,2,...\}$ is called a *stopping rule* $(\phi_j : \mathbb{R}^j \to \{0,1\})$ and $\underline{\delta}=\{\delta_j\,;\, j=0,1,2,...\}$ is called a *terminal decision rule*, δ_j being a decision rule on $(\mathbb{R}^j, B^j)$ for each $j \geqslant 0$.

The sequential decision rule $(\underline{\phi}, \underline{\delta})$ operates as follows: For an observation sequence $\{y_k\,;\, k=1,2,...\}$, the rule $(\underline{\phi}, \underline{\delta})$ makes the decision $\delta_N(y_1, y_2, ..., y_N)$, where N is the *stopping time* defined by $N=\min\{n \,|\, \phi_n(y_1, y_2, ..., y_n)=1\}$. That is, $\underline{\phi}$ tells us when to stop taking samples by the mechanism that when $\phi_n(y_1, ..., y_n)=0$, we take another sample [the $(n+1)^{\underline{st}}$] and when $\phi_n(y_1, ..., y_n) = 1$, we stop sampling and make a decision. In this way the number of samples, N, is random since it depends on the data sequence. The terminal decision rule $\underline{\delta}$ tells us what decision to make when we do stop sampling. For example, an ordinary fixed-sample-size decision rule δ operating with n samples is given by the sequential decision rule $(\underline{\phi}, \underline{\delta})$ defined by

† That is, $\mathbb{R}^\infty=\{y \,|\, y=\{y_k\}_{k=1}^\infty \text{ with } y_k \in \mathbb{R},\, k \geqslant 1\}$.

$$\phi_j(y_1, \ldots, y_j) = \begin{cases} 0 \text{ if } j \neq n \\ 1 \text{ if } j = n \end{cases}$$

(III.D.2)

$$\delta_j(y_1, \ldots, y_j) = \begin{cases} \delta(y_1, \ldots, y_n) \text{ if } j = n \\ \textit{arbitrary} \quad \text{if } j \neq n. \end{cases}$$

To derive optimum sequential decision rules for (III.D.1), we first consider the Bayesian version of this problem, in which priors π_1 and $\pi_0=(1-\pi_1)$ are assigned to the hypotheses H_1 and H_0, respectively, and costs C_{ij} are assigned to our decisions. For the sake of simplicity we will assume uniform costs, although other cost assignments are easily handled. Since we theoretically have an infinite number of i.i.d. observations at our disposal, in order to make the problem realistic we should also assign a cost to observation. Thus we will assign a cost $C > 0$ to each sample we take, so that the cost of taking n samples is nC.

With the foregoing cost assignments, the conditional risks for a given sequential decision rule are

$$R_0(\underline{\phi}, \underline{\delta}) = E_0\{\delta_N(Y_1, \ldots, Y_N)\} + CE_0\{N\}$$

and (III.D.3)

$$R_1(\underline{\phi}, \underline{\delta}) = 1 - E_1\{\delta_N(Y_1, \ldots, Y_N)\} + CE_1\{N\},$$

where the subscripts denote the hypothesis under which expectation is computed and N is the stopping time defined above operating on the random sequence $\{Y_k\}$. The Bayes risk is thus given by

$$r(\underline{\phi}, \underline{\delta}) = (1-\pi_1)R_0(\underline{\phi}, \underline{\delta}) + \pi_1 R_1(\underline{\phi}, \underline{\delta}), \qquad \text{(III.D.4)}$$

and a Bayesian sequential rule is one that minimizes $r(\underline{\phi}, \underline{\delta})$.

To see the structure of the optimum decision rule in this Bayesian framework, it is useful to consider the function

$$V^*(\pi_1) \triangleq \min_{\substack{\underline{\phi}, \underline{\delta} \\ \phi_0=0}} r(\underline{\phi}, \underline{\delta}), \; 0 \leq \pi_1 \leq 1. \tag{III.D.5}$$

Since $\phi_0=0$ means that the test does not stop with zero observations, (III.D.5) describes the minimum Bayes risk over all sequential tests that take at least one sample. It is straightforward to show that the function $V^*(\pi_1)$ is a concave, continuous function of π_1 with $V^*(0)=V^*(1)=C$, as illustrated in Fig. III.D.1 [see, e.g., Ferguson (1967)]. Also shown on Fig. III.D.1 are plots of the Bayes risk versus π_1 for two other sequential decision rules: The one that takes no samples and decides H_1 (i.e., $\phi_0=\delta_0=1$) and the one that takes no samples and decides H_0 (i.e., $\phi_0=1-\delta_0=1$). Note that

$$r(\underline{\phi}, \underline{\delta})\Big|_{\phi_0=\delta_0=1} = (1-\pi_1) \quad \text{and} \quad r(\underline{\phi}, \underline{\delta})\Big|_{\phi_0=1-\delta_0=1} = \pi_1.$$

The latter two decision rules represent the only possible Bayes

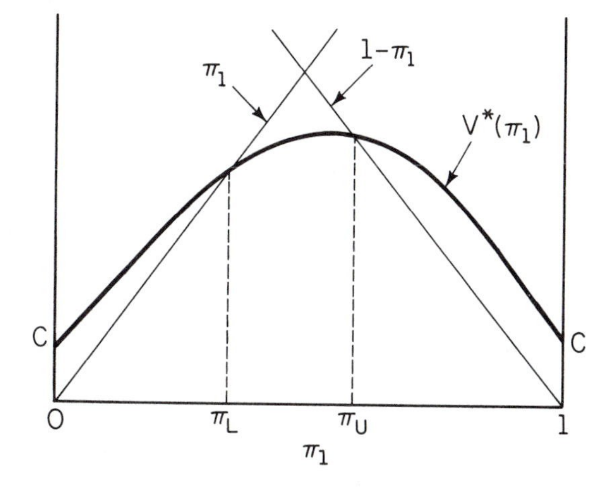

Fig. III.D.1: Relationships yielding the Bayes sequential rule for uniform costs of errors and cost C per sample.

rules that are not included in the minimization of (III.D.5.)[†]. Note that on Fig. III.D.1 we have denoted by π_U the abscissa of the intersection of $r(\underline{\phi},\underline{\delta})|_{\phi_0=\delta_0=1}$ and $V^*(\pi_1)$, and by π_L the intersection of $r(\underline{\phi},\underline{\delta})|_{\phi_0=1-\delta_0=1}$ and $V^*(\pi_1)$.

By inspection of Fig. III.D.1, we see that the Bayes rule for a fixed prior π_1 is $\phi_0=1-\delta_0=1$ if $\pi_1 \leqslant \pi_L$, it is $\phi_0=\delta_0=1$ if $\pi_1 \geqslant \pi_U$, and it is the decision rule with minimum Bayes risk among all $(\underline{\phi},\underline{\delta})$ with $\phi_0=0$ if $\pi_L < \pi_1 < \pi_U$. So if $\pi_1 \leqslant \pi_L$ we take no samples and choose H_0, if $\pi_1 \geqslant \pi_U$ we take no samples and choose H_1, and otherwise we take at least one sample.

Now suppose that we have the condition $\pi_L < \pi_1 < \pi_U$. Here we know that the optimum test takes at least one sample, but the test is otherwise unspecified. However, note that after having taken one sample, the problem of optimizing the test is conditionally the same as that with no samples, in the sense that we still have infinitely many i.i.d. samples at our disposal and the costs are the same. The one difference is that we now have taken a sample and so we have more information about which hypothesis is true. In particular, instead of having a prior probability π_1, we now have prior $\pi_1(y_1)$ that is actually the *posterior* probability of H_1 given our observation of Y_1; i.e., $\pi_1(y_1)=P(H_1 \text{ is true } | Y_1=y_1)$. Thus the picture after having taken one sample is exactly the same as Fig. III.D.1 except that the abscissa variable π_1 is replaced with $\pi_1(y_1)$. Because the samples are independent, knowledge of Y_1 does not affect the shape of V^* (which now represents minimum risk over all tests that take at least *two* samples). So we conclude that after taking one sample the optimum test stops and chooses H_0 if $\pi_1(y_1) \leqslant \pi_L$, it stops and chooses H_1 if $\pi_1(y_1) \geqslant \pi_U$, and it takes another sample if $\pi_L \leqslant \pi_1(y_1) \leqslant \pi_U$.

† It is straightforward to see that randomization cannot help with $\phi_0=1$, since if we choose $\delta_0=1$ with probability γ and $\delta_0=0$ with probability $(1-\gamma)$ we get a Bayes risk $\gamma(1-\pi_1)+(1-\gamma)\pi_1$, which is always larger than $\min\{\pi_1, 1-\pi_1\}$.

If both $\pi_L < \pi_1 < \pi_U$ and $\pi_L < \pi_1(y_1) < \pi_U$, then from the above we see that the optimum test takes at least two samples. In this case we start over with the new prior $\pi_1(y_1, y_2) = P(H_1 \text{ is true } | Y_1 = y_1, Y_2 = y_2)$ and make the same comparison again. Continuing this reasoning for an arbitrary number of samples taken, we see that the Bayes sequential test continues sampling until the quantity $\pi_1(y_1, \ldots, y_n) \triangleq P(H_1 \text{ is true } | Y_1 = y_1, \ldots, Y_n = y_n)$ falls out of the interval (π_L, π_U), and then it chooses H_0 if $\pi_1(y_1, \ldots, y_n) \leqslant \pi_L$ and H_1 if $\pi_1(y_1, \ldots, y_n) \geqslant \pi_U$. [For $n=0$, $\pi_1(y_1, \ldots, y_n)$ denotes π_1.] This test is described by the stopping rule

$$\phi_n(y_1, \ldots, y_n) = \begin{cases} 0 & \text{if } \pi_L < \pi_1(y_1, \ldots, y_n) < \pi_U \\ 1 & \text{otherwise} \end{cases} \tag{III.D.6}$$

and the terminal decision rule

$$\delta_n(y_1, \ldots, y_n) = \begin{cases} 1 & \text{if } \pi_1(y_1, \ldots, y_n) \geqslant \pi_U \\ 0 & \text{if } \pi_1(y_1, \ldots, y_n) \leqslant \pi_L . \end{cases} \tag{III.D.7}$$

The Bayes test is illustrated in Fig. III.D.2. Under mild conditions $\pi_1(Y_1, \ldots, Y_n)$ converges almost surely to 1 under H_1 and to 0 under H_0. Thus the test terminates with probability 1. All that is needed to specify the optimum test are the two probabilities π_L and π_U and a scheme to compute $\pi_1(y_1, \ldots, y_n)$. Unfortunately, it is difficult to obtain π_L and π_U exactly except in some special cases (e.g., when the Y_n's take on only discrete values). On the other hand, the computation of the posterior probability $\pi_1(y_1, \ldots, y_n)$ is quite easy. In particular, assuming that P_0 and P_1 have densities p_0 and p_1, Bayes' formula implies that $\pi_1(y_1, \ldots, y_n)$ can be written as

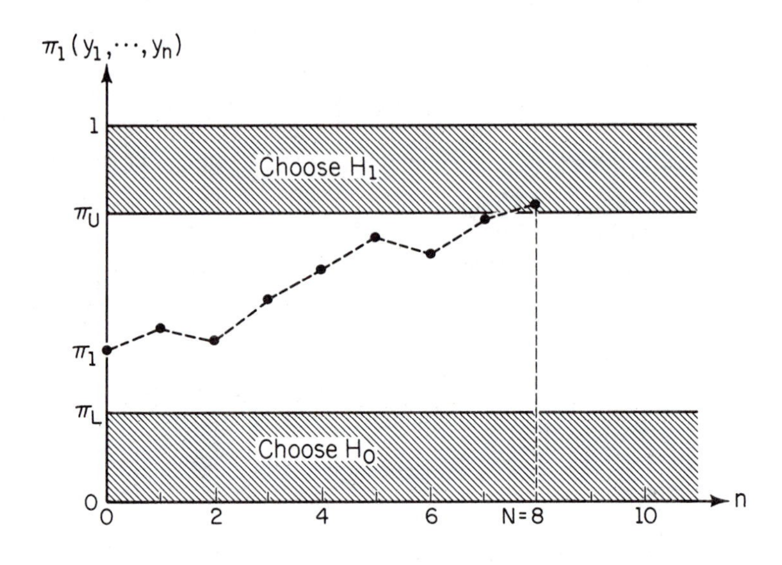

Fig. III.D.2: Depiction of a realization of a Bayes sequential test.

$$\pi_1(y_1, \ldots, y_n) = \frac{\pi_1 \prod_{k=1}^{n} p_1(y_k)}{\pi_0 \prod_{k=1}^{n} p_0(y_n) + \pi_1 \prod_{k=1}^{n} p_1(y_n)} \tag{III.D.8}$$

$$= \frac{\pi_1 \lambda_n(y_1, \ldots, y_n)}{\pi_0 + \pi_1 \lambda_n(y_1, \ldots, y_n)},$$

where λ_n is the likelihood ratio based on n samples given by

$$\lambda_n(y_1, \ldots, y_n) = \prod_{k=1}^{n} [p_1(y_n)/p_0(y_n)]. \tag{III.D.9}$$

(For consistency we define $\lambda_0 = 1$.) Since (III.D.8) is monotone increasing in λ_n, the test of (III.D.6) and (III.D.7) can be rewritten as

$$\phi_n(y_1, \ldots, y_n) = \begin{cases} 0 \text{ if } \underline{\pi} < \lambda_n(y_1, \ldots, y_n) < \overline{\pi} \\ \\ 1 \text{ otherwise} \end{cases} \tag{III.D.10}$$

and

$$\delta_n(y_1, \ldots, y_n) = \begin{cases} 1 \text{ if } \lambda_n(y_1, \ldots, y_n) \geqslant \overline{\pi} \\ \\ 0 \text{ if } \lambda_n(y_1, \ldots, y_n) \leqslant \underline{\pi}, \end{cases} \tag{III.D.11}$$

where $\underline{\pi} \triangleq \pi_0 \pi_L / \pi_1(1-\pi_L)$ and $\overline{\pi} \triangleq \pi_0 \pi_U / \pi_1(1-\pi_U)$. Thus the Bayes sequential test takes samples until the likelihood ratio falls outside the interval $(\underline{\pi}, \overline{\pi})$ and then it decides on H_0 or H_1, depending on whether λ_n falls below $\underline{\pi}$ or above $\overline{\pi}$.

The test of (III.D.10) and (III.D.11) is an example of a *sequential probability ratio test* (SPRT). In particular, for any real numbers A and B satisfying $0 < A \leqslant 1 \leqslant B < \infty$, the *SPRT with boundaries A and B* [denoted by *SPRT* (A, B)] is defined as in (III.D.10) and (III.D.11) with $\underline{\pi}$ replaced by A and $\overline{\pi}$ replaced by B, and with the decision rule left arbitrary if $A = B$. Thus the *SPRT* (A, B) continues sampling until the likelihood ratio λ_n falls outside the "boundaries" A and B and then chooses H_1 if $\lambda_n \geqslant B$ and H_0 if $\lambda_n \leqslant A$. Note that if $A = 1 < B$, we take no samples and choose H_0; if $A < B = 1$, we take no samples and choose H_1; and if $A = B = 1$, we take no samples and make an arbitrary choice.

Example III.D.1: Sequential Detection of a Constant Signal

Consider the problem of detecting a constant signal in additive i.i.d. noise.

$$H_0: Y_k = N_k, \quad k = 1, 2, ...,$$
$$H_1: Y_k = N_k + \theta, \quad k = 1, 2, ..., \qquad \text{(III.D.12)}$$

where $\theta > 0$ and $\{N_k\}_{k=1}^{\infty}$ is an i.i.d. sequence of $N(0, \sigma^2)$ noise samples. For this problem the likelihood ratio based on n samples is given by

$$\lambda_n(y_1, ..., y_n) = \exp\left\{\frac{\theta}{\sigma^2}\sum_{n=1}^{n}(y_k - \theta/2)\right\}, \qquad \text{(III.D.13)}$$

so the *SPRT* (A, B) continues sampling as long as this quantity lies between A and B. Equivalently, by taking logarithms, we see that the *SPRT* (A, B) computes $(\theta/\sigma^2)\sum_{k=1}^{n}(y_k - \theta/2)$ at each stage and compares it with $\log A$ and $\log B$, stopping and choosing H_1 when $(\theta/\sigma^2)\sum_{k=1}^{n}(y_k - \theta/2)$ exceeds $\log B$, and stopping and choosing H_0 when $\theta\sum_{k=1}^{n}(y_k - \theta/2)$ falls below $\log A$.

In addition to the optimality of the *SPRT* $(\underline{\pi}, \overline{\pi})$ in the Bayesian problem, the *SPRT* (A, B) has another optimality property that, from a signal detection viewpoint, is perhaps more important. This property is summarized in the Wald-Wolfowitz theorem (given below), which is analogous to the Neyman-Pearson lemma.

For a sequential decision rule $(\underline{\phi}, \underline{\delta})$, let $P_F(\underline{\phi}, \underline{\delta})$ denote the probability of a false alarm and let $P_M(\underline{\phi}, \underline{\delta})$ denote the probability of a miss; i.e.,

$$P_F(\underline{\phi}, \underline{\delta}) = P(\delta_N(Y_1, ..., Y_N) = 1 | H_0)$$

and

$$P_M(\underline{\phi}, \underline{\delta}) = P(\delta_N(Y_1, ..., Y_N) = 0 | H_1).$$

Also, let $N(\underline{\phi})$ denote the random stopping time associated with $\underline{\phi}$; i.e.,

$$N(\underline{\phi}) = \min\{n \,|\phi_n(Y_1, ..., Y_n) = 1\}.$$

$N(\underline{\phi})$ is also known as the *sample number* of $\underline{\phi}$. We then have the following result.

Proposition III.D.1: The Wald-Wolfowitz Theorem

Suppose that $(\underline{\phi}_0, \underline{\delta}_0)$ is the *SPRT* (A, B) and that $(\underline{\phi}, \underline{\delta})$ is any other sequential decision rule for which

$$P_F(\underline{\phi}, \underline{\delta}) \leqslant P_F(\underline{\phi}_0, \underline{\delta}_0)$$

and

$$P_M(\underline{\phi}, \underline{\delta}) \leqslant P_M(\underline{\phi}_0, \underline{\delta}_0).$$

Then

$$E\{N(\underline{\phi})|H_j\} \geqslant E\{N(\underline{\phi}_0)|H_j\} \text{ for } j = 0 \text{ and } 1.$$

Thus we see that for a given level of performance, no sequential decision rule has a smaller expected sample size than does the SPRT with that performance. Note also that since a fixed-sample-size detector can be characterized as a specific type of sequential decision rule, this theorem implies that *the average sample size of an SPRT is no larger than the sample size of a fixed-sample-size test with the same performance.* It is also true that for given expected sample sizes, no sequential decision rule has smaller error probabilities than does the SPRT [see, e.g., Ferguson (1967)]. The validity of the Wald-Wolfowitz theorem actually follows from the Bayes optimality of the SPRT. The proof will be omitted here.

The Wald-Wolfowitz theorem tells us that for a given P_F and P_M, the SPRT is the detector with the smallest average sample size. It turns out that the *SPRT* (A, B) allows us to choose P_F and P_M arbitrarily at the expense of larger expected sample sizes, as we will see below. The next question that arises is: How does one choose A and B to yield a desired level of performance? The answer to this question is given partially through the following analysis.

Suppose that $(\underline{\phi}, \underline{\delta})$ is the *SPRT* (A, B) with $A<1<B$, and let $\alpha=P_F(\underline{\phi}, \underline{\delta})$, $\gamma=(1-\beta)=P_M(\underline{\phi}, \underline{\delta})$, and $N=N(\underline{\phi})$. Note that the rejection region of $(\underline{\phi}, \underline{\delta})$ can be written as

$$\Gamma_1 = \{y \in \mathbb{R}^\infty | \lambda_N(y_1, \ldots, y_N) \geqslant B\} = \bigcup_{n=1}^{\infty} Q_n ,$$

where $Q_n = \{y \in \mathbb{R}^\infty | N=n \text{ and } \lambda_n(y_1, y_2, \ldots, y_n) \geqslant B\}$. Since Q_n and Q_m are mutually exclusive sets, we can write

$$\alpha = P(\lambda_N(Y_1, \ldots, Y_N) \geqslant B | H_0) = \sum_{n=1}^{\infty} \int_{Q_n} \prod_{k=1}^{n} [p_0(y_k)\mu(dy_k)].$$

On Q_n, we have $\Pi_{k=1}^n p_0(y_k) \leqslant B^{-1} \Pi_{k=1}^n p_1(y_k)$, so that

$$\alpha \leqslant B^{-1} \sum_{k=1}^{n} \int_{Q_n} \prod_{k=1}^{n} [p_1(y_k)\mu(dy_k)]$$

$$= B^{-1} P(\lambda_N(Y_1, \ldots, Y_N) \geqslant B | H_1)$$

$$= B^{-1}(1-\gamma).$$

Similarly,

$$\gamma = P(\lambda_N(Y_1, \ldots, Y_N) \leqslant A | H_1) \leqslant A P(\lambda_N(Y_1, \ldots, Y_N) \leqslant A | H_0)$$

$$= A(1-\alpha).$$

Thus we have

$$B \leqslant (1-\gamma)/\alpha \text{ and } A \geqslant \gamma/(1-\alpha). \qquad \text{(III.D.14)}$$

We can use the inequalities of (III.D.14) to get approximate values for boundaries A and B to give desired α and γ by assuming that when the likelihood ratio λ_n crosses a boundary, the excess over the boundary [i.e., $(\lambda_N(Y_1, ..., Y_N) - B)$ or $(A - \lambda_N(Y_1, ..., Y_N))$] is negligible. This approximation will be accurate if N is relatively large on the average. Thus we assume that either $\lambda_N(Y_1, ..., Y_N) \cong A$ or $\lambda_N(Y_1, ..., Y_N) \cong B$ and the inequalities of (III.D.14) become approximate equalities; i.e.,

$$B \cong (1-\gamma)/\alpha \text{ and } A \cong \gamma/(1-\alpha). \qquad \text{(III.D.15)}$$

These approximations are known as *Wald's approximations*.

Suppose that α_d and γ_d are desired error probabilities and that we use the approximation of (III.D.15) to choose the actual boundaries, i.e.,

$$A_a = \gamma_d/(1-\alpha_d) \text{ and } B_a = (1-\gamma_d)/\alpha_d. \qquad \text{(III.D.16)}$$

Then the actual error probabilities α_a and γ_a will satisfy the inequalities of (III.D.14), so that

$$\alpha_a/(1-\gamma_a) \leqslant B_a^{-1} = \alpha_d/(1-\gamma_d)$$

and

$$\gamma_a/(1-\alpha_a) \leqslant A_a = \gamma_d/(1-\alpha_d),$$

from which we have

$$\alpha_a \leqslant \alpha_d(1-\gamma_a)/(1-\gamma_d) \leqslant \alpha_d/(1-\gamma_d)$$

and

$$\gamma_a \leqslant \gamma_d(1-\alpha_a)/(1-\alpha_d) \leqslant \gamma_d/(1-\alpha_d). \qquad \text{(III.D.17)}$$

So, for example, if $\gamma_d = \alpha_d$, we have

$$\alpha_a \leqslant \alpha_d + O(\alpha_d^2)$$

and

$$\gamma_a \leqslant \gamma_d + O(\gamma_d^2).$$

Note that (III.D.17) guarantees that we can obtain arbitrarily good error-probability performance by proper choice of the boundaries A and B. Note also that the inequalities (III.D.17) are exact and do not rely [except for motivation for choosing A_a and B_a by (III.D.15)] on the approximation based on ignoring the excess over the boundaries. Perhaps the most surprising thing about all of the above is that the actual distribution of the Y_k's is irrelevant to the analysis. The latter fact is a significant advantage of using an SPRT over a fixed-sample-size test.

Although one can arbitrarily decrease the probabilities of error with an SPRT, this is done at the expense of increasing the expected sample size. To see this, assume that α and γ are small so that $A \cong \gamma$ and $B \cong 1/\alpha$. Then as γ and α decrease, A gets smaller and B gets larger. This widens the interval (A, B), so for a given realization, the *SPRT* (A, B) must take more samples. The analysis in the preceding paragraphs indicates that the error probabilities of the SPRT can be controlled without regard to the distributions of the observations. However, the expected sample size is dependent on the distribution of the data and is the primary performance indicator used to compare various sequential detectors. The next question that arises then is: How does one evaluate the expected sample size of a sequential detector? To answer this question we first consider a slightly more general sequential test for H_0 versus H_1 defined as follows:

For each $a < 0 < b$ and each function $g : \mathbb{R} \rightarrow \mathbb{R}$, define the sequential decision rule $ST(a, b; g)$ by the pair $(\underline{\phi}, \underline{\delta})$ given by

$$\phi_j(y_1, \ldots, y_j) = \begin{cases} 0 \text{ if } a < \sum_{k=1}^{j} g(y_k) < b \\ \\ 1 \text{ otherwise} \end{cases}$$

and

$$\delta_j(y_1, \ldots, y_j) = \begin{cases} 1 \text{ if } \sum_{k=1}^{j} g(y_k) \geqslant b \\ \\ 0 \text{ if } \sum_{k=1}^{j} g(y_k) \leqslant a . \end{cases}$$

Note that for $0 < A < 1 < B < \infty$ and $p_1 / p_0 < \infty$, the *SPRT* (A, B) is $ST(a, b; g)$ with $a = \log A$, $b = \log B$, and $g = \log(p_1/p_0)$. For the example of Example VI.B.1, *SPRT* (A, B) is $ST(a, b; g)$ with $g(x) = \theta(x - \theta/2)/\sigma^2$. The test $ST(a, b; g)$ is illustrated in Fig. III.D.3.

Within the framework of the $ST(a, b; g)$ we have the following result, which is due to Wald [see Ferguson (1968)].

Proposition III.D.2: The Fundamental Identity of Sequential Analysis

Suppose $(\underline{\phi}, \underline{\delta}) = ST(a, b; g)$. Define $N = \min\{n \mid \theta_n(Y_1, \ldots, Y_n) = 1\}$ and $S_n = \sum_{k=1}^{n} g(Y_k)$, and let M_j denote the moment generating function of the random variable $g(Y_1)$ under hypothesis H_j; i.e.,

$$M_j(t) = E\{\exp\{tg(Y_1)\} \mid H_j\},\ j = 0, 1.$$

Suppose that $j = 0$ or 1. Then, if $P(g(Y_1) = 0 \mid H_j) \neq 1$ and $P(|g(Y_1)| < \infty \mid H_j) = 1$, we have

$$E\{\exp\{tS_N\}[M_j(t)]^{-N} \mid H_j\} = 1$$

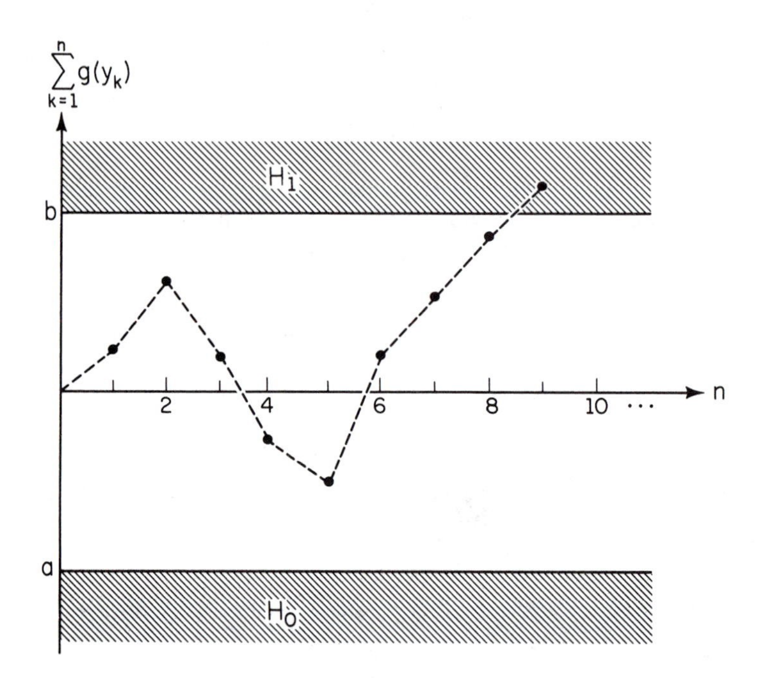

Fig. III.D.3: Illustration of $ST(a, b; g)$.

for all real t for which $M(t) < \infty$.

This result is also known as *Wald's identity*. It is a straightforward consequence of the optimal sampling theorem for martingales [Breiman (1968)] and will not be proved here. A result following from this Proposition III.D.2 is the following:

Proposition III.D.3: Corollary to Wald's Identity

Under the hypothesis of Proposition III.D.3, suppose that $M_j(t) < \infty$ in a neighborhood of $t = 0$. Define $\mu_j = E\{g(Y_1)|H_j\}$ and $\sigma_j^2 = Var(g(Y_1)|H_j)$. Then

$$(a)\ E\{S_N|H_j\} = \mu_j E\{N|H_j\}$$

and

$$(b)\ E\{(S_N - N\mu_j)^2 | H_j\} = \sigma_j^2 E\{N | H_j\}.$$

Using Wald's identity and its corollary, we can extend Wald's approximations [(III.D.15)] to the test $ST(a, b; g)$, as follows.

Suppose that we can find two nonzero numbers t_0 and t_1 such that $M_j(t_j)=1$ for $j=0$ and 1 (the existence of such a t_j is assured if $P_j(g(Y_1)<0)>0, P_j(g(Y_1)>0)>0$, and $\mu_j \neq 0$ [see Ferguson (1967)]. Then Wald's identity implies that

$$E\{\exp\{t_j S_N\}|H_j\} = 1 \text{ for } j = 0, 1. \qquad \text{(III.D.18)}$$

If we ignore the "excess over the boundaries," then under H_0, S_N is a discrete random variable taking the values b and a with probabilities $P_F(\underline{\phi}, \underline{\delta}) \triangleq \alpha$ and $(1-\alpha)$, respectively, and under H_1, S_N takes values a and b with probabilities $\gamma = P_M(\underline{\phi}, \underline{\delta})$ and $(1-\gamma)$, respectively. Thus (III.D.18) implies that

$$(1-\alpha)e^{t_0 a} + \alpha e^{t_0 b} \cong 1$$

and

$$\gamma e^{t_1 a} + (1-\gamma)e^{t_1 b} \cong 1,$$

from whence we have

$$\alpha \cong (1-e^{t_0 a})/(e^{t_0 b} - e^{t_0 a}) \qquad \text{(III.D.19)}$$

and

$$\gamma \cong (1-e^{t_1 b})/(e^{t_1 a} - e^{t_1 b}). \qquad \text{(III.D.20)}$$

As an example, suppose that we take $g = \log(p_1/p_0)$ so that $ST(a, b; g) = SPRT(e^a, e^b)$. Then we have

$$M_0(t) = \int_{\mathbb{R}} \exp\{t \log[p_1(y)/p_0(y)]\} p_0(y) \mu(dy)$$

$$= \int_{\mathbb{R}} [p_1(y)]^t [p_0(y)]^{1-t} \mu(dy).$$

Noting that

$$M_0(1) = \int_{\mathbb{R}} p_1(y) \mu(dy) = 1,$$

we have $t_0 = 1$. Similarly, for this case

$$M_1(t) = \int_{\mathbb{R}} [p_1(y)]^{t+1} [p_0(y)]^{-t} \mu(dy),$$

which yields $t_1 = -1$. On defining $A = e^a$ and $B = e^b$, (III.D.19) and (III.D.20) imply that

$$\alpha \cong (1-A)/(1-B) \text{ and } \gamma \cong (1-B^{-1})/(A^{-1}-B^{-1}),$$

which are equivalent to Wald's approximations of (III.D.15).

If $\mu_0 = 0$, then no nonzero t_0 exists such that $M_0(t_0) = 1$. But in this case, the corollary to Wald's identity implies that $E\{S_N | H_0\} = 0$, which yields

$$a(1-\alpha) + b\alpha \cong 0$$

or

$$\alpha \cong -a/(b-a). \tag{III.D.21}$$

Similarly, if $\mu_1 = 0$, we have

$$\gamma \cong -b(a-b). \qquad \text{(III.D.22)}$$

In a similar manner, we can approximate the expected sample sizes of $ST(g\,;a\,,b\,)$ using Proposition III.D.3. First, suppose that $\mu_0 \neq 0$. Then by ignoring the excess over the boundaries the corollary implies that

$$E\{N|H_0\} = \frac{1}{\mu_0}E\{S_N|H_0\} \cong \frac{1}{\mu_0}(a(1-\alpha)+b\,\alpha).$$

If $\mu_0=0$, we can use

$$E\{N|H_0\} = \frac{1}{\sigma_0^2}E\{S^2|H_0\} \cong \frac{1}{\sigma_0^2}(a^2(1-\alpha)+B^2\alpha).$$

Similar expressions hold for $E\{N|H_1\}$. Note that either $(a\,,b\,)$ or (α,γ) can be eliminated from these expressions via the approximations of (III.D.19) and (III.D.20) or (III.D.21) and (III.D.22). For example, if $\mu_j=0$, we have

$$E\{N|H_j\} \cong -ab\,/\,\sigma_j^2.$$

For the *SPRT* $(e^a\,,e^b\,)$ we have $\mu_j \neq 0$ for $j=0$ and 1, and Wald's approximations yield

$$E\{N|H_0\} \cong \frac{1}{\mu_0}\left[(1-\alpha)\log\frac{\gamma}{1-\alpha}+\alpha\log\frac{1-\gamma}{\alpha}\right]$$

and (III.D.23)

$$E\{N|H_1\} \cong \frac{1}{\mu_1}\left[\gamma\log\frac{\gamma}{1-\alpha}+(1-\gamma)\log\frac{1-\gamma}{\alpha}\right].$$

We now consider the following example.

Example III.D.2: A Comparison of Sequential and FSS Detection

To illustrate the savings in expected sample size caused by using on SPRT over using a fixed-sample-size (FSS) test, consider again the constant-signal detection problem of Example III.D.1.

Of course, the best FSS detector is the likelihood ratio detector, which for a given sample size n , has error probabilities α and γ related through the expression (III.B.31).

$$(1-\gamma) = 1-\Phi(\Phi^{-1}(1-\alpha)-n^{1/2}\theta/\sigma). \qquad \text{(III.D.24)}$$

Thus for given α and γ the sample size required for the FSS likelihood ratio test to produce these error probabilities is given by inverting (III.D.24); namely

$$n_{FSS} = \lceil \sigma^2[\Phi^{-1}(1-\alpha)-\Phi^{-1}(\gamma)]^2/\theta^2 \rceil, \qquad \text{(III.D.25)}$$

where $\lceil x \rceil$ denotes the smallest integer that is not smaller than x.

For the SPRT with boundaries chosen for given error probabilities, the expected sample sizes are given by (III.D.23). To evaluate the expression we must compute μ_0 and μ_1, given by

$$\begin{aligned}\mu_j &= E\{\log[p_1(Y_1)/p_0(Y_1)]|H_j\} \\ &= \frac{\theta}{\sigma^2}E\{(Y_1-\theta/2)|H_j\} \\ &= \begin{cases} -\theta^2/2\sigma^2 & \text{if } j=0 \\ +\theta^2/2\sigma^2 & \text{if } j=1. \end{cases}\end{aligned} \qquad \text{(III.D.26)}$$

For simplicity we assume that $\alpha=\gamma$, in which case

$$E\{N|H_1\} = E\{N|H_0\}$$

$$\cong 2\sigma^2 \left[\gamma \log \frac{\gamma}{1-\alpha} + (1-\gamma) \log \frac{1-\gamma}{\alpha} \right]. \tag{III.D.27}$$

For $\alpha=\gamma=0.1$ and $\theta^2/\sigma^2=1$, evaluation of (III.D.25) and (III.D.27) yields $n_{FSS} \cong 22$ and $E\{N|H_j\} \cong 9$. Thus, on the average, the SPRT uses less than half the samples of the FSS test in this case. Interestingly, by using (III.D.25) and (III.D.27), it can be shown that

$$\lim_{\alpha=\gamma\to 0} \frac{E\{N|H_j\}}{n_{FSS}} = 1/4. \tag{III.D.28}$$

Thus for vanishingly small error probabilities (with $\alpha=\gamma$), the SPRT requires only one-fourth as many samples, on the average, as does the FSS test.

Example III.D.2 illustrates that the SPRT can, on the average, offer substantial savings over the best FSS test in terms of the number of samples required to perform a test with a given level of performance. This is particularly advantageous in applications in which a large number of identical tests are to be performed. An example of such an application is search radar in which the radar performs a test (target present versus target absent) in each of many cells in a search area.

From the performance gains indicated in Example III.D.2 one might ask: Why not abandon the use of FSS likelihood-ratio tests in favor of SPRTs in all cases? Unfortunately, SPRTs have several practical disadvantages that are not readily apparent from the discussion above. One such disadvantage lies in the fact that although the sample size of an SPRT is finite with probability 1, it is not bounded. The SPRT saves samples by making quick decisions when the hypothesis is clear from observed data; but on the other hand, if the observed data are ambiguous, the SPRT can run on for a large number of samples. As the Wald-Wolfowitz theorem implies the average of these two effects is beneficial; however, the occasional long run may not be practical for many applications. Fortunately, this difficulty can be overcome quite easily by modifying the SPRT to stop sampling and make a hard (single-threshold) decision after some maximum number of samples. This type of test is known as a *truncated* SPRT, and truncated SPRTs retain the favorable properties of SPRTs provided that the truncation point is not taken to be too small.

Another practical disadvantage of SPRTs is that their implementation requires an exact knowledge of both p_0 and p_1. For example, in the constant-signal detection problem of Example III.D.1 it is necessary to know the signal value θ in order to implement the test. This is in contrast to the FSS Neyman-Pearson test for the same problem, which is uniformly most powerful for $\theta > 0$. An incorrect guess as to the actual signal value can result in a loss in performance of the SPRT in this case. For example, if the actual location parameter were only half of that assumed, the test statistic $\theta \sum_{k=1}^{n} (y_k - \theta/2)/\sigma^2$ would fluctuate around zero under H_1, possibly resulting in very long tests. This can be alleviated to a certain degree by truncation; however, this type of problem is one of the main limitations of sequential tests.

A third disadvantage of sequential tests is that the theory of these tests is limited when the i.i.d. assumption cannot be invoked. Note that our original derivation of the SPRT as a Bayes optimal test would not work if we did not have independence of the past and future, and if the future was not identical at each stage.

Despite these three disadvantages, the advantages of SPRTs makes them attractive for many applications. A number of additional properties of the SPRT and its practical modifications are discussed in a survey article by Tantaratana (1986).

III.E Nonparametric and Robust Detection

In Chapter II and in Section III.B we have considered hypothesis testing and signal detection problems under a variety of assumptions about the statistical behavior of available observations. One assumption common to all these problems is that the probability distribution of the data is known (perhaps only up to a set of unknown parameters) under each hypothesis. In practical situations, it is often unrealistic to assume that these distributions are known exactly, and sometimes it cannot even be assumed that they are known approximately. Without such knowledge, the techniques of the preceding sections cannot be applied directly, and thus alternative design criteria must be established. Two design philosophies that can be applied in this context are *non-parametric* and *robust* detection. Basically, nonparametric techniques address the problem of detecting signals with only very coarse information about the statistical behavior of the noise, while robust techniques are applicable to situations in which the noise statistics are known approximately but not exactly. In this section we give a brief overview of these two types of detection strategies.

Consider the following general composite binary hypothesis testing problem based on an independent and identically distributed (i.i.d.) observation sequence

$$H_0: Y_k \sim P \in \hat{P}_0, \quad k=1,2,\ldots,n$$
$$\text{versus} \tag{III.E.1}$$
$$H_1: Y_k \sim P \in \hat{P}_1, \quad k=1,2,\ldots,n\,,$$

where $\hat{P}_0$ and $\hat{P}_1$ are two nonoverlapping classes of possible marginal distributions for the observations. This problem is said to be a *parametric* hypothesis testing problem if the classes $\hat{P}_0$ and $\hat{P}_1$ can be parameterized by a real or vector parameter. For example, the composite hypothesis testing problems discussed in Section II.E are parametric problems. If $\hat{P}_0$ or $\hat{P}_1$ cannot be parameterized in this way, (III.E.1) is said to be a *nonparametric* hypothesis testing problem. The general idea in nonparametric problems is that $\hat{P}_0$ and $\hat{P}_1$ are too broad to be parameterized by a finite-dimensional parameter.

An example of a nonparametric hypothesis testing problem is the location testing problem

$$H_0: Y_k = N_k\,, \quad k=1,2,\ldots,n$$
$$\text{versus} \tag{III.E.2}$$
$$H_1: Y_k = N_k + \theta, \quad k=1,2,\ldots,n\,,$$

in which $\{N_k\}_{k=1}^n$ is an i.i.d. sequence whose marginal distribution is known only to be symmetric about zero. Such a model might arise, for example, in the problem of detecting a constant signal in a noise environment that is completely unknown statistically except for identicality, independence, and symmetry properties. The problem is nonparametric since the class of all symmetric distributions is certainly not finite-dimensional.

Both robust and nonparametric hypothesis tests are designed within the context of nonparametric hypotheses. We begin our discussion with nonparametric tests. Robust tests will be discussed subsequently.

III.E.1 Nonparametric Detection

Generally speaking, a *nonparametric test* is one designed to operate over wide classes $\hat{P}_0$ and $\hat{P}_1$ with some performance characteristic being invariant over the classes. These tests usually tend to be simple, using rough information about the data (e.g., signs, ranks, etc.) rather than the exact values of the data. Almost always, the performance characteristic that is to be kept invariant in nonparametric problems is the false-alarm probability. Thus, the standard definition of a nonparametric test (or detector) for (III.E.1) is one whose false-alarm probability is constant over $\hat{P}_0$. For situations in which Y consists of a sequence of observations $Y_1, \ldots, Y_n$, we are also interested in sequences of tests, $\{\delta_n(y_1, \ldots, y_n)\}_{n=1}^{\infty}$, that are *asymptotically nonparametric* for (III.E.1); i.e., sequences of tests for which $\lim_{n\to\infty} P_F(\delta_n)$ is constant for all $P \in \hat{P}_0$.

Nonparametric tests and detectors have found many applications in areas such as radar and sonar. In such applications, nonparametric detectors are sometimes called *constant-false-alarm-rate (CFAR) detectors*. In the following paragraphs, we describe some of the most commonly used nonparametric methods.

The Sign Test

Suppose that we have a sequence $Y_1, \ldots, Y_n$ of independent and identically distributed (i.i.d.) real-valued observations. Define the parameter p by $p = P(Y_1 > 0)$, and consider the hypothesis pair

$$H_0: p = \tfrac{1}{2}$$

$$\text{versus} \tag{III.E.3}$$

$$H_1: \tfrac{1}{2} < p < 1.$$

In words, (III.E.3) is the hypothesis that the Y_k's have zero median versus the hypothesis that the median of the Y_k's is greater than zero. With various interpretations, this model arises in a number of applications. For example, the model of (III.E.2) is a subset of this model.

The hypotheses of (III.E.3) are nonparametric since, in terms of (III.E.1), we have

$$\hat{P}_0 = \{P \in M \mid P((0, \infty)) = \tfrac{1}{2}\} \tag{III.E.4}$$

and

$$\hat{P}_1 = \{P \in M \mid 1 > P((0, \infty)) > \tfrac{1}{2}\}, \tag{III.E.5}$$

where M denotes the class of all distributions on $(\mathbb{R}, B)$. Neither of these classes can be parameterized by (i.e., put into one-to-one correspondence with) a finite-dimensional parameter.

To derive an optimum test for (III.E.3), let us first choose an arbitrary distribution Q_1 in $\hat{P}_1$. For purposes of illustration, we will assume that Q_1 has a density q_1, although the following development can be carried out without this assumption. Define two functions

$$q_1^+(x) = \begin{cases} q_1(x) & \text{if } x > 0 \\ 0 & \text{if } x \leqslant 0 \end{cases}$$

and (III.E.6)

$$q_1^-(x) = \begin{cases} 0 & \text{if } x > 0 \\ q_1(x) & \text{if } x \leqslant 0 \end{cases}$$

and define a density q_0 on $(\mathbb{R}, B)$ by

$$q_0(x) = \frac{q_1^+(x)}{2\int_0^\infty q_1(t)\,dt} + \frac{q_1^-(x)}{2\int_{-\infty}^0 q_1(t)\,dt}. \tag{III.E.7}$$

Note that the distribution Q_0 corresponding to the density q_0 is a member of $\hat{P}_0$ since

$$Q_0\Big((0, \infty)\Big) = \int_0^\infty q_0(x)\,dx = \frac{\int_0^\infty q_1^+(x)\,dx}{2\int_0^\infty q_1(x)\,dx} = 1/2. \tag{III.E.8}$$

Consider the simple hypothesis pair

$$\begin{aligned} &H_0': Y_k \sim Q_0, \quad k=1, \ldots, n \\ \text{versus} \\ &H_1': Y_k \sim Q_1, \quad k=i, \ldots, n. \end{aligned} \tag{III.E.9}$$

By the Neyman-Pearson lemma, a most powerful α-level test of H_0' versus H_1' is the likelihood ratio test based on comparison of the statistic

$$L(\underline{y}) = \prod_{k=1}^n \frac{q_1(y_k)}{q_0(y_k)} \tag{III.E.10}$$

to a threshold. Note that

$$\frac{q_1(y_k)}{q_0(y_k)} = \begin{cases} 2Q_1^+ & \text{if } y_k > 0 \\ 2(1-Q_1^+) & \text{if } y_k \leq 0, \end{cases} \tag{III.E.11}$$

where $Q_1^+ \triangleq Q_1((0, \infty))$, so that $L(y)$ can be written as

$$L(\underline{y}) = 2^n [Q_1^+]^n [Q_1^+/(1-Q_1^+)]^{t(\underline{y})}, \qquad \text{(III.E.12)}$$

where

$$t(\underline{y}) \triangleq \sum_{k=1}^{n} u(y_k), \qquad \text{(III.E.13)}$$

with u denoting the unit-step function defined by, $u(x)=1$ if $x>0$ and $u(x)=0$ if $x \leqslant 0$. Note that $t(\underline{y})$ is the number of the observed y_k's that are positive.

By hypothesis $\frac{1}{2}<Q_1^+<1$, so that $Q_1^+/(1-Q_1^+)>1$. This implies that $L(\underline{y})$ is a monotone increasing function of $t(\underline{y})$, and so a most-powerful α-level test of H_0' versus H_1' is given by

$$\tilde{\delta}_s(\underline{y}) = \begin{cases} 1 & > \\ \gamma \text{ if } t(\underline{y}) & = \tau, \\ 0 & < \end{cases} \qquad \text{(III.E.14)}$$

where γ and τ are the randomization and threshold for false-alarm probability α. It is easy to see that $t(\underline{Y})$ is a binomial random variable with parameters (n, Q_1^+) under H_1' and $(n, \frac{1}{2})$ under H_0'. Thus, for size α, the threshold τ is the smallest integer such that

$$2^{-n} \sum_{k=\tau}^{n} \frac{n!}{(n-k)!k!} \leqslant \alpha \qquad \text{(III.E.15)}$$

and the randomization constant is

$$\gamma = \frac{\alpha - 2^{-n} \sum_{k=\tau+1}^{n} \frac{n!}{(n-k)!k!}}{2^{-n} \frac{n!}{(n-\tau)!\tau!}}. \qquad \text{(III.E.16)}$$

The distribution of $t(\underline{Y})$ is binomial with parameters $(n, 1/2)$ for any $Q_0 \epsilon \hat{P}_0$, so (III.E.14) has size α for the entire class $\hat{P}_0$. Thus the test (III.E.14) is nonparametric for this problem. This also implies that (III.E.14) is the most powerful α-level test of H_0 versus H_1'. Furthermore, by noting that the test of (III.E.14) does not depend on the choice of Q_1, we see that it is a uniformly most powerful α-level test of H_0 versus H_1.

Since $t(\underline{Y})$ is binomial (n, p) under H_1, the detection probability of (III.E.14) is given by

$$P_D = \sum_{k=\tau+1}^{n} \frac{n!}{(n-k)!k!} p^k (1-p)^{n-k}$$

$$+ \gamma \frac{n!}{(n-\tau)!\tau!} p^\tau (1-p)^{n-\tau}. \tag{III.E.17}$$

From (III.E.17) we see that although the false-alarm probability of $\tilde{\delta}_s$ is constant under H_0, its detection probability depends on p and thus is not independent of the choice $P \epsilon \hat{P}_1$. It can be shown that P_D increases monotonically from α to 1 as p increases from 1/2 to 1.

The test of (III.E.14) uses only the (algebraic) signs of the observations $y_1, \ldots, y_n$ and so it is known as the *sign test*. Although the sign test is α-level UMP for $\hat{P}_0$ versus $\hat{P}_1$, we could do better than the sign test if we knew the exact distribution of the observations under the two hypotheses by using the likelihood ratio test between those two distribution. That is, the sign test is not a UMP α-level test for a particular $P \epsilon \hat{P}_0$ versus the class $\hat{P}_1$. An interesting question is: how much performance do we lose by assuming nothing about the distribution other than the very coarse assumptions made in the hypotheses of (III.E.3)?

As a partial answer to this question, we consider an asymptotic $(n \to \infty)$ analysis based on the Pitman asymptotic relative efficiency (ARE) introduced in Section III.C. Recall that the asymptotic efficiency of one detector relative to another is a measure of the relative number of samples that one needs to achieve the same performance as the other in the limit as the number of samples increases without bound.

In order to analyze the sign test via the ARE, we need a specific model and test to use as a basis for comparison. To do this we consider again the model

$$
\begin{aligned}
&H_0: Y_k = N_k, \quad k=1, 2, \ldots, \\
&\text{versus} \qquad\qquad\qquad\qquad\qquad\qquad \text{(III.E.18)} \\
&H_1: Y_k = N_k + \theta, \quad k=1, 2, \ldots, n
\end{aligned}
$$

where $N_1, N_2, \ldots, N_n$ is an i.i.d. sequence with zero mean. As discussed above, (III.E.18) is, of course, the problem of detecting a constant signal in i.i.d. additive noise.

If we assume that the noise distribution in (III.E.18) is $N(0, \sigma^2)$ with σ^2 unknown, then it can be shown [see, e.g., Lehmann (1986)] that a UMP (among unbiased tests) α-level test of H_0 versus H_1 is given by

$$
\tilde{\delta}_t(y) = \begin{cases} 1 & > \\ \gamma \text{ if } \dfrac{\bar{y}}{[\overline{s^2}]^{1/2}} & = \tau \\ 0 & > \end{cases} \qquad \text{(III.E.19)}
$$

where $\bar{y}$ is the sample mean $[\bar{y} \triangleq 1/n \sum_{k=1}^{n} y_k]$ and $\overline{s^2}$ is the sample variance $[\overline{s^2}=1/n \sum_{k=1}^{n} (y_k - \bar{y})^2]$.

The test of (III.E.19) is known as the *t*-*test*. Not only is this test UMP for the Gaussian case of (III.E.18), but also by choosing the threshold $\tau=\Phi^{-1}(1-\alpha)/\sqrt{n}$ and the randomization γ arbitrarily the t-test becomes asymptotically nonparametric at $P_F=\alpha$ for (III.E.18) with any noise distribution having zero mean and finite variance. To see this, we note that the false-alarm probability of this test is given by

$$P_F(\tilde{\delta}_t) = P_0(\bar{Y}/(\overline{S^2})^{1/2} > \tau) + \gamma P_0(\bar{Y}/(\overline{S^2})^{1/2} = \tau)$$

$$= P_0\left(\frac{1}{\sqrt{n}} \sum_{k=1}^{n} Y_k / (\overline{S^2})^{1/2} > \Phi^{-1}(1-\alpha)\right) \qquad \text{(III.E.20)}$$

$$+ \gamma P_0\left(\frac{1}{\sqrt{n}} \sum_{k=1}^{n} Y_k / (\overline{S^2})^{1/2} = \Phi^{-1}(1-\alpha)\right).$$

By the weak law of large numbers, $\overline{S^2}$ converges in probability to $Var(N_1)$, and by the central limit theorem, $n^{-1/2} \sum_{k=1}^{n} Y_k / (\overline{S^2})^{1/2}$ converges in distribution to a $N(0, 1)$ random variable, so

$$\lim_{n \to \infty} P_F(\tilde{\delta}_t) = \frac{1}{\sqrt{2\pi}} \int_{\Phi^{-1}(1-\alpha)}^{\infty} e^{-x^2/2} dx = \alpha. \qquad \text{(III.E.21)}$$

Equation (III.E.21) implies that the t-test is asymptotically nonparametric. (Note that γ is irrelevant since the limiting distribution is continuous.)

From the above, we see that the t-test is optimal for (III.E.18) with Gaussian noise and is asymptotically nonparametric for (III.E.18) with finite-variance noise. Note that the first of these problems corresponds to the testing a subset of the distributions from (III.E.3) and, if we impose the additional constraint that the noise have zero median in addition to zero mean, the second problem also corresponds to testing a subset of the distributions for (III.E.3). It is of interest to compare the sign test and the t-test under these latter conditions.

If we assume that the noise (III.E.18) has a pdf f that has zero mean, variance $\sigma^2 < \infty$, and that is continuous at zero, then it follows straightforwardly from the Pitman-Noether theorem (see Section III.C) that the asymptotic efficiency of the sign test relative to the t-test under (III.E.18) is given by

$$ARE_{s,t} = 4\sigma^2 f^2(0). \qquad \text{(III.E.22)}$$

For the particular case of Gaussian noise, in which f is the

$N(0, \sigma^2)$ density, (III.E.22) becomes

$$ARE_{s,t} = 4\sigma^2 \left(\frac{1}{\sqrt{2\pi}\sigma} \right)^2 = \frac{2}{\pi} \cong 0.64,$$

so that the t-test requires 64% of the samples required by an equivalent sign test. Alternatively for the Laplacian noise case ($f(x)=\frac{\alpha}{2} e^{-\alpha|x|}$), we have $\sigma^2=2/\alpha^2$ and

$$ARE_{s,t} = \frac{8}{\alpha^2} \left(\frac{\alpha}{2} \right)^2 = 2.$$

Thus, for this case, the t-test requires twice as many samples as the equivalent sign test. It should be noted that the sign test is optimum in terms of asymptotic efficiency for the Laplacian noise case.

It can be shown [see Kendall and Stuart (1961)] that for any symmetric unimodal density [i.e., $f(x)=f(-x)$ and $f(|x_1|)>f(|x_2|)$ if $|x_2|>|x_1|$], $ARE_{s,t}$ satisfies the inequality

$$ARE_{s,t} \geqslant 1/3. \tag{III.E.23}$$

Thus, the t-test requires at least 1/3 of the number of samples required by an equivalent sign test under these conditions. Since there is no corresponding upper bound on $ARE_{s,t}$, the sign test is preferable to the t-test when the class of possible noise distributions is quite broad. Furthermore, the sign test is exactly nonparametric over a very broad class of distributions while the t-test is only asymptotically nonparametric over a somewhat narrower class. These factors, added to the computational simplicity of the sign test, make the sign test a very useful alternative to the optimum tests of preceding sections for signal detection problems. Both the sign test and the t-test are used quite frequently in applications such a CFAR radar detection (in this particular application, the sign test is sometimes termed a *binary integrator*).

Rank Tests

We see from the discussion above that, although it uses only very coarse information about the observed data, the sign test is fairly efficient even in its worst case compared to a competitive test that uses much more information about the data. However, by using more information about the data, the nonparametric character of the sign test can be retained while improving on the worst case efficiency relative to the t-test.

For example, suppose that we replace the sign test statistic $t(\underline{y})=\sum_{k=1}^{n} u(y_k)$ of (III.E.13) with a weighted version $\sum_{k=1}^{n} \lambda_k u(y_k)$ where λ_k is the rank of y_k in the sample $y_1, \ldots, y_n$ when reordered in increasing order of absolute value. That is, suppose we rank $y_1, \ldots, y_n$ as $y_{k_1}, \ldots, y_{k_n}$ where $|y_{k_1}| \leqslant |y_{k_2}| \leqslant \ldots | y_{k_n}|$, and perform a threshold test based on the statistic

$$t_W(\underline{y}) = \sum_{i=1}^{n} i u(y_{k_i}). \tag{III.E.24}$$

The resulting test is known as the *Wilcoxon test* and it is an example of a *rank test* since it is based on the ranks of the individual observations within the entire observation sample.

The Wilcoxon test statistic of (III.E.24) can be rewritten as

$$t_W(\underline{y}) = \sum_{k=1}^{n} \sum_{j=1}^{k} u(y_k + y_j), \tag{III.E.25}$$

the derivation of which is left as an exercise. It can be shown from (III.E.25) that the Wilcoxon test is nonparametric for the hypothesis that $Y_1, \ldots, Y_n$ are i.i.d. with a symmetric marginal distribution [*i.e.*, $F_{Y_k}(b)=1-F_{Y_k}(-b)$ for all real b]. Note that this is a smaller class of models than the class of all distributions with zero median (for which the sign test is nonparametric). The asymptotic efficiency of the Wilcoxon test relative to the t-test in the hypothesis pair of (III.E.18) is given by the Pitman-Noether theorem as

$$ARE_{w,t} = 12\sigma^2[\int_{-\infty}^{\infty} f^2(x)\,dx]^2, \qquad \text{(III.E.26)}$$

where we have assumed that the noise variables have a symmetric density f. For the case of Gaussian noise $[N_k \sim N(0, \sigma^2)]$ computation of (III.E.26) gives $ARE_{w,t} = 3/\pi = 0.955$. Thus the Wilcoxon test is nearly optimum for the Gaussian case. For the Laplacian case $ARE_{w,t} = 1.5$, which indicates a loss in efficiency of 25% relative to the sign test in this case. However, it can be shown by minimizing $\int_{-\infty}^{\infty} f^2(x)\,dx$ subject to the constraint $\int_{-\infty}^{\infty} x^2 f(x)\,dx = \sigma^2$ [see Kendall and Stuart (1961) for details] that

$$ARE_{w,t} \geqslant 0.864 \qquad \text{(III.E.27)}$$

for any symmetric noise density. Thus, the Wilcoxon test is never less than 86.4% as efficient as the t-test and, since there is no corresponding upper bound on $ARE_{w,t}$ (the variance σ^2 is not bounded), the Wilcoxon test offers substantial advantages over the t-test. However, a disadvantage of the Wilcoxon test is that all samples must be stored in order to compute its test statistic. This is not true of either the sign test or the t-test.

Even better performance against the t-test can be obtained by using rank tests that are more complicated than the Wilcoxon test. One such test is the *Fisher-Yates* or *normal scores* test, which uses the test statistic

$$t_{FY}(\underline{y}) = \sum_{i=1}^{n} h_n(iu(y_{k_i})), \qquad \text{(III.E.28)}$$

where $y_{k_1}, \ldots, y_{k_n}$ is the ordered sample as in the Wilcoxon test, and where h_n is a function defined by

$$h_n(i) = \begin{cases} 0 \text{ if} & i=0 \\ E\{X_{(i)}\} \text{ if} & i=1, \ldots, n, \end{cases} \quad \text{(III.E.29)}$$

where $X_{(1)} < X_{(2)} < \ldots < X_{(n)}$ are ordered values of i.i.d. $N(0, 1)$ random variables $X_1, X_2, \ldots, X_n$. The Fisher-Yates test has an efficiency relative to the t -test that satisfies

$$ARE_{FY,t} \geqslant 1$$

in the model of (III.E.18) with symmetric noise. Thus the Fisher-Yates test is always at least as efficient as the t -test in this model. Again, this efficiency is gained at the expense of complexity.

For further discussion of rank tests, the reader is referred to the books by Hajek and Sidak (1967) and Kendall (1948).

Two-Channel Tests

A number of applications involve observation sets that consist of samples taken from two or more sensors or channels. Such applications arise in sonar, seismology, and radio astronomy problems in which arrays of sensors are often used to detect signals. Several important nonparametric tests have been developed for this type of problem, and some of these will be discussed here for the particular case in which two observation channels are available.

We consider an observation sequence consisting of n independent pairs of random variables; i.e., $\underline{Y} = [(U_1, V_1), (U_2, V_2), \ldots, (U_n, V_n)]$, where (U_k, V_k), $k=1, \ldots, n$, are mutually independent.

One type of problem within this framework is that of detecting the presence or absence of a common random signal in two sensors. This problem can be modeled by the following hypothesis pair

$$H_0: \begin{array}{l} U_k = N_k \\ V_k = W_k \end{array}, \quad k = 1, 2, ..., n$$

versus (III.E.30)

$$H_1: \begin{array}{l} U_k = N_k + S_k \\ V_k = W_k + S_k \end{array}, \quad k = 1, 2, ..., n$$

where $\{N_k\}_{k=1}^n$, $\{W_k\}_{k=1}^n$, and $\{S_k\}_{k=1}^n$, are independent sequences of i.i.d. random variables with marginal distribution functions F_N, F_W, and F_S, respectively.

By the independence assumptions, it is easily seen that, under H_0, each pair (U_k, V_k) has joint distribution function

$$Q_0(u, v) = F_N(u) F_W(v) \qquad \text{(III.E.31)}$$

and, under H_1, (U_k, V_k) has joint distribution

$$Q_1(u, v) = \int_{-\infty}^{\infty} F_N(u-s) F_W(v-s)\, dF_S(s). \qquad \text{(III.E.32)}$$

From these distributions the likelihood-ratio for optimum detection in (III.E.30) can be obtained if F_N, F_W, and F_S are known.

Suppose, for example, that F_N, F_W, and F_S are all Gaussian distributions with zero means, $Var(N_k) = Var(W_k) = \sigma^2$, and $Var(S_k) = \sigma_S^2$. Then Q_0 is the bivariate Gaussian density with both means zero, both variances σ^2, and zero correlation coefficient; and Q_1 is the bivariate Gaussian density with both means zero, both variances $\sigma^2/(1-\rho)$, and correlation coefficient ρ, where $\rho \triangleq \sigma_S^2/(\sigma^2+\sigma_S^2)$. For known σ^2, an α-level UMP test for this Gaussian problem is given by an energy detector of the form

$$\delta_{ED}(\underline{y}) = \begin{cases} 1 & > \\ \gamma \text{ if } \sum_{k=1}^{n} (u_k + v_k)^2 & = \tau, \\ 0 & > \end{cases} \tag{III.E.33}$$

where τ is chosen for size α. The analysis of this test is virtually identical to that for the single-channel radiometer of Section III.B.

If, as commonly occurs in practice, the distributions F_N, F_W, and F_S are all unknown, then an alternative to an optimum detector such as that of (III.E.33) must be sought. One such detector is that is widely used in practice is the *polarity coincidence correlator* (PCC), which is given by

$$\delta_{PCC}(\underline{y}) = \begin{cases} 1 & > \\ \gamma \text{ if } \sum_{k=1}^{n} u(u_k v_k) & = \tau, \\ 0 & < \end{cases} \tag{III.E.34}$$

where u again denotes the unit-step function

$$u(x) = \begin{cases} 1 \text{ if } x > 0 \\ 0 \text{ if } x \leqslant 0. \end{cases} \tag{III.E.35}$$

Note that this detector makes its decision based on the number of "polarity coincidences" in the observations; i.e., the PCC decision is based on the number of times the outputs of both channels have the same sign. Since the signal is common to both channels under H_1 and absent from both channels under

H_0 in the model of (III.E.30), one would expect more polarity coincidences under H_1 than under H_0, and so this is a reasonable way of detecting the signal here.

Under either hypothesis of (III.E.30), the PCC statistic, $t_{PCC}(\underline{Y}) \triangleq \sum_{k=1}^{n} u(U_k V_k)$ is the sum of n i.i.d. Bernoulli random variables. Thus, under H_j, $t_{PCC}(\underline{Y})$ is a binomial (n, λ_j) random variable, where $\lambda_j = P_j(U_1 V_1 \geqslant 0)$, $j=0, 1$. We can write

$$\lambda_j = P_j(U_1 \geqslant 0, V_1 \geqslant 0) + P_j(U_1 \leqslant 0, V_1 \leqslant 0),$$

which straightforwardly reduces to

$$\lambda_j = 1 - Q_j(0, \infty) - Q_j(\infty, 0) + 2Q_j(0, 0). \qquad \text{(III.E.36)}$$

Using (III.E.31) and (III.E.32), we have that

$$\begin{aligned} \lambda_0 &= 1 - F_N(0) - F_W(0) + 2F_N(0) \\ &= \tfrac{1}{2} + 2[F_N(0) - \tfrac{1}{2}][F_W(0) - \tfrac{1}{2}] \end{aligned} \qquad \text{(III.E.37)}$$

and

$$\begin{aligned} \lambda_1 &= 1 - \int_{-\infty}^{\infty} F_N(-s)\, dF_S(s) - \int_{-\infty}^{\infty} F_W(-s)\, dF_S(s) \\ &\quad + 2\int_{-\infty}^{\infty} F_N(-s) F_W(-s) F_S(s) \\ &= \tfrac{1}{2} + 2\int_{-\infty}^{\infty} [F_N(s) - \tfrac{1}{2}][F_W(s) - \tfrac{1}{2}]\, dF_S(s). \end{aligned} \qquad \text{(III.E.38)}$$

Note from (III.E.36) that if either noise process has zero median (i.e., if $F_N(0)=\tfrac{1}{2}$ or $F_W(0)=\tfrac{1}{2}$), then $\lambda_0=\tfrac{1}{2}$. Thus the PCC is nonparametric against the class of noise models in which at least one of the noise processes has zero median, since the false-alarm probability can be fixed at α for this class by choosing τ and γ to give size α for t_{PCC} being binomially distributed with parameters $(n, 1/2)$.

It is interesting to compare the PCC to the optimum detector for Gaussian channels given by (III.E.33). Assuming that F_N and F_W have probability density functions f_N and f_W, respectively, and that f_N and f_W are continuous at zero and have finite second and fourth moments, it can be shown that the asymptotic efficiency of δ_{PCC} relative to δ_{ED} for the hypothesis pair of (III.E.30) is given by

$$ARE_{PCC,ED} =$$

$$[\gamma_N^4 + \gamma_W^4 + 4\sigma_N^2 \sigma_W^2 - \sigma_N^2 - \sigma_W^4] f_N^2(0) f_W^2(0),$$

where

$$\sigma_N^2 \triangleq E\{N_1^2\},\ \sigma_W^2 \triangleq E\{W_1^2\},\ \gamma_N^2 \triangleq E\{N_1^4\}, \text{ and } \gamma_W^4 = E\{W_1^4\}.$$

For example, for identical Gaussian channels with

$$N_1 \sim N(0, \sigma^2) \text{ and } W_1 \sim N(0, \sigma^2),$$

we have

$$\sigma_N^2 = \sigma_W^2 = \sigma^2,$$

which gives

$$ARE_{PCC,ED} = \frac{2}{\pi^2} = 0.202.$$

Alternatively, for identical Laplacian channels with

$$f_{N_1}(x) = f_{W_1}(x) = \frac{\alpha}{2} e^{-\alpha|x|}$$

we have $\sigma_N^2=\sigma_W^2=2/\alpha^2$ and $\gamma_N^4=\gamma_W^4=24/\alpha^2$, from whence

$$ARE_{PCC,ED} = 3.5.$$

Thus, the PCC performs relatively poorly when compared to power detector for Gaussian channels, but the PCC performs quite well for Laplacian channels. The poor efficiency for Gaussian channels in the price paid here for the simplicity of implementation and nonparametric performance exhibited by the PCC. As in the single-channel case, the improved efficiency can be obtained for two- and multi-sample problems while retaining nonparametric performance by considering more complex detectors based on ranks [see, for example, Carlyle (1968)].

The hypotheses of (III.E.30) describe one common type of two-channel problem. Another type of two-channel problem that arises frequently is the problem of detecting a signal with a reference noise source. This problem is described by the hypothesis pair

$$H_0: \begin{array}{l} U_k = N_k \\ V_k = W_k \end{array}, \quad k=1, 2, ..., n$$

$$\text{versus} \qquad \text{(III.E.39)}$$

$$H_1: \begin{array}{l} U_k = N_k \\ V_k = W_k + S_k \end{array}, \quad k=1, 2, ..., n$$

where $\{N_k\}_{k=1}^n$ and $\{W_k\}_{k=1}^n$ are independent i.i.d. noise sequences with the same marginal and $\{S_k\}_{k=1}^n$ is a signal sequence independent of the noise. Thus this problem corresponds to the usual single-sample problem in the V-channel, with the additional observation of a sample of noise only in the U-channel. Note that both sets of samples might actually be taken from a single channel with the noise-only samples being taken at a time when it is known that no signal is present and then stored for later processing with the additional samples. [For simplicity, we have assumed in (III.E.39) that the same number of samples are taken from each channel although this is not necessarily what would happen in practice.]

There are a number of tests that can be applied to detect signals with observations modeled as in (III.E.39). First we note that if the statistics of the noise are known exactly, then the information provided by the noise-only channel is useless under the given independence assumptions. This is easily seen from the likelihood ratio, from which all terms involving the U_k's disappear. Thus, it is only when there is something unknown about the noise distribution that the signal-free channel can be useful.

One test that is useful for (III.E.39) when the signal is a positive constant is a two-sample version of the Wilcoxon test, which is based on the statistic

$$\sum_{k=1}^{n} r_k , \qquad \text{(III.E.40)}$$

where r_k is the rank of v_k when all observation $u_1, ..., u_n$ and $v_1, ..., v_n$ are grouped together and put in increasing order. This test is also known as the *Mann-Whitney test*. A statistic equivalent to that of (III.E.40) is

$$\sum_{k=1}^{n} \sum_{j=1}^{n} u(v_k - u_j),$$

from which it can be seen that the Mann-Whitney detector is nonparametric for symmetrically-distributed noise. As with the single-channel Wilcoxon detector, this two-version performs quite well in comparison with other detectors for (III.E.39).

The Mann-Whitney test is one of several useful tests for (III.E.39) fall under the category of *Kolmogorov-Smirnov tests*, which are based on functionals evaluated at the function

$$\hat{F}_V(x) - \hat{F}_U(x), \; -\infty < x < \infty,$$

where $\hat{F}_V$ and $\hat{F}_U$ are the so-called *empirical distribution functions* of the sample $\{u_k\}_{k=1}^{n}$ and $\{v_k\}_{k=1}^{n}$, respectively; i.e., $\hat{F}_V$

is defined as

$$\hat{F}_V(x) = \frac{1}{n} \sum_{k=1}^{n} u(x - v_k), \; -\infty < x < \infty, \qquad \text{(III.E.41)}$$

and $\hat{F}_U$ is defined analogously.† As $n \to \infty$, $\hat{F}_U$ and $\hat{F}_V$ converge to the respective marginal distributions of the U_k's and V_k's respectively, and so functions of the difference $(\hat{F}_V - \hat{F}_U)$ can be useful in deciding whether $\{U_k\}_{k=1}^n$ and $\{V_k\}_{k=1}^n$ have the same distribution or not [i.e., $(\hat{F}_V - \hat{F}_U)$ is useful in testing for *homogeneity*]. Aside from the Mann-Whitney test, other useful tests of this type are those based on comparing the statistics

$$\sup_{-\infty < x < \infty} [\hat{F}_V(x) - \hat{F}_U(x)],$$

or

$$\sup_{-\infty < x < \infty} |\hat{F}_V(x) - \hat{F}_U(x)|$$

to a threshold. These tests are nonparametric for the hypothesis that $\{U_k\}_{k=1}^n$ and $\{V_k\}_{k=1}^n$ have the same marginal [i.e., for H_0 in (III.E.39)] and can outperform the Mann-Whitney for some ranges of false-alarm probability.

For further discussion and details of the theory and applications of nonparametric detection, the reader is referred to the book by Kassam and Thomas (1980).

III.E.2 Robust Detection

In Section III.B we discussed the design of detection systems under the assumption that a complete statistical description of the observation data is available. Alternatively, in the paragraphs above, we considered detection systems for situations in which very little is known about the observation statistics. Between these two extremes is the situation in which a reasonably accurate nominal model is available for

† Note that for each x, $\hat{F}_V(x)$ is simply the number of v_k's that are smaller than or equal to x divided by the total number of v_k's.

the data statistics but in which some small deviations from this model may occur.

Consider, for example, the problem of testing between two possible marginal distributions, P_0 and P_1, for an i.i.d. sequence $Y_1, ..., Y_n$. Assuming densities p_0 and p_1, optimum tests are based on the likelihood ratio

$$L(\underline{y}) = \prod_{k=1}^{n} \frac{p_1(y_k)}{p_0(y_k)}. \tag{III.E.42}$$

Note that the likelihood ratio is very sensitive to observations for which either $p_1(y_k) >> p_0(y_k)$ or $p_0(y_k) >> p_1(y_k)$. Since the condition $p_1(y_k) >> p_0(y_k)$ is much more likely to occur under H_1 than under H_0, [and vice versa for the condition $p_0(y_k) >> p_1(y_k)$] this sensitivity is simply part of the intended action of the test under the assumed model.

Suppose, however, that the actual marginal distribution of the data is not exactly P_0 or P_1 but rather is only approximately P_0 or P_1. For example, suppose the actual distribution is of the form

$$(1-\epsilon)P_j + \epsilon M_j, \quad j = 0, 1, \tag{III.E.43}$$

where P_0 and P_1 are the nominal distributions, M_0 and M_1 are unknown and arbitrary "contaminating" distributions, and ϵ is a number between 0 and 1 representing the degree of uncertainty to be placed on the model. Such a model might arise, for example, in a communications or radar channel in which an extraneous interferer is present for a fraction ϵ of the time or in which impulsive noise (lightning, etc.) occurs with probability ϵ. Also, intermittent sensor faults and other measurement or data-recording errors can be modeled in this way. The key idea here is that the M_j 's represent an aspect of the established model that is completely unknown to the designer.

Suppose that $p_1(y_k)/p_0(y_k)$ is an unbounded function of y_k. Since M_0 is arbitrary, it could place all of its probability in regions where $p_1(y_k) >> p_0(y_k)$. This would tend to cause the test based on the nominal likelihood ratio (III.E.42) to make false alarms (i.e., errors under H_0) more often that it should. As we shall see below, this would make the false-alarm probability (and overall error probability) roughly on the order of $1-(1-\epsilon)^n$, which increases with n, and equals ϵ for $n=1$. Since many detection systems are designed to operate at false-alarm probabilities in the range 10^{-4} to 10^{-6}, even 1% of uncertainty in this model (i.e., $\epsilon=0.01$) could potentially destroy the detector's false-alarm performance. A similar phenomenon could occur under H_1 if $p_1(y_k)/p_0(y_k)$ is not bounded away from zero since M_1 might place its probability in observation regions where $p_1(y_k) << p_0(y_k)$, thus tending to drive the likelihood ratio below the threshold.

The above discussion, although heuristic, points to a certain lack of robustness in performance of the likelihood ratio tests in situations where $p_1(y_k)/p_0(y_k)$ is not bounded from above and (away from zero) from below. In particular, we see that even relatively small deviations in the model might result in substantial performance loss in this situation. A question that arises in whether anything reasonable can be done to alleviate this lack of robustness. One fairly obvious possible way of stabilizing the performance of the likelihood ratio test is to replace the likelihood ratio p_1/p_0 with a version that is limited from above and below. That is, suppose we replace $l \triangleq p_1/p_0$ in the product $\Pi_{k=1}^n p_1(y_k)/p_0(y_k)$ with the function

$$[l]_a^b(y) = \begin{cases} b & \text{if } l(y) > b \\ l(y) & \text{if } a \leqslant l(y) \leqslant b \\ a & \text{if } l(y) < a, \end{cases} \tag{III.E.44}$$

where $0 < a < b < \infty$. Then for properly chosen a and b, a test based on $\Pi_{k=1}^n [l]_a^b(y_k)$ would not exhibit the difficulties noted above for unbounded l. However, by introducing the limiting of (III.E.44), some performance (and certainly optimality) at the nominal is lost, since those observations for which $l(y)$ is very large or very close to zero are, in a sense, the most informative ones.

It is of interest to find a criterion by way of which tests can be optimized for performance under modeling uncertainty. The discussion above concerning the lack of robustness of the likelihood-ratio test with unbounded likelihood ratio points to one such criterion. In particular, that discussion centered around the worst-case performance of the test over the class of statistics possible under the uncertainty model. Thus a reasonable design criterion might be to replace the usual error probabilities with their worst-case values over some reasonable neighborhood of the nominal model [such as (III.E.43)] and to choose a test to optimize a corresponding criterion.

Recall that if P_0 is the true marginal of the Y_k, the false-alarm probability of a test δ is

$$P_F(\delta, P_0) = \int_\Gamma \delta(\underline{y}) \left[\prod_{k=1}^{n} p_0(y_k) \right] \mu(d\underline{y}), \tag{III.E.45}$$

where p_0 is the marginal density corresponding to P_0. Similarly, the miss probability when P_1 is the true marginal is

$$P_M(\delta, P_1) = \int_\Gamma [1-\delta(\underline{y})] \left[\prod_{k=1}^{n} p_1(y_k) \right] \mu(d\underline{y}). \tag{III.E.46}$$

Assuming for simplicity that costs are uniform, the three usual criteria for simple binary hypothesis testing are then:

(i) $\min_\delta [\pi_0 P_F(\delta, P_0) + \pi_1 P_M(\delta, P_1)]$ (Bayes).

(ii) $\min_\delta [\max\{P_F(\delta, P_0), P_M(\delta, P_1)\}]$ (Minimax).

(iii) $\min_\delta P_M(\delta, P_1)$ subject to $P_F(\delta, P_0) \leq \alpha$ (Neyman-Pearson).

If instead of assuming that the marginal distribution of the Y_k is exactly P_0 or P_1, we assume that the marginal lies either in a neighborhood $\hat{P}_0$ of P_0 or in a neighborhood $\hat{P}_1$ of P_1 [as, e.g, in (III.E.43)], then by replacing $P_F(\delta, P_0)$ and $P_M(\delta, P_1)$ in (i) - (iii) with their worst-case values

$$P_F(\delta, \hat{P}_0) \triangleq \sup_{P \epsilon \hat{P}_0} P_F(\delta, P) \qquad \text{(III.E.47)}$$

and

$$P_M(\delta, \hat{P}_1) \triangleq \sup_{P \epsilon \hat{P}_1} P_M(\delta, P), \qquad \text{(III.E.48)}$$

we arrive at the alternative design problems

(i') $\min_\delta [\pi_0 P_F(\delta, \hat{P}_0) + \pi_1 P_M(\delta, \hat{P}_1)]$.

(ii') $\min_\delta [\max \{P_F(\delta, \hat{P}_0), P_M(\delta, \hat{P}_1)]$.

(iii') $\min_\delta \hat{P}_M(\delta, \hat{P}_1)$ subject to $\hat{P}_F(\delta, \hat{P}_0) \leqslant \alpha$.

Solutions to (i') - (iii') will have the best worst-case performance (over the neighborhoods $\hat{P}_0$ and $\hat{P}_1$) of all possible tests. Of course, there is a danger that such tests might be overly conservative, and this certainly would be true if $\hat{P}_0$ and $\hat{P}_1$ are too large. However, the idea here is that $\hat{P}_0$ and $\hat{P}_1$ are small neighborhoods of a nominal model and the goal is to avoid the possible performance instability for such neighborhoods noted above. It turns out that solutions to (i')-(iii') do achieve this goal for uncertainty models such as (III.E.43).

Although there is a general approach to solving problems (i')-(iii') due to Huber and Strassen (1973), it is somewhat involved and we will not discuss it here. Instead, we will focus on the solutions to (i')-(iii') for the particular uncertainty neighborhoods described in (III.E.43), which are known as *ϵ-contaminated mixtures* . This solution is found in Huber (1965), in which the formulation (i')-(iii') was first proposed as a design technique for robust tests.

It turns out that the solutions to (i')-(iii') for ϵ-contaminated mixtures are the corresponding optimum tests for (i)-(iii) when P_0 and P_1 in (i)-(iii) are replaced by a pair $Q_0 \epsilon \hat{P}_0$ and $Q_1 \epsilon \hat{P}_1$ of *least-favorable* distributions. Q_0 and Q_1 are given in terms of their densities by

$$q_0(y_k) = \begin{cases} (1-\epsilon)p_0(y_k) & \text{if } p_1(y_k) < c''p_0(y_k) \\ \frac{1-\epsilon}{c''} p_1(y_k) & \text{if } p_1(y_k) \geqslant c''p_0(y_k) \end{cases} \tag{III.E.49a}$$

and

$$q_1(y_k) = \begin{cases} (1-\epsilon)p_1(y_k) & \text{if } p_1(y_k) > c'p_0(y_k) \\ c'(1-\epsilon)\, p_0(y_k) & \text{if } p_1(y_1) \leqslant c'p_0(y_k), \end{cases} \tag{III.E.49b}$$

where $0 < c' < 1 < c'' < \infty$ are two constants chosen so that Q_0 and Q_1 are probably distributions, i.e., so that their total probability equals 1. This condition is given by

$$(1-\epsilon)[P_0(l(Y_k) < c'') + P_1(l(Y_k) \geqslant c'')/c''] = 1 \tag{III.E.50a}$$

and

$$(1-\epsilon)[P_1(l(Y_k) > c') + c'P_0(l(Y_k) \leqslant c')] = 1. \tag{III.E.50b}$$

Since each of (i)-(iii) is solved by a likelihood-ratio test, the solutions to (i')-(iii') are likelihood-ratio tests between Q_0 and Q_1; namely, they are based on the likelihood ratio, $\Pi_{k=1}^{n} q_1(y_k)/q_0(y_k)$. Using (III.E.49) we have that

$$\frac{q_1(y_k)}{q_0(y_k)} = \begin{cases} c' & \text{if } l(y_k) < c' \\ l(y_k) & \text{if } c' \leqslant l(y_k) \leqslant c'' \\ c'' & \text{if } l(y_k) > c''. \end{cases} \tag{III.E.51}$$

Thus the solutions to (i')-(iii') are threshold tests based on

$$\prod_{k=1}^{n} [l]_{c'}^{c''}(y_k), \qquad \text{(III.E.52)}$$

and we have arrived analytically at a test of the type proposed earlier as an ad hoc robust test. This test, however, has a specific optimality property in solving (i')-(iii'). Moreover, in addition to solving (i')-(iii'), it can be shown that

$$P_F(\delta_R, \hat{P}_0) = P_F(\delta_R, Q_0) \qquad \text{(III.E.53a)}$$

and

$$P_M(\delta_R, \hat{P}_1) = P_M(\delta_R, Q_1) \qquad \text{(III.E.53b)}$$

for any threshold test δ_R based on the likelihood ratio (III.E.52). This implies that the worst-case performance of δ_R is in fact its performance at the pair of distributions (Q_0, Q_1) for which it is optimum. This allows us to compute upper bounds on the error probabilities of δ_R for the entire classes $\hat{P}_0$ and $\hat{P}_1$ simply by evaluating the error probabilities of δ_R for marginals Q_0 and Q_1.

Remarks

1. In a sense, the least-favorable densities Q_0 and Q_1 are as close in shape to one another as is possible within the constraints $Q_0 \epsilon \hat{P}_0$ and $Q_1 \epsilon \hat{P}_1$. It can be shown that if $P_0 \neq P_1$, and ϵ is small enough (and positive), then equations (III.E.50) have solutions satisfying $0 < c' < 1 < c'' < \infty$. This implies that $Q_0 \neq Q_1$ and that δ_R is not a trivial test. On the other hand, if ϵ is too large, then $\hat{P}_0$ and $\hat{P}_1$ will overlap, $c' = c'' = 1$, and Q_0 will equal Q_1. In this case, $q_1(y_k)/q_0(y_k) \equiv 1$ and the tests solving (i')-(iii') simply ignore the observations and guess at the hypothesis. Thus, if the neighborhoods $\hat{P}_0$ and $\hat{P}_1$ are too large so as to overlap, (i')-(iii') are not good design criteria. However, as noted above, they are really intended for small ϵ.

2. Even for very small ϵ, the difference in performance between the tests based on l and $[l]_{c'}^{c''}$ can be quite dramatic. Consider, for example, the Bayesian formulation (i) with

$\pi_0=\pi_1=1/2$. Then, for either test, the threshold is unity and the randomization is arbitrary. Suppose that $0<l(y_k)<\infty$ and $\sup_{y_k \in \mathbb{R}} l(y_k)=\infty$. Then, as noted above, since M_0 is arbitrary, it can be chosen to put all of its probability on a value of y_k for which $l(y_k)$ is arbitrarily large. In this way, any of the observations can cause δ_0 to commit a false-alarm with probability ϵ, where δ_0 is the test based on l. Since there are n observations, this implies that

$$P_F(\delta_0, \hat{P}_0) \geqslant 1-(1-\epsilon)^n . \tag{III.E.54}$$

Similarly, if $\inf_{y_k \in \mathbb{R}}(y_k)=0$, the miss probability of δ_0 over $\hat{P}_1$ satisfies

$$P_M(\delta_0, \hat{P}_1) \geqslant 1-(1-\epsilon)^n ,$$

so the worst-case average error probability of δ_0 satisfies

$$\begin{aligned} \sup P_e(\delta_0) &= 1/2 P_F(\delta_0, \hat{P}_0) + 1/2 P_M(\delta_0, \hat{P}_1) \\ &\geqslant 1-(1-\epsilon)^n , \end{aligned} \tag{III.E.55}$$

where the supremum is taken over $\hat{P}_0$ and $\hat{P}_1$. For any $\epsilon>0$, $\lim_{n\to\infty}(1-\epsilon)^n=0$. Thus

$$\lim_{n\to\infty} [\sup P_e(\delta_0)] = 1;$$

so the performance of δ_0 can be arbitrarily bad, and in fact it can be worse than simply guessing at the hypothesis on the basis of a coin toss, since guessing would cause an error probability of 1/2.

In contrast, consider the test δ_R based on q_1/q_0. From the property (III.E.50), it follows that

$$\sup P_e(\delta_R) = 1/2\, P_F(\delta_R, Q_0) + 1/2 P_M(\delta_R, Q_1). \tag{III.E.56}$$

Applying the Chernoff bound (III.C.18) to the right-hand side of (III.E.56), we have that

$$\sup P_e(\delta_R) \leq \tfrac{1}{2}\left[\int [q_0 q_1]^{1/2}\right]^n . \qquad \text{(III.E.57)}$$

As noted in Section III.C, if $Q_0 \neq Q_1$ we will have $\int [q_0 q_1]^{1/2} < 1$, so that

$$\lim_{n \to \infty} [\sup P_e(\delta_R)] = 0. \qquad \text{(III.E.58)}$$

From (III.E.55) and (III.E.56), we see that in terms of worst-case performance, $P_e(\delta_0)$ converges exponentially to unity while $P_e(\delta_R)$ converges exponentially to zero. Although the behavior that drives sup $P_e(\delta_0)$ to unity is somewhat extreme, this nevertheless points to a limitation of such tests. It should be noted that under nominal conditions, $P_e(\delta_0)$ also converges exponentially to zero via the Chernoff bound. Thus δ_R achieves, in its worst case, behavior similar to the nominal behavior of δ_0, while δ_0 in its worst case behaves radically differently.

3. Solutions to (i')-(iii') are known for a number of uncertainty models other than the ϵ-contaminated mixture. (The ϵ-contaminated mixture itself can be generalized slightly to allow different values of ϵ under each hypothesis. For simplicity, we chose them to be equal here.) For example, several interesting types of neighborhoods that can be treated in this context are of the form $\hat{P}_0 = \{P | \rho(P, P_0) \leq \epsilon_0]$ and $\hat{P}_1 = \{P | \rho(P, P_1) \leq \epsilon_1\}$, where ρ is some measure of distance between probability distributions. In most cases, other known solutions to (i')-(iii') are similar to those discussed here for the ϵ-contaminated case in that they are solutions to (i)-(iii) for least-favorable pairs and that they usually involve some form of amplitude limiting of the likelihood ratio. The ϵ-contaminated model can also be generalized to allow for time-varying nominals and ϵ's. In this case, $\Pi_{k=1}^n [l]_{c'}^{c''}(y_k)$ is simply replaced by $\Pi_{k=1}^n [l_k]_{c_k'}^{c_k''}(y_k)$ where l_k is the nominal likelihood ratio for the kth sample and c_k' and c_k'' solve (III.E.50) for the kth nominal distributions.

We illustrate the application of the results discussed in the paragraph above by the following example.

Example III.E.1: The Correlator-Limiter

Consider the coherent signal detection problem

$$\begin{aligned} &H_0: Y_k = N_k, \qquad k=1, \ldots, n \\ \text{versus} \\ &H_1: Y_k = N_k + \theta s_k, \quad k=1, \ldots, n, \end{aligned} \tag{III.E.59}$$

where $s_1, \ldots, s_n$ is a known signal sequence, $N_1, \ldots, N_n$ is an i.i.d. sequence of $N(0, 1)$ noise samples, and θ is a known positive amplitude. The kth-sample likelihood ratio for this problem is given [see (III.B.9)] by $l_k(y_k) = \exp[\theta s_k(y_k - \theta s_k/2)]$, and so the log-likelihood ratio is

$$\log L(\underline{y}) =. \theta \sum_{k=1}^{n} s_k(y_k - \theta s_k/2), \tag{III.E.60}$$

which leads to the correlation detector illustrated by Fig. III.E.1.

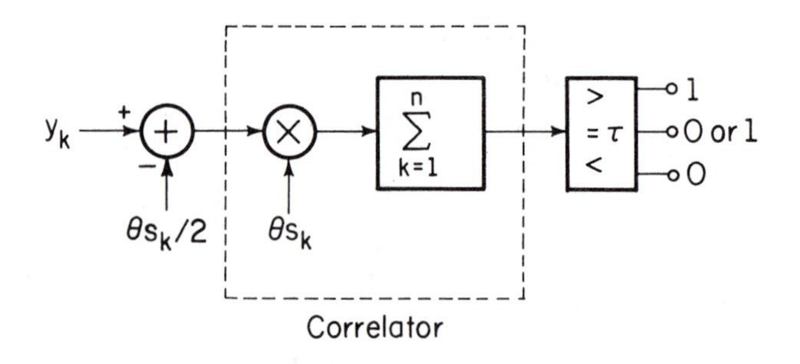

Fig. III.E.1: Nominally optimum detector for coherent signals in Gaussian noise.

In a practical situation, the model of (III.E.59) can only be assumed to be approximately correct. In particular, the noise distribution is unlikely to be exactly $N(0, 1)$, and the measurement model in which signal and noise are assumed to be additive is not completely accurate due to nonlinearities in the observation mechanism. Thus, to have a more realistic model, we could modify (III.E.59) to

$$H_0: Y_k \sim (1-\epsilon)P_0 + \epsilon M_0, \quad k=1, \dots, n$$

versus (III.E.61)

$$H_1: Y_k \sim (1-\epsilon)P_1^{(k)} + \epsilon M_1^{(k)}, \quad k=1, \dots, n,$$

where P_0 is the $N(0, 1)$ distribution, $P_1^{(k)}$ is the $N(\theta s_k, 1)$ distribution, M_0 and $M_1^{(k)}, k=1, \dots, n$ are arbitrary (note: M_0 could also be allowed to change with k; this would not change the solution given below), and ϵ is between 0 and 1.

Note that with $l_k(y_k) = \exp[\theta s_k(y_k - \theta s_k)/2)]$, we have $0 < l_k(y_k) < \infty$, $\sup_{y_k \in \mathbb{R}} l_k(y_k) = \infty$, and $\inf_{y_k \in \mathbb{R}} l_k(y_k) = 0$, so that the correlation detector will suffer the performance degradation discussed above in the presence of uncertainty modeled as in (III.E.61). Huber's robust likelihood ratio test δ_R is thus preferred in this situation. For the model of (III.E.61) the logarithm of the robust likelihood ratio

$$L_R(\underline{y}) = \prod_{k=1}^{n} [l_k]_{c_k'}^{c_k''}(y_k),$$

can be written as

$$\log L_R(\underline{y}) = \sum_{k=1}^{n} [\theta s_k(y_k - \theta s_k/2)]_{d_k'}^{d_k''}, \qquad \text{(III.E.62)}$$

where $d_k' \triangleq \log c_k'$ and $d_k'' \triangleq \log c_k''$. Using the symmetry properties of the $N(0, 1)$ distribution and of the likelihood ratio, it can be shown that $d_k' = -d_k''$ and that d_k'' is the solution to

$$\Phi\left(\frac{d_k''}{\theta|s_k|}+\frac{\theta|s_k|}{2}\right)+e^{-d_k''}\left[1+\Phi\left(\frac{d_k''}{\theta|s_k|}-\frac{\theta|s_k|}{2}\right)\right]$$
(III.E.63)

$$=(1-\epsilon)^{-1},$$

where Φ is the $N(0, 1)$ cumulative distribution function. d_k'' decreases monotonically from ∞ to zero as ϵ increases from zero.

The detector of (III.E.62) is depicted in Fig. III.E.2. Note that it is identical to the correlation receiver of Fig. III.E.1 except that there is a (time-varying) limiter with limits $\pm d_k''$ placed between the multiplier and the accumulator in the correlator. This detector structure is called a *correlator-limiter*, and it is originally derived and analyzed by Martin and Schwartz (1971). It will be robust provided that ϵ is small enough to prevent overlap of the classes under H_0 and H_1. For fixed ϵ, this corresponds to the signal strength θ being large enough to prevent overlap. In particular, if $|s_k|>0$ for $k=1, ..., n$ the uncertainty classes under H_0 and H_1 do not overlap for any k if $\theta>\theta_\epsilon/\min|s_k|$, where θ_ϵ is the solution to

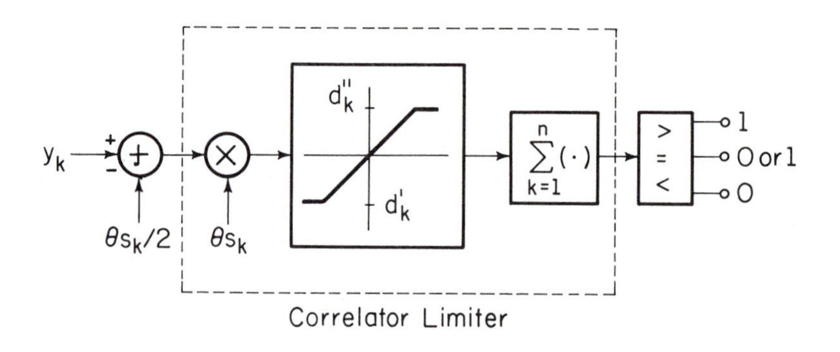

Fig. III.E.2: Robust detector for coherent signals in nominally Gaussian noise.

$(1-\epsilon)\Phi(\theta_\epsilon/2)=\frac{1}{2}$. For example, if $\epsilon=0.012$, then $\theta_\epsilon=0.03$; if $\epsilon=0.055$, then $\theta_\epsilon=0.15$; and if $\epsilon=0.138$, then $\theta_\epsilon=0.4$. Thus robustness is retained for fairly small signal-to-noise ratios. It should be noted that unlike the linear correlator, the correlator-limiter is not uniformly optimum in θ even for $\theta>\theta_\epsilon$ since d_k'' depends on θ. This is a disadvantage for many applications since θ is not necessarily known. Methods for dealing with this difficulty and the small-θ robustness problems have been developed. These and other aspects of robust detection are discussed in the review article, Kassam and Poor (1985).

III.F Exercises

1. Show that the filter with impulse response

$$\tilde{h}_k = \begin{cases} \tilde{s}_{n-k} & , 0\leqslant k \leqslant n-1 \\ 0 & , \text{otherwise} \end{cases}$$

with $\underline{\tilde{s}}=\mathbf{\Sigma}_N^{-1}\underline{s}$ has maximum output signal-to-noise ratio at time n among all linear filters, when the input signal is $\underline{s}=(s_1, ..., s_n)^T$ and the input noise has zero mean and covariance $\mathbf{\Sigma}_N$.

2. Suppose the random observation vector $\underline{Y}$ is given by

$$Y_k = N_k + \theta S_k, \quad k = 1, ..., n$$

where $\underline{N}$ is a zero-mean Gaussian random vector with $E\{N_k N_l\}=\sigma^2\rho^{|k-l|}$ for all $0\leqslant k, l \leqslant n$, $|\rho|<1$ and where $\underline{s}$ is a known signal vector.

(a) Show that the test

$$\delta(\underline{y}) = \begin{cases} 1 \text{ if } \sum_{k=1}^{n} b_k z_k \geq \tau' \\ \\ 0 \text{ if } \sum_{k=1}^{n} b_k z_k < \tau' \end{cases}$$

is equivalent to the likelihood ratio test for $\theta=0$ versus $\theta=1$, where

$$b_1 = s_1/\sigma$$

$$b_k = (s_k - \rho s_{k-1})/\sigma\sqrt{1-\rho^2}, \quad k = 2, ..., n$$

$$z_1 = y_1/\sigma$$

$$z_k = (y_k - \rho y_{k-1})/\sigma\sqrt{1-\rho^2}, \quad k = 2, ..., n .$$

(b) Find the ROCs of the detector from (a) as a function of $\theta/\sigma, \rho, n$, and the false-alarm probability α.

3. Consider the M-ary decision problem: $(\Gamma=\mathbb{R}^n)$

$$H_0: \underline{Y} = \underline{N} + \underline{s}_0$$

$$H_1: \underline{Y} = \underline{N} + \underline{s}_1$$

$$\vdots$$

$$H_{M-1}: \underline{Y} = \underline{N} + \underline{s}_{M-1},$$

where $\underline{s}_0, \underline{s}_1, ..., \underline{s}_{M-1}$ are known signals with equal energies, $\|\underline{s}_0\|^2=\|\underline{s}_1\|^2=...=\|\underline{s}_{M-1}\|^2$.

(a) Assuming $\underline{N} \sim N(\underline{0}, \sigma^2 \mathbf{I})$, find the decision rule achieving minimum error probability when all hypotheses are equally likely.
(b) Assuming further that the signals are orthogonal, show that the minimum error probability is given by

$$P_e = 1 - \frac{1}{\sqrt{2\pi}} \int_{-\infty}^{\infty} [\Phi(x)]^{M-1} e^{-(x-d)^2/2} dx$$

where $d^2 = \|\underline{s}_0\|^2 / \sigma^2$.

4. Quaternary Phase-shift Keying (QPSK) is an example of the situation in Exercise 3 with four signals ($M=4$) given by

$$s_{lk} = E_0 \sin(\omega_c T(k-1) + (l + 1/2)\pi/2),\ k = 1, \ldots, n,$$

$$, l = 0, \ldots, 3.$$

Assuming ω_c, T, and n are as in Example III.B.5, find the minimum error probability for equally like signals in i.i.d. $N(0, \sigma^2)$ noise. (Note: that these signals are not orthogonal).

5. Suppose $\underline{Y} \sim N(\underline{\mu}, \Sigma)$. For each $k \geq 2$, define $\hat{Y}_k = E\{Y_k | Y_1, \ldots, Y_{k-1}\}$ and $\hat{\sigma}^2_{Y_k} = Var(Y_k | Y_1, \ldots, Y_{k-1})$. Also define $\hat{Y}_1 = E\{Y_1\}$ and $\hat{\sigma}^2_{Y_1} = Var(Y_1)$. Define a sequence $I_1, I_2, \ldots, I_n$ by

$$I_k = (Y_k - \hat{Y}_k) / \hat{\sigma}_{Y_k}.$$

Show that $\underline{I} \sim N(0, \mathbf{I})$, and thus that the above scheme provides whitening of $\underline{Y}$.

6. Consider the hypothesis pair

$$H_0: Y_k = N_k \qquad , k = 1, ..., n$$

versus

$$H_1: Y_k = N_k + \Theta S_k \ , k = 1, ..., n$$

where $\underline{N} \sim N(\underline{0}, \Sigma), \underline{s}$ is known, and Θ is a random variable independent of $\underline{N}$.
(a) Find the α-level Neyman-Pearson detector and ROC's assuming that Θ is a discrete random variable taking the values +1 and -1 with equal probabilities (i.e., $P(\Theta=+1)=P(\Theta=-1)=½$).
(b) Suppose that $\Theta \sim N(0, \sigma_\theta^2)$. Assuming $\Sigma=\sigma^2\mathbf{I}$, show that the likelihood ratio is of the form

$$L(\underline{y}) = k_1 e^{k_2 \|\underline{s}^T \underline{y}\|^2}$$

where k_1 and k_2 are positive constants. Find k_2.

7. Suppose we have observations $Y_k = N_k + \theta S_k$, $k = 1, ..., n$, where $\underline{N} \sim N(\underline{0}, \mathbf{I})$ and where $S_1, ..., S_n$ are i.i.d. random variables, independent of $\underline{N}$, and each taking on the values +1 and −1 with equal probabilities of ½.
(a) Find the likelihood ratio for testing $H_0: \theta=0$ versus $H_1: \theta=A$, where A is a known constant.
(b) For the case $n=1$, find the Neyman-Pearson rule and corresponding detection probability for false-alarm probability $\alpha \in (0, 1)$, for the hypotheses of (a).
(c) Is there a UMP test of $H_0: \theta=0$ versus $H_1: \theta \neq 0$ in this model? If so, why and what is it? If not, why not? Consider the cases $n=1$ and $n>1$ separately.

8. Derive Equation (III.B.82).

9. Differential Phase-shift Keying (DPSK) is a binary signaling scheme using modulated sinusoids in which a "zero" is transmitted in a given bit interval by sending the same signal as that sent in the immediately preceding bit interval, and a "one" is transmitted by sending a signal whose

carrier is phase-shifted by 180° from that of the previous bit interval. Under the same signal and noise assumptions as in Example III.B.5, find the Bayes detector for DPSK when the costs are uniform and the priors are equal. Find the resulting probability of error.

10. Consider the model

$$Y_k = \theta^{1/2} s_k R_k + N_k, \quad k = 1, \ldots, n$$

where $s_1, s_2, \ldots, s_n$ is a known signal sequence, $\theta \geq 0$ is a constant and $R_1, R_2, \ldots, R_n, N_1, N_2, \ldots, N_n$ are i.i.d. $N(0, 1)$ random variables

(a) Consider the hypothesis pair

$$H_0: \theta = 0$$

versus

$$H_1: \theta = A$$

where A is a known positive constant. Describe the structure of the Neyman-Pearson detector.

(b) Consider now the hypothesis pair

$$H_0: \theta = 0$$

versus

$$H_1: \theta > 0.$$

Under what conditions on $s_1, s_2, \ldots, s_n$ does a UMP test exist?

(c) For the hypothesis pair of part (b) with $s_1, s_2, \ldots, s_n$ general, is there a *locally* optimum detector? If so, find it. If not, describe the generalized likelihood ratio test.

11. Repeat Exercise 10 under the alternate assumption that $R_1 = R_1 = \ldots = R_n \sim N(0, 1)$. Retain the assumption that $\underline{R}$ and $\underline{N}$ are independent.

12. Consider the problem of Example III.B.5 in which the amplitude sequence $a_1, a_2, \ldots, a_n$ is given by

$$a_k = Ab_k, \quad k = 1, 2, \ldots, n,$$

where $\sum_{k=1}^n b_k^2 = n$, and A is a positive random variable, independent of the phase Θ, having the Rayleigh density with parameter A_0; i.e.,

$$p_A(a) = (a/A_0^2)\exp\{-a^2/2A_0^2\}, \quad a \geq 0.$$

Find the Neyman-Pearson detector, including the threshold for size α, and derive an expression for the ROC's.

13. Find the $\underline{\hat{s}}$ solving

$$e^{(\underline{\hat{s}}^T \underline{y} - \frac{1}{2}\|\underline{\hat{s}}\|^2)/\sigma^2} = \int_{\mathbb{R}^n} e^{(\underline{s}^T \underline{y} - \frac{1}{2}\|\underline{s}\|^2)/\sigma^2} p_{\underline{S}}(\underline{s}) d\underline{s}$$

for the case in which $p_{\underline{S}}$ is the $N(\underline{0}, \Sigma_S)$ density.

14. Derive Eq (III.B.112) from (III.B.111).

15. Let $\mu_{T,0}(s)$ denote the cumulant generating function of the log-likelihood ratio under H_0. Assume $\mu_{T,0}(s)$ is twice differentiable.
(a) Show that $(\mu_{T,0}(s) - s\tau)$ is a convex function of s.
(b) Show that

$$\mu'_{T,0}(j) = E\{\log L(Y) | H_j\} \text{ for } j = 0, 1,$$

where

$$\mu'_{T,0}(s) = \frac{d}{ds}\mu_{T,0}(s).$$

(c) Show that, if $\min_{s \geq 0}[\mu_{T,0}(s) - s\tau]$ occurs for $s \geq 1$, then $\min_{s \leq 1}[\mu_{T,0}(s) + (1-s)\tau]$ occurs for $s = 1$ and the corresponding minimum value is zero.

(d) Show that, if $\min_{s \leqslant 1}[\mu_{T,0}(s)-s\tau]$ occurs for $s \leqslant 0$, then $\min_{s \geqslant 0}[\mu_{T,0}(s)-s\tau]$ occurs for $s=0$ and the corresponding minimum value is zero.

16 Derive Eq. (III.C.15).

17. Compute the Chernoff bound for the binary symmetric channel with equal priors ($\pi_0=\pi_1=\frac{1}{2}$), and compare it to the actual minimum error probability.

18. Consider the hypothesis pair

$$H_0: Y_k = N_k - S_k, \quad k = 1, ..., n$$

versus

$$H_1: Y_k = N_k + S_k, \quad k = 1, ..., n$$

where $N_1, ..., N_n$ are i.i.d. Laplacian random variables and where $s_1, ..., s_n$ is a known signal satisfying $s_k \geqslant \Delta > 0$ for all k and some constant Δ. Show that the minimum error probability in deciding H_0 versus H_1 approaches zero as $n \to \infty$. (Δ is independent of n).

19. Consider the problem of detecting a $N(\underline{0}, \Sigma_S)$ signal in $N(\underline{0}, \sigma^2 \mathbf{I})$ noise with $n=2$ and

$$\Sigma_S = \sigma_S^2 \begin{pmatrix} 1 & \rho \\ \rho & 1 \end{pmatrix}.$$

For equally likely priors compute and compare the exact error probability and the Chernoff bound on the error probability for $\rho=0.0$, $\rho=-0.5$, and $\rho=+0.5$, and for $\sigma_S^2/\sigma^2=0.1$, $\sigma_S^2/\sigma^2=1.0$, and $\sigma_S^2/\sigma^2=10.0$.

20. Investigate the Chernoff bound for testing between the two marginal densities

$$p_0(y) = \begin{cases} 1 \text{ if } 0 \leqslant y \leqslant 1 \\ 0 \text{ otherwise} \end{cases}$$

and

$$p_1(y) = \begin{cases} 2y \text{ if } 0 \leqslant y \leqslant 1 \\ 0 \text{ otherwise,} \end{cases}$$

for a sequence of i.i.d. observations, $Y_1, Y_2, \ldots, Y_n$.

21. Consider a sequence of i.i.d. Bernoulli observations, $Y_1, Y_2, \ldots$, with distribution

$$P(Y_k = 1) = 1 - P(Y_k = 0) = 1/3$$

under hypothesis H_0, and

$$P(Y_k = 1) = 1 - P(Y_k = 0) = 2/3$$

under hypothesis H_1.
(a) Use Wald's approximations to suggest values of A and B so that the $SPRT(A, B)$ has maximum error probability $p^* = \max(P_F, P_M)$ approximately equal to 0.01. Describe the resulting test in detail. Also, using Wald's approximations, give an approximation to the expected sample sizes $E\{N|H_0\}$ and $E\{N|H_1\}$.
(b) Find an integer n as small as you can so that the maximum error probability for the optimal test with fixed sample size n is no more than 0.01. Compare n to the expected sample sizes found in part (a) (**Note**: You may use a Chernoff bound to find n, rather than finding the smallest possible n.)

(c) Compute $p^*, E\{N|H_0\}$ and $E\{N|H_1\}$ exactly for the test you found in part (a), and compare with the approximate values you found in part (a). (**Hint**: Use the fact that the SPRT you found in part (a) is equivalent to $SPRT(A', B')$ where A' and B' are integer powers of 2.)

22. Let $N_1, N_2, \ldots$ be independent Gaussian random variables with means 0 and variances 1, and let $S_1, S_2, \ldots$ be independent Gaussian random variables with means 0 and variances 3. Assume the N_k's and S_k's are independent of each other and consider the hypothesis pair

$$H_0: Y_k = N_k, \quad k = 1, 2, \cdots$$

versus

$$H_1: Y_k = S_k + N_k, \quad k = 1, 2, \cdots$$

(a) Repeat (a) of Exercise 21 for this new model.
(b) Repeat (b) of Exercise 21 for this new model.

23. Derive Eq. (III.E.22).

24. Show that Eqs. (III.E.24) and (III.E.25) are equivalent.

25. Derive Eq. (III.E.26).

26. Verify Eq. (III.E.54).

IV ELEMENTS OF PARAMETER ESTIMATION

IV.A Introduction

In Chapters II and III we have considered the design of optimum procedures for deciding between two possible statistical situations on the basis of a random observation Y. In many situations arising in practice we are interested not in making a choice between two (or among several) discrete situations, but rather in making a choice among a continuum of possible states of nature. In particular, as in the composite hypothesis testing problems discussed in Chapter II, we can think of a family of distributions on the observation space, indexed by a parameter or set of parameters. But unlike the case of composite hypothesis testing in which we wish to make a binary decision about the parameter, we wish here to determine as accurately as possible the actual value of the parameter from the observation.

Such problems are known as *parameter* (or *point*) *estimation problems*, and in this chapter we discuss the basic ideas relating to the design of optimum procedures for estimating parameters. As with the hypothesis-testing problem (which incidentally can be thought of as a special case of the parameter estimation problem), a variety of estimation design philosophies can be used, these differing primarily in the amount of prior information known about the parameter and in the performance criteria applied.

In this chapter we discuss two basic approaches to parameter estimation--one, the Bayesian, in which the parameter is assumed to be a random quantity related statistically to the observation, and a second in which the parameter is assumed to be unknown but without being endowed with any probabilistic structure. Of these two approaches, the Bayesian is the most straightforward and so is considered first, in Section IV.B, with nonrandom parameter estimation being considered in the remainder of the chapter.

It should be noted that in this treatment, we consider only the estimation of parameters that are static, i.e., that are constant in time. The estimation of dynamic parameters (i.e., signals) is considered in Chapter V.

IV.B Bayesian Parameter Estimation

Throughout this chapter we assume as a model a family of distributions for the random parameter Y, indexed by a parameter θ taking values in a parameter set Λ; i.e., we have the family $\{P_\theta; \theta \in \Lambda\}$, where P_θ denotes a distribution on the observation space (Γ, G). We also assume that the parameter set Λ is a subset of $\mathbb{R}^m$ for some m. Within this model the goal of the parameter estimation problem is to find a function $\hat{\theta}: \Gamma \to \Lambda$ such that $\hat{\theta}(y)$ is the "best" guess of the true value of θ (i.e., the value of θ for which $Y \sim P_\theta$) based on the observation $Y = y$.

Of course, the solution to this problem depends on the criterion of goodness by which we measure estimation performance; so, as in the hypothesis-testing problem, we begin by assigning costs to our decisions about the parameter. In particular, we suppose that there is a function $C: \Lambda \times \Lambda \to \mathbb{R}$ such that $C[a, \theta]$ is the cost of estimating a true value of θ as a, for a and θ in Λ. Given such a function C we can then associate with an estimator $\hat{\theta}$ a conditional risk or cost averaged over Y for each $\theta \in \Lambda$; i.e., we have

$$R_\theta(\hat{\theta}) = E_\theta\{C[\hat{\theta}(Y), \theta]\}. \qquad \text{(IV.B.1)}$$

If we now adopt the interpretation that the actual parameter value θ is the realization of a random variable Θ, we can define an average or Bayes risk as

$$r(\hat{\theta}) \triangleq E\{R_\Theta(\hat{\theta})\}, \qquad \text{(IV.B.2)}$$

and the appropriate design goal is to find an estimator minimizing $r(\hat{\theta})$. Such an estimator is known as a *Bayes estimate* of θ.

Noting that $R_\theta(\hat{\theta}) = E\{C[\hat{\theta}(Y), \Theta] \big| \Theta = \theta\}$, we have

$$r(\hat{\theta}) = E\{C[\hat{\theta}(Y), \Theta]\} = E\{E\{C[\hat{\theta}(Y), \Theta]|Y\}\}. \tag{IV.B.3}$$

By inspection of (IV.B.3) we see that the Bayes estimate of θ can be found (if it exists) by minimizing, for each $y \in \Gamma$, the posterior cost given $Y=y$:

$$E\{C[\hat{\theta}(y), \Theta]|Y = y\}. \tag{IV.B.4}$$

This is the same procedure as that followed in the Bayesian hypothesis-testing problem (see Section II.E). Note that if we assume that Θ has a conditional density $w(\theta|y)$ given $Y=y$ for each $y \in \Gamma$, then the Bayes estimate $\hat{\theta}(y)$ corresponding to $y \in \Gamma$ can be sought by minimizing

$$\int_{\Lambda} C[\hat{\theta}(y), \theta] w(\theta|y) \mu(d\theta). \tag{IV.B.5}$$

The following cases illustrate the application of this criterion.

Case IV.B.1: Minimum-Mean-Squared-Error (MMSE) Estimation

For situations in which $\Lambda = \mathbb{R}$ and $E\{\Theta^2\} < \infty$, a commonly used cost function is that given by

$$C[a, \theta] = (a - \theta)^2, \quad (a, \theta) \in \mathbb{R}^2. \tag{IV.B.6}$$

This cost function is a natural one for many situations since it measures the performance of an estimator in terms of the square of the estimation error, $\hat{\theta}(y) - \theta$. The Bayes risk here is $E\{(\hat{\theta}(y) - \Theta)^2\}$, a quantity known as the *mean-squared-error* (MSE). Thus the Bayes estimate in this case is a *minimum-mean-squared-error* (MMSE) *estimator*.

The posterior cost given $Y=y$ is given in this case by

$$E\{(\hat{\theta}(y)-\Theta)^2|Y=y\} = E\{[\hat{\theta}(y)]^2|Y=y\}$$

$$-2E\{\hat{\theta}(y)\Theta|Y=y\}$$

$$+E\{\Theta^2|Y=y\} \qquad \text{(IV.B.7)}$$

$$=[\hat{\theta}(y)]^2-2\hat{\theta}(y)E\{\Theta|Y=y\}$$

$$+E\{\Theta^2|Y=y\}.$$

The expression in (IV.B.7) is a quadratic function of $\hat{\theta}(y)$, so it achieves its unique minimum at the point where its derivative with respect to $\hat{\theta}(y)$ is zero. On differentiating (IV.B.7) we have that the Bayes estimate, denoted by $\hat{\theta}_{MMSE}$, is given by

$$\hat{\theta}_{MMSE}(y) = E\{\Theta|Y=y\}. \qquad \text{(IV.B.8)}$$

Thus the MMSE estimate of Θ given $Y=y$ is the conditional mean of Θ given $Y=y$. This is a very basic result to which we will return in subsequent chapters. This estimate is sometimes termed the *conditional mean estimate* (CME).

Case IV.B.2: Minimum-Mean-Absolute-Error (MMAE) Estimation

Another cost function that is sometimes applied in the case $\Lambda=\mathbb{R}$ is the *absolute error*, given by

$$C[a,\theta]=|a-\theta|,\ (a,\theta)\in\mathbb{R}^2 \qquad \text{(IV.B.9)}$$

The Bayes risk here is $E\{|\hat{\theta}(Y)-\Theta|\}$, a quantity known as the *mean-absolute-error*, so the corresponding Bayes estimate is known as the *minimum-mean-absolute-error* (MMAE) *estimate*.

To derive the MMAE estimate we make use of the fact that if X is a random variable with $P(X\geqslant 0)=1$, then $E\{X\}=\int_0^\infty P(X>x)dx$. This result follows essentially by integrating by parts (see, e.g., Breiman (1968)).

Since $|\hat{\theta}(y)-\Theta|\geqslant 0$, we have from the result above that

$$E\{|\hat{\theta}(y) - \Theta||Y = y\}$$

$$= \int_0^\infty P(|\hat{\theta}(y) - \Theta| > x\,|Y = y)dx$$

$$= \int_0^\infty P(\Theta > x + \hat{\theta}(y)|Y = y)dx \qquad \text{(IV.B.10)}$$

$$+ \int_0^\infty P(\Theta < -x + \hat{\theta}(y)|Y = y)dx\,.$$

Substituting $t = x + \hat{\theta}(y)$ in the first integral and $t = -x + \hat{\theta}(y)$ in the second integral on the right of (IV.B.10), we have

$$E\{|\hat{\theta}(y) - \Theta||Y = y\} = \int_{\hat{\theta}(y)}^{\infty} P(\Theta > t\,|Y = y)dt \qquad \text{(IV.B.11)}$$

$$+ \int_{-\infty}^{\hat{\theta}(y)} P(\Theta < t\,|Y = y)dt.$$

With $E\{|\hat{\theta}(y) - \Theta||Y = y\}$ in the form (IV.B.11) we see that it is a differentiable function of $\hat{\theta}(y)$. On differentiating we get

$$\frac{\partial}{\partial\hat{\theta}(y)} E\{|\hat{\theta}(y) - \Theta||Y = y\} = P(\Theta < \hat{\theta}(y)|Y = y) \qquad \text{(IV.B.12)}$$

$$- P(\Theta > \hat{\theta}(y)|Y = y).$$

From (IV.B.12) we note that this derivative is a nondecreasing function of $\hat{\theta}(y)$ that approaches -1 as $\hat{\theta}(y) \to -\infty$ and +1 as $\hat{\theta}(y) \to +\infty$. Thus $E\{|\hat{\theta}(y) - \Theta||Y = y\}$ achieves its minimum over $\hat{\theta}(y)$ at the point (or on the set of points) where its derivative changes sign. That is, the Bayes estimate in this case, denoted by $\hat{\theta}_{ABS}(y)$, is any point such that

$$P(\Theta<t\,|Y=y)\leqslant P(\Theta>t\,|Y=y),\, t<\hat{\theta}_{ABS}(y)$$

and (IV.B.13)

$$P(\Theta<t\,|Y=y)\geqslant P(\Theta>t\,|Y=y),\, t>\hat{\theta}_{ABS}(y).$$

Note that a point $\hat{\theta}_{ABS}(y)$ satisfying (IV.B.13) is a *median* of the conditional distribution of Θ given $Y=y$. Thus the MMAE estimate is a *conditional median estimate*. This estimate coincides with the MMSE estimate only when the distribution of Θ given $Y=y$ has the same value for its mean and median. Which of these two is the "better" estimate of Θ depends, of course, on which criterion one adopts.

Case IV.B.3: Maximum A Posteriori Probability (MAP) Estimation

Another estimation method which, although not properly a Bayes estimate, fits within the Bayesian framework is maximum *a posteriori* probability (MAP) estimation.

To motivate this method, we assume the case $\Lambda=\mathbb{R}$ and consider the so-called *uniform cost function*,

$$C[a,\theta]=\begin{cases}0 \text{ if } |a-\theta|\leqslant\Delta\\ 1 \text{ if } |a-\theta|>\Delta,\end{cases} \qquad \text{(IV.B.14)}$$

where $\Delta>0$. For an estimator $\hat{\theta}$ the average posterior cost given $Y=y$ in this case is given by

$$E\{C[\hat{\theta}(y),\Theta]|Y=y\}=P(|\hat{\theta}(y)-\Theta|>\Delta|Y=y)$$

(IV.B.15)

$$=1-P(|\hat{\theta}(y)-\Theta|\leqslant\Delta|Y=y).$$

In considering the minimization of (IV.B.15) suppose first that Θ is a discrete random variable taking values in a finite set $\Lambda=\{\theta_0,\ldots,\theta_{M-1}\}$ with $|\theta_i-\theta_j|>\Delta$ for $i\neq j$. Then we have

$$E\{C[\hat{\theta}(y),\Theta]|Y=y\} = 1-P(\Theta=\hat{\theta}(y)|Y=y) \tag{IV.B.16}$$
$$= 1-w(\hat{\theta}(y)|y) \text{ for } \hat{\theta}(y)\in\Lambda,$$

where $w(\theta|y)$ is the conditional probability mass function of Θ given $Y=y$. We see from (IV.B.16) that the Bayes estimate in this case is given for each $y\in\Gamma$ by any value of θ that maximizes $w(\theta|y)$ over $\theta\in\Lambda$. That is, the Bayes estimate is the value of Θ that has the maximum *a posteriori* probability of occurring given $Y=y$.†

Now suppose that $\Lambda=\mathbb{R}$ and Θ is a continuous random variable with conditional density function $w(\theta|y)$ given $Y=y$. In this case the posterior risk becomes

$$E\{C[\hat{\theta}(y),\Theta]|Y=y\} = 1 - \int_{\hat{\theta}(y)-\Delta}^{\hat{\theta}(y)+\Delta} w(\theta|y)d\theta. \tag{IV.B.17}$$

The quantity in (IV.B.17) is minimized over $\hat{\theta}(y)$ by maximizing the area under $w(\theta|y)$ over the interval $(\hat{\theta}(y)-\Delta, \hat{\theta}(y)+\Delta)$. Referring to Fig. IV.B.1, we see that if $w(\theta|y)$ is a smooth function of θ and if Δ is sufficiently small, this area will be approximately maximized by choosing $\hat{\theta}(y)$ to be a point of maximum of $w(\theta|y)$. That is, for small Δ and smooth $w(\theta|y)$, we have

$$\int_{\hat{\theta}(y)-\Delta}^{\hat{\theta}(y)+\Delta} w(\theta|y)d\theta \cong 2\Delta w(\theta|y)\Big|_{\theta=\hat{\theta}(y)}, \tag{IV.B.18}$$

and the right-hand side is maximized by choosing $\hat{\theta}(y)$ to be the value of θ maximizing $w(\theta|y)$ over Λ.

In either of the cases above, the uniform cost criterion leads to the procedure for estimating Θ as that value maximizing the *a posteriori* (discrete or continuous) density $w(\theta|y)$. [Similarly, with θ discrete but taking on infinitely many values, it can be argued that (IV.B.15) is minimized

† Of course, this case in which Λ is a finite set is simply an M-ary hypothesis testing problem, and the cost criterion of (IV.B.14) reduces here to $C[a,\theta]=1$ if $a\neq\theta$ and $C[a,\theta]=0$ if $a=\theta$, since $|\theta_i-\theta_j|>\Delta$ for $i\neq j$. The Bayes estimate in this case is thus the M-ary Bayes decision rule for uniform cost (as in Exercise 15 of Chapter II.)

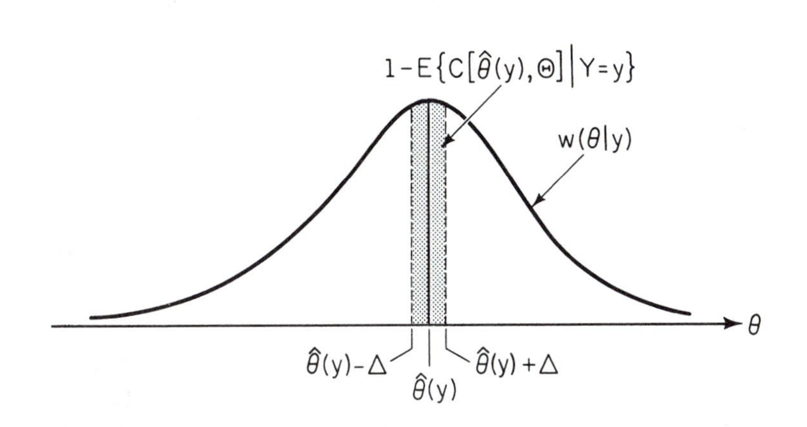

Fig. IV.B.1: Illustration of MAP estimation.

approximately by choosing $\hat{\theta}(y)$ to maximize the conditional mass function $w(\theta|y)$.] This estimate is known as the *maximum a posteriori probability* (MAP) *estimate* and is denoted by $\hat{\theta}_{MAP}$. Although this estimate often only approximates the Bayes estimate for uniform cost with small Δ, the MAP criterion is widely used to design estimates. A principal reason for this is that MAP estimates are often easier to compute than MMSE, MMAE, or other estimates.

Note that a point at which a density achieves its maximum value is termed a *mode* of the corresponding probability distribution. Thus since $\hat{\theta}_{MAP}$ estimates Θ by the mode of its conditional distribution, it is a *conditional mode estimate.*

From the Cases IV.B.1 through IV.B.3 [and from (IV.B.4)] we see that Bayes estimates for a given situation are determined from the conditional distribution of the parameter given the observations. In particular, the MMSE, MMAE, and MAP estimates are the mean, median, and mode of this distribution, respectively.

In modeling a given statistical situation we usually start with the family $\{P_\theta; \theta \epsilon \Lambda\}$ of conditional distributions of Y given $\Theta=\theta$, and for the Bayesian formulation we also have a prior distribution for Θ. To obtain the conditional distribution of Θ given Y from the prior and the conditional of Y given Θ we only need to apply Bayes' formula. In particular, supposing that P_θ has density p_θ for each $\theta \epsilon \Lambda$ and that the prior distribution of Θ has density $w(\theta)$, we have that the conditional distribution of Θ given $Y=y$ has density

$$w(\theta|y) = \frac{p_\theta(y)w(\theta)}{\int_\Lambda p_\theta(y)w(\theta)\mu(d\theta)}. \qquad \text{(IV.B.19)}$$

Note that the denominator of (IV.B.19) is $p(y)$, the unconditioned density of Y.

The Bayes estimates for the three cases above can be obtained straightforwardly from (IV.B.19). Note that the MAP estimate can be obtained without the computation of $p(y)$ since this term will not affect the maximization over θ. That is, $\hat{\theta}_{MAP}(y)$ is found by maximizing $p_\theta(y)w(\theta)$ over $\theta \epsilon \Lambda$. Since the logarithm is an increasing function, $\hat{\theta}_{MAP}(y)$ also maximizes $[log p_\theta(y) + \log w(\theta)]$ over $\theta \epsilon \Lambda$. If Θ is a continuous random variable given $Y=y$, then for sufficiently smooth p_θ and w, a necessary condition for this maximization is

$$\frac{\partial}{\partial\theta} \log p_\theta(y)\Big|_{\theta=\hat{\theta}_{MAP}(y)} = -\frac{\partial}{\partial\theta} \log w(\theta)\Big|_{\theta=\hat{\theta}_{MAP}(y)}. \qquad \text{(IV.B.20)}$$

Equation (IV.B.20) is known as the *MAP equation*.

The following two examples serve to illustrate the computation of the MMSE, MMAE, and MAP estimates.

Example IV.B.1: Estimation of the Parameter of an Exponential Distribution

Consider the situation $\Lambda=(0,\infty)$ and $\Gamma=\mathbb{R}$, in which the observations have the following conditional probability density function given $\Theta=\theta$:

$$p_\theta(y) = \begin{cases} \theta e^{-\theta y} & \text{if } y \geq 0 \\ 0 & \text{if } y < 0. \end{cases} \qquad \text{(IV.B.21)}$$

This is the *exponential density* with parameter θ. The exponential density models many physical phenomena. It is particularly useful in modeling the time intervals between successive events occurring randomly in time, such as messages or data packets arriving at a communications switching station, vehicles arriving at an intersection of roads, photons emitting from a coherent light source, or devices failing in a logic circuit. The parameter θ in this model can be interpreted as the rate of such occurrences, and thus we can think of the estimation problem here as that of estimating the rate of occurrences of such events from an observation of the time between successive occurrences of them.

Suppose that our prior information about Θ is that it also has an exponential distribution with density

$$w(\theta) = \begin{cases} \alpha e^{-\alpha\theta} & \text{if } \theta \geq 0 \\ 0 & \text{if } \theta < 0, \end{cases} \tag{IV.B.22}$$

where $\alpha > 0$ is known. We can then find the posterior distribution of Θ given $Y = y$ from (IV.B.19). We have

$$\begin{aligned} w(\theta|y) &= \frac{\alpha\theta e^{-(\alpha+y)\theta}}{\int_0^\infty \alpha\theta e^{-(\alpha+y)\theta} d\theta} \\ &= (\alpha + y)^2 \theta e^{-\theta(\alpha + y)}, \end{aligned} \tag{IV.B.23}$$

for $\theta \geq 0$ and $y \geq 0$, and $w(\theta|y) = 0$ otherwise.

The MMSE estimate is the mean of (IV.B.23) and thus is given by

$$\hat{\theta}_{MMSE}(y) = \int_0^\infty \theta w(\theta|y)d\theta = (\alpha + y)^2 \int_0^\infty \theta^2 e^{-\theta(\alpha+y)} d\theta \qquad \text{(IV.B.24)}$$

$$= \frac{2}{\alpha + y}.$$

Note that for fixed α, this estimate of Θ varies inversely with y. This is intuitively reasonable from the foregoing interpretation of the exponential model since a large interarrival time (large y) would be evidence a low rate (small θ). This behavior is tempered to a degree depending on the value of α since the estimate is never greater than $2/\alpha$. Note that a small value of α implies that Θ is distributed diffusely [i.e., $w(\theta)$ is relatively spread out] and the corresponding estimate allows larger values of Θ if implied by the observation. Alternatively, a large value of α implies that Θ is close to zero with high probability, so the estimate is never large in this case.

The minimum value of the MSE can be computed straightforwardly in this case. First, we note from (IV.B.3) that the Bayes risk is the average of the posterior cost, so that

$$MMSE = r(\hat{\theta}_{MMSE}) = E\{E\{(\hat{\theta}_{MMSE}(Y) - \Theta)^2|Y\}\}$$

$$= E\{E\{(\Theta - E\{\Theta|Y\})^2|Y\}\} \qquad \text{(IV.B.25)}$$

$$= E\{\,Var(\Theta|Y)\}.$$

Thus the minimum MSE is the average of the conditional variance of Θ given Y. Since

$$Var\,(\Theta|Y = y) = E\{\Theta^2|Y = y\} - E^2\{\Theta|Y = y\},$$

we have

$$Var\ (\Theta|Y=y) = \int_0^\infty \theta^2 w(\theta|y)d\,\theta - [\hat{\theta}_{MMSE}(y)]^2$$

$$= (\alpha+y)^2 \int_0^\infty \theta^3 e^{-\theta(\alpha+y)} d\,\theta - \frac{4}{(\alpha+y)^2}$$

$$= \frac{2}{(\alpha+y)^2}.$$

Thus

$$MMSE = E\{\frac{2}{(\alpha+Y)^2}\} = \int_0^\infty \frac{2}{(\alpha+y)^2} p(y)dy$$

$$= \int_0^\infty \frac{2\alpha}{(\alpha+y)^4} dy \qquad \text{(IV.B.26)}$$

$$= \frac{2}{\alpha^2},$$

where we have used $p(y) = \int_0^\infty \alpha\theta e^{-(\alpha+y)\theta} d\,\theta = \alpha/(\alpha+y)^2$, as in (IV.B.23).

The MMAE estimate, $\hat{\theta}_{ABS}(y)$, is the median of $w(\theta|y)$. Since Θ is continuous given $Y=y$, we can find $\hat{\theta}_{ABS}(y)$ by solving the equation

$$\int_{\hat{\theta}_{ABS}(y)}^\infty w(\theta|y)d\,\theta = \tfrac{1}{2}. \qquad \text{(IV.B.27)}$$

Inserting (IV.B.23) and integrating yields

$$(1 + (\alpha+y)\hat{\theta}_{ABS}(y))e^{(\alpha+y)\hat{\theta}_{ABS}(y)} = \tfrac{1}{2}, \qquad \text{(IV.B.28)}$$

so that we have

$$\hat{\theta}_{ABS}(y) = \frac{T_o}{\alpha + y}, \tag{IV.B.29}$$

with T_o being the solution to $(1+T_o)e^{-T_o}=½$, which is given by $T_o \cong 1.68$. Comparing (IV.B.29) with (IV.B.24), we see the same general behavior as the MMSE estimate, these differing only in the constant in the numerator. The minimum Bayes risk for this situation can be computed similarly to that for the MMSE estimate, and this computation is left as an exercise.

The MAP estimate of Θ can also be obtained easily in this case. Noting that

$$\frac{\partial}{\partial\theta}[\log p_\theta(y) + \log w(\theta)] = \frac{\partial}{\partial\theta}(\log\theta - \theta y + \log\alpha - \alpha\theta)$$

$$= \theta^{-1} - (\alpha + y)$$

and

$$\frac{\partial^2}{\partial\theta^2}[\log p_\theta(y) + \log w(\theta)] = -\theta^{-2} < 0,$$

we see that $w(\theta|y)$ has its unique maximum at

$$\hat{\theta}_{MAP}(y) = \frac{1}{\alpha + y}. \tag{IV.B.30}$$

Thus we again get an estimate differing from the MMSE estimate by only a scale factor.

In Example IV.B.1 the three estimation criteria considered lead to three different estimators for Θ. To decide which one to use, one must decide which of the three corresponding cost functions penalizes the estimation error in the way most suitable for the application of interest. In many problems of interest one does not need to make such a choice because the three estimates coincide. The following is an example of such a situation.

Example IV.B.2: Estimation of Signal Amplitude

Consider the case $\Gamma=\mathbb{R}^n$ and $\Lambda=\mathbb{R}$ with

$$Y_k = N_k + \Theta s_k , \quad k = 1, ..., n , \tag{IV.B.31}$$

where $\underline{N} \sim N(0, \Sigma), \underline{s}$ is known, $\Theta \sim N(\mu, v^2)$, and $\underline{N}$ and Θ are independent. Note that this problem corresponds to the estimation of the unknown amplitude of an otherwise known signal observed in the presence of additive noise.

Given $\Theta=\theta$, we have that $\underline{Y} \sim N(\theta \underline{s}, \Sigma)$. Thus the posterior density for Θ is

$$w(\theta | \underline{y}) =$$

$$\frac{\frac{1}{(2\pi)^{n/2}|\Sigma|^{1/2}} e^{-(\underline{y}-\theta\underline{s})^T \Sigma^{-1} (\underline{y}-\theta\underline{s})} \frac{1}{\sqrt{2\pi} v} e^{-(\theta-\mu)^2/2v^2}}{\int_{-\infty}^{\infty} \frac{1}{(2\pi)^{n/2}|\Sigma|^{1/2}} e^{-(\underline{y}-\theta\underline{s})^T \Sigma^{-1} (\underline{y}-\theta\underline{s})} \frac{1}{\sqrt{2\pi} v} e^{-(\theta-\mu)^2/2v^2} d\theta} \tag{IV.B.32}$$

$$= K(\underline{y}) \exp\left\{ -\frac{\theta^2}{2}(d^2 + 1/v^2) + \theta(\underline{s}^T \Sigma^{-1} \underline{y} + \frac{\mu}{v^2}) \right\},$$

where as in Chapter III we define $d^2 = \underline{s}^T \Sigma^{-1} \underline{s}$ and where $K(\underline{y})$ is a function depending on $\underline{y}$ but not on θ. Note from (IV.B.32) that $w(\theta | \underline{y})$ is the exponential of a quadratic term in θ, so that it must be a Gaussian density. If $w(\theta | \underline{y})$ were $N(m, q^2)$, we would have

$$\begin{aligned} w(\theta | \underline{y}) &= \frac{1}{\sqrt{2\pi} q} e^{-(\theta-m)^2/2q^2} \\ &= \frac{e^{-m^2/2q^2}}{\sqrt{2\pi} q} e^{-\frac{\theta^2}{2} q^{2} + \theta m q^{-2}} . \end{aligned} \tag{IV.B.33}$$

Comparing (IV.B.32) with (IV.B.33), we see that given $\underline{Y} = \underline{y}$, $\Theta \sim N(m, q^2)$ with

$$q^2 = (d^2 + 1/v^2)^{-1}$$

and

$$m = (d^2 + 1/v^2)^{-1}(\underline{s}^T \Sigma^{-1} \underline{y} + \mu/v^2),$$

and $K(\underline{y})$ becomes $e^{-m^2/2q^2}/\sqrt{2\pi}q$.

Since the first parameter of the Gaussian density is its mean, we immediately have that the conditional mean estimate of Θ is

$$\hat{\theta}_{MMSE}(\underline{y}) = \frac{\underline{s}^T \Sigma^{-1} \underline{y} + \mu/v^2}{d^2 + 1/v^2}$$
$$= \frac{v^2 d^2 \hat{\theta}_1(\underline{y}) + \mu}{v^2 d^2 + 1}, \quad \text{(IV.B.34)}$$

where $\hat{\theta}_1(\underline{y}) \triangleq \underline{s}^T \Sigma^{-1} \underline{y} / d^2$. Moreover, the minimum-mean-squared-error is

$$MMSE = E\{ Var(\Theta | \underline{Y})\} = \frac{1}{d^2 + 1/v^2} = \frac{v^2}{v^2 d^2 + 1}, \quad \text{(IV.B.35)}$$

since $Var(\Theta|\underline{Y}) = (d^2 + 1/v^2)^{-1}$ which does not depend on $\underline{Y}$. Also, since the Gaussian density is symmetric about its mean and it achieves its maximum at its mean, the conditional median and conditional mode both equal the conditional mean; i.e., we have $\hat{\theta}_{ABS} = \hat{\theta}_{MAP} = \hat{\theta}_{MMSE}$.

The behavior of this estimate well illustrates the nature of Bayesian estimation. Note that v^2 determines the accuracy of our prior knowledge about Θ; that is, the smaller v^2 is, the more accurately we know θ in the absence of observations. On the other hand, in view of the discussion of coherent detection in Gaussian noise in Chapter III, the quantity d^2 is a measure of the quality with which $\underline{s}$ can be distinguished from the $N(0, \Sigma)$ noise. That is, d^2 is a measure of the accuracy of our observations in terms of producing information about the signal - large d^2 corresponds to high-quality observations and small d^2 to low-quality observations in this sense.

With these ideas in mind, consider the estimate $\hat{\theta}_{MMSE}$ of (IV.B.34). If v^2d^2 is very small relative to the other quantities in this estimate, we have $\hat{\theta}_{MMSE}(\underline{y}) \cong \mu$. This occurs when the prior knowledge is very accurate relative to the observations (i.e., v^2 is small relative to $1/d^2$), so the estimator ignores the observations and chooses the mean of the prior distribution as its estimate. Note that the MMSE in this case is approximately v^2, the prior variance. On the other hand, if v^2d^2 is large, then $\hat{\theta}_{MMSE}(\underline{y}) \cong \hat{\theta}_1(\underline{y})$, an estimate that depends only on the observations and does not incorporate the prior information at all. The latter situation is also reasonable since with v^2 large relative to $1/d^2$, we are better off trusting the observations rather than the prior information. The MMSE in the latter case is approximately $1/d^2$.† Between these two extremes the optimum estimator balances the prior knowledge and the observations, and the corresponding MMSE reflects this balance.

It is interesting to consider the particular case $\mathbf{\Sigma}=\sigma^2\mathbf{I}$ and $\underline{s}=\underline{1} \triangleq (1, 1, ..., 1)^T$ in the context of the discussion above. In this case our observations are

$$Y_k = N_k + \Theta, \quad k = 1, ..., n,$$

with $N_1, ..., N_n$ i.i.d. $N(0, \sigma^2)$. The quantity $v^2d^2 = nv^2/\sigma^2$ and $\hat{\theta}_1(\underline{y}) = \bar{y} \triangleq (1/n)\sum_{k=1}^n y_k$, the sample mean. If we have no observations ($n=0$), we simply estimate Θ as its prior mean μ, but as we take more observations (increase n) the sample mean $\bar{y}$ becomes more reliable and we place more weight on it. In the limit as $n \to \infty$ we disregard the prior mean entirely and adopt the sample mean as our estimate. The scale of this behavior is controlled by the ratio v^2/σ^2 (note that σ^2 determines the accuracy of each observation).

In the discussion above we have concentrated on the estimation of a single real parameter. However, in many

† Note that in the absence of any observations the MMSE estimate is μ and the MMSE is v^2, which corresponds to the approximate conditions for v^2d^2 small. It turns out (as we shall see in the following sections) that the estimate $\hat{\theta}_1(\underline{y})$ and the accuracy $1/d^2$ are optimum in the absence of any prior information. Thus these two extremes are quite reasonable.

problems arising in practice we wish to estimate several parameters simultaneously. The Bayesian formulation, of course, applies equally well to the vector-parameter situation, and in the following discussion we treat this case.

Case IV.B.4: Estimation of Vector Parameters

We consider now the case in which $\Lambda = \mathbb{R}^m$. To follow the Bayesian procedure for designing an estimate of $\underline{\Theta}$ we must specify a cost function $C : \mathbb{R}^m \times \mathbb{R}^m \to \mathbb{R}$. It is sometimes meaningful to use a cost function of the form

$$C[\underline{a}, \underline{\theta}] = \sum_{i=1}^{m} C_i[a_i, \theta_i], \tag{IV.B.36}$$

where C_i is a cost function associated with the estimation of the $i^{\underline{th}}$ component of the parameter. If we have a cost function of this form, the conditional posterior cost for an estimate $\underline{\hat{\theta}}$ is given by

$$E\{C[\underline{\hat{\theta}}(y), \underline{\Theta}] | Y = y\}$$
$$= \sum_{i=1}^{m} E\{C_i[\hat{\theta}_i(y), \Theta_i] | Y = y\}, \tag{IV.B.37}$$

so that we essentially have m scalar estimation problems to solve. That is, $\hat{\theta}_i(y)$ [the $i^{\underline{th}}$ component $\hat{\theta}(y)$] is chosen to minimize $E\{C_i[\hat{\theta}_i(y), \Theta_i] | Y = y\}$.

An example of a useful cost function that decomposes as in (IV.B.36) is the square of the Euclidean norm of the error:

$$C[\underline{a}, \underline{\theta}] = \|\underline{a} - \underline{\theta}\|^2 = \sum_{i=1}^{m} (a_i - \theta_i)^2. \tag{IV.B.38}$$

It follows from Case IV.B.1 that for this cost function the $i^{\underline{th}}$ component of the Bayes estimate is $E\{\Theta_i | Y = y\}$; i.e., the Bayes estimate is

$$\hat{\underline{\theta}}_B(y) = E\{\underline{\Theta}|Y = y\}, \tag{IV.B.39}$$

the conditional mean of $\underline{\Theta}$ given $Y=y$.

Another example of a cost function satisfying (IV.B.36) is the following

$$C[\underline{a}, \underline{\theta}] = \sum_{i=1}^{m} |a_i - \theta_i|. \tag{IV.B.40}$$

This function provides an alternative to $\|\underline{a} - \underline{\theta}\|$ as a measure of the distance between $\underline{a}$ and $\underline{\theta}$. From Case IV.B.2 we see that this cost function leads to the estimate whose $i^{\underline{th}}$ component is the conditional median of Θ_i given $Y=y$.

To extend the concept of MAP estimation to vector parameters we might consider a cost function of the form (IV.B.36), in which $C_i[a_i, \theta_i]$ is the uniform cost function of (IV.B.14). This leads to the vector estimator that has as its $i^{\underline{th}}$ component the conditional mode of Θ_i given $Y=y$. However, this decomposed cost function is not the most meaningful extension of the uniform cost function to the vector case. A more meaningful one is

$$C[\underline{a}, \underline{\theta}] = \begin{cases} 1 \text{ if } \max_{1 \leqslant i \leqslant m} |a_i - \theta_i| > \Delta \\ 0 \text{ if } \max_{1 \leqslant i \leqslant m} |a_i - \theta_i| \leqslant \Delta, \end{cases} \tag{IV.B.41}$$

for which we have

$$E\{C[\hat{\underline{\theta}}(y), \underline{\Theta}]|Y = y\} = 1 - P(|\hat{\theta}_1(Y) - \Theta_1| \leqslant \Delta, \ldots, |\hat{\theta}_m(Y) - \Theta_m| \leqslant \Delta | Y = y). \tag{IV.B.42}$$

From (IV.B.42) we can argue the approximate optimality of estimating $\underline{\Theta}$ as its conditional mode given $Y=y$, a quantity that differs in general from the vector whose $i^{\underline{th}}$ component is the conditional mode of Θ_i given $Y=y$ obtained from decomposing the cost. The estimate that chooses the conditional mode of $\underline{\Theta}$ given $Y=y$ is the MAP estimate for the vector-parameter case. Note that the region where $\max_{1 \leq i \leq m} |a_i - \theta_i| \leq \Delta$ is an m-dimensional cube centered at $\underline{\theta}$ with side length 2Δ. We could define similar cost functions by replacing this cube with other shapes (e.g., an m-dimensional ball, $\|\underline{a}-\underline{\theta}\| \leq \Delta$); however, the approximate optimality of the MAP estimate would still be implied within the appropriate smoothness conditions.

A further useful cost function of interest in estimating vector parameters is a generalization of the squared-error norm. In particular, it is of interest to consider cost functions of the form

$$C[\underline{a}, \underline{\theta}] = (\underline{a} - \underline{\theta})^T \mathbf{A} (\underline{a} - \underline{\theta}), \qquad \text{(IV.B.43)}$$

where $\mathbf{A}$ is a symmetric, positive-definite matrix. Note that this cost function allows for joint weightings of errors in different parameters, a desirable feature for some applications since the accuracy of our estimate of one of the real parameters forming $\underline{\Theta}$ may have an impact on how well we need to know other parameters.

To derive the Bayes estimate for (IV.B.43), we write

$$\begin{aligned} &E\{(\underline{\hat{\theta}}(y) - \underline{\Theta})^T \mathbf{A} (\underline{\hat{\theta}}(y) - \underline{\Theta}) | Y = y\} \\ &\quad = [\underline{\hat{\theta}}(y)]^T \mathbf{A} \underline{\hat{\theta}}(y) - 2[\underline{\hat{\theta}}(y)]^T \mathbf{A} E\{\underline{\Theta} | Y = y\} \qquad \text{(IV.B.44)} \\ &\qquad + E\{\underline{\Theta}^T \mathbf{A} \underline{\Theta} | Y = y\}. \end{aligned}$$

Since the function of (IV.B.44) is quadratic in $\underline{\hat{\theta}}(y)$, it achieves its minimum at the point at which its gradient with respect to $\underline{\hat{\theta}}(y)$ vanishes. We have straightforwardly that

$$\begin{aligned} &\nabla_{\underline{\hat{\theta}}(y)} E\{C[\underline{\hat{\theta}}(y), \underline{\Theta}] | Y = y\} \\ &\quad = 2\mathbf{A} \underline{\hat{\theta}}(y) - 2\mathbf{A} E\{\underline{\Theta} | Y = y\}. \end{aligned} \qquad \text{(IV.B.45)}$$

So the Bayes estimate, $\hat{\underline{\theta}}_B$, for (IV.B.43) satisfies

$$2A\hat{\underline{\theta}}_B(y) = 2A\,E\{\underline{\Theta}|Y = y\}. \qquad \text{(IV.B.46)}$$

Premultiplying (IV.B.46) by $(½)A^{-1}$ yields that $\hat{\underline{\theta}}_B(y)=E\{\underline{\Theta}|Y=y\}$.

Thus we see that the quadratic cost criterion of (IV.B.43) yields the conditional mean vector as a Bayes estimate regardless of choice of A. The resulting Bayes risk of course does depend on A and it is straightforward to show (see Exercise 10) that for this case

$$r(\hat{\underline{\theta}}_B) = tr\{A\,E\{Cov\,(\underline{\Theta}|Y)\}\}, \qquad \text{(IV.B.47)}$$

where $tr\{\cdot\}$ denotes the trace operator (i.e., summation of the diagonal terms) and where $Cov\,(\underline{\Theta}|Y)$ is the conditional covariance matrix of $\underline{\Theta}$ given $Y=y$. Note that the squared error norm is the special case of (IV.B.47) with $A=I$, so in the latter case $r(\hat{\underline{\theta}}_B)$ is simply the trace of $E\{Cov\,(\underline{\Theta}|Y)\}$.

Example IV.B.3: Estimation of a Gaussian Vector from a Jointly Gaussian Observation

Consider the situation in which $\Gamma=\mathbb{R}^n$, $\Lambda=\mathbb{R}^m$, and Y and $\underline{\Theta}$ are jointly Gaussian with mean vectors $\underline{\mu}_Y$ and $\underline{\mu}_\Theta$, covariance matrices Σ_Y and Σ_Θ, and cross-covariance matrix $\Sigma_{Y\Theta} \triangleq E\{(\underline{Y}-\underline{\mu}_Y)(\underline{\Theta}-\underline{\mu}_\theta)^T\}$; that is, we assume that

$$\begin{pmatrix} \underline{Y} \\ \underline{\Theta} \end{pmatrix} \sim N\left(\begin{pmatrix} \underline{\mu}_Y \\ \underline{\mu}_\Theta \end{pmatrix}, \begin{pmatrix} \Sigma_Y & \Sigma_{Y\Theta} \\ \Sigma_{\Theta Y} & \Sigma_\Theta \end{pmatrix}\right) \qquad \text{(IV.B.48)}$$

with $\Sigma_{\Theta Y}=\Sigma_{Y\Theta}^T$.

Within this model it is straightforward to show that the conditional distribution of $\underline{\Theta}$ given $\underline{Y}=\underline{y}$ is also Gaussian, with conditional mean $\hat{\underline{\mu}}(y)$ given by

$$\underline{\hat{\mu}}(\underline{y}) = \underline{\mu}_{\Theta} + \Sigma_{\Theta Y} \Sigma_Y^{-1} (\underline{y} - \underline{\mu}_Y) \qquad \text{(IV.B.49a)}$$

and with conditional covariance matrix $\hat{\Sigma}$ given by

$$\hat{\Sigma} = \Sigma_{\theta} - \Sigma_{\Theta Y} \Sigma_Y^{-1} \Sigma_{Y\Theta}. \qquad \text{(IV.B.49b)}$$

From this property we can find all of the optimum estimates discussed in Case IV.B.4. In particular, we note immediately that the conditional-mean estimate is equal to $\underline{\hat{\mu}}(\underline{y})$ of (IV.B.49a). Also, since the multivariate Gaussian density has its mode at its mean, the MAP estimate is given by $\underline{\hat{\mu}}(\underline{y})$ as well. Moreover, since $\underline{\Theta}$ being Gaussian given $\underline{Y}=\underline{y}$ implies that Θ_i is marginally Gaussian given $\underline{Y}=\underline{y}$, the marginal mode and median of Θ_i given $\underline{Y}=\underline{y}$ occur at $\hat{\mu}_i(\underline{y})$, the ith component of $\underline{\hat{\mu}}(\underline{y})$. Thus $\underline{\hat{\mu}}(\underline{y})$ provides the optimum estimate in all the senses discussed under Case IV.B.4. It should be noted that this estimate is linear (or, more properly, *affine*) in $\underline{y}$, so that it is easily computed if Σ_Y^{-1} can be determined efficiently. We will comment further on this issue later.

The minimum Bayes risk can also be computed easily for the quadratic cost function of (IV.B.43) via (IV.B.47). In particular we note that $Cov(\underline{\Theta}|\underline{Y})=\hat{\Sigma}$, which does not depend on $\underline{Y}$. Thus $E\{Cov(\underline{\Theta}|\underline{Y})\}=\hat{\Sigma}$ and the minimum Bayes risk becomes

$$r(\underline{\hat{\theta}}_B) = tr\{\mathbf{A}\,\hat{\Sigma}\} = tr\{\mathbf{A}\,\Sigma_{\theta}\} - tr\{\mathbf{A}\,\Sigma_{\Theta Y} \Sigma_Y^{-1} \Sigma_{Y\Theta}\}. \qquad \text{(IV.B.50)}$$

Note also that $\hat{\Sigma}=E\{(\underline{\Theta}-\underline{\hat{\theta}}_B(\underline{Y}))(\underline{\Theta}-\underline{\hat{\theta}}_B(\underline{Y}))^T\}$, so that $\hat{\Sigma}$ is the covariance matrix of the estimation error, $\underline{\Theta}-\underline{\hat{\theta}}_B(\underline{Y})$.

A special case of interest of this general Gaussian problem arises from the so-called *linear observation model:*

$$\underline{Y} = \mathbf{H}\,\underline{\Theta} + \underline{N}, \qquad \text{(IV.B.51)}$$

where $\underline{\Theta} \sim N(\underline{\mu}_{\theta}, \Sigma_{\Theta})$, $\underline{N} \sim N(\underline{0}, \Sigma)$, $\mathbf{H}$ is a fixed $n \times m$ matrix, and $\underline{\Theta}$ and $\underline{N}$ are independent. Such models arise in many applications. For example, the model of Example IV.B.2 in which we wish to estimate signal amplitude is of this form with $m=1$ and $\mathbf{H}=\underline{s}$. Furthermore, if we think of $\Theta_1, \ldots, \Theta_m$

as being samples of a stochastic signal, then

$$Y_k = \sum_{j=1}^{m} h_{k,j}\,\Theta_j + N_k, \quad k = 1, ..., n \qquad \text{(IV.B.52)}$$

is an observation sequence consisting of linearly filtered signal plus additive noise - a situation arising, for example, when a signal is observed through a channel with finite bandwidth or other linearly distorting characteristic. In this case the estimation of $\underline{\Theta}$ is known as the problem of *equalizing* the channel. A further applications of the model is discussed in Chapter V in the context of Kalman-Bucy filtering.

In this model it is straightforward to show that $\underline{Y}$ and $\underline{\Theta}$ are jointly Gaussian with $\underline{\mu}_\Theta$ and $\mathbf{\Sigma}_\Theta$ given, $\underline{\mu}_Y = \mathbf{H}\,\underline{\mu}_\theta$, $\mathbf{\Sigma}_Y = \mathbf{H}\,\mathbf{\Sigma}_\theta \mathbf{H}^T + \mathbf{\Sigma}$, and $\mathbf{\Sigma}_{\Theta Y} = \mathbf{\Sigma}_\theta \mathbf{H}^T$. Thus we get the Bayes estimate

$$\hat{\underline{\mu}}(\underline{y}) = \underline{\mu}_\theta + \mathbf{\Sigma}_\Theta \mathbf{H}^T (\mathbf{H}\,\mathbf{\Sigma}_\theta \mathbf{H}^T + \mathbf{\Sigma})^{-1} (\underline{y} - \mathbf{H}\,\underline{\mu}_\theta) \qquad \text{(IV.B.53)}$$

and the error covariance matrix

$$\hat{\mathbf{\Sigma}} = \mathbf{\Sigma}_\theta - \mathbf{\Sigma}_\theta \mathbf{H}^T (\mathbf{H}\,\mathbf{\Sigma}_\theta \mathbf{H}^T + \mathbf{\Sigma})^{-1} \mathbf{H}\,\mathbf{\Sigma}_\theta. \qquad \text{(IV.B.54)}$$

With regard to the computation of (IV.B.53) we note that it involves the inversion of an $n \times n$ matrix, a computation whose complexity is of the order of n^3 unless the matrix has some special structure. This computational complexity can sometimes be reduced by making use of the following simple matrix identity.

$$\begin{aligned} &\mathbf{\Sigma}_\theta \mathbf{H}^T (\mathbf{H}\,\mathbf{\Sigma}_\theta \mathbf{H}^T + \mathbf{\Sigma})^{-1} \\ &= (\mathbf{H}^T \mathbf{\Sigma}^{-1} \mathbf{H} + \mathbf{\Sigma}_\theta^{-1}) \mathbf{H}^T \mathbf{\Sigma}^{-1}. \end{aligned} \qquad \text{(IV.B.55)}$$

If $\mathbf{\Sigma}^{-1}$ is known (e.g., if $\mathbf{\Sigma} = \sigma^2 \mathbf{I}$) and $m < n$, the matrix on the right-hand side of (IV.B.55) is easier to compute than that on the left.

In Chapter V, (IV.B.53) and (IV.B.54) will be used to derive the Kalman-Bucy filter. It is also interesting to rework Example IV.B.2 in this general context. In this case we have $m=1, \mathbf{H}=\underline{s}, \underline{\mu}_\theta=\mu$, and $\mathbf{\Sigma}_\theta=v^2$. Inserting these quantities into (IV.B.53) and (IV.B.54) and applying (IV.B.55), we get

$$\hat{\underline{\mu}}(\underline{y}) = \mu + (\underline{s}^T\mathbf{\Sigma}^{-1}\underline{s} + 1/v^2)^{-1}\underline{s}^T\mathbf{\Sigma}^{-1}(\underline{y} - \underline{s}\mu)$$

$$= \frac{v^2 d^2 \hat{\theta}_1(\underline{y}) + \mu}{v^2 d^2 + 1}$$

and

$$r(\hat{\underline{\mu}}) = \hat{\mathbf{\Sigma}} = v^2 - (\underline{s}^T\mathbf{\Sigma}^{-1}\underline{s} + v^{-2})^{-1}\underline{s}^T\mathbf{\Sigma}^{-1}\underline{s}v^2$$

$$= \frac{v^2}{v^2 d^2 + 1}$$

as in (IV.B.34) and (IV.B.35).

IV.C Nonrandom Parameter Estimation - General Structure

In Section IV.B we considered the problem of estimating a random parameter indexing a class of distributions on the observation space. A related problem is that in which we have a parameter (indexing the class of observation statistics) which is not modeled as a random variable but, nevertheless, is unknown. In particular, we may not have enough prior information about the parameter to assign a prior probability distribution to it, but yet we wish to treat the estimation of such parameters in an organized manner.

Suppose, then, that we have an observation $Y \in \Gamma$ and that the distribution of Y is a member of a class of distributions on (Γ, G) indexed by a parameter θ lying in some set Λ. As before, we denote this set of distributions by $\{P_\theta; \theta \in \Lambda\}$. Assume for now that the parameter θ is real-valued. We do not know anything about the true value of θ other than the fact that it lies in Λ, and simply stated, the problem we would like to solve is: Given the observation $Y=y$, what is the best estimate of θ? In view of the procedures developed in Section IV.B, we might begin to answer this question by seeking an estimate $\hat{\theta}(y)$ that minimizes some average performance criterion. Throughout the remainder of this chapter we consider exclusively the squared-error cost, although some results discussed here apply straightforwardly to other cost assignments as well. In the absence if a prior on Λ, the only averaging of cost that can be done is with respect to the distribution of Y given θ; i.e., we can use only the conditional risk function $R_\theta(\hat{\theta}) \triangleq E_\theta\{(\hat{\theta}(Y)-\theta)^2\}$, $\theta \epsilon \Lambda$.

As was seen in the hypothesis-testing case in Chapter II, we cannot generally expect to minimize $R_\theta(\hat{\theta})$ uniformly for $\theta \epsilon \Lambda$. This is easily seen for the squared-error cost since for any particular value of θ, say θ_o, the conditional mean-squared-error can be made zero by choosing $\hat{\theta}(y)$ to be identically θ_o for all observations $y \epsilon \Gamma$; but such an estimate would perform poorly if θ_o were not near the true value of θ. Thus it is obvious that the conditional mean-squared-error is not by itself a suitable design criterion for an estimator of a nonrandom parameter unless the class of estimators is somehow restricted to contain only reasonable estimators [e.g., to exclude estimators such as $\hat{\theta}(y) \equiv \theta_o$].

A reasonable restriction to place on an estimate of θ is that its expected value equal the true parameter value; i.e., that

$$E_\theta\{\hat{\theta}(Y)\} = \theta, \quad \theta \epsilon \Lambda. \tag{IV.C.1}$$

Such an estimate is termed *unbiased*. Within this restriction, the conditional mean-squared-error becomes the variance of the estimate under P_θ, and an unbiased estimate minimizing the mean-squared-error for each $\theta \epsilon \Lambda$ is termed a *minimum-variance unbiased estimator* (MVUE).

In this section we consider the general structure of nonrandom parameter estimation problems with a goal of characterizing MVUEs.

We begin with the concept of sufficiency, defined as follows (until otherwise noted, we now assume that Λ is general, i.e., not necessarily a subset of $\mathbb{R}$).

Definition IV.C.1: Sufficiency

Suppose that Δ is an arbitrary set and D is an event class on Δ. A function $T:(\Lambda, G) \rightarrow (\Delta, D)$ is said to be a *sufficient statistic* for $\{P_\theta; \theta \epsilon \Lambda\}$ if the distribution of Y conditioned on $T(Y)$ when $Y \sim P_\theta$ does not depend on θ for $\theta \epsilon \Lambda$. (When $\{P_\theta; \theta \epsilon \Lambda\}$ is understood, we may simply say that T *is sufficient for* θ.)

Note that θ affects the observations only through its distribution P_θ. So we can only learn about θ by viewing the statistical behavior of Y. Thus if knowing $T(Y)$ removes any further dependence on θ of the distribution of Y, we can conclude that $T(Y)$ contains all the information in Y that is useful for estimating θ. Thus the origin of the term "sufficient."

Note that any one-to-one mapping of the observations is trivially sufficient for θ, so there are always many sufficient statistics for any given estimation model. However, it is desirable to find a sufficient statistic that reduces the observations as much as possible. In this context we make the following definition.

Definition IV.C.2: Minimal Sufficiency

A function T on (Γ, G) is said to be *minimal sufficient* for $\{P_\theta; \theta \epsilon \Lambda\}$ if it is a function of every other sufficient statistic for $\{P_\theta; \theta \epsilon \Lambda\}$.

In other words, a minimal sufficient statistic represents the furthest that the observation can be reduced without destroying information about θ. Unfortunately, minimal sufficient statistics do not exist for many estimation problems, and they are often difficult to identify when they do exist.

On the other hand, it is often very easy to find useful (although not necessarily minimal) sufficient statistics by way of the following result.

Proposition IV.C.1: The Factorization Theorem

Suppose that $\{P_\theta; \theta \epsilon \Lambda\}$ has a corresponding family of densities $\{p_\theta; \theta \epsilon \Lambda\}$. A statistic T is sufficient for θ if and only if there are functions g_θ and h such that

$$p_\theta(y) = g_\theta[T(y)]h(y) \qquad \text{(IV.C.2)}$$

for all $y \epsilon \Lambda$ and $\theta \epsilon \Lambda$.

Proof: We prove this result only for the case in which Γ is discrete. This case illustrates the general idea of this proposition without introducing technicalities required for the general case. A proof of the general case can be found in Lehmann (1986).

Suppose that Γ is discrete and $\{p_\theta; \theta \epsilon \Lambda\}$ satisfies (IV.C.2) for a function T. Let $p_\theta(y|t)$ denote the density of Y given $T(Y)=t$ when $Y \sim P_\theta$. By Bayes formula we have

$$\begin{aligned} p_\theta(y|t) &\triangleq P_\theta(Y=y|T(Y)=t) \\ &= \frac{P_\theta(T(Y)=t|Y=y)P_\theta(Y=y)}{P_\theta(T(Y)=t)}. \end{aligned} \qquad \text{(IV.C.3)}$$

Since $p_\theta(T(Y)=t|Y=y)$ equals 1 if $T(y)=t$ and 0 if $T(y)\neq t$, and since $P_\theta(Y=y)=p_\theta(y)$, (IV.C.3) becomes

$$p_\theta(y|t) = \begin{cases} p_\theta(y)/P_\theta(T(Y)=t) \text{ if } T(y)=t \\ \\ 0 \text{ if } T(y)\neq t. \end{cases} \qquad \text{(IV.C.4)}$$

Now $P_\theta(T(Y)=t)=\sum_{\{y|T(y)=t\}} p_\theta(y)$. Thus from (IV.C.2), we have

$$P_\theta(T(Y)=t) = \sum_{\{y|T(Y)=t\}} g_\theta[T(y)]h(y)$$

$$= g_\theta(t) \sum_{\{y|T(y)=t\}} h(y),$$

and we also have $p_\theta(y)=g_\theta[T(y)]h(y)=g_\theta(t)h(y)$. From (IV.C.4) we then have

$$p_\theta(y|t) = \begin{cases} h(y)/\sum_{\{y|T(y)=t\}} h(y), & \text{if } T(y)=t \\ 0, & \text{if } T(y) \neq t. \end{cases}$$

Since this expression does not depend on θ, T is a sufficient statistic for $\{P_\theta; \theta \epsilon \Lambda\}$. This proves that T is sufficient if (IV.C.2) holds.

To prove that T is sufficient only if (IV.C.2) holds, let T be any sufficient statistic for θ. From (IV.C.4) we can write

$$p_\theta(y) = p_\theta[y|T(y)]P_\theta[T(Y)=T(y)]. \qquad \text{(IV.C.5)}$$

Since T is sufficient for $\theta, p_\theta[y|T(y)]$ depends only on y and not on θ. Also, $P_\theta[T(Y)=T(y)]$ is a function only of $T(y)$ and θ. On defining $h(y) \triangleq p_\theta[y|T(y)]$ and $g_\theta[T(y)] \triangleq P_\theta[T(Y)=T(y)]$, we see that (IV.C.5) implies the factorization of (IV.C.2). This completes the proof of this proposition for the discrete Γ case.

□

To illustrate Proposition IV.C.1, we consider the following simple example.

Example IV.C.1: A Sufficient Statistic for Hypothesis Testing

Consider the hypothesis-testing problem $\Lambda=\{0, 1\}$ with densities p_0 and p_1. Noting that

$$p_\theta(y) = \begin{cases} p_0(y) \text{ if } \theta = 0 \\ \\ \dfrac{p_1(y)}{p_0(y)} p_0(y) \text{ if } \theta = 1, \end{cases}$$

we can see the factorization $p_\theta(y) = g_\theta[T(y)]h(y)$ with $h(y) = p_0(y)$, $T(y) = p_1(y)/p_0(y) \triangleq L(y)$, and $g_\theta(t)$ defined by

$$g_\theta(t) = \begin{cases} 1 \text{ if } \theta = 0 \\ \\ t \text{ if } \theta = 1. \end{cases}$$

Thus we see that the likelihood ratio $L(y)$ is a sufficient statistic for the binary hypothesis-testing problem. It is a very useful sufficient statistic because it is one-dimensional regardless of the nature of Γ. Of course, we have already seen that all of the optimum tests for $\Lambda = \{0, 1\}$ defined in Chapter II depend on the observation y only through this sufficient statistic $L(y)$.

The usefulness of sufficient statistics in seeking good unbiased estimators of real parameters can be seen partly from the following result. Here we allow Λ to be arbitrary, but we suppose that we wish to estimate some real-valued function g of θ.

Proposition IV.C.2: The Rao-Blackwell Theorem

Suppose that $\hat{g}(y)$ is an unbiased estimate of $g(\theta)$ and that T is sufficient for θ. Define $\tilde{g}[T(y)]$ by

$$\tilde{g}[T(y)] = E_\theta\{\hat{g}(Y) | T(Y) = T(y)\}.$$

Then $\tilde{g}[T(Y)]$ is also an unbiased estimate of $g(\theta)$. Furthermore,

$$Var_\theta(\tilde{g}[T(Y)]) \leqslant Var_\theta[\hat{g}(Y)],$$

with equality if and only if $P_\theta(\hat{g}(Y)=\tilde{g}[T(Y)])=1$.

Proof: We remark first that the expectation defining $\tilde{g}$ does not depend on θ by virtue of the sufficiency of T [i.e., given $T(Y)$, the distribution of Y, and hence the mean of $\hat{g}(Y)$, does not depend on θ]. To see that $\tilde{g}$ is unbiased, we note that

$$E_\theta\{\tilde{g}[T(Y)]\} = E_\theta\{E_\theta\{\hat{g}(Y)|T(Y)\}\}$$

$$= E_\theta\{\hat{g}(Y)\} = g(\theta),$$

where we have used the fact that $E\{E\{X|Z\}\}=E\{X\}$ to get the second equality and the unbiasedness of $\hat{g}$ to get the third equality.

To see that $Var_\theta(\tilde{g}[T(Y)])\leqslant Var_\theta[\hat{g}(Y)]$, we first note that

$$Var_\theta(\tilde{g}[T(Y)]) = E_\theta\{[\tilde{g}[T(Y)]]^2\} - g^2(\theta)$$

and

$$Var_\theta(\hat{g}(Y)) = E_\theta\{[\hat{g}(Y)]^2\} - g^2(\theta).$$

So we only need to show that $E_\theta\{[\tilde{g}[T(Y)]]^2\}\leqslant E_\theta\{[\hat{g}(Y)]^2\}$. We have

$$E_\theta\{(\tilde{g}[T(Y)])^2\} = E_\theta\{[E_\theta\{\hat{g}(Y)|T(Y)\}]^2\}$$

$$\leqslant E_\theta\{E_\theta\{[\hat{g}(Y)]^2|T(Y)\}\} = E_\theta\{[\hat{g}(Y)]^2\}, \tag{IV.C.6}$$

where the inequality follows from applying Jensen's inequality to get $[E_\theta\{\hat{g}(Y)|T(Y)\}]^2\leqslant E_\theta\{[\hat{g}(Y)]^2|T(Y)\}$, and the final equality follows from interated expectations. Note that we have equality in Jensen's inequality here if and only if $P_\theta[\hat{g}(Y)=E_\theta\{[\hat{g}(Y)|T(Y)\}|T(Y)]=1$. Since $\tilde{g}[T(Y)] \triangleq E_\theta\{\hat{g}(Y)|T(Y)\}$, this is equivalent to the condition $P_\theta[\hat{g}(Y)=\tilde{g}[T(Y)]]=1$. This completes the proof of

Proposition IV.C.2.

□

From the Rao-Blackwell theorem we see that with a sufficient statistic T we can improve any unbiased estimator that is not already a function of T by conditioning it on $T(Y)$. Furthermore, this theorem implies that if T is sufficient for θ and if there is only one function of T that is an unbiased estimate of $g(\theta)$, that function is an MVUE for $g(\theta)$. To see this, suppose that $g^*[T(y)]$ is the only function of $T(y)$ for which $E_\theta\{g^*[T(Y)]\}=g(\theta)$. Let $\hat{g}(y)$ be any unbiased estimator of $g(\theta)$. Then, by the Rao-Blackwell theorem, $\tilde{g}[T(y)] \triangleq E_\theta\{\hat{g}(Y)|T(Y)=T(y)\}$ is unbiased for $g(\theta)$ and it is a function of $T(y)$. So by uniqueness of g^*, we must have $g^*=\tilde{g}$. The Rao-Blackwell theorem also asserts that $Var_\theta(\tilde{g}[T(Y)]) \leqslant Var_\theta[\hat{g}(Y)]$. Since $\hat{g}$ is arbitrary, we see that $Var_\theta(g^*[T(Y)]) \leqslant Var_\theta[\hat{g}(Y)]$ for any unbiased estimate of $g(\theta)$; in other words, $g^*[T(y)]$ is an MVUE of $g(\theta)$.

Thus we see that an MVUE of $g(\theta)$ can be constructed if we can find a sufficient statistic T with such a unique unbiased estimate $g^*[T(y)]$. Toward the end of finding such a statistic, we introduce the notion of completeness.

Definition IV.C.3: Completeness

The family $\{P_\theta; \theta \epsilon \Lambda\}$ is said to be *complete* if the condition $E_\theta\{f(Y)\}=0$ for all $\theta \epsilon \Lambda$ implies that $P_\theta[f(Y)=0]=1$ for all $\theta \epsilon \Lambda$.

This notion of completeness is very similar to the notion of completeness of a set of vectors in $\mathbb{R}^n$. To see this, consider the situation in which Γ is a finite set $\{\gamma_1, ..., \gamma_n\}$. In this case, for any function f on Γ we can write

$$E_\theta\{f(Y)\} = \underline{f}^T \underline{p}_\theta,$$

where $\underline{f}=[f(\gamma_1), f(\gamma_2), ..., f(\gamma_n)]^T$ and $\underline{p}_\theta=[p_\theta(\gamma_1), p_\theta(\gamma_2), ..., p_\theta(\gamma_n)]^T$. Assuming that $p_\theta(\gamma_i)>0$ for all $\theta \epsilon \Lambda$ and $i=1, ..., n$, the completeness of $\{p_\theta; \theta \epsilon \Lambda\}$ is defined by the condition that $\underline{f}^T \underline{p}_\theta=0$ for all $\theta \epsilon \Lambda$ implies that $\underline{f}$ is the n-vector of all zeros. That is, $\{P_\theta; \theta \epsilon \Lambda\}$ is complete if $\underline{0}$ is the only vector that is orthogonal to all the vectors $\{\underline{p}_\theta; \theta \epsilon \Lambda\}$. This, of course, is the ordinary notion of completeness of the set of vectors $\{\underline{p}_\theta; \theta \epsilon \Lambda\}$ in $\mathbb{R}^n$. (Recall that a complete set of vectors in $\mathbb{R}^n$ is said to *span* $\mathbb{R}^n$.) Similar analogies hold for more general observation spaces.

To illustrate the notion of completeness further, consider the following example.

Example IV.C.2: Completeness of the Binomial Distribution

Suppose that $\Gamma=\{0, 1, ..., n\}$, $\Lambda=(0, 1)$, and

$$p_\theta(y) = \frac{n!}{y!(n-y)!}\theta^y (1-\theta)^{n-y}, \; y = 0, ..., n, \; 0<\theta<1.$$

For any function f on Γ we have

$$E_\theta\{f(Y)\} = \sum_{y=0}^{n} \frac{n!}{y!(n-y)!} f(y)\theta^y (1-\theta)^{n-y}$$

$$= (1-\theta)^n \sum_{y=0}^{n} a_y x^y,$$

where

$$a_y \triangleq \frac{n!}{y!(n-y)!} f(y) \text{ for } y = 0, ..., n,$$

and

$$x \triangleq \theta/(1-\theta).$$

The condition $E_\theta\{f(Y)\}=0$ for all $\theta \epsilon \Lambda$ is equivalent to the condition

$$\sum_{y=0}^{n} a_y x^y = 0, \text{ for all } x > 0. \qquad \text{(IV.C.7)}$$

The function $\sum_{y=0}^{n} a_y x^y$ is an $n^{\underline{th}}$-order polynominal and thus has at most n zeros unless all its coefficients are zero. It follows that (IV.C.7) can be satisfied only with $f(y)=0, y=0, \ldots, n$. So $\{p_\theta;\ \theta \epsilon \Lambda\}$ is complete. Note that completeness is retained here for any Λ containing at least $(n+1)$ nonzero parameter values.

The notions of completeness and sufficiency are closely related. To see this, suppose that T is sufficient for the complete family $\{P_\theta; \theta \epsilon \Lambda\}$, and for convenience assume that $E_\theta\{|Y|\}<\infty$ for each $\theta \epsilon \Lambda$. Define a function $f(y)$ by

$$f(y) = y - E_\theta\{Y | T(Y) = T(y)\}.$$

Note that f does not depend on θ since T is sufficient. For each $\theta \epsilon \Lambda$ we have

$$E_\theta\{f(Y)\} = E_\theta\{Y\} - E_\theta\{E_\theta\{Y | T(Y)\}\}$$

$$= E_\theta\{Y\} - E_\theta\{Y\} = 0.$$

Thus the completeness of $\{P_\theta; \theta \epsilon \Lambda\}$ implies that $P_\theta[Y=E_\theta\{Y|T(Y)\}]=1$ for all $\theta \epsilon \Lambda$ or, in effect, that $y=E_\theta\{Y|T(Y)=T(y)\}$. Since $E_\theta\{Y|T(Y)=T(y)\}$ is a function of $T(y)$, the latter condition implies that y itself is a function of $T(y)$. Since $T(y)$ is obviously a function of y, we see that $T(y)$ must be a one-to-one function of y; that is T is a trivial sufficient statistic. We conclude then that *if* $\{P_\theta; \theta \epsilon \Lambda\}$ *is complete, then there is no nontrivial sufficient statistic for* θ; i.e., the observation Y cannot be reduced without destroying information about θ.

Completeness is a useful concept in characterizing MVUEs. To see this, suppose that T is sufficient for θ, and let Q_θ denote the distribution of $T(Y)$ when $Y \sim P_\theta$. If $\{Q_\theta; \theta \epsilon \Lambda\}$ is complete, then T is said to be a *complete sufficient statistic*.†

---

† In view of the discussion above, we see that the observation cannot be reduced beyond $T(Y)$ without destroying information about θ. Thus if there is a minimal sufficient statistic for $\{P_\theta; \theta \epsilon \Lambda\}$, T must be it. We cannot, however, claim the converse; i.e., a complete sufficient statistic is not necessarily minimal since there may be other sufficient statistics of which T is not a function.

Suppose that T is complete and let $\tilde{g}[T(y)]$ and $g^*[T(y)]$ be any functions of $T(y)$ that are unbiased estimators of $g(\theta)$. We have

$$E_\theta\{\tilde{g}[T(Y)] - g^*[T(Y)]\} = E_\theta\{\tilde{g}[T(Y)]\} - E_\theta\{g^*[T(Y)]\}$$

$$= g(\theta) - g(\theta) = 0$$

for all $\theta \epsilon \Lambda$. Thus, by the completeness of T, we see that $P_\theta(\tilde{g}[T(Y)]=g^*[T(Y)])=1$ for all $\theta \epsilon \Lambda$, i.e., that $\tilde{g}[T(y)]$ and $g^*[T(y)]$ are the same estimator. Thus since $\tilde{g}$ and g^* were chosen arbitrarily, we see that *any unbiased estimator that is a function of a complete sufficient statistic is unique in this respect and thus is an MVUE.*

We thus see a procedure for seeking MVUEs:

1. Find a complete sufficient statistic T for $\{P_\theta; \theta \epsilon \Lambda\}$.
2. Find *any* unbiased estimator $\hat{g}(y)$ of $g(\theta)$.
3. Then $\tilde{g}[T(y)] \triangleq E_\theta\{\hat{g}(Y)|T(Y)=T(y)\}$ is an MVUE of $g(\theta)$.

Of these stages the first appears to be the least straightforward, since the second step is often fairly easy and the third is accomplished directly by probability calculus. However, for many models of interest in practice the first step turns out to be quite easy. To develop this, we first present the following definition.

Definition IV.C.4: Exponential Families

A class of distributions $\{P_\theta; \theta \epsilon \Lambda\}$ is said to be an *exponential family* if there are real-valued functions $C, Q_1, \ldots, Q_m, T_1, \ldots, T_m$, and h such that P_θ has density

$$p_\theta(y) = C(\theta) \exp\left\{\sum_{l=1}^{m} Q_l(\theta) T_l(y)\right\} h(y), \qquad \text{(IV.C.8)}$$

for all $\theta \epsilon \Lambda$ and $y \epsilon \Gamma$.

Many distributions encountered in practice can be put into the form of exponential families, including Gaussian,

Poisson, Laplacian, binomial, geometric, and certain multivariate forms of these. Exponential families play an important role in the theory of minimum-variance unbiased estimation by virtue of the following result.

Proposition IV.C.3: The Completeness Theorem for Exponential Families

Suppose that $\Gamma=\mathbb{R}^n$, $\Lambda=\mathbb{R}^m$ and that each P_θ has density p_θ given by

$$p_\theta(y) = C(\theta)\exp\left\{\sum_{l=1}^{m}\theta_l T_l(y)\right\}h(y), \qquad \text{(IV.C.9)}$$

where $C, T_1, \ldots, T_n$, and h are real-valued functions.† Then $T(y)=[T_1(y), \ldots, T_m(y)]$ is a complete sufficient statistic for $\{P_\theta; \theta \in \Lambda\}$ if Λ contains a m-dimensional rectangle.

Outline of Proof: A complete proof of Proposition IV.C.3 can be found in Lehmann (1986). The steps in this proof can be outlined as follows.

We first note that T is sufficient for θ by the factorization theorem (Proposition IV.C.1), so we need only show completeness of T. With Y distributed according to (IV.C.9), it is straightforward to show that $T(Y)$ will have a density (on $\mathbb{R}^m$) of the form

$$g_\theta(t) = C(\theta)\exp\left\{\sum_{l=1}^{m}\theta_l t_l\right\}h_T(t), \qquad \text{(IV.C.10)}$$

where h_T is a real-valued function of t. Suppose that f is a real-valued function on $\mathbb{R}^m$ such that $E_\theta\{f[T(Y)]\}=0$. We have

---

† Note that (IV.C.8) can be reparameterized to be put into the form of (IV.C.9).

$$E_\theta\{f\,[T(Y)]\}$$

$$= C(\theta)\int_{\mathbb{R}^m} f(t)\exp\left\{\sum_{l=1}^{m}\theta_l t_l\right\} h_T(t)\mu(dt). \qquad \text{(IV.C.11)}$$

Suppose that Λ contains an m-dimensional rectangle $J=\{\theta | a_1 \leqslant \theta_1 \leqslant b_1, a_2 \leqslant \theta_2 \leqslant b_2, \ldots, a_m \leqslant \theta_m \leqslant b_m\}$. By simple translation of the parameters we can always choose this rectangle to be of the form $J'=\{\theta | -1 \leqslant \theta_1 \leqslant 1, -1 \leqslant \theta_2 \leqslant 1, \ldots, -1 \leqslant \theta_m \leqslant 1\}$. Consider (IV.C.11) as a function of a complex variable by replacing θ_l with $\theta_l + iu_l$, $l=1, \ldots, m$. It can be shown that this function is analytic in the region $C=\{\theta+iu | -1 \leqslant \theta_l \leqslant 1, -\infty < u_l < \infty, l=1, \ldots, m\}$, and thus the condition that it be zero for all real arguments in J' implies that it is zero throughout the strip C. In particular, this function is zero in the region $C'=\{\theta+iu | \theta_l=0, -\infty < u_l < \infty, l=1, \ldots, m\}$, i.e., we have

$$C(\theta)\int_{\mathbb{R}^m} f(t)\exp\left\{\sum_{l=1}^{m}u_l t_l\right\} h_T(t)\mu(dt) = 0, \qquad \text{(IV.C.12)}$$

for all $u \in \mathbb{R}^m$. Note that the function on the left of (IV.C.12) is a multidimensional Fourier transform. This being identically zero for all $\theta \in \Lambda$ implies that the function being transformed is zero for all $\theta \in \Lambda$, or equivalently that $P_\theta\{f(Y)=0\}=1$ for all $\theta \in \Lambda$. This implies in turn that T is complete, and thus completes the proof of the proposition.

□

To illustrate the use of Proposition IV.C.3, we consider the following example.

Example IV.C.3: Minimum-Variance Unbiased Estimation of Signal Amplitude

Consider the model

$$Y_k = N_k + \mu s_k, \quad k = 1, \ldots, n$$

where $N_1, \ldots, N_n$ are i.i.d. $N(0, \sigma^2)$ noise samples, $\underline{s}=(s_1, \ldots, s_n)^T$ is a known signal, and μ is a signal amplitude parameter. Assume for now that σ^2 is known and that we wish to estimate the amplitude parameter μ. The density of $\underline{Y}$ is given by

$$\frac{1}{(2\pi\sigma^2)^{n/2}} \exp\left\{\frac{1}{2\sigma^2}\sum_{k=1}^{n}(y_k - \mu s_k)^2\right\} \tag{IV.C.13}$$

$$= C(\theta_1)\exp\{\theta_1 T_1(\underline{y})\}h(\underline{y}),$$

where we have defined

$$\theta_1 = \mu/\sigma^2,$$

$$T_1(\underline{y}) = \sum_{k=1}^{n} s_k y_k,$$

$$C(\theta_1) = \exp\{(\theta_1/2)\sum_{k=1}^{n} s_k^2\}/(2\pi\sigma^2)^{n/2},$$

and

$$h(\underline{y}) = \exp\{-(1/2\sigma^2)\sum_{k=1}^{n} y_k^2\}.$$

Assuming that μ is an arbitrary real number, the parameter set is $\Lambda=\{\theta_1 | -\infty<\theta_1<\infty\}=\mathbb{R}$. A one-dimensional rectangle is an interval, and Λ obviously contains an interval, so from Proposition IV.C.3 and (IV.C.13), we see that $T_1(\underline{y})$ is a complete sufficient statistic for θ_1.

We wish to estimate $\mu=g(\theta)=\sigma^2\theta_1$. Note that $E_\theta\{Y_1\}=\mu s_1$. So, assuming that $s_1 \neq 0$, the estimate $\hat{g}(\underline{y})=y_1/s_1$ is an unbiased estimator of $g(\theta)$. Thus since T_1 is complete, the estimate

$$\tilde{g}[T_1(\underline{y})] = E_\theta\{\hat{g}(\underline{Y}) | T_1(\underline{Y}) = T_1(\underline{y})\}$$

is an MVUE. To compute (IV.C.14) we note that $\hat{g}(\underline{Y})$ and

$T_1(\underline{Y})$ are both linear functions of $\underline{Y}$, which is Gaussian. Thus $\hat{g}(\underline{Y})$ and $T_1(\underline{Y})$ are jointly Gaussian. It is easy to see that

$$E_\theta\{\hat{g}(\underline{Y})\} = \mu,$$

$$E_\theta\{T_1(\underline{Y})\} = n\,\mu\overline{s^2}$$

$$Var_\theta\{\hat{g}(\underline{Y})\} = \sigma^2/s_1^2,$$

$$Var_\theta\{T_1(\underline{Y})\} = n\,\sigma^2\overline{s^2},$$

and

$$Cov_\theta[\hat{g}(\underline{Y}), T_1(\underline{Y})] = \sigma^2,$$

where we have defined $\overline{s^2} \triangleq (1/n)\sum_{k=1}^n s_k^2$. So, applying the results of Section IV.B, we can write this conditional mean of (IV.C.14) as

$$\begin{aligned}\tilde{g}[T_1(\underline{y})] &= E_\theta\{\hat{g}(\underline{Y})\} + Cov_\theta[\hat{g}(\underline{Y}), T_1(\underline{Y})] \\ &\quad \times [Var_\theta[T_1(\underline{Y})]]^{-1}[T_1(\underline{y}) - E_\theta\{T_1(\underline{Y})\}] \\ &= \mu + \sigma^2(n\,\sigma^2\overline{s^2})^{-1}[T_1(\underline{y}) - n\,\mu\overline{s^2}] \\ &= T_1(\underline{y})/n\overline{s^2} = (\sum_{k=1}^n s_k y_k)/n\overline{s^2}.\end{aligned} \quad \text{(IV.C.15)}$$

Thus we have constructed an MVUE for the signal amplitude μ. The variance of this estimator is

$$Var_\theta(\tilde{g}[T_1(\underline{Y})]) = \sigma^2/n\overline{s^2}. \quad \text{(IV.C.16)}$$

Suppose now that both μ and σ^2 are unknown, with μ ranging over $\mathbb{R}$ and σ^2 ranging over $(0,\infty)$, and that we would like to estimate both of these parameters. We see from (IV.C.16) that estimating σ^2 gives us an estimate of the accuracy of our amplitude estimate. Note that $h(\underline{y})$ as defined in (IV.C.13) is a function of σ^2, so that (IV.C.13) as written is not a correct exponential family if σ^2 is not known. However, we can rewrite the density as

$$\frac{1}{(2\pi\sigma^2)^{n/2}} \exp\left\{-\frac{1}{2\sigma^2}\sum_{k=1}^{n}(y_k - \mu s_k)^2\right\}$$

$$= C(\theta) \exp\left\{\theta_1 T_1(\underline{y}) + \theta_2 T_2(\underline{y})\right\} h(\underline{y}), \tag{IV.C.17}$$

where θ_1 and T_1 are as in (IV.C.13), but, we now define $\theta = (\theta_1, \theta_2)$,

$$\theta_2 = \sigma^2/2,$$

$$T_2(\underline{y}) = \sum_{k=1}^{n} y_k^2,$$

$$C(\theta) = \exp\{(\theta_1/2)\sum_{k=1}^{n} s_k^2\}/(-2\pi/\theta_2)^{n/2},$$

and

$$h(\underline{y}) \equiv 1.$$

The range $\{(\mu, \sigma^2) | \mu \in \mathbb{R}, \sigma^2 < 0\}$ corresponds to $\Lambda = \{(\theta_1, \theta_2) | \theta_1 \in \mathbb{R}, \theta_2 < 0)$, which certainly contains a rectangle. Thus $T = (T_1, T_2)$ is a complete sufficient statistic for θ.

We wish to estimate $\mu = g_1(\theta) \triangleq -\theta_1/\theta_2$ and $\sigma^2 = g_2(\theta) \triangleq -1/\theta_2$. Note that the estimate found in (IV.C.15) is computed without knowledge of σ^2, it is unbiased, and it is a function of $T_1(\underline{y})$ [and hence of $T(\underline{y})$]. Thus it is an MVUE of μ even when σ^2 is not known.

To find an MVUE of σ^2 we can first seek an unbiased estimator of σ^2 and then condition it on $T(\underline{y})$. It is simpler in this case, however, to look directly for an unbiased function of T. In particular, we note that since $T_1(\underline{Y}) \sim N(n\mu\overline{s^2}, n\sigma^2\overline{s^2})$, we have

$$E_\theta\{T_1^2(\underline{Y})\} \equiv Var_\theta[T_1(\underline{Y})] + (E_\theta\{T_1(\underline{Y})\})^2$$

$$= n\sigma^2\overline{s^2} + n^2\mu^2(\overline{s^2})^2.$$

Also, we have that

$$E_\theta\{T_2(\underline{Y})\} = \sum_{k=1}^{n} E_\theta\{Y_k^2\} = \sum_{k=1}^{n} (\sigma^2 + \mu^2 s_k^2)$$

$$= n\sigma^2 + n\mu^2\overline{s^2}.$$

From these two results we see that the quantity $[T_2(\underline{Y}) - T_1^2(\underline{Y})/n\overline{s^2}]$ has mean

$$E_\theta\{T_2(\underline{Y})\} - E_\theta\{T_1^2(\underline{Y})\}/n\overline{s^2} = (n-1)\sigma^2. \quad \text{(IV.C.18)}$$

Thus the function $\tilde{g}_2[T(y)] = [T_2(\underline{y}) - T_1^2(\underline{y})/n\overline{s^2}]/(n-1)$ is an unbiased estimator of σ^2, and by the completeness of T it is an MVUE. We can rewrite $\tilde{g}_2$ as

$$\tilde{g}_2[T(\underline{y})] = \frac{1}{n-1}\sum_{k=1}^{n}(y_k - \hat{\mu}s_k)^2 \triangleq \hat{\sigma}^2, \quad \text{(IV.C.19)}$$

where $\hat{\mu}$ is the MVUE of μ from (IV.C.15). Note that $\hat{n}_k \triangleq y_k - \hat{\mu}s_k$ is an estimate of the noise in the kth sample, so $\hat{\sigma}^2$ estimates the variance (which equals the second moment) of the noise by $[1/(n-1)]\sum_{k=1}^{n}(\hat{n}_k)^2$. Note that a more natural estimator for the second moment would be $(1/n)\sum_{k=1}^{n}(\hat{n}_k)^2$; but as we see from the analysis above, the latter estimate is biased. Further discussion of this point is included in Section IV.D.

The theory outlined in the paragraphs above provides a means for seeking minimum-variance unbiased estimators. For many models of interest, however, the structure required for applying results such as Proposition IV.C.3 is not present. Thus we are often faced with the problem of proposing an estimator and evaluating its performance (i.e., its bias and variance) in the absence of any knowledge about the optimality of the estimator. In such cases it is useful to have a standard to which estimators can be compared; i.e., it would be useful to know the fundamental limitations on estimator performance imposed by a given model. Such a standard is provided in part by the following result.

Proposition IV.C.4: The Information Inequality

Suppose that $\hat{\theta}$ is an estimate of the parameter θ in a family $\{P_\theta; \theta \epsilon \Lambda\}$ and that the following conditions hold:

(1) Λ is an open interval.

(2) The family $\{P_\theta; \theta \epsilon \Lambda\}$ has a corresponding family of densities $\{p_\theta; \theta \epsilon \Lambda\}$, all of the members of which have the same support.†

(3) $\partial p_\theta(y)/\partial\theta$ exists and is finite as for all $\theta \epsilon \Lambda$ and all y in the support of p_θ.

(4) $\partial \int_\Gamma h(y) p_\theta(y) \mu(dy)/\partial\theta$ exists and equals $\int_\Gamma h(y)[\partial p_\theta(y)/\partial\theta]\mu(dy)$ for all $\theta \epsilon \Lambda$ for $h(y)=\hat{\theta}(y)$ and $h(y)=1$.

Then

$$Var_\theta[\hat{\theta}(Y)] \geqslant \frac{\left[\frac{\partial}{\partial\theta} E_\theta\{\hat{\theta}(Y)\}\right]^2}{I_\theta} \qquad \text{(IV.C.20)}$$

where

† That is, the set $\{y \mid p_\theta(y) > 0\}$ is the same for all $\theta \epsilon \Lambda$.

$$I_\theta \triangleq E_\theta\left\{\left(\frac{\partial}{\partial\theta}\log p_\theta(Y)\right)^2\right\}. \quad \text{(IV.C.21)}$$

Furthermore, if the following condition also holds:

(5) $\partial^2 p_\theta(y)/\partial\theta^2$ exists for all $\theta \in \Lambda$ and y in the support of p_θ and

$$\int \frac{\partial^2}{\partial\theta^2} p_\theta(y)\mu(dy) = \frac{\partial^2}{\partial\theta^2}\int p_\theta(y)\mu(dy),$$

then I_θ can be computed via

$$I_\theta = -E_\theta\left\{\frac{\partial^2}{\partial\theta^2}\log p_\theta(Y)\right\}. \quad \text{(IV.C.22)}$$

Proof: The proof of this result follows straightforwardly from the Schwarz inequality. In particular, we have that

$$E_\theta\{\hat{\theta}(Y)\} = \int_\Gamma \hat{\theta}(y) p_\theta(y)\mu(dy). \quad \text{(IV.C.23)}$$

On differentiating (IV.C.23) and applying condition (4), we have

$$\frac{\partial}{\partial\theta}E_\theta\{\hat{\theta}(Y)\} = \int_\Gamma \hat{\theta}(y)\frac{\partial}{\partial\theta}p_\theta(y)\mu(dy).$$

Condition (4) also implies that

$$\int_\Gamma \frac{\partial}{\partial\theta}p_\theta(y)\mu(dy) = \frac{\partial}{\partial\theta}\int_\Gamma p_\theta(y)\mu(dy) = \frac{\partial}{\partial\theta}(1) = 0,$$

so that we have

$$\frac{\partial}{\partial\theta}E_\theta\{\hat{\theta}(Y)\} = \int_\Gamma (\hat{\theta}(y) - E_\theta\{\hat{\theta}(Y)\})\frac{\partial}{\partial\theta}p_\theta(y)\mu(dy)$$

$$= \int_\Gamma (\hat{\theta}(y) - E_\theta\{\hat{\theta}(Y)\})\left[\frac{\partial}{\partial\theta}\log p_\theta(y)\right]p_\theta(y)\mu(dy) \quad \text{(IV.C.24)}$$

$$= E_\theta\left\{[\hat{\theta}(Y) - E_\theta\{\hat{\theta}(Y)\}]\left[\frac{\partial}{\partial\theta}\log p_\theta(Y)\right]\right\},$$

where the second inequality follows from the fact that $\partial\log p_\theta(y)/\partial\theta = [\partial p_\theta(y)/\partial\theta]/p_\theta(y)$. Applying the Schwarz inequality to (IV.C.24), we have

$$\left(\frac{\partial}{\partial\theta}E_\theta\{\hat{\theta}(Y)\}\right)^2 \leqslant E_\theta\{[\hat{\theta}(Y) - E_\theta\{\hat{\theta}(Y)\}]^2\}I_\theta, \quad \text{(IV.C.25)}$$

where I_θ is from (IV.C.21). Noting that $E_\theta\{[\hat{\theta}(Y) - E_\theta\{\hat{\theta}(Y)\}]^2\} = Var_\theta[\hat{\theta}(Y)]$, (IV.C.20) follows.

To see (IV.C.22), we note that

$$\frac{\partial^2}{\partial\theta^2}\log p_\theta(Y) = \left(\frac{\partial^2}{\partial\theta^2}p_\theta(Y)/p_\theta(Y)\right) - \left(\frac{\partial}{\partial\theta}\log p_\theta(Y)\right)^2. \quad \text{(IV.C.26)}$$

Taking $E_\theta\{\cdot\}$ of both sides of (IV.C.26) and rearranging yields

$$I_\theta = -E_\theta\left(\frac{\partial^2}{\partial\theta^2}\log p_\theta(Y)\right) - \int_\Gamma \frac{\partial^2}{\partial\theta^2}p_\theta(y)\mu(dy).$$

Using condition (5) we have

$$\int_{\Gamma} \frac{\partial^2}{\partial \theta^2} p_\theta(y) \mu(dy) = \frac{\partial^2}{\partial \theta^2} \int_{\Gamma} p_\theta(y) \mu(dy) = \frac{\partial^2}{\partial \theta^2}(1) = 0,$$

and (IV.C.22) follows.

□

The quantity I_θ defined in (IV.C.21) is known as *Fisher's information* for estimating θ from Y, and (IV.C.20) is called the *information inequality.* The higher this information measure is for a given model, the better is the lower bound on estimation accuracy provided by the information inequality. The existence of an estimate that achieves equality in the information inequality is possible only under special circumstances [see, e.g., Lehmann (1983) and the discussion below]. For the particular case in which $\hat{\theta}$ is unbiased [$E_\theta\{\hat{\theta}(Y\}=\theta$], the information inequality reduces to

$$Var_\theta[\hat{\theta}(Y)] \geqslant \frac{1}{I_\theta}, \qquad \text{(IV.C.27)}$$

a result known as the *Cramér-Rao Lower Bound* (CRLB).

Examples illustrating the information inequality in specific estimation problems will be discussed in the following section. The following general example illustrates further the role of exponential families in parameter estimation.

Example IV.C.4: The Information Inequality for Exponential Families

Suppose that Λ is an open interval and $p_\theta(y)$ is given by

$$p_\theta(y) = C(\theta) e^{g(\theta) T(y)} h(y), \qquad \text{(IV.C.28)}$$

where C, g, T, and h are real-valued functions of their arguments and where $g(\theta)$ has derivative $g'(\theta)$. Assuming that $E_\theta\{|T(Y)|\} < \infty$ and

$$\frac{\partial}{\partial \theta} \int_{\Gamma} e^{g(\theta) T(y)} h(y) \mu(dy) = \int_{\Gamma} \frac{\partial}{\partial \theta} e^{g(\theta) T(y)} h(y) \mu(dy),$$

conditions (1)-(4) of Proposition IV.C.4 hold. Since $p_\theta(y)$

must integrate to unity, we can write $C(\theta)=[\int_\Gamma e^{g(\theta)T(y)}h(y)\mu(dy)]^{-1}$.

To compute I_θ for this family of densities, we write

$$\log p_\theta(y) = g(\theta)T(y) + \log h(y)$$

$$- \log[\int_\Gamma e^{g(\theta)T(y)}h(y)\mu(dy)].$$

On differentiating we have

$$\frac{\partial}{\partial\theta}\log p_\theta(y) = g'(\theta)T(y) - \frac{g'(\theta)\int_\Gamma T(y)e^{g(\theta)T(y)}h(y)\mu(dy)}{\int_\Gamma e^{g(\theta)T(y)}h(y)\mu(dy)}$$

$$= g'(\theta)[T(y) - E_\theta\{T(Y)\}].$$

Thus

$$I_\theta \triangleq E_\theta\left\{\left(\frac{\partial}{\partial\theta}\log p_\theta(Y)\right)^2\right\} = [g'(\theta)]^2 E_\theta\{[T(Y) - E_\theta\{T(Y)\}])^2\}$$

$$= [g'(\theta)]^2 Var_\theta[T(Y)],$$

and the information inequality in this case is

$$Var_\theta[\hat{\theta}(Y)] \geqslant \frac{\left[\frac{\partial}{\partial\theta}E_\theta\{\hat{\theta}(Y)\}\right]^2}{[g'(\theta)]^2 Var_\theta[T(Y)]}. \quad \text{(IV.C.29)}$$

Suppose that we consider $T(y)$ itself as an estimator of θ. Then we have

$$E_\theta\{T(Y)\} = \frac{\int_\Gamma T(y) e^{g(\theta)T(y)} h(y)\mu(dy)}{\int_\Gamma e^{g(\theta)T(y)} h(y)\mu(dy)}. \qquad \text{(IV.C.30)}$$

On differentiating (IV.C.30) we have straightforwardly that

$$\frac{\partial}{\partial\theta} E_\theta\{T(Y)\} = g'(\theta) Var_\theta[T(Y)],$$

and thus (IV.C.29) implies that the lower bound in the information inequality equals

$$\frac{[\frac{\partial}{\partial\theta} E_\theta\{T(Y)\}]^2}{[g'(\theta)]^2 Var_\theta[T(Y)]} = Var_\theta[T(Y)]. \qquad \text{(IV.C.31)}$$

From (IV.C.31) we see that $T(Y)$ achieves the information lower bound, so it has minimum variance among all estimators $\hat{\theta}$ satisfying $\partial E_\theta\{\hat{\theta}(Y)\}/\partial\theta = \partial E_\theta\{T(Y)\}/\partial\theta$. In particular, if T is unbiased for θ, then it is an MVUE, a fact that we know already from the fact that T is a complete sufficient statistic for θ in this case.

We see that the exponential form (IV.C.28) is sufficient for the variance of T to achieve the information lower bound within the regularity assumed above. It turns out that this form is also necessary for achieving the lower bound for all $\theta \epsilon \Lambda$, again within regularity conditions. In particular, we note that an estimator $\hat{\theta}$ has variance equal to the information lower bound for all $\theta \epsilon \Lambda$ if and only if we have equality in the Schwarz inequality applied in (IV.C.25). This, in turn, will happen if and only if

$$\frac{\partial}{\partial\theta} \log p_\theta(y) = k(\theta)[\hat{\theta}(y) - E_\theta\{\hat{\theta}(Y)\}]$$

with probability 1 under P_θ, for some $k(\theta)$. Letting (a, b)

denote Λ and $f(\theta)$ denote $E_\theta\{\hat{\theta}(Y)\}$, we thus conclude that $\hat{\theta}$ achieves the information bound if and only if

$$p_\theta(y) = \frac{\exp\{\int_a^\theta k(\sigma)[\hat{\theta}(y) - f(\sigma)]d\sigma\}}{\int_a^b \exp\{\int_a^\theta k(\sigma)[\hat{\theta}(y) - f(\sigma)]d\sigma\}d\theta}, \; y \in \Gamma, \quad \text{(IV.C.32)}$$

Equation (IV.C.32) will be recognized as the exponential form of (IV.C.28) with

$$C(\theta) = \exp\{-\int_a^\theta k(\sigma)f(\sigma)d\sigma\},$$

$$g(\theta) = \int_a^\theta k(\sigma)d\sigma,$$

$$T(y) = \hat{\theta}(y),$$

and

$$h(y) = \left[\int_a^b \exp\{\int_a^\theta k(\sigma)[\hat{\theta}(y) - f(\sigma)]d\sigma d\theta\right]^{-1}.$$

[Note that $k(\theta)$ must be equal to $I_\theta / \partial E_\theta\{\hat{\theta}(\underline{Y})\}/\partial\theta$ in this situation, as can be seen from substituting (IV.C.32) into (IV.C.24).] Thus we conclude that within regularity, *the information lower bound is achieved by* $\hat{\theta}$ *if and only if* $\hat{\theta}(y)=T(y)$ *in a one-parameter exponential family.*

IV.D Maximum-Likelihood Estimation

For many observation models arising in practice, it is not possible to apply the results of Section IV.C to find MVUEs, either because of intractability of the required analysis or because of the lack of a useful complete sufficient statistic. For such models, an alternative method for seeking good estimators is needed. One very commonly used method of designing estimators is the maximum-likelihood method, which is the subject of this section.

To motivate maximum-likelihood estimation, we first consider MAP estimation in which we seek $\hat{\theta}_{MAP}(y)$ given by

$$\hat{\theta}_{MAP}(y) = \arg\{\max_{\theta \in \Lambda} p_\theta(y) w(\theta)\}. \qquad \text{(IV.D.1)}$$

In the absence of any prior information about the parameter, we might assume that it is uniformly distributed in its range [i.e., $w(\theta)$ is constant on Λ] since this represents more or less a worst-case prior. In this case, the MAP estimate for a given $y \in \Gamma$ is any value of θ that maximizes $p_\theta(y)$ over Λ. Since $p_\theta(y)$ as a function of θ is sometimes called the *likelihood function* [hence, $p_1(y)/p_0(y)$ is the likelihood ratio], this estimate is called the *maximum likelihood estimate* (MLE). Denoting this estimate by $\hat{\theta}_{ML}$, we have

$$\hat{\theta}_{ML}(y) = \arg\{\max_{\theta \in \Lambda} p_\theta(y)\}. \qquad \text{(IV.D.2)}$$

There are two things wrong with the above argument. First, it is not always possible to construct a uniform distribution on Λ, since Λ may not be a bounded set. Second, and more important, assuming a uniform prior for the parameter is different from assuming that the prior is unknown or that the parameter is not a random variable. However, the maximum-likelihood estimate turns out to be very useful in many situations, and as we will see in this section, its use can be motivated in other, more direct, ways. Moreover, finding the value of θ that makes the observations most likely is a legitimate criterion on its own.

Maximizing $p_\theta(y)$ is equivalent to maximizing $\log p_\theta(y)$, and assuming sufficient smoothness of this function, a necessary condition for the maximum-likelihood estimate is

$$\frac{\partial}{\partial\theta}\log p_\theta(y)\Big|_{\theta=\hat{\theta}_{ML}(y)} = 0. \qquad \text{(IV.D.3)}$$

Equation (IV.D.3) is known as the *likelihood equation*, and we will see that its solutions have useful properties even when they are not maxima of $p_\theta(y)$.

For example, suppose we have equality in the Cramér-Rao lower bound (IV.C.27); i.e., suppose that $\hat{\theta}$ is an unbiased estimate of θ with $Var_\theta[\hat{\theta}(Y)]=1/I_\theta$. (Note that such a $\hat{\theta}$ is an MVUE of θ.) Then, from (IV.C.32), we see that log $p_\theta(y)$ must be of the form

$$\log p_\theta(y) = \int_a^\theta I_\sigma[\hat{\theta}(y) - \sigma]d\sigma + \log h(y), \qquad \text{(IV.D.4)}$$

where we have used the facts that $f(\theta)=\theta$ and $k(\theta)=f'(\theta)/I_\theta$. From (IV.D.4), the likelihood equation becomes

$$\frac{\partial}{\partial\theta}\log p_\theta(y)\Big|_{\theta=\hat{\theta}_{ML}(y)} = I_\theta[\hat{\theta}(y) - \theta]\Big|_{\theta=\hat{\theta}_{ML}(y)} = 0, \qquad \text{(IV.D.5)}$$

which has the solution $\hat{\theta}_{ML}(y)=\hat{\theta}(y)$. Thus we conclude that *if* $\hat{\theta}$ *achieves the CRLB, it is the solution to the likelihood equation*. In other words, only solutions to the likelihood can achieve the CRLB. Unfortunately, it is not always true that solutions to the likelihood equation will achieve the CRLB or even that they are unbiased. [However, when log p_θ has the form (IV.D.4), this will happen.] Also, when the solution to the likelihood equation does not satisfy the CRLB, there may be other estimators with the same bias that have smaller variance then $\hat{\theta}_{ML}$.

From the above discussion we see that the solution to the likelihood equation can sometimes be an MVUE. For the case in which the observation space is $\mathbb{R}^n$ with Y consisting of i.i.d. components, it happens that within regularity the solution to the likelihood equation is unbiased and achieves the CRLB asymptotically as $n \to \infty$. Before studying these asymptotic properties we give the following two examples to illustrate the maximum-likelihood approach.

Example IV.D.1: Maximum-Likelihood Estimation of the Parameter of the Exponential Distribution

Suppose that $\Gamma=\mathbb{R}^n$, $\Lambda=(0,\infty)$, and $Y_1,\ldots,Y_n$ are i.i.d. exponential random variables with parameter θ, i.e., $p_\theta(\underline{y})=\prod_{k=1}^n f_\theta(y_k)$ with

$$f_\theta(y_k)=\begin{cases}\theta e^{-\theta y_k} & \text{if } y_k \geqslant 0\\ 0 & \text{if } y_k<0.\end{cases} \tag{IV.D.6}$$

We have $p_\theta(y)=\theta^k \exp\{-\theta n\bar{y}\}$ with $\bar{y} \triangleq (1/n)\sum_{k=1}^n y_k$, so the likelihood equation is

$$\frac{\partial}{\partial\theta}\log p_\theta(\underline{y})\Big|_{\theta=\hat{\theta}_{ML}(y)}=\frac{n}{\theta}-n\bar{y}\Big|_{\theta=\hat{\theta}_{ML}(\underline{y})}=0, \tag{IV.D.7}$$

which has the unique solution $\hat{\theta}_{ML}(\underline{y})=1/\bar{y}$. Since $\partial^2\log p_\theta(\underline{y})/\partial\theta^2=-n/\theta^2<0$, this solution gives the unique maximum of $p_\theta(\underline{y})$. Note that $E_\theta\{Y_k\}=1/\theta$, so that $E_\theta\{\bar{Y}\}=1/\theta$ and thus it makes sense to estimate θ as $1/\bar{y}$. In fact, the weak law of large numbers implies that $\bar{Y}\to 1/\theta$ in probability under P_θ, which in turn implies that $1/\bar{Y}\to\theta$ in probability under P_θ; i.e., the MLE converges in probability to the true parameter value, a property known as *consistency*. This property of MLEs is not specific to this example but rather is true in a very general context as we shall see below.

Fisher's information for this case can be computed via

$$I_\theta=-E_\theta\left[\frac{\partial^2}{\partial\theta^2}\log p_\theta(\underline{Y})\right]=-E_\theta\{-n/\theta^2\}=n/\theta^2,$$

and so the CRLB is θ^2/n. Since $\partial \log p_\theta(\underline{y})/\partial\theta$ is not of the form $k(\theta)[\hat{\theta}_{ML}(\underline{y})-f(\theta)]$, we know that the information inequality is not achieved in this problem. However, we can compute the mean and variance of $\hat{\theta}_{ML}$ directly. In particular, by using characteristic functions it is straightforward to show that the sample mean $\bar{Y}$ has pdf

$$p_Y(y) = \begin{cases} \frac{1}{n!\theta^n}(y/n)^{n-1}e^{-y/n\theta} & \text{if } y \geq 0 \\ 0 & \text{if } y < 0, \end{cases}$$

from which we can compute (for $n > 1$)

$$E_\theta\{\hat{\theta}_{ML}(\underline{Y})\} = E_\theta\left\{\frac{1}{\bar{Y}}\right\} = \frac{n\theta}{n-1} \tag{IV.D.8}$$

and (for $n > 2$)

$$Var_\theta[\hat{\theta}_{ML}(\underline{Y})] = \frac{\theta^2 n^2}{(n-1)^2(n-2)}. \tag{IV.D.9}$$

We see from (IV.D.8) that although $\hat{\theta}_{ML}(\underline{Y})$ is biased, it does have the property that $\lim_{n\to\infty}E_\theta\{\hat{\theta}_{ML}(\underline{Y})\}=\theta$; that is, it is *asymptotically unbiased*. Also, we note that

$$Var_\theta[\hat{\theta}_{ML}(\underline{Y})]I_\theta = \frac{n^3}{(n-1)^2(n-2)} \to 1$$

as $n \to \infty$; and thus $\hat{\theta}_{ML}$ has variance asymptotically equal to the CRLB, a property known as *asymptotic efficiency*. As we shall see, these two properties of asymptotic unbiasedness and efficiency are characteristic of MLEs under general conditions for i.i.d. observations.

As a final comment on this example, we note from Proposition IV.C.3 that $\bar{Y}$ is a complete sufficient statistic for θ in this model. Also, from (IV.D.8) we see that

$$\frac{n-1}{n}\hat{\theta}_{ML}(\underline{y}) \equiv \left(\frac{1}{n-1}\sum_{k=1}^{n} y_k\right)^{-1}$$

is an unbiased estimator of θ depending on $\bar{Y}$. Thus

$$\frac{n-1}{n}\hat{\theta}_{ML}(\underline{Y}) \triangleq \hat{\theta}_{MV}(\underline{y})$$

is an MVUE of θ in this problem. From (IV.D.9), its variance is seen to be given by

$$Var_\theta[\hat{\theta}_{MV}(\underline{Y})] = \frac{\theta^2}{n-2}, \qquad \text{(IV.D.10)}$$

a quantity that is larger than the CRLB (as it must be since we know the CRLB cannot be achieved here), but that approaches the CRLB as n becomes large.

The variance of (IV.D.10) equals the MSE of $\hat{\theta}_{MV}$ since it is unbiased. For the MLE, the MSE is

$$E_\theta\{[\hat{\theta}_{ML}(\underline{Y}) - \theta]^2\} = Var_\theta[\hat{\theta}_{ML}(\underline{Y})] + b^2(\theta), \qquad \text{(IV.D.11)}$$

where $b(\theta) \triangleq E_\theta\{\hat{\theta}_{ML}(\underline{Y})\} - \theta$ is the *bias* of $\hat{\theta}_{ML}$. Using (IV.D.8) and (IV.D.9), we have

$$E_\theta\{[\hat{\theta}_{ML}(\underline{Y}) - \theta]^2\} = \frac{\theta^2(n+2)}{(n-1)(n-2)},$$

a quantity that is strictly greater than $\theta^2/(n-2)$, the MSE of $\hat{\theta}_{MV}$. Thus in this case the MVUE is preferable to the MLE, although they are asymptotically equivalent.

Example IV.D.2: Maximum-Likelihood Estimation of Signal Amplitude

Consider the model treated in Example IV.C.3:

$$y_k = N_k + \mu s_k, \quad k = 1, \ldots, n$$

with $N_1, \ldots, N_n$ i.i.d. $N(0, \sigma^2)$ and $\underline{s} = (s_1, \ldots, s_n)^T$ known. The likelihood equation for estimating μ with σ^2 known is given by

$$-\frac{\partial}{\partial\mu}\left(\sum_{k=1}^{n} \tfrac{1}{2}\log(2\pi\sigma^2) + \frac{1}{2\sigma^2}\sum_{k=1}^{n}(y_k - \mu s_k)^2\right)\Bigg|_{\mu=\hat{\mu}_{ML}(\underline{y})} \tag{IV.D.12}$$

$$= \frac{1}{\sigma^2}\sum_{k=1}^{n} s_k\,[y_k - \hat{\mu}_{ML}(\underline{y})s_k] = 0,$$

which implies that

$$\hat{\mu}_{ML}(\underline{y}) = \frac{1}{n}\sum_{k=1}^{n} s_k\, y_k / \overline{s^2}, \tag{IV.D.13}$$

where, as before, $\overline{s^2} \triangleq (1/n)\sum_{k=1}^{n} s_k^2$. Since

$$-\frac{\partial^2}{\partial\mu^2}\sum_{k=1}^{n}\left[\tfrac{1}{2}\log(2\pi\sigma^2) + \frac{1}{2\sigma^2}(y_k - \mu s_k)^2\right] \tag{IV.D.14}$$

$$= -n\overline{s^2}/\sigma^2 < 0,$$

we see that $\log p_\theta(\underline{y})$ is concave in μ, so the solution to the likelihood equation does give a global maximum here.

Note that $\hat{\mu}_{ML}$ is the same as the MVUE of μ (see Example IV.C.3), so that $E_\theta\{\hat{\mu}_{ML}(\underline{y})\}=\mu$ and $Var_\theta[\hat{\mu}_{ML}(\underline{Y})]=\sigma^2/n\overline{s^2}$. From (IV.D.14), we see that $I_\theta = n\overline{s^2}/\sigma^2$, so

$$CRLB = \frac{\sigma^2}{n\overline{s^2}} = Var_\theta[\hat{\mu}_{ML}(\underline{Y})]. \tag{IV.D.15}$$

Note that with $\theta=\mu$, we can write

$$\frac{\partial}{\partial\theta}\log p_\theta(\underline{y}) = k(\theta)[\hat{\theta}_{ML}(\underline{y}) - \theta]$$

with $k(\theta)=1/I_\theta=n\overline{s^2}/\sigma^2$, as is required for achievement of the CRLB.

Suppose that now that μ is known but we wish to estimate σ^2. The likelihood equation becomes

$$\frac{\partial}{\partial \sigma^2} \log p_\theta(\underline{y})\Big|_{\sigma^2 = \hat{\sigma}^2_{ML}(\underline{y})}$$

$$= \frac{n}{2\hat{\sigma}^2_{ML}(\underline{y})} + \frac{1}{2[\hat{\sigma}^2_{ML}(\underline{y})]^2} \sum_{k=1}^{n} (y_k - \mu s_k)^2 = 0, \tag{IV.D.16}$$

which has the unique solution

$$\hat{\sigma}^2_{ML}(\underline{y}) = \frac{1}{n} \sum_{k=1}^{n} (y_k - \mu s_k)^2. \tag{IV.D.17}$$

Since

$$\frac{\partial}{\partial \sigma^2} \log p_\theta(\underline{y}) = \frac{n}{2\sigma^4} \left[\hat{\sigma}^2_{ML}(\underline{y}) - \sigma^2\right], \tag{IV.D.18}$$

we see that $\log p_\theta(\underline{y})$ is increasing in σ^2 for $\sigma^2 < \hat{\sigma}^2_{ML}(\underline{y})$ and decreasing in σ^2 for $\sigma^2 > \hat{\sigma}^2_{ML}(\underline{y})$. Thus $\log p_\theta(\underline{y})$ achieves its absolute maximum at $\hat{\sigma}^2_{ML}(\underline{y})$. We also see from (IV.D.18) that with $\theta = \sigma^2$,

$$\frac{\partial}{\partial \theta} \log p_\theta(\underline{y}) = \frac{n}{2\theta^2} [\hat{\theta}_{ML}(\underline{y}) - \theta], \tag{IV.D.19}$$

which from Example IV.C.4 implies that $\hat{\sigma}^2_{ML}(\underline{y})$ is unbiased and achieves the CRLB, and thus that $\hat{\sigma}^2_{ML}$ is an MVUE of σ^2. By inspection of (IV.D.19) we have $I_\theta = n/2\theta^2 \equiv n/2\sigma^4$, so

$$CRLB = \frac{2\sigma^4}{n} = Var_\theta[\hat{\sigma}^2_{ML}(\underline{Y})]. \tag{IV.D.20}$$

Now suppose that both μ and σ^2 are unknown. Putting $\theta = (\mu, \sigma^2)$, the MLE of θ is found by maximizing $p_\theta(\underline{y})$ over μ and σ^2. Since the maximum $\hat{\mu}_{ML}(\underline{y})$ from (IV.D.13) does not depend on σ^2, we have that

$$\max_{(\mu,\sigma^2)} \log p_\theta(\underline{y}) = \max_{\sigma^2} \{\max_{\mu} \log p_\theta(\underline{y})\}$$

$$= \max_{\sigma^2} \sum_{k=1}^{n} \left\{ -\tfrac{1}{2}\log(2\pi\sigma^2) - \frac{1}{2\sigma^2} \sum_{k=1}^{n} [y_k - \hat{\mu}_{ML}(\underline{y}) s_k]^2 \right\}.$$

But the right-hand side of this equation is the same maximization problem as for estimating σ^2 with known μ, with μ set equal to $\hat{\mu}_{ML}(\underline{y})$. Thus the maximum is achieved by (IV.D.17) with $\hat{\mu}_{ML}$ substituted for μ and the MLE for $\theta=(\mu, \sigma^2)$ is $\hat{\theta}_{ML}(\underline{y})=[\hat{\mu}_{ML}(\underline{y}), \hat{\sigma}^2_{ML}(\underline{y})]$, where

$$\hat{\mu}_{ML}(\underline{y}) = \frac{1}{n} \sum_{k=1}^{n} s_k y_k / \overline{s^2} \qquad \text{(IV.D.21a)}$$

and

$$\hat{\sigma}^2_{ML}(\underline{y}) = \frac{1}{n} \sum_{k=1}^{n} [y_k - \hat{\mu}_{ML}(\underline{y}) s_k]^2. \qquad \text{(IV.D.21b)}$$

$\hat{\mu}_{ML}(\underline{y})$ is still on MVUE of μ in this case. However, from (IV.C.19) we see that $\hat{\sigma}^2_{ML}(\underline{y})$ is $[(n-1)/n]\hat{\sigma}^2_{MV}(\underline{y})$. Thus

$$E_\theta\{\hat{\sigma}^2_{ML}(\underline{Y})\} = \frac{n-1}{n}\sigma^2,$$

and the MLE of σ^2 is biased here (although it is asymptotically unbiased). Note that

$$Var_\theta[\hat{\sigma}^2_{ML}(\underline{Y})] = [(n-1)^2/n^2] Var_\theta[\hat{\sigma}^2_{MV}(\underline{Y})],$$

so that $\hat{\sigma}^2_{ML}(\underline{y})$ has lower variance than the MVUE. It can be shown that (see Exercise 14.)†

$$Var_\theta(\hat{\sigma}^2_{MV}(\underline{Y})) = \frac{2\sigma^4}{(n-1)}. \qquad \text{(IV.D.22)}$$

Thus for the MVUE of σ^2 the MSE, $E_\theta\{[\hat{\sigma}^2_{MV}(\underline{Y}) - \sigma^2]^2\}$, is $2\sigma^4/(n-1)$. Alternatively, for the MLE of σ^2, the MSE is given by

$$E_\theta\{[\sigma^2_{ML}(\underline{Y}) - \sigma^2]^2\}$$

$$= Var_\theta[\sigma^2_{ML}(\underline{Y})] + [E_\theta\{\hat{\sigma}^2_{ML}(\underline{Y})\} - \sigma^2]^2$$

$$= \frac{(n-1)^2}{n^2}\frac{2\sigma^4}{n-1} + \left(\frac{n-1}{n}\sigma^2 - \sigma^2\right)^2 \qquad \text{(IV.D.23)}$$

$$= \sigma^4\frac{(2n-1)}{n^2}.$$

The ratio of these two quantities is

$$\frac{E_\theta\{[\hat{\sigma}^2_{MV}(\underline{Y}) - \sigma^2]^2\}}{E_\theta\{[\hat{\sigma}^2_{ML}(\underline{Y}) - \sigma^2]^2\}} = \left(\frac{n}{n-1}\right)\left(\frac{2n}{2n-1}\right) > 1. \qquad \text{(IV.D.24)}$$

We see from (IV.D.34) that *the MLE in this case has a uniformly lower MSE than the MVUE.* This is because the increase in MSE due to the bias of the MLE is more than offset by the increase in variance of the MVUE needed to achieve unbiasedness. Thus, achieving the goal of minimum-variance unbiased estimation does not always lead to an optimum estimate in terms of mean-squared error.

---

† It is interesting to note that the MVUE of σ^2 with μ known has variance $2\sigma^4/n$ [from (IV.D.20)] and the MVUE with μ unknown has variance $2\sigma^4/(n-1)$ (from IV.D.22). Thus for unbiased estimation of σ^2, there is a "penalty" of one observation when μ is unknown.

One of the principal motivations for using maximum-likelihood estimation is illustrated by the two examples above; namely, estimates based on independent samples have good asymptotic properties as the number of samples increases without bound. The reason for this asymptotic behavior can be seen from the arguments in the following paragraphs.

Suppose that we have a sequence of i.i.d. observations $Y_1, Y_2, \ldots, Y_n$, each with marginal density f_θ coming from the family $\{f_\theta; \theta \epsilon \Lambda\}$. Let $\hat{\theta}_n$ denote a solution to the likelihood equation for sample size n, i.e.,

$$\frac{\partial}{\partial\theta} \log p_\theta(\underline{y})\Big|_{\theta=\hat{\theta}_n(\underline{y})} = \sum_{k=1}^{n} \psi[y_k; \hat{\theta}_n(\underline{y})] = 0,$$

where $\psi(y_k; \theta) \triangleq \partial \log f_\theta(y_k)/\partial\theta$. Equivalently, we can write

$$\frac{1}{n}\sum_{k=1}^{n} \psi[y_k; \hat{\theta}_n(\underline{y})] = 0. \tag{IV.D.25}$$

For a fixed parameter value $\theta' \epsilon \Lambda$, consider the quantity $\sum_{k=1}^{n}\psi(Y_k; \theta')/n$. Assuming that θ is the true parameter value (i.e., $Y_k \sim f_\theta$), the weak law of large numbers implies that

$$\frac{1}{n}\sum_{k=1}^{n} \psi(Y_k; \theta') \overset{(i.p.)}{\rightarrow} E_\theta\{\psi(Y_1; \theta')\}.$$

We have

$$E_\theta\{\psi(Y_1; \theta')\} = \int_{\mathbb{R}} \frac{\partial}{\partial\theta} \log f_\theta(y_1)\Big|_{\theta=\theta'} f_\theta(y_1)\mu(dy_1) \tag{IV.D.26}$$

$$\triangleq J(\theta; \theta').$$

Assuming that order of integration and differentiation can be interchanged in (IV.D.26), $J(\theta; \theta)$ can be written as

$$J(\theta;\theta) = \int [\frac{\partial}{\partial\theta} \log f_\theta(y_1)] f_\theta(y_1)\mu(dy_1)$$

$$= \int \frac{\partial}{\partial\theta} f_\theta(y_1)\mu(dy_1)$$

$$= \frac{\partial}{\partial\theta} \int f_\theta(y_1)\mu(dy_1) = \frac{\partial}{\partial\theta}(1) = 0.$$

Thus the equation $J(\theta;\theta')=0$ has a solution $\theta'=\theta$. Suppose that this is the unique root of $J(\theta;\theta')$, and suppose that that $J(\theta;\theta')$ and $\sum_{k=1}^{n}\psi(Y_k;\theta')/n$ are both smooth functions of θ'. Then, since $\sum_{k=1}^{n}\psi(Y_k;\theta')/n$ is close to $J(\theta;\theta')$ for large n, we would expect the roots of these two functions to be close when n is large. That is, $\hat{\theta}_n(\underline{Y})$ should be close to the true parameter value θ when n is large. And as $n \to \infty$, we would expect that $\hat{\theta}_n(\underline{Y}) \to \theta$ in some statistical sense. In fact, within the appropriate smoothness and uniqueness conditions, the solutions to the likelihood equation are *consistent*; that is, they converge in probability to the true parameter value:

$$\lim_{n\to\infty} P_\theta(|\hat{\theta}_n(\underline{Y}) - \theta| > \epsilon) = 0 \text{ for all } \epsilon > 0.$$

One set of conditions under which solutions to the likelihood are consistent is summarized in the following.

Proposition IV.D.1: Consistency of MLEs

Suppose that $\{Y_k\}_{k=1}^{\infty}$ is an i.i.d. sequence of random variables each with density f_θ, and assume that τ and ψ are well defined as above. Suppose that further that the following conditions hold:

(1) $J(\theta;\theta')$ is a continuous function of θ' and has a unique root at $\theta'=\theta$, at which point it changes sign.

(2) $\psi(Y_k;\theta')$ is a continuous function of θ' (with probability 1).

(3) For each n, $\sum_{k=1}^{n}\psi(Y_k;\theta')/n$ has a unique root $\hat{\theta}_n$ (with probability 1).

Then $\hat{\theta}_n \to \theta$ (i.p.).

Proof: Choose $\epsilon > 0$. By condition (1), $J(\theta;\theta+\epsilon)$ and $J(\theta;\theta-\epsilon)$ must have opposite signs. Define $\delta = \min\{|J(\theta;\theta+\epsilon)|, |J(\theta;\theta-\epsilon)|\}$ and for each n, define the events

$$A_n^+ = \{|J(\theta;\theta+\epsilon) - \frac{1}{n}\sum_{k=1}^{n}\psi(Y_k;\theta+\epsilon)| \leqslant \delta\},$$

$$A_n^- = \{|J(\theta;\theta-\epsilon) - \frac{1}{n}\sum_{k=1}^{n}\psi(Y_k;\theta-\epsilon)| \leqslant \delta\}, \qquad \text{(IV.D.27)}$$

and $A_n = A_n^+ \cap A_n^-$.

Now, on A_n^+, $\sum_{k=1}^{n}\psi(Y_k, \theta+\epsilon)/n$ must have the same sign as $J(\theta;\theta+\epsilon)$, and on A_n^-, $\sum_{k=1}^{n}(Y_k;\theta-\epsilon)/n$ must have the same sign as $J(\theta;\theta-\epsilon)$. Thus on A_n, $\sum_{k=1}^{n}\psi(Y_k;\theta+\epsilon)/n$ and $\sum_{k=1}^{n}\psi(Y_k;\theta-\epsilon)/n$ have opposite signs. By the continuity assumption (2), $\sum_{k=1}^{n}\psi(Y_k;\theta')/n$ can change sign only by passing through zero. Thus on A_n, the root $\hat{\theta}_n$ is between $\theta-\epsilon$ and $\theta+\epsilon$. This implies that A_n is a subset of $\{|\hat{\theta}_n - \theta| \leqslant \epsilon\}$, so that $P(|\hat{\theta}_n - \theta| \leqslant \epsilon) \geqslant P(A_n)$.

By the weak law of large numbers,

$$\frac{1}{n}\sum_{k=1}^{n}\psi(Y_k;\theta+\epsilon) \to J(\theta;\theta+\epsilon) \ (i.p.)$$

and (IV.D.28)

$$\frac{1}{n}\sum_{k=1}^{n}\psi(Y_k;\theta-\epsilon) \to J(\theta;\theta-\epsilon) \ (i.p.).$$

Thus $P(A_n^+) \to 1$ and $P(A_n^-) \to 1$ as $n \to \infty$. We have

$$\begin{aligned} 1 &\geqslant P(|\hat{\theta}_n - \theta| \leqslant \epsilon) \\ &\geqslant P(A_n) = P(A_n^+) + P(A_n^-) - P(A_n^+ \cap A_n^-) \qquad \text{(IV.D.29)} \\ &\geqslant P(A_n^+) + P(A_n^-) - 1 \to 1. \end{aligned}$$

Thus $P(|\hat{\theta}_n - \theta| \leqslant \epsilon) \to 1$, and since ϵ was chosen arbitrarily we have the desired result.

□

Remarks: The conditions on this proposition can be relaxed in various ways. First, the continuity of the functions $J(\theta;\theta')$ and $\psi(Y_k;\theta')$ can be relaxed to continuity in a neighborhood of $\theta'=\theta$. Also, it is not necessary to assume the existence of the roots $\hat{\theta}_n$, since the development above shows that there must be a root to the likelihood equation on A_n, which has probability tending to 1. In fact, with only the assumption of local continuity, the proof above can be used to show that *with probability tending to 1, there is a sequence of roots to the likelihood equation converging to any isolated root of* $J(\theta;\theta')$. Thus if $J(\theta;\theta')$ has multiple roots, inconsistent sequences can arise by solving the likelihood equation.

In addition to consistency, we saw in the examples above that the solutions to the likelihood equation may also be asymptotically unbiased and efficient. We know that under the conditions of Proposition IV.D.1, $\hat{\theta}_n$ converges to θ in probability. Thus if we would write

$$\lim_{n\to\infty} E_\theta\{\hat{\theta}_n\} = E_\theta\{\lim_{n\to\infty}\hat{\theta}_n\} \tag{IV.D.30}$$

for this type of convergence, then asymptotic unbiasedness would follow. The interchange of limits and expectations in (IV.D.30) is not always valid for convergence in probability. However, under various conditions on ψ, this interchange can be shown to be valid. [A sufficient condition for the validity of this interchange is the existence of a random variable X such that $|\hat{\theta}_n| \leqslant X$ for each n and $E_\theta\{X\}<\infty$. This is known as the *dominated convergence theorem*.] Thus asymptotic unbiasedness is not an unreasonable property to expect in view of the consistency of $\hat{\theta}_n$.

It is less clear why MLEs might be asymptotically efficient. To see why this might be so, we consider the related question of finding the asymptotic distribution of the error, $\hat{\theta}_n-\theta$. In particular, we prove the following proposition.

Proposition IV.D.2: Asymptotic Normality of MLEs

Suppose that that $\{Y_k\}_{n=1}^{\infty}$ is a sequence of i.i.d. random variables each with density f_θ, and that $\{\hat{\theta}_n\}_{n=1}^{\infty}$ is a consistent sequence of roots of the likelihood equation. Suppose further that ψ satisfies the following regularity conditions.

(1) $0 < i_\theta \triangleq E_\theta\{[\psi(Y_1;\theta)]^2\} < \infty$.

(2) The derivatives $\psi'(Y_1;\theta') \triangleq \partial\psi(Y_1;\theta')/\partial\theta'$ and $\psi''(Y_k;\theta') \triangleq \partial^2\psi(Y_k;\theta')/(\partial\theta')^2$ exist (with probability 1).

(3) There is a function $M(Y_1)$ such that $|\psi''(Y_1;\theta')| \leq M(Y_1)$ for all $\theta' \epsilon \Lambda$ and $E_\theta\{M(Y_1)\} < \infty$.

(4) $J(\theta;\theta)=0$, where $J(\theta;\theta')$ is defined as in (IV.D.26).

(5) Condition (5) of Proposition IV.C.4 holds.

Then

$$P_\theta(\sqrt{ni_\theta}(\hat{\theta}_n - \theta) \leq x) \rightarrow \Phi(x) \text{ for all } x \in \mathbb{R},$$

where Φ is the standard Gaussian distribution function. That is, $\sqrt{n}(\hat{\theta}_n - \theta)$ converges in distribution to a $N(0, 1/i_\theta)$ random variable.

Proof: Using Taylor's theorem, we can expand the left-hand side of the likelihood equation, $(1/n)\sum_{k=1}^{n}\psi(Y_k;\hat{\theta}_k)=0$, about θ to yield

$$\frac{1}{n}\sum_{k=1}^{n}\psi(Y_k;\theta) + (\hat{\theta}_n - \theta)\frac{1}{n}\sum_{k=1}^{n}\psi'(Y_k;\theta)$$

$$+ \tfrac{1}{2}(\hat{\theta}_n - \theta)^2\frac{1}{n}\sum_{k=1}^{n}\psi''(Y_k;\bar{\theta}_n) = 0, \quad \text{(IV.D.31)}$$

where $\bar{\theta}_n$ is between θ and $\hat{\theta}_n$. Rearranging (IV.D.31) gives an expression for the quantity $\sqrt{n}(\hat{\theta}_n - \theta)$:

$$\sqrt{n}\,(\hat{\theta}_n - \theta)$$

$$= \frac{-\dfrac{1}{\sqrt{n}}\displaystyle\sum_{k=1}^{n}\psi(Y_k\,;\theta)}{\dfrac{1}{n}\displaystyle\sum_{k=1}^{n}\psi'(Y_k\,;\theta)+\tfrac{1}{2}(\hat{\theta}_n - \theta)\sum_{k=1}^{n}\psi''(Y_k\,;\bar{\theta}_n)}. \qquad \text{(IV.D.32)}$$

Consider the denominator on the right-hand side of (IV.D.32). By the weak law of large numbers, the first term, $\sum_{k=1}^{n}\psi'(Y_k\,;\theta)/n$, converges to $E_\theta\{\psi'(Y_1;\theta)]$ in probability. By condition (3), the second term $\tfrac{1}{2}(\hat{\theta}_n-\theta)\sum_{k=1}^{n}\psi''(Y_k\,;\bar{\theta}_n)/n$ is bounded as

$$\left|\tfrac{1}{2}(\hat{\theta}_n - \theta)\frac{1}{n}\sum_{k=1}^{n}\psi''(Y_k\,;\hat{\theta}_n)\right|$$

$$\leqslant \tfrac{1}{2}|\hat{\theta}_n - \theta|\frac{1}{n}\sum_{k=1}^{n}M(Y_k). \qquad \text{(IV.D.33)}$$

Now, $|\bar{\theta}_n-\theta|\to 0$ (i.p.) and the weak law of large numbers implies that $(1/n)\sum_{k=1}^{n}M(Y_k)\to E_\theta\{M(Y_1)\}<\infty$. Thus the second term converges in probability to zero and the denominator then converges in probability to $E_\theta\{\psi'(Y_1;\theta\}$.

The numerator sum $\sum_{k=1}^{n}\psi(Y_k\,;\theta)$ in (IV.D.32) is the sum of n i.i.d. random variables, each with mean $E_\theta\{\psi(Y_1;\theta)\}=J(\theta;\theta)=0$ and variance $E_\theta\{\psi^2(Y_1;\theta)\}=i_\theta<\infty$. Thus by the central limit theorem, $-(1/\sqrt{n})\sum_{k=1}^{n}\psi(Y_k\,;\theta)$ converges in distribution to a $N(0, i_\theta)$ random variable.

The two results above imply that $\sqrt{n}\,(\hat{\theta}_n-\theta)$ converges in distribution to a $N(0, v^2)$ random variable with

$$v^2 = i_\theta / E_\theta^2\{\psi'(Y_1;\theta)\}. \qquad \text{(IV.D.34)}$$

But using the argument used in deriving (IV.C.22), $E_\theta\{\psi'(Y_k\,;\theta\}=-E_\theta\{\psi^2(Y_1;\theta\}=-i_\theta$, so $v^2=1/i_\theta$. This completes the proof.

□

Remarks: It is easy to see that Fisher's information is given by $I_\theta = n i_\theta$ for this i.i.d. case. Heuristically, we can think of the conclusion of this proposition as the condition that $\hat{\theta}_n$ is asymptotically $N(\theta, 1/ni_\theta)$; that is, asymptotically $\hat{\theta}_n$ has mean θ and variance equal to $1/ni_\theta$, the Cramér-Rao lower bound. Actually, what we have proved is that the asymptotic distribution of $\sqrt{n}\,(\hat{\theta}_n - \theta)$ has zero mean and variance $1/i_\theta$, which is not the same as $E_\theta\{\sqrt{n}\,(\hat{\theta}_n - \theta)\} \to 0$ and $Var_\theta[\sqrt{n}\,(\hat{\theta}_n - \theta)] \to 1/i_\theta$. The latter two conditions (the second of which is asymptotic efficiency back 30 {)} may, in fact, hold; however, additional conditions are required to assume this. [These properties can be examined via (IV.D.32).] Nevertheless, the conclusion of Proposition IV.D.2 is sufficient practical justification for considering the MLE to be an asymptotically optimum (MVUE) estimator. And, in fact, asymptotic unbiasedness and efficiency are often alternatively defined in terms of the mean and variance of the asymptotic error distribution.

IV.E Further Aspects and Extensions of Maximum-Likelihood Estimation

IV.E.1 Estimation of Vector Parameters

It should be noted that all of the analysis of the preceding section can be generalized to the case in which the parameter is a vector, say of dimension m. In this case, the likelihood equation is a vector equation

$$\begin{gathered} \frac{\partial}{\partial \theta_1} \log p_{\underline{\theta}}(y)\Big|_{\underline{\theta}=\underline{\hat{\theta}}} = 0 \\ \vdots \\ \frac{\partial}{\partial \theta_m} \log p_{\underline{\theta}}(y)\Big|_{\underline{\theta}=\underline{\hat{\theta}}} = 0, \end{gathered} \qquad \text{(IV.E.1)}$$

which for i.i.d. models becomes

$$\sum_{k=1}^{n} \psi_1(Y_k ; \underline{\hat{\theta}}_n) = 0$$

$$\vdots \qquad \text{(IV.E.2)}$$

$$\sum_{k=1}^{n} \psi_m(Y_k ; \underline{\hat{\theta}}_n) = 0,$$

where $\psi_j(Y_k ; \underline{\theta}) = \partial \log f_{\underline{\theta}}(Y_k) / \partial \theta_j$ and where $f_{\underline{\theta}}$ is the marginal of Y_k.

The information inequality (Proposition IV.C.4) can, within regularity, be extended to the vector case. For example, the Cramér-Rao lower bound in the variance of unbiased estimates becomes

$$Cov_{\underline{\theta}}(\underline{\hat{\theta}}) \geqslant \mathbf{I}_{\underline{\theta}}^{-1}, \qquad \text{(IV.E.3)}$$

where $Cov_{\underline{\theta}}(\underline{\hat{\theta}}) \triangleq E_{\underline{\theta}}\{(\underline{\hat{\theta}}-\underline{\theta})(\underline{\hat{\theta}}-\underline{\theta})^T\}$, and $\mathbf{I}_{\underline{\theta}}$ is the $m \times m$ *Fisher information matrix* with $j-l^{\underline{th}}$ element

$$(\mathbf{I}_{\underline{\theta}})_{j,l} = E_{\underline{\theta}}\left\{\left[\frac{\partial}{\partial \theta_j} \log p_{\underline{\theta}}(Y)\right]\left[\frac{\partial}{\partial \theta_l} \log p_{\underline{\theta}}(Y)\right]\right\}. \qquad \text{(IV.E.4)}$$

Note that $\mathbf{I}_{\underline{\theta}}$ is the covariance matrix of the zero-mean vector

$$\left(\frac{\partial}{\partial \theta_1} \log p_{\underline{\theta}}(Y), \frac{\partial}{\partial \theta_2} \log p_{\underline{\theta}}(Y), \ldots, \frac{\partial}{\partial \theta_m} \log p_{\underline{\theta}}(Y)\right)^T,$$

and so it is at least nonnegative definite. Equation (IV.E.3) assumes that it is positive definite. The inequality $\mathbf{A} \geqslant \mathbf{B}$ for matrices mean that $(\mathbf{A}-\mathbf{B})$ is nonnegative definite. For the i.i.d. case, (IV.E.3) becomes

$$Cov_{\underline{\theta}}(\underline{\hat{\theta}}) \geqslant \frac{1}{n} \mathbf{i}_{\underline{\theta}}^{-1} \qquad \text{(IV.E.5)}$$

where

$$(\mathbf{i}_{\underline{\theta}})_{j,l} = E_{\underline{\theta}}\{\psi_j(Y_1;\underline{\theta})\psi_l(Y_1;\underline{\theta})\}. \qquad \text{(IV.E.6)}$$

Within conditions similar to those of Proposition IV.D.1, solutions to the likelihood equation are consistent, i.e.,

$$\|\underline{\hat{\theta}}_n - \underline{\theta}\| \triangleq \left[\frac{1}{m}\sum_{j=1}^{m}[(\underline{\hat{\theta}}_n)_j - \theta_j)]^2\right]^{1/2} \to 0 \ (i.p.); \qquad \text{(IV.E.7)}$$

and within conditions similar to those of Proposition IV.D.2,

$$\sqrt{n}\,(\underline{\hat{\theta}}_n - \underline{\theta}) \to N(\underline{0}, \mathbf{i}_{\underline{\theta}}^{-1}) \qquad \text{(IV.E.8)}$$

in distribution. Thus the vector parameter case is very similar to the scalar one.

Details of this and other aspects of the behavior of MLEs for i.i.d. models can be found in the book by Lehmann (1983).

IV.E.2 Estimation of Signal Parameters

The asymptotic properties of MLEs can also be extended to some time varying problems. Of particular interest is the situation in which we have real-valued observations of the form

$$Y_k = s_k(\theta) + N_k, \quad k = 1, ..., n, \qquad \text{(IV.E.9)}$$

where $\{s_k(\theta)\}_{k=1}^n$ is a signal sequence that is a known function of the unknown parameter θ, and where $\{N_k\}_{k=1}^n$ is an i.i.d. noise sequence with marginal probability density f. We assume for simplicity that θ is a scalar parameter lying in an interval Λ.

The maximum-likelihood estimate of θ in (IV.E.9) solves the equation

$$\hat{\theta}_n = \arg \max_{\theta \in \Lambda} \left[\sum_{k=1}^{n} \log f \,[Y_k - s_k(\theta)] \right],$$

or equivalently,

$$\hat{\theta}_n = \arg \min_{\theta \in \Lambda} \left[- \sum_{k=1}^{n} \log f \,[Y_k - s_k(\theta)] \right], \quad \text{(IV.E.10)}$$

and the likelihood equation is thus

$$\sum_{k=1}^{n} s_k'(\hat{\theta}_n)\psi[Y_k - s_k(\hat{\theta}_n)] = 0, \quad \text{(IV.E.11)}$$

where $\psi \triangleq -f'/f$, $f'(x) \triangleq df(x)/dx$, and $s_k'(\theta) \triangleq \partial s_k(\theta)/\partial\theta$. For example, when f is a $N(0, \sigma^2)$ density, (IV.E.10) and (IV.E.11) are equivalent to

$$\hat{\theta}_n = \arg \left[\min_{\theta \in \Lambda} \sum_{k=1}^{n} [Y_k - s_k(\theta)]^2 \right] \quad \text{(IV.E.12)}$$

and

$$\sum_{k=1}^{n} s_k'(\hat{\theta}_n)[Y_k - s_k(\hat{\theta}_n)] = 0, \quad \text{(IV.E.13)}$$

respectively. The particular estimator (IV.E.12) is sometimes known as the *least-squares estimate* of θ, since it chooses that value of θ for which $\{s_k(\theta)\}_{k=1}^{n}$ is the least-squares fit to the data. That is, it chooses θ to minimize the sum of the squared errors between the data and the signal that arises from that choice of θ. Least squares is a classical estimation technique and is used frequently in models such as (IV.E.9) even when the errors cannot be assumed to be Gaussian.

Solutions to the likelihood equation (IV.E.11) can have asymptotic properties similar to those for MLEs in i.i.d. models. However, the time variation of the signal adds different considerations to the asymptotic analysis. For example, if the signal becomes identically zero (or otherwise independent of θ) after some finite number of samples, it would be unrealistic to expect consistency in this model. To illustrate the types of conditions needed on the signal for the solutions to the likelihood equation (IV.E.11) to enjoy the properties of their i.i.d. counterparts, we will analyze the particular case of the least squares estimate (IV.E.13). Similar results will hold for the general case (IV.E.11) within sufficient regularity on ψ.

The equation (IV.E.13) satisfied by the least-squares estimate can be written using the observation model (IV.E.9) as

$$\sum_{k=1}^{n} s_k'(\hat{\theta}_n)n_k + \sum_{k=1}^{n} s_k'(\hat{\theta}_n)[s_k(\theta) - s_k(\hat{\theta}_n)] = 0. \quad \text{(IV.E.14)}$$

To analyze the behavior of $\hat{\theta}_n$, let us consider for each $\theta' \epsilon \Lambda$ the sequence of random variables

$$J_n(\theta;\theta') \triangleq \sum_{k=1}^{n} s_k'(\theta')N_k + \sum_{k=1}^{n} s_k'(\theta')[s_k(\theta) - s_k(\theta')]. \quad \text{(IV.E.15)}$$

Note that in the absence of noise ($N_k \equiv 0$), $\hat{\theta}_n = \theta$ is a solution to the likelihood equation (IV.E.14). However, unless $\theta' = \theta$ is the only root of

$$K_n(\theta;\theta') \triangleq \sum_{k=1}^{n} s_k'(\theta')[s_k(\theta) - s_k(\theta')], \quad \text{(IV.E.16)}$$

Equation (IV.E.14) may not lead to a perfect estimate even in the noiseless case. Thus for consistency in (IV.E.14), we would expect that we need the noise term, $\sum_{k=1}^{n} s_k'(\theta')N_k$, in (IV.E.15) to be asymptotically negligible relative to the term, $K_n(\theta;\theta')$, and for the latter term to have a unique root asymptotically. Since the solution to (IV.E.14) is unchanged if we divide each side by some $d_n > 0$, we can modify the statements above to apply to the corresponding terms in $J_n(\theta;\theta')/d_n$; i.e., if we can find a sequence $\{d_n\}_{n=1}^{\infty}$ such that $\sum_{k=1}^{n} s_k'(\theta')N_k/d_n$ is asymptotically negligible and $K_n(\theta;\theta')/d_n$ has a unique root asymptotically, then we can expect the roots of (IV.E.14) to be consistent by analogy with what happens in the i.i.d. case.

Note that, on assuming $N(0, \sigma^2)$ noise, we have

$$\frac{1}{d_n} J_n(\theta; \theta') \sim N\left(\frac{1}{d_n} K_n(\theta; \theta'), \frac{\sigma^2}{d_n^2} \sum_{k=1}^{n} [s_k'(\theta')]^2 \right). \tag{IV.E.17}$$

It is easily seen from this that for given $\theta, \theta' \epsilon \Lambda$, $J_n(\theta; \theta')/d_n$ converges in probability to a constant if and only if

$$\lim_{n \to \infty} \frac{1}{d_n^2} \sum_{k=1}^{n} [s_k'(\theta')]^2 = 0 \tag{IV.E.18}$$

and

$$\lim_{n \to \infty} \frac{1}{d_n} K_n(\theta; \theta') \text{ exists.} \tag{IV.E.19}$$

From this result we can prove the following proposition, which is analogous to Proposition IV.D.1.

Proposition IV.E.1: Consistency of Least Squares

Suppose that we have the model of (IV.E.9) with $N(0, \sigma^2)$ noise and that there exists a sequence of scalars $\{d_n\}_{n=1}^{\infty}$ such that (IV.E.18) and (IV.E.19) hold for all $\theta' \epsilon \Lambda$. Suppose further that $s_k(\theta')$, $s_k'(\theta')$, and

$$J(\theta; \theta') \triangleq \lim_{n \to \infty} \frac{1}{d_n} K_n(\theta; \theta') \tag{IV.E.20}$$

are all continuous functions of θ', and that $J(\theta; \theta')$ has a unique root at $\theta'=\theta$. Then, with probability tending to 1, the likelihood equation (IV.E.13) has a sequence of roots converging in probability to θ. In particular, if (IV.E.13) has a unique root $\hat{\theta}_n$ for each n, then $\hat{\theta}_n \to \theta$ (i.p.).

The proof of this result is virtually identical to that of Proposition IV.D.1, and is left as an exercise. As an example, consider the problem of signal-amplitude estimation (see Example IV.D.2), in which

$$s_k(\theta) = \theta s_k, \quad k = 1, 2, \ldots, n, \tag{IV.E.21}$$

for a known sequence $\{s_k\}_{k=1}^{\infty}$. In this case, we have $s_k'(\theta) = s_k$, so that $\sum_{k=1}^{n}[s_k'(\theta)]^2 = \sum_{k=1}^{n} s_k^2$ and $K_n(\theta;\theta') = (\theta-\theta')\sum_{k=1}^{n} s_k^2$. Thus a sufficient condition for consistency following from the proposition is the existence of a divergent sequence $\{d_n\}_{n=1}^{\infty}$ such that

$$0 < \lim_{n\to\infty} \frac{1}{d_n} \sum_{k=1}^{n} s_k^2 < \infty. \tag{IV.E.22}$$

Asymptotic normality can also be assured for the least-squares estimate in (IV.E.9) under regularity conditions on the signal sequence. Note that if $s_k(\theta)$ has third derivatives, the likelihood equation can be expanded in a Taylor series about θ, to give

$$\sum_{k=1}^{n} s_k'(\theta)[Y_k - s_k(\theta)]$$

$$+ (\hat{\theta}_n - \theta) \sum_{k=1}^{n} [s_k''(\theta)[Y_k - s_k(\theta)] - [s_k'(\theta)]^2] \tag{IV.E.23}$$

$$+ \tfrac{1}{2}(\hat{\theta}_n - \theta)^2 \sum_{k=1}^{n} [s_k'''(\bar{\theta}_n)[Y_k - s_k(\bar{\theta}_k)] - 3 s_k''(\bar{\theta}_n) s_k'(\bar{\theta}_n)] = 0.$$

with $\bar{\theta}_n$ between θ and $\hat{\theta}_n$. On rearranging we have

$$\hat{\theta}_n - \theta$$

$$= \frac{-\sum_{k=1}^{n} s_k'(\theta) N_k}{\sum_{k=1}^{n} s_k''(\theta) N_k - \sum_{k=1}^{n} [s_k'(\theta)]^2 + \tfrac{1}{2}(\hat{\theta}_n - \theta) \sum_{k=1}^{n} Z_k(\bar{\theta}_n)} \tag{IV.E.24}$$

where

$$Z_k(\theta') \triangleq [s_k'''(\theta')[N_k + s_k(\theta) - s_k(\theta')] - 3s_k''(\theta')s_k'(\theta')].$$

From this expression for the error, the following result can be proven.

Proposition IV.E.2: Asymptotic Normality of Least Squares

Suppose that we have the model of (IV.E.9) with $N(0, \sigma^2)$ noise, and $\{\hat{\theta}_n\}_{n=1}^{\infty}$ is a consistent sequence of least-squares estimates of θ. Suppose that further than the following regularity conditions hold:

(1) There exists a function M such that $|Z_k(\theta')| \leqslant M(N_k)$ uniformly in θ', and $E_\theta\{M(N_k)\} < \infty$. [The existence of the relevant derivatives of $s_k(\theta)$ is also assumed.]

(2) $\lim_{n\to\infty}(1/n)\sum_{k=1}^{n}[s_k'(\theta)]^2 > 0$.

(3) $\lim_{n\to\infty}\sum_{k=1}^{n}[s_k''(\theta)]^2/[\sum_{k=1}^{n}[s_k'(\theta)]^2]^2 = 0$.

Then,

$$\left(\sum_{k=1}^{n}[s_k'(\theta)]^2\right)^{1/2}(\hat{\theta}_n - \theta) \to N(0, \sigma^2) \qquad \text{(IV.E.25)}$$

in distribution.

The proof of this result is similar to that for the analogous i.i.d. case and is left as an exercise. Note that Fisher's information is given here by

$$I_\theta = \sum_{k=1}^{n}[s_k'(\theta)]^2/\sigma^2. \qquad \text{(IV.E.26)}$$

Thus in the same sense as in the i.i.d. case, the least-squares estimate is asymptotically efficient for (IV.E.9) with $N(0, \sigma^2)$ errors.

The signal-amplitude estimation problem, $s_k(\theta)=\theta s_k$, again provides a straightforward example. In this case, the differentiability conditions are trivial, $Z_k(\theta')\equiv 0$, and $s_k''(\theta)=0$; thus the only condition needed for asymptotic normality is that $\lim_{n\to\infty} \sum_{k=1}^{n} s_k^2/n > 0$. Recall, however, that the desirable properties of the MLE in this particular case follow by direct analysis (even for finite n), as was seen in Example IV.D.2.

A less obvious example is given by the following.

Example IV.E.1: Identification of a First-Order Linear System

An important class of applications of parameter estimation problems falls within the context of *system identification*, in which we wish to infer the structure of some input/output system by putting in an input and observing the output. One of the simplest possible identification problems is that of identifying a stable first-order time-invariant linear system. This type of system can be described by the signal model

$$s_k(\theta) = \theta s_{k-1}(\theta) + u_k, \quad k = 1, 2, ..., n, \tag{IV.E.27}$$

where $|\theta|<1$ and $\{u_k\}_{k=1}^{n}$ is the known input sequence. Note that θ here is the coefficient of the homogeneous equation $s_k(\theta)=\theta s_{k-1}(\theta)$, and thus this parameter completely determines the system once we have made the assumptions of linearity, time invariance, and unit order. The observation of the system output is usually corrupted by measurement noise, so assuming that this noise is i.i.d., the estimation of θ is a problem in the form of IV.E.9. We consider the case of $N(0, \sigma^2)$ errors and the least-squares estimate of θ.

Assume that the system (IV.E.27) is initially at rest $[s_0(\theta)=0]$, in which case the solution to (IV.E.27) is given by

$$s_k(\theta) = \sum_{l=1}^{k} \theta^{k-l} u_l. \tag{IV.E.28}$$

Whether or not θ can be identified (as $n \to \infty$) depends on the input sequence $\{u_k\}_{k=1}^n$. Consider, for example, a constant input signal $u_k = 1$ for all $k \geqslant 1$. The output is then

$$s_k(\theta) = \sum_{l=1}^{k} \theta^{i-l} = \sum_{m=0}^{k-1} \theta^m = \frac{1-\theta^k}{1-\theta},$$

and

$$s_k'(\theta) = \frac{(1-k\theta^{k-1})(1-\theta) + (1-\theta^k)}{(1-\theta)^2}.$$

This implies that

$$\lim_{n\to\infty} \frac{1}{n} \sum_{k=1}^{n} [s_k'(\theta)]^2 = (2-\theta)^2/(1-\theta)^4 \tag{IV.E.29}$$

and

$$\lim_{n\to\infty} \left| \frac{1}{n} \sum_{k=1}^{n} [s_k'(\theta')[s_k(\theta) - s_k(\theta')] \right| = \frac{(2-\theta')(\theta-\theta')}{(1-\theta')^2(1-\theta)}. \tag{IV.E.30}$$

Since (IV.E.30) has a unique root at $\theta' = \theta$ and the relevant quantities are continuous for $|\theta'| < 1$, (IV.E.29) and (IV.E.30) imply that the hypothesis of Proposition IV.E.1 is satisfied with $d_n = n$. Thus we have a consistent sequence of roots to the likelihood equation. [In fact, since $J(\theta; \theta')$ is bounded away from zero off a neighborhood of $\theta' = \theta$, it can be shown that any sequence of roots is consistent.]

It is not difficult to see why the consistent estimation of θ is possible in this case. Note that the asymptotic value of $s_k(\theta)$ is $1/(1-\theta)$. Thus the system achieves a unique steady-state value for each value of parameter θ. From this we would expect to be able to determine the parameter value perfectly by observing the noisy output for $k = 1, 2, \ldots, \infty$, since the noise can be averaged out in infinite time. On the other hand, suppose that we use an input with only finite duration. Then, since the system is stable, the steady-state output of the system is zero for every parameter value. It is easy to see that the hypothesis of Proposition IV.E.1 fails to hold in this case. If the measurement noise were not present, it might be possible to determine the parameter perfectly in this case from the transient behavior; however, the presence of the noise makes it necessary that the parameter be identifiable in the steady state as well. The quality of an input that produces this effect is sometimes known as *persistence of excitation*. (A related quality that is sometimes required of an input in linear-system identification problems is *sufficient richness*. Basically, this property means that the frequency content of the input signal is sufficiently rich to excite all oscillatory modes of the system.)

For the constant input signal, Proposition IV.E.2 cannot be applied directly to this model with $\Lambda=(-1,1)$ because $Z_k(\theta')$ cannot be uniformly bounded on this set. However, if we assume that θ is bounded away from unity [i.e., if we take $\Lambda=(-1,\theta_u)$ with $\theta_u<1$], then the regularity conditions of Proposition IV.D.4 do hold, and asymptotic normality and efficiency of the consistent roots of the likelihood equation follow. Note that the asymptotic variance of $\sqrt{n}\,(\hat{\theta}_n-\theta)$ in this case is $\sigma^2(1-\theta)^4/(2-\theta)^2$.

Some additional aspects of maximum-likelihood and least squares estimates of signal parameters are discussed below and in Chapter VII. However, before leaving this subject for now, we note that the properties of least squares summarized in Propositions IV.D.3 and IV.D.4 hold more generally. In particular, we have the following.

Proposition IV.E.3: Consistency and Asymptotic Normality of Least-Squares with Non-Gaussian Noise

Propositions IV.E.1 and IV.E.2 remain valid if the assumption $N_k \sim N(0,\sigma^2)$ is replaced by the assumption $E\{N_k\}=0$ and $E\{N_k^2\}=\sigma^2<\infty$.

Note, however, that this result does not imply that least squares is asymptotically efficient when the noise is not Gaussian, since Fisher's information is no longer given by (IV.E.26) in the non-Gaussian case.

IV.E.3 Robust Estimation of Signal Parameters

Consider again the model of (IV.E.9), in which we have noted that MLEs are asymptotically optimum in the sense of minimum asymptotic variance. As we discussed in Section III.E, statistical models such as this are only approximately valid in practice, and an important question arising in such situations is whether or not procedures designed for a particular model are *robust*; i.e., whether their performance is insensitive to small changes in the model.

Consider, for example, a nominal model in which the noise samples have the $N(0, 1)$ distribution. Then, within regularity, and assuming that $e_\theta \triangleq \lim_{n\to\infty}\sum_{k=1}^{n}[s_k'(\theta)]^2/n$ exists and is positive, the least-squares estimate is asymptotically $N(\theta, 1/ne_\theta)$. Suppose, however, that the actual statistical behavior of the noise is described by a pdf that is only approximately $N(0, 1)$. For example, suppose that the noise density f is of the form

$$f(x) = (1-\epsilon)\frac{1}{\sqrt{2\pi}}e^{-x^2/2} + \epsilon h(x),\ x \in \mathbb{R}, \tag{IV.E.31}$$

where $h(x)$ is an arbitrary density, symmetric about zero, and with variance

$$\sigma_h^2 \triangleq \int_{-\infty}^{\infty} x^2 h(x) dx$$

finite but not bounded. Then, by Proposition IV.E.3, the least-squares estimate will have asymptotic variance

$$v_h^2 \sim \frac{(1-\epsilon) + \epsilon\sigma_h^2}{ne_\theta}. \tag{IV.E.32}$$

Note that v_h^2 can be arbitrarily large for any $\epsilon > 0$ since σ_h^2 is not bounded. In particular, the worst-case asymptotic variance over the class of densities (IV.E.31) is

$$\sup_h [(1-\epsilon) + \epsilon \sigma_h^2] = \infty \tag{IV.E.33}$$

for any $\epsilon > 0$.

This points to a lack of robustness of the least-squares estimate for situations in which a small fraction of the noise samples may come from a high variance distribution. (This may happen, for example, in radar measurements, in which very high-variance impulsive interference may be present in a small fraction ϵ of the measurements. Observations that are improbably large for a given nominal model are sometimes termed *outliers*.) As in the signal detection problems treated in Section III.E, an alternative to asymptotic variance at a nominal model is needed as a design criterion for such situations.

Suppose that the noise density f in (IV.E.9) is an even symmetric function. Consider estimates of θ of the form

$$\sum_{k=1}^{n} s_k'(\hat{\theta}_n)\psi(Y_k - s_k(\hat{\theta}_n))] = 0, \tag{IV.E.34}$$

where ψ is a general odd-symmetric function. With $\psi(x) = x$, (IV.E.34) gives the least-squares estimate, and with $\psi(x) = -f'(x)/f(x)$, (IV.E.34) gives the MLE. Estimates of this form are known as M-estimates. Assuming that $0 < e_\theta < \infty$ and within regularity on ψ, f, and $\{s_k(\theta)\}_{k=1}^{\infty}$, it can be shown, using the techniques developed above, that M-estimates are consistent and asymptotically $N[\theta, V(\psi, f)/ne_\theta]$, where

$$V(\psi, f) \triangleq \frac{\int \psi^2 f}{(\int \psi' f)^2} \tag{IV.E.35}$$

with $\psi'(x) = d\psi(x)/dx$.

In view of these properties, one possible way of designing a robust estimator for an uncertainty class F of noise densities is to seek a function ψ that minimizes the worst case M-

estimate variance, $\sup_{f \in F} V(\psi, f)$. That is, one possible design method is to restrict attention to M-estimates and solve

$$\min_{\psi} \sup_{f \in F} V(\psi; f). \qquad \text{(IV.E.36)}$$

The problem (IV.E.37) has been studied by Huber (1981) for general sets F. Within appropriate conditions, its solution is basically as follows.

Consider the functional

$$I(f) \triangleq \int (f')^2/f\,, \qquad \text{(IV.E.37)}$$

and let f_L be a density in F that minimizes $I(f)$ over F; i.e.,

$$I(f_L) = \min_{f \in F} I(f). \qquad \text{(IV.E.38)}$$

Then the M-estimate with ψ-function $\psi_R(x) = -f'_L(x)/f_L(x)$ solves (IV.E.36). Note that for any f,

$$V(\psi, f)\Big|_{\psi=-f'/f} = 1/I(f), \qquad \text{(IV.E.39)}$$

so that $[ne_\theta I(f)]^{-1}$ is the asymptotic variance of the MLE in our model with given f. [Fisher's information here is $ne_\theta I(f)$.] Thus f_L is the member of F whose corresponding optimum estimate (the MLE) has the worst optimum performance. For this reason f_L can be considered a *least-favorable density*, and the robust M-estimate is the best estimate for this least-favorable model.

The problem $\min_{f \in F} I(f)$ has been solved for a number of uncertainty models of [see Huber (1981)]. For example, for the ϵ-contaminated $N(0, 1)$ model of (IV.E.31), the least favorable density is given by

$$f_L(x) = \begin{cases} (1-\epsilon)\frac{1}{\sqrt{2\pi}}e^{-x^2/2} & \text{if } |x| \leqslant k' \\ (1-\epsilon)e^{-k'(|x|-k')}\frac{1}{\sqrt{2\pi}}e^{(-k')^2/2} & \text{if } |x| > k', \end{cases} \quad \text{(IV.E.40)}$$

where k' is a constant given by the solution to

$$(1-\epsilon)^{-1} = 2\Phi(k') - 1 + \frac{1}{k'}(2/\pi)^{1/2}e^{-(k')^2/2}. \quad \text{(IV.E.41)}$$

The corresponding robust ψ function is

$$\psi_k(x) = \begin{cases} x & \text{if } |x| \leqslant k' \\ k' \, sgn(x) & \text{if } |x| > k'. \end{cases} \quad \text{(IV.E.42)}$$

Thus, as in the analogous hypothesis testing problem, robustness is brought about by limiting the effects of outliers.

For further discussion of this and other approaches to robust estimation, the reader is referred to the survey article by Kassam and Poor (1985) and the books by Huber (1981) and Hampel, *et al.* (1986).

IV.E.4 Recursive Parameter Estimation

We see from the preceding discussions that maximum-likelihood estimates often have nice properties, particularly when the sample size is large. However, they sometimes have the disadvantages of being cumbersome to compute. For example, with n i.i.d. samples drawn from the density f_θ, computation of the MLE requires the maximization of the function $\sum_{k=1}^{n} \log f_\theta(y_k)$. Unless the maximizing θ can be found as a closed-form function of $\underline{y}$, an iterative technique must be used to find $\hat{\theta}_{ML}(\underline{y})$. This requires the storage and simultaneous manipulation of all n samples (unless a lower-dimensional sufficient statistic is available), a task that is undesirable if n is very large. It is thus sometimes desirable to consider alternatives to maximum likelihood that can be implemented in a recursive or sequential manner so that the contribution of each sample to the estimate is computed as the sample is taken.

One such estimation technique is suggested by the MLE. In particular, consider a consistent sequence $\{\hat{\theta}_n\}_{n=1}^{\infty}$ solving the likelihood equation

$$\sum_{k=1}^{n} \psi(Y_k\,;\hat{\theta}_n)=0 \tag{IV.E.43}$$

with $\psi(Y_k\,;\theta)=\partial \log f_\theta(Y_k)/\partial\theta$, as before. Since $\{\hat{\theta}_n\}_{n=1}^{\infty}$ is consistent, the difference, $\hat{\theta}_n-\hat{\theta}_{n-1}$, converges to zero as $n\to\infty$. Thus (IV.E.43) can be approximated by expanding about $\hat{\theta}_{n-1}$ to give

$$\sum_{k=1}^{n} \psi(Y_k\,;\hat{\theta}_{n-1})+(\hat{\theta}_n-\hat{\theta}_{n-1})\sum_{k=1}^{n} \psi'(Y_k\,;\hat{\theta}_{n-1})\sim 0, \tag{IV.E.44}$$

with $\psi'(Y_k\,;\theta)=\partial\psi(Y_k\,;\theta)/\partial\theta$. Rearranging (IV.E.44) gives

$$\hat{\theta}_n \sim \hat{\theta}_{n-1}-\frac{\sum_{k=1}^{n} \psi(Y_k\,;\hat{\theta}_{n-1})}{\sum_{k=1}^{n} \psi'(Y_k\,;\hat{\theta}_{n-1})}. \tag{IV.E.45}$$

Since $\hat{\theta}_{n-1}$ solves $\sum_{k=1}^{n-1}\psi(Y_k\,;\hat{\theta}_{n-1})=0$, the numerator sum on the right side of (IV.E.45) has only one term, $\psi(Y_n\,;\hat{\theta}_{n-1})$. Let us write the denominator sum as

$$n\left[\frac{1}{n}\sum_{k=1}^{n} \psi'(Y_k\,;\hat{\theta}_{n-1})\right]. \tag{IV.E.46}$$

Now, the weak law of large numbers implies that

$$-\frac{1}{n}\sum_{k=1}^{n} \psi'(Y_k\,;\theta)\to i_\theta\;(i.p.),$$

where $i_\theta = -E_\theta\{\psi'(Y_k;\theta)\} = E\{\psi^2(Y_k;\theta)\}$ is Fisher's information per sample. Since $\hat{\theta}_{n-1} \to \theta$, we can approximate

$$\frac{1}{n}\sum_{k=1}^{n}\psi'(Y_k;\hat{\theta}_{n-1}) \sim i_{\hat{\theta}_{n-1}}. \qquad \text{(IV.E.47)}$$

On combining (IV.E.45) and (IV.E.47) we have that, asymptotically, a consistent sequence of solutions to the likelihood equation will satisfy

$$\hat{\theta}_n \sim \hat{\theta}_{n-1} + \frac{\psi(Y_n;\theta_{n-1})}{n i_{\hat{\theta}_{n-1}}}. \qquad \text{(IV.E.48)}$$

This is an asymptotic recursive equation for $\hat{\theta}_n$, since $\hat{\theta}_n$ is computed from $\hat{\theta}_{n-1}$ and Y_n only.

It turns out that the (nonasymptotic) recursion

$$\hat{\theta}_n = \hat{\theta}_{n-1} + \frac{\psi(Y_n;\hat{\theta}_{n-1})}{n i_{\hat{\theta}_{n-1}}}, \quad n = 1, \ldots, \qquad \text{(IV.E.49)}$$

(with $\hat{\theta}_0$ arbitrary back 30 {)} suggested by (IV.E.48) has the same desirable asymptotic properties (i.e., consistency and efficiency back 30 {)} as the MLE within regularity on the model. This recursion is an example of a more general class of recursive parameter estimation algorithm known as *stochastic approximation* algorithms. Because of their recursive nature, such algorithms are of considerable interest in applications in which on-line or real-time parameter estimation is necessary. In modified form they are also useful in real-time tracking of slowly varying parameters. The reader interested in further aspects of such algorithms is referred to the book by Nevel'son and Has'minskii (1973). Similar recursive modifications of the MLE and least-squares estimates for time-varying problems such as (IV.E.9) have also been developed. The reader is referred to Ljung and Soderstrom (1982) and Goodwin and Sin (1984) for the development of these ideas.

IV.F Exercises

1. Suppose Θ is a random parameter and that, given $\Theta=\theta$, the real observation Y has density

$$p_\theta(y) = (\theta/2)\, e^{-\theta|y|}, \quad y \in \mathbb{R}.$$

Suppose further that Θ has prior density

$$w(\theta) = \begin{cases} 1/\theta, & 1 \leqslant \theta \leqslant e \\ 0, & \text{otherwise.} \end{cases}$$

(a) Find the MAP estimate of Θ based on Y.
(b) Find the MMSE estimate of Θ based on Y.

2. Suppose we have a real observation Y given by

$$Y = N + \Theta S$$

where $N \sim N(0, 1)$, $P(S=1)=P(S=-1)=½$, and Θ has pdf

$$w(\theta) = \begin{cases} K\, e^{\theta^2/2}, & 0 \leqslant \theta \leqslant 1 \\ 0, & \text{otherwise} \end{cases}$$

where $K=[\int_0^1 e^{\theta^2/2} d\theta]^{-1}$. Assume that N, Θ, and S are independent.
(a) Find the MMSE estimate of Θ given $Y=y$.
(b) Find the MAP estimate of Θ given $Y=y$.

3. Suppose Θ is a random parameter with prior density

$$w(\theta) = \begin{cases} \alpha e^{-\alpha\theta}, & \theta \geqslant 0 \\ 0, & \theta < 0 \end{cases}$$

where $\alpha > 0$ is known. Suppose our observation Y is a Poisson random variable with rate Θ; i.e., that

$$p_\theta(y) \equiv P(Y = y | \Theta = \theta) = \frac{\theta^y e^{-\theta}}{y!}, \quad y = 0, 1, 2, \ldots$$

Find the MMSE and MAP estimates of Θ based on Y. How would you find the MMAE estimate?

4. Suppose we have a single observation y of a random variable Y given by

$$Y = N + \Theta$$

where N is a Gaussian random variable with mean zero and variance σ^2. The parameter Θ is a random variable, independent of N, with probability mass function

$$w(\theta) = P(\Theta = \theta) = \begin{cases} 1/2, & \theta = -1 \\ 1/2, & \theta = +1. \end{cases}$$

(a) Find $\hat{\theta}_{MMSE}$ and $\hat{\theta}_{MAP}$.
(b) Under what conditions are the two estimates in (a) approximately equal?

5. Suppose Θ is a random parameter with prior density

$$w(\theta) = \begin{cases} e^{-\theta}, & \theta \geqslant 0 \\ 0, & \theta < 0, \end{cases}$$

and that Y has conditional density

$$p_\theta(y) = \tfrac{1}{2} e^{-|y-\theta|}, \quad -\infty < y < \infty.$$

Find $\hat{\theta}_{MMSE}$ and $\hat{\theta}_{MAP}$.

6. Suppose that N_1 and N_2 are two jointly-Gaussian random variables with zero means, unit variances, and correlation coefficient ρ ($|\rho|<1$). Suppose further that we observe Y_1 and Y_2 given by

$$Y_k = \frac{N_k}{\sqrt{\Theta}}, \quad k = 1, 2$$

where Θ is a random parameter, independent of N_1 and N_2, with prior density

$$w(\theta) = \begin{cases} 1/\alpha, & \theta \in [0, \alpha] \\ 0, & \theta \notin [0, \alpha] \end{cases}$$

where $\alpha > 0$ is known.
(a) Find the minimum-mean-squared-error estimate of Θ.
(b) Find the MAP estimate of Θ.
(c) Find the minimum-mean-absolute-error estimate of Θ.

7. Suppose Θ is uniformly distributed on the interval (0, 1) and that we observe $Y = N + \Theta$ where N is a random variable, independent of Θ, with density

$$p_N(n) = \begin{cases} e^{-n} & , n \geqslant 0 \\ 0 & , n < 0. \end{cases}$$

Find $\hat{\theta}_{MMSE}$, $\hat{\theta}_{ABS}$, and $\hat{\theta}_{MAP}$.

8. (a) Consider the observation model of Exercise 7 but with the prior of Exercises 5. (I.e., N and Θ both have the unit exponential distribution). Find the MMSE and MMAE estimates of Θ based on Y.
(b) Find the minimum mean-squared-error for (a).
(c) Consider now the observation model

$$Y_k = N_k + \Theta, \quad k = 1, ..., n,$$

where $N_1, N_2, ..., N_n$, and Θ are i.i.d random variables with the unit exponential distribution. Find the MAP estimate of Θ based on $Y_1, Y_2, ..., Y_n$.

9. Repeat Exercise 1 for the situation in which we have a sequence of observations $Y_1, Y_2, ..., Y_n$, which are conditionally i.i.d. with the given pdf p_θ given $\Theta=\theta$.

10. Derive Eq. (IV.B.47).

11. Suppose that we observe a sequence

$$Y_k = X_k + N_k, \quad k = 1, ..., n$$

where $N_1, ..., N_n$ is a sequence of independent Gaussian random variables, each with zero mean and variance σ^2, and $X_1, ..., X_n$ are defined by the equations

$$X_0 = \Theta$$

$$X_k = \alpha X_{k-1}, \quad k = 1, ..., n$$

where α is known and Θ is a Gaussian random parameter with zero mean and variance q^2.
(a) Assuming that Θ and $\underline{N}$ are independent, find the MMSE estimate of Θ based on $Y_1, \ldots, Y_n$.
(b) For each $n=1, 2, \ldots$, let $\hat{\theta}_n$ denote the MMSE estimate of Θ based on $Y_1, \ldots, Y_n$. Show that $\hat{\theta}_n$ can be computed recursively by

$$\hat{\theta}_n = K_n^{-1}[K_{n-1}\hat{\theta}_{n-1} + \alpha^n y_n], \quad n = 1, 2, \ldots$$

where $\hat{\theta}_0=0$ and the coefficients K_n are defined by

$$K_0 = q^{-2} \text{ and } K_n = K_{n-1} + \alpha^n, \quad n = 1, 2, \ldots .$$

Draw a block-diagram of this implementation.
(c) Find an expression for the mean squared error

$$e_n = E\{(\hat{\theta}_n - \Theta)^2\}, \quad n = 1, 2, \ldots$$

What happens when $n \to \infty$; $q^2 \to \infty$; $\sigma^2 \to 0$; $\alpha < 1$; $\alpha = 1$; $\alpha > 1$?

12. Suppose θ is a nonrandom parameter satisfying $\theta > 1$. Suppose further that, given $\theta, Y_1, Y_2, \ldots, Y_n$ are i.i.d. observations with each density

$$f_\theta(y) = \begin{cases} (\theta - 1)y^{-\theta}, & y \geq 1 \\ 0, & y \leq 1. \end{cases}$$

Find a sufficient statistic for θ which has a complete family of distributions. Justify your answer.

13. Suppose we toss a coin n independent times and define an observation sequence

$$Y_k = \begin{cases} 1 \text{ if the } k^{th} \text{ outcome is heads} \\ \\ 0 \text{ if the } k^{th} \text{ outcome is tails} \end{cases}$$

$k=1, 2, \ldots, n$. Let $\theta = P(Y_k = 1), k = 1, \ldots, n$.
(a) Find an MVUE of θ.
(b) Find the ML estimate of θ. Find its bias and variance.
(c) Compute the Cramér-Rao lower bound and compare with results from (a) and (b).

14. Derive Eq. (IV.D.22).

15. Suppose Y is Poisson. Find the ML estimate of its rate. Compute the bias, variance, and Cramér-Rao lower bound.

16. Suppose θ is a positive (nonrandom) parameter. Suppose further that we have a sequence of observations $Y_1, \ldots, Y_n$ where, given θ, $Y_1, \ldots, Y_n$ are i.i.d. each with pdf

$$f_\theta(y) = \begin{cases} \dfrac{(y)^M e^{-y/2\theta}}{2^M \theta^M M!}, & y \geq 0 \\ \\ 0, & y < 0 \end{cases}$$

where M is a known positive integer.
(a) Find the ML estimate of θ.
(b) Compute the bias and variance of the estimate from part (a).
(c) Compute the Cramér-Rao lower bound on the variance of unbiased estimates of θ.
(d) Is the ML estimate consistent? Is it efficient?

17. Suppose we observe two jointly Gaussian random variables Y_1 and Y_2, each of which has zero mean and unit

variance. We want to estimate the correlation coefficient $\rho = E\{Y_1 Y_2\}$.
(a) Find the equation for the maximum-likelihood estimate of ρ based on observation of (Y_1, Y_2).
(b) Compute the Cramér-Rao lower bound for unbiased estimates of ρ.

18. Suppose we observe a sequence $Y_1, Y_2, ..., Y_n$ given by

$$Y_k = N_k + \theta s_k, \quad k = 1, ..., n$$

where $\underline{N} = (N_1, ..., N_n)^T$ is a zero-mean Gaussian random vector with covariance matrix $\Sigma > 0$; $s_1, s_2, ..., s_n$ is a known signal sequence; and θ is a (real) nonrandom parameter.
(a) Find the maximum-likelihood estimate of the parameter θ.
(b) Compute the bias and variance of your estimate.
(c) Compute the Cramér-Rao lower bound for unbiased estimates of θ and compare with your result from (b).
(d) What can be said about the consistency of $\hat{\theta}_{ML}$ as $n \to \infty$? Suppose, for example, that there are positive constants a and b such that

$$\frac{1}{n} \sum_{k=1}^{n} s_k^2 > a \text{ for all } n$$

and

$$\lambda_{\min}(\Sigma^{-1}) > b \text{ for all } n$$

where $\lambda_{\min}(\Sigma^{-1})$ denotes the minimum eigenvalue of the matrix Σ^{-1}.

19. Suppose θ is a positive nonrandom parameter and that we

have a sequence $Y_1, ..., Y_n$ of observations given by

$$Y_k = \theta^{1/2} N_k, \quad k = 1, 2, ..., n$$

where $\underline{N} = (N_1, ..., N_n)^T$ is a Gaussian random vector with zero mean and covariance matrix $\mathbf{\Sigma}$. Assume that $\mathbf{\Sigma}$ is positive definite.

(a) Find the maximum-likelihood estimate of θ based on $Y_1, ..., Y_n$.

(b) Show that the maximum-likelihood estimate is unbiased.

(c) Compute the Cramér-Rao lower bound on the variance of unbiased estimates of θ.

(d) Compute the variance of the maximum-likelihood estimate of θ and compare to the Cramér-Rao lower bound.

20. Consider the observation model

$$Y_k = \theta^{1/2} s_k R_k + N_k, \quad k = 1, 2, ..., n$$

where $s_1, s_2, ..., s_n$ is a known signal, $N_1, N_2, ..., N_n, R_1, R_2, ..., R_n$ are i.i.d. $N(0, 1)$ random variables, and $\theta \geq 0$ is an unknown parameter.

(a) Find the likelihood equation for estimating θ from $Y_1, Y_2, ..., Y_n$.

(b) Find the Cramér-Rao lower bound on the variance of unbiased estimates of θ.

(c) Suppose $s_1, s_2, ..., s_n$ is a sequence of +1's and -1's. Find the MLE of θ explicitly.

(d) Compute the bias and variance of your estimate from (c), and compare the latter with the Cramér-Rao lower bound.

21. Suppose Y_1 and Y_2 are independent Poisson random variables each with parameter λ. Define the parameter θ by

$$\theta = e^{-\lambda}.$$

(a) Show that y_1+y_2 is a complete sufficient statistic for θ. (Assume λ ranges over $(0, \infty)$.)
(b) Define an estimate $\hat{\theta}$ by

$$\hat{\theta}(y) = \tfrac{1}{2}[f(y_1) + f(y_2)]$$

where f is defined by

$$f(y) = \begin{cases} 1 \text{ if } y = 0 \\ 0 \text{ if } y \neq 0. \end{cases}$$

Show that $\hat{\theta}$ is an unbiased estimate of θ.
(c) Find an MVUE of θ. (Hint: Y_1+Y_2 is Poisson with parameter 2λ.)
(d) Find the maximum-likelihood estimate of θ. Is the MLE unbiased; if so, why; if not, why not?
(e) Compute the Cramér-Rao bound on the variance of unbiased estimates of θ.

22. Suppose $\theta>0$ is a parameter of interest and that given θ, $Y_1, \ldots, Y_n$ is a set of i.i.d. observations with marginal distribution function

$$F_\theta(y) = [F(y)]^{1/\theta}, \quad -\infty<y<\infty,$$

where F is a known distribution function with pdf f.
(a) Show that

$$\hat{\theta}_{MV}(\underline{y}) = -\frac{1}{n}\sum_{k=1}^{n} \log F(y_k)$$

is an MVUE of θ.
(b) Suppose now that θ is replaced by a random variable Θ drawn at random using the prior density

$$w(\theta) = c^m \exp(-c/\theta)/(\Gamma(m)\theta^{m+1}), \quad \theta > 0,$$

where c and m are positive constants. Use the fact that $E\{\Theta\}=c/m$ to show that the MMSE estimator of Θ from $Y_1, ..., Y_n$ is

$$\hat{\theta}_{MMSE}(\underline{y}) = (k - \sum_{k=1}^{n} \log F(y_k))/(m+n).$$

(c) Compare $\hat{\theta}_{MV}$ and $\hat{\theta}_{MMSE}$ with regard to the role of the prior information.

23. Suppose we observe

$$Y_k = A \sin\left(\frac{k\pi}{2} + \Phi\right) + N_k, \quad k = 1, ..., n$$

where $\underline{N} \sim N(\underline{0}, \sigma^2 I)$ and n is even.
(a) Suppose A and Φ are nonrandom with $A \geqslant 0$ and $\Phi \in [-\pi, \pi]$. Find their ML estimates.
(b) Suppose A and Φ are random and independent with priors

$$w_\Phi(\phi) = \begin{cases} \frac{1}{2\pi}, & -\pi \leqslant \phi \leqslant \pi \\ 0, & \text{otherwise} \end{cases}$$

$$w_A(a) = \begin{cases} (a/\beta^2)e^{-a^2/2\beta^2} & \text{if } a \geqslant 0 \\ 0 & \text{if } a < 0 \end{cases}$$

where β is known. Assuming A and Φ are independent of $\underline{N}$, find the MAP estimates of A and Φ.
(c) Under what conditions are the estimates from (a) and (b) approximately equal?

24. Suppose that, given $\Theta=\theta$, $Y_1, \ldots, Y_n$ are i.i.d. real observations with marginal densities

$$f_\theta(y) = \begin{cases} \theta^{-1}e^{-y/\theta}, & y \geq 0 \\ 0, & y < 0. \end{cases}$$

(a) Find the maximum-likelihood estimate of θ based on $Y_1, \ldots, Y_n$. Compute its mean and variance.
(b) Compute the Cramér-Rao lower bound for the variance of unbiased estimates of θ.
(c) Suppose Θ is uniformly distributed on $(0, 1]$. Find the MAP estimate of Θ
(d) For $n=3$, find the MMSE estimate of Θ. Assume the same prior as in part (c).
(e) For $n=2$, find the MMAE estimate of Θ. Assume the same prior as in part (c).

25. Suppose that, given $\Theta=\theta$, the real observation Y has pdf

$$p_\theta(y) = \begin{cases} \dfrac{6(y^2+\theta y)}{2+3\theta}, & 0 \leq y \leq 1 \\ 0, & \text{otherwise.} \end{cases}$$

(a) Suppose Θ is uniformly distributed on [0, 1]. Find the MMSE estimate and corresponding minimum Bayes risk.
(b) With Θ as in (a), find the MAP estimate and the MMAE estimate of Θ.
(c) Find the maximum-likelihood estimate of θ and compute its bias.
(d) Compute the Cramér-Rao lower bound on the variance of unbiased estimates of θ.

V ELEMENTS OF SIGNAL ESTIMATION

V.A Introduction

In Chapter IV we discussed methods for designing estimators for static parameters, that is, for parameters that are not changing with time. In many applications we are interested in the related problem of estimating dynamic or time-varying parameters. In the traditional terminology, a dynamic parameter is usually called a *signal*, so the latter problem is known as *signal estimation* or *tracking*.

Such problems arise in many applications. For example, one function of many radar systems is to track targets as they move through the radar's scanning area. This means that the radar must estimate the position of the target (and perhaps its velocity) at successive times. Since the targets of interest are usually moving and the position measurements are noisy, this is a signal estimation problem. Another application is that of analog communications, in which analog information (e.g., audio or video) is transmitted by modulating the amplitude, frequency, or phase of a sinusoidal carrier. The receiver's function in this situation is to determine the transmitted information with as high a fidelity as possible on the basis of a noisy observation of the received waveform. Again, since the transmitted information is time varying, this problem is one of signal estimation.

The dynamic nature of the parameter in signal estimation problems adds a new dimension to the statistical modeling of these problems. In particular, the dynamic properties of the signal (i.e., how fast and in what manner it can change) must be modeled at least statistically in order to obtain meaningful signal estimation procedures. Also, performance expectations for estimators of dynamic parameters should be different from those for static parameters. In particular, unlike the static case, we cannot expect an estimator of a signal to be perfect as the number of observations becomes infinite because of the time variation in the signal.

In this chapter we discuss the basic ideas behind some of the signal estimation techniques used most often in practice. In Section V.B we discuss *Kalman-Bucy filtering*, which provides a very useful algorithm for estimating signals that are generated by finite-dimensional linear dynamical models. In Section V.C the general problem of estimating signals as linear transformations of the observations is developed, and in Section V.D a particular case of linear estimation, *Wiener-Kolmogorov filtering*, which is a method of estimating signals whose statistics are stationary in time, is considered.

V.B Kalman-Bucy Filtering

Many time-varying physical phenomena of interest can be modeled as obeying equations of the type

$$\underline{X}_{n+1} = \underline{f}_n(\underline{X}_n, \underline{U}_n), \quad n = 0, 1, \ldots, \qquad \text{(V.B.1)}$$

where $\underline{X}_0, \underline{X}_1, \ldots$ is a sequence of vectors in $\mathbb{R}^m$ representing the phenomenon under study; $\underline{U}_0, \underline{U}_1, \ldots$ is a sequence of vectors in $\mathbb{R}^s$ "acting" on $\{\underline{X}_n\}_{n=1}^{\infty}$; and where $\underline{f}_0, \underline{f}_1, \ldots$ is a sequence of functions (or, in other words, a time-varying function), each mapping $\mathbb{R}^m \times \mathbb{R}^s$ to $\mathbb{R}^m$. Equation (V.B.1) is an example of a *dynamical system*, with $\underline{X}_n$ representing the *state* of the system at time n and with $\underline{U}_n$ representing the *input* to the system at time n. A dynamical system is a system having the property that for any fixed times l and k, $\underline{X}_l$) is determined completely from the state at time k (i.e., $\underline{X}_k$) and the inputs from times k up through $l-1$ (i.e., $\{\underline{U}_n\}_{n=k}^{l-1}$). Note that complete determination of $\{\underline{X}_n\}_{n=1}^{\infty}$ from (V.B.1) requires not only the specification of the input sequence but also the specification of the *initial condition* $\underline{X}_0$. If the input sequence or the initial condition is random, the states $\underline{X}_0, \underline{X}_1, \ldots$ form a sequence of random vectors and (V.B.1) is referred to as *stochastic system*.

Equation (V.B.1) describes the evolution of the states of a system, so it is usually known as the *state equation* of the system. The system may also have associated with it on *output* sequence $\underline{Z}_0, \underline{Z}_1, \ldots$ of vectors in $\mathbb{R}^k$, possibly different from the state sequence, and given by the *output equation*

$$\underline{Z}_n = \underline{h}_n(\underline{X}_n), \quad n = 0, 1, ..., \qquad \text{(V.B.2)}$$

where $\underline{h}_n$ maps $\mathbb{R}^m$ to $\mathbb{R}^k$. Thus the overall system is a mapping from the initial condition $\underline{X}_0$ and input sequence $\{\underline{U}_n\}_{n=n}^{\infty}$ to the output sequence $\{\underline{Z}_n\}_{n=0}^{\infty}$.

An example of a system described by equations of the type (V.B.1) and (V.B.2) is the following.

Example V.B.1: One-Dimensional Motion

Suppose that we wish to model the one-dimensional motion of a particle that is subjected to an acceleration A_t for $t \geqslant 0$. Note that the position, P_t, and velocity, V_t, of the particle at each time t satisfy the equations $V_t = dP_t/dt$ and $A_t = dV_t/dt$. Assume that we look at the position of the particle every T_s seconds, and we wish to write a model of the form (V.B.1) and (V.B.2) describing the particle's motion from observation time to observation time. Assuming that T_s is small, a Taylor series approximation allows us to write

$$P_{(n+1)T_s} \cong P_{nT_s} + T_s V_{nT_s} \qquad \text{(V.B.3a)}$$

and

$$V_{(n+1)T_s} \cong V_{nT_s} + T_s A_{nT_s}. \qquad \text{(V.B.3b)}$$

We see from (V.B.3) that two states are needed to describe the motion of the particle, namely, position and velocity. On defining $Z_n = X_{1,n} = P_{nT_s}$, $X_{2,n} = V_{nT_s}$, and $U_n = A_{nT_s}$, the motion can be described approximately by the state equation

$$\underline{X}_{n+1} = \mathbf{F}\underline{X}_n + \mathbf{G}U_n, \quad n = 0, 1, ... \qquad \text{(V.B.4)}$$

and the output equation

$$Z_n = \mathbf{H}\underline{X}_n, \quad n = 0, 1, ..., \tag{V.B.5}$$

where $\mathbf{F}$ is the 2×2 matrix

$$\mathbf{F} = \begin{pmatrix} 1 & T_s \\ 0 & 1 \end{pmatrix}, \tag{V.B.6}$$

$\mathbf{G}$ is the 2×1 matrix

$$\mathbf{G} = \begin{pmatrix} 0 \\ T_s \end{pmatrix}, \tag{V.B.7}$$

and $\mathbf{H}$ is the 1×2 matrix

$$\mathbf{H} = (1 \vdots 0). \tag{V.B.8}$$

Thus in this case $m=2, s=1, k=1$, and $\underline{f}_n$ and $\underline{h}_n$ are given, respectively, by

$$\underline{f}_n(\underline{X}, \underline{U}) = \mathbf{F}\underline{X} + \mathbf{G}\underline{U} \tag{V.B.9}$$

and

$$\underline{h}_n(\underline{X}) = \mathbf{H}\underline{X}. \tag{V.B.10}$$

This particular model is discussed further below.

In many applications we are faced with the following problem. We observe the output of a stochastic system in the presence of observation noise (or *measurement noise*) up to some time, say t, and we wish to estimate the state of the system at some time u. That is, we have an observation sequence

$$\underline{Y}_n = \underline{Z}_n + \underline{V}_n, \quad n = 0, 1, ..., t \qquad \text{(V.B.11)}$$

from which we wish to estimate $\underline{X}_u$. In (V.B.11), the sequence $\underline{V}_0, \underline{V}_1, \ldots$ represents measurement noise, and (V.B.11) is sometimes known as the *measurement equation*. If $u=t$, this estimation problem is known as the *filtering* problem; for $u<t$, it is known as the *smoothing* problem; and for $u>t$, it is known as the *prediction* problem. Also, the term *state estimation* is applied to all such problems.

As noted above, state estimation problems arise in many applications. For example, in so-called track-while-scan (TWS) radar, radar measurements of the position of a target are made on each scan of a scanning radar. These measurements are noisy observations of a stochastic system similar to that of Example V.B.1 (with random acceleration), and the radar on each scan would like to estimate the current position of the target and also to predict the position the target will occupy on the next scan. At each scanning time t, then, a TWS radar estimates states at $u=t$ and $u=t+1$ based on the past observation record of the position of the target. (This particular application is discussed further below.)

Other applications of state estimation arise in automatic control systems such as those for aircraft flight control or chemical process control. In flight control the states of interest are the positional coordinates of the aircraft and also the attitudinal coordinates (roll, pitch, and yaw) describing the angular orientation of the aircraft. The state equation in this case describes the dynamics of the aircraft, and the inputs may consist of both control forces and random forces (such as turbulence) operating on the aircraft. In chemical process control the states may be quantities such as temperatures and concentrations of various chemicals, and the state equation describes the dynamics of the chemical reactions involved. Of course, many other applications fit within the context of the general model discussed here.

If we adopt the mean-norm-squared-error performance measure $E\{\|\hat{\underline{X}}_u - \underline{X}_u\|^2\}$ for state estimates $\hat{\underline{X}}_u$ in the model above, we know from Chapter IV (see Case IV.B.4) that the optimum estimate is the conditional mean

$$\hat{\underline{X}}_u = E\{\underline{X}_u | \underline{Y}_0, ..., \underline{Y}_t\}. \tag{V.B.12}$$

Of course, for fixed u and t, this problem is no different from the vector estimation problems discussed in Chapter IV. However, we are usually interested in producing estimates in real time as t increases. Since the data set grows linearly with t, the conditional-mean estimates of (V.B.12) will not be practical unless the system model has a structure that makes (V.B.12) computationally efficient. Thus before considering (V.B.12) further, we will first place suitable restrictions on the model of (V.B.1), (V.B.2), and (V.B.11).

One such restriction which we now impose is that the system be a *linear stochastic system*; i.e., that the state and observation equations are of the form

$$\underline{X}_{n+1} = \mathbf{F}_n \underline{X}_n + \mathbf{G}_n \underline{U}_n, \quad n = 0, 1, ... \tag{V.B.13a}$$

$$\underline{Y}_n = \mathbf{H}_n \underline{X}_n + \underline{V}_n, \quad n = 0, 1, ..., \tag{V.B.13b}$$

where, for each n, $\mathbf{F}_n$, $\mathbf{G}_n$, and $\mathbf{H}_n$ are matrices of appropriate dimensions ($m \times m$, $m \times s$, and $k \times m$, respectively). The linear model of (V.B.13) is appropriate for many applications. For example, the one-dimensional motion model (and its two- and three-dimensional analogs) of Example V.B.1 gives rise to a linear stochastic system when the acceleration acting on the particle is random. Also, many nonlinear systems can be approximated by linear systems when the states of interest represent deviations of the system trajectory from some nominal trajectory. In particular, many systems can be linearized about a nominal state trajectory by use of Taylor series expansions of the nonlinearities $\underline{f}_n$.

A further assumption that allows great simplification of the estimate (V.B.12) is that the input sequence $\{\underline{U}_n\}_{n=0}^{\infty}$ and the observation noise $\{\underline{V}_n\}_{n=0}^{\infty}$ are independent sequences of independent zero-mean Gaussian random vectors. It is also convenient to assume that the initial condition $\underline{X}_0$ is a Gaussian random vector independent of $\{\underline{U}_n\}_{n=0}^{\infty}$ and $\{\underline{V}_n\}_{n=0}^{\infty}$. As is discussed briefly below, the independence assumptions on the sequences $\{\underline{U}_n\}_{n=0}^{\infty}$ and $\{\underline{V}_n\}_{n=0}^{\infty}$ can be relaxed. Also, the assumption of zero mean is primarily for convenience. The

Gaussian assumption, on the other hand, is crucial. However, this assumption is not unrealistic in many models since the observation noise is often due to Gaussian thermal noise in the sensor electronics, and the random inputs to the system are often due to phenomena such as turbulence which can be modeled accurately as having Gaussian statistics. Moreover, it turns out that the Gaussian assumption can be relaxed if one is willing to accept the best estimator among the class of all *linear* estimators, as will be discussed below.

Within the assumptions above, the conditional-mean state estimator (V.B.12) takes on a very nice form from the viewpoint of computational efficiency. Although this form appears in several other state estimation problems, we will consider the particular problems of filtering ($u=t$) and one-step prediction ($u=t+1$), as these are the most common cases arising in applications. The simultaneous solution to these two problems is given by the following.

Proposition V.B.1: The Discrete-Time Kalman-Bucy Filter

For the linear stochastic system (V.B.13) with $\{\underline{U}_n\}_{n=0}^{\infty}$ and $\{\underline{V}_n\}_{n=0}^{\infty}$ being independent sequences of independent zero-mean Gaussian vectors independent of the Gaussian initial condition $\underline{X}_0$, the estimates $\hat{\underline{X}}_{t|t} \triangleq E\{\underline{X}_t | \underline{Y}_0^t\}$† and $\hat{\underline{X}}_{t+1|t} \triangleq E\{\underline{X}_{t+1} | \underline{Y}_0^t\}$ are given recursively by the following equations.

$$\hat{\underline{X}}_{t|t} = \hat{\underline{X}}_{t|t-1} + \mathbf{K}_t(\underline{Y}_t - \mathbf{H}_t \hat{\underline{X}}_{t|t-1}), \quad t = 0, 1, \ldots \qquad \text{(V.B.14a)}$$

and

$$\hat{\underline{X}}_{t+1|t} = \mathbf{F}_t \hat{\underline{X}}_{t|t}, \quad t = 0, 1, \ldots, \qquad \text{(V.B.14b)}$$

with the initialization $\hat{\underline{X}}_{0|-1} = \underline{m}_0 \triangleq E\{\underline{X}_0\}$, where the matrix $\mathbf{K}_t$ is given by

† For compactness of notation we will use the symbol $\underline{Y}_a^b$ to denote the set $\underline{Y}_a, \ldots, \underline{Y}_b$ for $b > a$.

$$\mathbf{K}_t = \Sigma_{t|t-1}\mathbf{H}_t^T(\mathbf{H}_t\Sigma_{t|t-1}\mathbf{H}_t^T + \mathbf{R}_t)^{-1} \quad \text{(V.B.15)}$$

with $\Sigma_{t|t-1} \triangleq Cov\,(\underline{X}_t | \underline{Y}_0^{t-1})$ and $\mathbf{R}_t \triangleq Cov\,(\underline{V}_t)$. Note that since $\hat{\underline{X}}_{t|t-1} = E\{\underline{X}_t | \underline{Y}_0^{t-1}\}$, $\Sigma_{t|t-1}$ is the covariance matrix of the prediction error, $\underline{X}_t - \hat{\underline{X}}_{t|t-1}$, conditioned on $\underline{Y}_0^{t-1}$. This matrix can be computed jointly with the filtering error covariance, $\Sigma_{t|t} \triangleq Cov\,(\underline{X}_t | Y_0^t)$ from the following recursion.

$$\Sigma_{t|t} = \Sigma_{t|t-1} - \mathbf{K}_t\mathbf{H}_t\Sigma_{t|t-1}, \quad t = 0, 1, \ldots \quad \text{(V.B.16a)}$$

$$\Sigma_{t+1|t} = \mathbf{F}_t\Sigma_{t|t}\mathbf{F}_t^T + \mathbf{G}_t\mathbf{Q}_t\mathbf{G}_t^T, \quad t = 0, 1, \ldots \quad \text{(V.B.16b)}$$

with the initialization $\Sigma_{0|-1} = \Sigma_0 \triangleq Cov\,(\underline{X}_0)$, where $\mathbf{Q}_t$ is the covariance matrix of the tth-state input $[\mathbf{Q}_t \triangleq Cov\,(\underline{U}_t)]$.

Proof: To prove the proposition, we first show (V.B.14b) and (V.B.16b) directly, and then prove (V.B.14a) and (V.B.16a) by induction. To see (V.B.14b), we note from the state equation that

$$\begin{aligned} \hat{\underline{X}}_{t+1|t} &= E\{\underline{X}_{t+1} | \underline{Y}_0^t\} = E\{\mathbf{F}_t\underline{X}_t + \mathbf{G}_t\underline{U}_t | \underline{Y}_0^t\} \\ &= \mathbf{F}_t E\{\underline{X}_t | \underline{Y}_0^t\} + \mathbf{G}_t E\{\underline{U}_t | \underline{Y}_0^t\} \qquad \text{(V.B.17)} \\ &= F_t\hat{\underline{X}}_{t|t} + \mathbf{G}_t E\{\underline{U}_t | \underline{Y}_0^t\}, \end{aligned}$$

where the third equality follows from the linearity of the expectation and the final equality follows from the definition of $\hat{\underline{X}}_{t|t}$. Note that $\underline{Y}_0$ is determined by $\underline{X}_0^t$ and $\underline{V}_0^t$ or in turn by $\underline{X}_0, \underline{U}_0^{t-1}$, and $\underline{V}_0^t$, all of which are independent of $\underline{U}_t$. Thus the conditioning in the second term of (V.B.17) is irrelevant and $E\{\underline{U}_t | \underline{Y}_0^t\} = E\{\underline{U}_t\} = \underline{0}$. Equation (V.B.14b) then follows from (V.B.17). Similarly, we have

$$\Sigma_{t+1|t} = Cov\ (\underline{X}_{t+1}|\underline{Y}_0^t)$$

$$= Cov\ (\mathbf{F}_t\underline{X}_t + \mathbf{G}_t\underline{U}_t|\underline{Y}_0^t)$$

$$= Cov\ (\mathbf{F}_t\underline{X}_t|\underline{Y}_0^t) + Cov\ (\mathbf{G}_t\underline{U}_t|\underline{Y}_0^t) \qquad \text{(V.B.18)}$$

$$= Cov\ (\mathbf{F}_t\underline{X}_t|\underline{Y}_0^t) + Cov\ (\mathbf{G}_t\underline{U}_t),$$

since $\underline{U}_t$ is independent of $\underline{X}_t$ and $\underline{Y}_0^t$. Using the property that $Cov\ (\mathbf{A}\underline{X})=\mathbf{A}\,Cov\ (\underline{X})\mathbf{A}^T$ and the definitions of $\Sigma_{t|t}$ and $\mathbf{Q}_t$, we have

$$\Sigma_{t+1|t} = \mathbf{F}_t\ Cov\ (\underline{X}_t|Y_0^t)\mathbf{F}_t^T + \mathbf{G}_t\ Cov\ (\underline{U}_t)\mathbf{G}_t^T$$
$$= \mathbf{F}_t\Sigma_{t|t}\mathbf{F}_t^T + \mathbf{G}_t\mathbf{Q}_t\mathbf{G}_t^T, \qquad \text{(V.B.19)}$$

which is (V.B.16b).

Thus we have shown that (V.B.14b) and (V.B.16b) hold. We now use induction to show that the other two equations [(V.B.14a) and (V.B.16a)] in the recursion are valid. To do this we must show that they are valid for $t=0$ and that for arbitrary $t_0>0$, their validity for $t=t_0-1$ implies their validity for $t=t_0$. For $t=0$ the measurement equation is given by

$$\underline{Y}_0 = \mathbf{H}_0\underline{X}_0 + \underline{V}_0. \qquad \text{(V.B.20)}$$

Since $\underline{X}_0$ and $\underline{V}_0$ are independent Gaussian vectors, we see that the estimation of $\underline{X}_0$ from $\underline{Y}_0$ fits the linear estimation model discussed as Example IV.B.3. In particular, since $\underline{X}_0 \sim N(\underline{m}_0, \Sigma_0)$ and $\underline{V}_0 \sim N(0, \mathbf{R}_0)$, we see from (IV.B.53) that

$$\underline{\hat{X}}_{0|0} \triangleq E\{\underline{X}_0 | \underline{Y}_0\}$$

$$= \underline{m}_0 + \Sigma_0 \mathbf{H}_0^T (\mathbf{H}_0 \Sigma_0 \mathbf{H}_0^T + \mathbf{R}_0)^{-1} (\underline{Y}_0 - \mathbf{H}_0 \underline{m}_0) \quad \text{(V.B.21)}$$

$$= \underline{\hat{X}}_{0|-1} + \mathbf{K}_0 (\underline{Y}_0 - \mathbf{H}_0 \underline{\hat{X}}_{0|-1}),$$

where we have used the following definitions from the proposition: $\underline{\hat{X}}_{0|-1} = \underline{m}_0$, $\mathbf{K}_0 = \Sigma_{0|-1} \mathbf{H}_0^T (\mathbf{H}_0 \Sigma_{0|-1} \mathbf{H}_0^T + \mathbf{R}_0)^{-1}$, and $\Sigma_{0|-1} = \Sigma_0$. Equation (V.B.21) is (V.B.14a) for $t=0$. The error covariance from (V.B.21) is given from (IV.B.54) as

$$\Sigma_{0|0} = \Sigma_0 - \Sigma_0 \mathbf{H}_0^T (\mathbf{H}_0 \Sigma_0 \mathbf{H}_0^T + \mathbf{R}_0)^{-1} \mathbf{H}_0 \Sigma_0$$
$$\text{(V.B.22)}$$
$$= \Sigma_{0|-1} - \mathbf{K}_0 \mathbf{H}_0 \Sigma_{0|-1},$$

which is (V.B.16a) for $t=0$.

To complete the proof, we now assume that (V.B.14a) and (V.B.16a) are valid for $t = t_0 - 1$. Note that $\underline{X}_{t_0}$ and $\underline{Y}_0^{t_0-1}$ are derived by linear transformation of the Gaussian vectors $\underline{X}_0, \underline{U}_0^{t_0-1}$, and $\underline{V}_0^{t_0-1}$. This implies that $\underline{X}_{t_0}$ and $\underline{Y}_0^{t_0-1}$ are jointly Gaussian and thus that $\underline{X}_{t_0}$ is conditionally Gaussian given $\underline{Y}_0^{t_0-1}$. In particular, the condition distribution of $\underline{X}_{t_0}$ given $\underline{Y}_0^{t_0-1}$ is $N(\underline{\hat{X}}_{t_0|t_0-1}, \Sigma_{t_0|t_0-1})$. Also note that $\underline{V}_{t_0}$ is Gaussian and independent of $\underline{Y}_0^{t_0-1}$, so it is also conditionally Gaussian given $\underline{Y}_0^{t_0-1}$ with distribution $N(\underline{0}, \mathbf{R}_{t_0})$. Since $\underline{V}_{t_0}$ is independent of all of $\underline{X}_0, \underline{V}_0^{t_0-1}$ and $\underline{U}_0^{t_0-1}$, it is conditionally independent of $\underline{X}_{t_0}$ given $\underline{Y}_0^{t_0-1}$. From the remarks above we see that, given $\underline{Y}_0^{t_0-1}$, the observation equation

$$\underline{Y}_{t_0} = \mathbf{H}_{t_0} \underline{X}_{t_0} + \underline{V}_{t_0} \quad \text{(V.B.23)}$$

is a Gaussian linear equation of the form discussed in Example IV.B.3. Now, if we compute the conditional expectation of $\underline{X}_{t_0}$ given $\underline{Y}_{t_0}$ under the conditional model (V.B.23) given $\underline{Y}_0^{t_0-1}$ we will get $\underline{\hat{X}}_{t_0|t_0}$, the conditional expectation of $\underline{X}_{t_0}$ given $\underline{Y}_0^{t_0}$.

we will get $\hat{\underline{X}}_{t_0|t_0}$, the conditional expectation of $\hat{\underline{X}}_{t_0}$ given $\underline{Y}_0^{t_0}$. From (IV.B.53) we thus have

$$\hat{\underline{X}}_{t_0|t_0} = \hat{\underline{X}}_{t_0|t_0-1}$$

$$+ \Sigma_{t_0|t_0-1} \mathbf{H}_{t_0}^T (\mathbf{H}_{t_0} \Sigma_{t_0|t_0-1} \mathbf{H}_{t_0}^T + \mathbf{R}_{t_0})^{-1} \qquad \text{(V.B.24)}$$

$$\times (\underline{Y}_{t_0} - \mathbf{H}_{t_0} \hat{\underline{X}}_{t_0|t_0-1}),$$

where we have used the fact that $\hat{\underline{X}}_{t_0}$ has the $N(\hat{\underline{X}}_{t_0|t_0-1}, \Sigma_{t_0|t_0-1})$ distribution conditioned on $\underline{Y}_0^{t_0-1}$. Using the definition of $\mathbf{K}_{t_0}$, we see that (V.B.24) is (V.B.16a) for $t = t_0$. Similarly, by applying (IV.B.54) and the argument above, we arrive at (V.B.16b). We thus have shown that $\hat{\underline{X}}_{t_0|t_0}$ [resp. $\Sigma_{t_0|t_0}$] is given in terms of $\hat{\underline{X}}_{t_0|t_0-1}$ [resp. $\Sigma_{t_0|t_0-1}$] by (V.B.14a) [resp. (V.B.16a)]. We have already shown that $\hat{\underline{X}}_{t_0|t_0-1}$ [resp. $\Sigma_{t_0|t_0-1}$] is obtained from $\hat{\underline{X}}_{t_0-1|t_0-1}$ [resp. $\Sigma_{t_0-1|t_0-1}$] via (V.B.14b) [resp. (V.B.16b)], and thus assuming the validity of (V.B.14a) [resp. (V.B.16a)] for $t = t_0 - 1$ implies its validity for $t = t_0$. This completes the proof of the proposition.

□

The estimator structure described by Proposition V.B.1 is known as the *discrete-time Kalman-Bucy filter* because it is the discrete-time version of a continuous-time recursive state estimator developed principally by R. E. Kalman and R. S. Bucy in the late 1950s. This estimator is depicted in Fig. V.B.1. The computational simplity of this structure is evident from the figure. In particular, although the estimators $\hat{\underline{X}}_{t+1|t}$ or $\hat{\underline{X}}_{t|t}$ depend on all the data $\underline{Y}_0^t$, they are computed at each stage from only the latest observation $\underline{Y}_t$ and the previous prediction $\hat{\underline{X}}_{t|t-1}$. Thus rather than having to store the $(t+1)k$-dimensional vectors $\underline{Y}_0$ (and hence having a linearly growing memory and computational burden), we need only to store and update the single m-vector $\hat{X}_{t|t-1}$. All other parts of the estimator (including the *Kalman gain matrix*, $\mathbf{K}_t$) are determined completely from the parameters of the model and

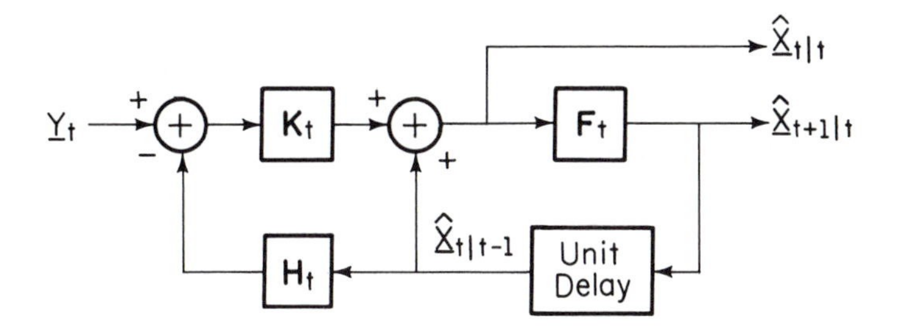

Fig. V.B.1: The discreie-time Kalman-Bucy filter

are independent of the data.

Note that the recursions (V.B.14) and (V.B.16) each consist of two basic steps. The first of these steps is the *measurement update* [(V.B.14a) and (V.B.16a)], which provides the means of updating the estimate and covariance of $\underline{X}_t$ given $\underline{Y}_0^{t-1}$ to incorporate the new observation $\underline{Y}_t$. The second basic step is the *time update* [(V.B.14b) and (V.B.16b)], which provides the means for projecting the state estimate and covariance based on the observation $\underline{Y}_0^t$ to the next time $(t+1)$ before the $(t+1)^{\underline{st}}$ measurement is taken. Examination of the proof of Proposition V.B.1 reveals that the time update is derived exclusively from the state equation, whereas the measurement update is derived from the measurement equation.

It is interesting to consider the measurement update equation (V.B.14a) further. In particular, the estimate $\hat{\underline{X}}_{t|t}$, which is the best estimate of $\underline{X}_t$ based on $\underline{Y}_0^t$, can be viewed as the combination of the best estimate of $\underline{X}_t$ based on the past data, $\hat{\underline{X}}_{t|t-1}$, and a correction term, $\mathbf{K}_t(\underline{Y}_t - \mathbf{H}_t\hat{\underline{X}}_{t|t-1})$. The vector $\underline{I}_t \triangleq (\underline{Y}_t - \mathbf{H}_t\hat{\underline{X}}_{t|t-1})$ appearing in the correction term has an interesting interpretation. In particular, since $\underline{Y}_t = \mathbf{H}_t\underline{X}_t + \underline{V}_t$, we note that $\hat{\underline{Y}}_{t|t-1} \triangleq E\{\underline{Y}_t | \underline{Y}_0^{t-1}\} = \mathbf{H}_t E\{\underline{X}_t | \underline{Y}_0^{t-1}\} + E\{\underline{V}_t | \underline{Y}_0^{t-1}\} = \mathbf{H}_t\hat{\underline{X}}_{t|t-1}$, where we have used the facts that $\underline{V}_t$ is independent of $\underline{Y}_0^{t-1}$ and has zero mean. Thus $\underline{I}_t = \underline{Y}_t - \hat{\underline{Y}}_{t|t-1}$ represents an error signal; it is the error in the prediction of $\underline{Y}_t$ from its past $\underline{Y}_0^{t-1}$. This error is sometimes known as the (prediction) *residual* on the *innovation*. This latter term comes from the fact that we can write $\underline{Y}_t$ as

$$\underline{Y}_t = \hat{\underline{Y}}_{t|t-1} + \underline{I}_t , \qquad \text{(V.B.25)}$$

with the interpretation that $\hat{\underline{Y}}_{t|t-1}$ is the part of $\underline{Y}_t$ that can be predicted from the past, and $\underline{I}_t$ is the part of $\underline{Y}_t$ that cannot be predicted. Thus $\underline{I}_t$ contains the *new* information that is gained by taking the tth observation; hence the term "innovation". (Recall that this sequence arose in the Gaussian detection problems of Chapter III.)

It is not hard to show that the innovation sequence $\{\underline{I}_t\}_{t=0}^{\infty}$ is a sequence of independent zero-mean Gaussian random vectors. First, the fact that $\{\underline{I}_t\}_{t=0}^{\infty}$ is a Gaussian sequence follows from the fact that $\{\underline{Y}_t\}_{t=0}^{\infty}$ is a Gaussian sequence and that $\{\underline{I}_t\}_{t=0}^{\infty}$ is a linear transformation on $\{\underline{Y}_t\}_{t=0}^{\infty}$. The mean of $\underline{I}_t$ is

$$E\{\underline{I}_t\} = E\{\underline{Y}_t - E\{\underline{Y}_t | \underline{Y}_0^{t-1}\}\}$$

$$= E\{\underline{Y}_t\} - E\{\underline{Y}_t\} = 0,$$

where we have used the iterated expectation property of conditional expectations ($E\{Y\}=E\{E\{Y|X\}\}$). Also, we note that because $E\{\underline{I}_t\}=0$,

$$Cov\ (\underline{I}_t , \underline{I}_s) = E\{\underline{I}_t\ \underline{I}_s^T\}.$$

Assuming that $s < t$, we have

$$E\{\underline{I}_t \underline{I}_s^T\} = E\{E\{\underline{I}_t \underline{I}_s^T | \underline{Y}_0^s\}\} = E\{E\{\underline{I}_t | \underline{Y}_0^s\} \underline{I}_s^T\}, \qquad \text{(V.B.26)}$$

where the second equality follows from the fact that $\underline{I}_s$ is constant given $\underline{Y}_0^s$. Noting that

$$E\{\underline{I}_t | \underline{Y}_0^s\} = E\{\underline{Y}_t | \underline{Y}_0^s\} - E\{E\{\underline{Y}_t | \underline{Y}_0^{t-1}\} \underline{Y}_0^s\}$$

$$= E\{\underline{Y}_t | \underline{Y}_0^s\} - E\{\underline{Y}_t | \underline{Y}_0^s\} = \underline{0},$$

(V.B.26) implies that $Cov(\underline{I}_t, \underline{I}_s) = \underline{0}$. For $t > s$, a symmetrical argument yields the same result. Thus the innovation vectors are mutually uncorrelated, and since they are jointly Gaussian, this implies that they are mutually independent.

From the discussion above and (V.B.25) we can reiterate the interpretation that $\underline{Y}_t$ consists of a part, $\hat{\underline{Y}}_{t|t-1}$, completely dependent on the past and a part, $\underline{I}_t$, completely independent of the past. This implies that the innovations sequence provides a set of independent observations that is equivalent to the original set $\{\underline{Y}_t\}_{t=0}^{\infty}$. Thus the formation of the innovations sequence is a prewhitening operation as discussed in Chapter III.†

The following examples illustrate various properties of the Kalman filter.

Example V.B.2: The Time-Invariant Single-Variable Case

The simplest model with which the Kalman filter can be illustrated is the one-dimensional ($m = k = 1$) case in which all parameters of the model are independent of time. In particular, consider the model

$$X_{n+1} = f X_n + U_n, \quad n = 0, 1, \ldots$$
$$Y_n = h X_n + V_n, \quad n = 0, 1, \ldots, \tag{V.B.27}$$

where $\{U_n\}_{n=0}^{\infty}$ and $\{V_n\}_{n=0}^{\infty}$ are independent sequences of i.i.d. $N(0, q)$ and $N(0, r)$ random variables, respectively, $X_0 \sim N(m_0, \Sigma_0)$, and where f, h, q, r, and Σ_0 are scalars.

† Note that the vectors $\underline{I}_t$ are not identically distributed. However, it is easy to see that $Cov(\underline{I}_t) = \mathbf{H}_t \mathbf{\Sigma}_{t|t-1} \mathbf{H}_t^T + \mathbf{R}_t \triangleq \mathbf{D}_t$, so $\mathbf{D}_t^{-1/2} \underline{I}_t$ will give a sequence of i.i.d. $N(\underline{0}, \mathbf{I})$ observations equivalent to $\{\underline{I}_t\}_{t=0}^{\infty}$, where $\mathbf{D}^{1/2}$ denotes the square root of the matrix $\mathbf{D}$ as discussed in Section III.B. Note that the gain $\mathbf{K}_t$ can be written as $\mathbf{\Sigma}_{t|t-1} \mathbf{H}_t \mathbf{D}_t^{-1/2} \mathbf{D}_t^{-1/2}$, so that the Kalman-Bucy filter is actually providing a white (i.i.d.) sequence equivalent to the observation.

The estimation recursions for this case are

$$\hat{X}_{t+1|t} = f\hat{X}_{t|t}, \quad t = 0, 1, \ldots \tag{V.B.28a}$$

and

$$\hat{X}_{t|t} = \hat{X}_{t|t-1} + K_t(Y_t - h\hat{X}_{t|t-1}), \quad t = 0, 1, \ldots, \tag{V.B.28b}$$

with K_t given by

$$K_t = \frac{\Sigma_{t|t-1}h}{(h^2\Sigma_{t|t-1} + r)} = \frac{1}{h}\frac{\Sigma_{t|t-1}}{\Sigma_{t|t-1} + r/h^2}. \tag{V.B.29}$$

The role of the Kalman gain in the measurement update (and hence the operation of the Kalman filter) is easily seen from the expression of (V.B.29). In particular, we note that $\Sigma_{t|t-1}$ is the MSE incurred in the estimation of X_t from $\underline{Y}_0^{t-1}$, and the ratio r/h^2 is a measure of the "noisiness" of the observations. The latter observation follows from the fact that $Y_t/h = X_t + V_t/h$ is an equivalent measurement to Y_t (assuming that $h \neq 0$), and the variance of V_t/h is r/h^2. From these observations on (V.B.29) we see that if the previous prediction of X_t is of much higher quality than the current observation (i.e., $\Sigma_{t|t-1} << r/h^2$), then the gain $K_t \cong 0$ and $\hat{X}_{t|t} \cong \hat{X}_{t|t-1}$. That is, in this case we trust our previous estimate of X_t much more than we trust our observation, so we retain the former estimate. In the opposite situation in which our previous estimate is much noiser than our observation (i.e., $\Sigma_{t|t-1} >> r/h^2$), the Kalman gain $K_t \cong 1/h$, and $\hat{X}_{t|t} \cong Y_t/h$. Thus in the second case we simply ignore our previous measurements and invert the current measurement equation. Of course, between these two extremes the measurement update balances these two ways of updating. The update in the vector case has a similar interpretation, although it cannot be parameterized as easily as in this scalar case.

It is interesting to compare the measurement update here with the Bayesian estimation of signal amplitude as discussed in Example IV.B.2. In particular, we can write the measurement update equation as

$$\hat{X}_{t|t} = \frac{v^2 d^2 \hat{\theta}_1 + \mu}{v^2 d^2 + 1}, \tag{V.B.30}$$

where we have identified $\hat{\theta}_1 = Y_t / h$, $\mu = \hat{X}_{t|t-1}$, $v^2 = \Sigma_{t|t-1}$, and $d^2 = h^2 / r$. Comparing (V.B.30) with (IV.B.34), we see that the distribution of X_t conditioned on Y_0^{t-1} can be interpreted as a prior distribution for X_t [it is $N(\hat{X}_{t|t-1}, \Sigma_{t|t-1})$], and the update balances this prior knowledge with the knowledge gained by the observation Y_t, according to the value of $v^2 d^2$. (Of course, this fact is the essence of the derivation of the measurement update given in the proof of Proposition V.B.1.)

For this scalar time-invariant model, the time and measurement updates for the estimation covariance become

$$\Sigma_{t+1|t} = f^2 \Sigma_{t|t} + q \tag{V.B.31a}$$

and

$$\Sigma_{t|t} = \frac{\Sigma_{t|t-1}}{\frac{h^2}{r}\Sigma_{t|t-1} + 1}. \tag{V.B.31b}$$

Note that we can eliminate the coupling between these equations to get separate recursions for each quantity. For example, inserting (V.B.31b) into (V.B.31a) yields the recursion

$$\Sigma_{t+1|t} = \frac{f^2 \Sigma_{t|t-1}}{\frac{h^2}{r}\Sigma_{t|t-1} + 1} + q, \quad t = 0, 1, \ldots . \tag{V.B.32}$$

(Of course the initialization is $\Sigma_{0|-1} = \Sigma_0$.)

In examining (V.B.32), the question arises as to whether the sequence generated by this recursion approaches a constant as t increases. If so, the Kalman gain approaches a constant also and the Kalman-Bucy filter becomes time-invariant asymptotically in t. Note that if $\Sigma_{t+1|t}$ does approach a constant, say Σ_∞, then Σ_∞ must satisfy

$$\Sigma_\infty = \frac{f^2 \Sigma_\infty}{\frac{h^2}{r}\Sigma_\infty + 1} + q \qquad \text{(V.B.33)}$$

since both $\Sigma_{t+1|t}$ and $\Sigma_{t|t-1}$ are approaching Σ_∞. Equation (V.B.33) is a quadratic equation and it has the unique positive solution

$$\Sigma_\infty = \tfrac{1}{2}\left\{ [\frac{r}{h^2}(1-f^2) - q]^2 + \frac{4rq}{h^2} \right\}^{1/2} - \frac{r}{2h^2}(1-f^2) + q . \qquad \text{(V.B.34)}$$

On combining (V.B.32) and (V.B.33), we have†

$$|\Sigma_{t+1|t} - \Sigma_\infty| = f^2 \left| \frac{\Sigma_{t|t-1}}{\frac{h^2}{r}\Sigma_{t|t-1} + 1} - \frac{\Sigma_\infty}{\frac{h^2}{r}\Sigma_\infty + 1} \right| \qquad \text{(V.B.35)}$$

$$\leqslant f^2 |\Sigma_{t|t-1} - \Sigma_\infty|, \quad t = 0, 1, ...,$$

which implies that

$$|\Sigma_{t+1|t} - \Sigma_\infty| \leqslant f^{2(t+1)} |\Sigma_0 - \Sigma_\infty|. \qquad \text{(V.B.36)}$$

If $|f|<1$, then (V.B.36) implies that $\Sigma_{t+1|t} \to \Sigma_\infty$ as $t \to \infty$. Thus the condition $|f|<1$ is sufficient for the Kalman-Bucy filter and its performance to approach a steady state for this model. [Note that $|f|<1$ is also the condition for asymptotic

---

† To see the inequality in (V.B.35), define $g(x)=x/(ax+1)$, with $a=h^2/r$. By Taylor's theorem, we have for each real x and y,

$$|g(x)-g(y)| = |x-y| \, |g'(\xi)|$$

for some ξ between x and y. We have $g'(\xi)=1/(a\xi+1)^2$ which satisfies $|g'(\xi)| \leqslant 1$ for $\xi \geqslant 0$. Since $\Sigma_{t|t-1}>0$, and $\Sigma_\infty>0$, we have $|g(\Sigma_{t|t-1})-g(\Sigma_\infty)| \leqslant |\Sigma_{t|t-1}-\Sigma_\infty|$.

stability of the original system (V.B.27).]

Example V.B.3: Track-While-Scan (TWS) Radar

A commonly used type of radar is one that regularly scans some area (say an airfield) and keeps track of the trajectories of various targets in the scanning area by processing position measurements taken once each scan. The radar also predicts the positions the targets will occupy on the next scan. Since the maneuver strategies of the targets are usually unknown to the radar, one way of modeling target motion for the purposes of devising optimum tracking schemes is to assume that the targets of interest undergo random accelerations. A simple model for this type of motion is to assume that these accelerations are i.i.d. from scan to scan and are Gaussian. Although the target motion is three-dimensional, it is simpler to discuss this tracking problem in a single dimension only. The assumptions above lead to a state/measurement model of the form described in Example V.B.1. In particular, we can use the model

$$\begin{pmatrix} P_{n+1} \\ V_{n+1} \end{pmatrix} = \begin{pmatrix} 1 & T_s \\ 0 & 1 \end{pmatrix} \begin{pmatrix} P_n \\ V_n \end{pmatrix} + \begin{pmatrix} 0 \\ T_s \end{pmatrix} A_n$$

$$Y_n = (1 \vdots 0) \begin{pmatrix} P_n \\ V_n \end{pmatrix} + \epsilon_n , \qquad \text{(V.B.37)}$$

where P_n and V_n represent the target position and velocity, respectively, on the nth scan, T_s is the time the radar takes to complete each scan, A_n is the target acceleration during the nth scanning period, Y_n is the position measurement at the nth sighting, and ϵ_n is the error in this measurement. (To track in all three dimensions we would have a six-state, three-measurement model. However, if the accelerations and measurement noises in the three dimensions are independent of one another, the three dimensions can be tracked independently.)

Thus assuming that all statistics are Gaussian and time-invariant, the optimum tracker/predictor equations are

$$\begin{pmatrix} \hat{P}_{t+1|t} \\ \hat{V}_{t+1|t} \end{pmatrix} = \begin{pmatrix} \hat{P}_{t|t} + T_s \hat{V}_{t|t} \\ \hat{V}_{t|t} \end{pmatrix} \quad \text{(V.B.38)}$$

and

$$\begin{pmatrix} \hat{P}_{t|t} \\ \hat{V}_{t|t} \end{pmatrix} = \begin{pmatrix} \hat{P}_{t|t-1} \\ \hat{V}_{t|t-1} \end{pmatrix} + \begin{pmatrix} K_{t,1} \\ K_{t,2} \end{pmatrix} (Y_t - \hat{P}_{t|t-1}), \quad \text{(V.B.39)}$$

where in this case, the gain matrix $\mathbf{K}_t$ is a 2×1 vector. This gain vector is given by

$$\begin{pmatrix} K_{t,1} \\ K_{t,2} \end{pmatrix} = \begin{pmatrix} \Sigma_{t|t-1}(1, 1)/(\Sigma_{t|t-1}(1, 1)+r) \\ \Sigma_{t|t-2}(2, 1)/(\Sigma_{t|t-1}(1, 1)+r) \end{pmatrix}, \quad \text{(V.B.40)}$$

where $\Sigma_{t|t-1}(k, l)$ is the $(k-l)^{\underline{th}}$ component of the matrix $\Sigma_{t|t-1}$, and where r is the variance of the measurement noise. The matrix $\Sigma_{t|t-1}$, of course, is computed through the recursions of Proposition V.B.1.

To reduce the computational burden of this tracker, the time-varying filter (V.B.39) is sometimes replaced in practical systems with a time-invariant filter

$$\begin{pmatrix} \hat{P}_{t|t} \\ \hat{V}_{t|t} \end{pmatrix} = \begin{pmatrix} \hat{P}_{t|t-1} \\ \hat{V}_{t|t-1} \end{pmatrix} + \begin{pmatrix} \alpha \\ \beta/T_s \end{pmatrix} (Y_t - \hat{P}_{t|t-1}), \quad \text{(V.B.41)}$$

where α and β are constants. The constants α and β can be chosen to trade-off various performance characteristics, such as speed of response and accuracy of track. This type of tracker is sometimes known as an *$\alpha-\beta$ tracker*.

The TWS radar problem will be discussed further below.

Returning to the general Kalman-Bucy filter of Proposition V.B.1, we note that the coupled recursions in each of (V.B.1) and (V.B.16) can be separated to give recursions for the prediction quantities $\hat{\underline{X}}_{t+1|t}$ and $\Sigma_{t+1|t}$ not involving the filtering quantities $\hat{\underline{X}}_{t|t}$ and $\Sigma_{t|t}$, and vice versa (as was noted in Example V.B.2). For example on substituting the measurement updates into the time updates we have

$$\hat{\underline{X}}_{t+1|t} = \mathbf{F}_t \hat{\underline{X}}_{t|t-1} + \mathbf{F}_t \mathbf{K}_t \underline{I}_t, \quad t = 0, 1, \ldots \quad \text{(V.B.42a)}$$

and

$$\Sigma_{t+1|t} = \mathbf{F}_t \Sigma_{t|t-1} \mathbf{F}_t^T - \mathbf{F}_t \mathbf{K}_t \mathbf{H}_t \Sigma_{t|t-1} \mathbf{F}_t^T + \mathbf{G}_t \mathbf{Q}_t \mathbf{G}_t, \quad t = 0, 1, \ldots . \quad \text{(V.B.42b)}$$

Note that the prediction filter (V.B.42a) is a linear stochastic system driven by the innovations sequence. This system has the same dynamics (i.e., $\mathbf{F}_t$'s) as the system we are trying to track. Thus to track $\underline{X}_t$ we are building a system comprising a duplicate of the dynamics that govern $\underline{X}_t$ and then driving it with the innovations through the matrix sequence $\mathbf{F}_t \mathbf{K}_t$.

The covariance update (V.B.42b) is a dynamical system with a matrix state. It is a nonlinear system since the $\mathbf{K}_t$ term in the second term on the right depends on $\Sigma_{t|t-1}$. This equation is known as a (discrete-time) *Riccati equation*. As in the scalar case of Example V.B.2, the time-invariant version of this equation (in which $\mathbf{F}_t, \mathbf{G}_t, \mathbf{H}_t, \mathbf{Q}_t$, and $\mathbf{R}_t$ are all independent of t) can be studied for possible convergence to steady state. A sufficient (but not necessary) condition for $\Sigma_{t+1|t}$ to converge to a steady state is that all eigenvalues of $\mathbf{F}$ have less than unit magnitude. (This condition is necessary and sufficient for the original system to be asymptotically stable.) Another issue relating to (V.B.42b) is that numerical problems sometimes arise in the computation of the matrix inverse $(\mathbf{H}_t \Sigma_{t|t-1} \mathbf{H}_t^T + \mathbf{R}_t)^{-1}$ appearing in the $\mathbf{K}_t$ term of this equation. Thus it is sometimes convenient to replace (V.B.42b) with an equivalent equation for propagating the square root of $\Sigma_{t+1|t}$ which leads to fewer numerical problems. See Anderson and Moore (1979) for a discussion of these and related issues.

All of the assumptions regarding the system and measurement models that we have made here were used in the derivation of the Kalman-Bucy filter. All of these assumptions are necessary, but as mentioned earlier in this section, some of them can be circumvented by appropriately redefining the model or performance objectives. For example, the independence assumptions on the input and noise sequences $\{\underline{U}_k\}_{k=0}^{\infty}$ and $\{\underline{V}_k\}_{k=0}^{\infty}$ can be relaxed by modeling these processes as themselves being derived from linear stochastic systems driven by independent sequences. The states of the original stochastic system can then be augmented with the states of these additional systems to give an overall higher-dimensional model, but one driven by and observed in independent sequences. The standard Kalman-Bucy filter can then be applied to this augmented system. The disadvantage of this approach, of course, is that it requires a higher-dimensional filter because the noise and input states must also be tracked.

To illustrate this approach we consider the following modification of Example V.B.3.

Example V.B.4: TWS Radar with Dependent Acceleration Sequences

In this example we reconsider the track-while-scan (TWS) radar application discussed in Example V.B.3. For the scanning speeds and target types of interest in many applications, it is often unrealistic to assume that the target acceleration is independent from scan to scan. (For example, the inertial characteristics of the target may preclude such motion.) A simple yet useful model for target acceleration that allows for dependence between accelerations on different scans is that the acceleration sequence $\{A_n\}_{n=0}^{\infty}$ is generated by the stochastic system

$$A_{n+1} = \rho A_n + W_n, \quad n = 0, 1, \ldots \tag{V.B.43}$$

with a Gaussian initial condition A_0 and an i.i.d. Gaussian input sequence $\{W_n\}_{n=0}^{\infty}$, where ρ is a parameter satisfying $0 \leqslant \rho < 1$. Note that if $\rho=0$, there is no dependence in the acceleration sequence, whereas larger values of ρ imply more highly correlated accelerations.

With accelerations satisfying (V.B.43), the model of (V.B.37) no longer satisfies the assumptions required for the Kalman-Bucy filter. However, we can augment this model to include the acceleration dynamics (V.B.43) by treating the acceleration as a state rather than as an input. In particular, we have the model

$$\begin{pmatrix} P_{n+1} \\ V_{n+1} \\ A_{n+1} \end{pmatrix} = \begin{pmatrix} 1 & T_s & 0 \\ 0 & 1 & T_s \\ 0 & 0 & \rho \end{pmatrix} \begin{pmatrix} P_n \\ V_n \\ A_n \end{pmatrix} + \begin{pmatrix} 0 \\ 0 \\ 1 \end{pmatrix} W_n, \quad n = 0, 1, \ldots \quad \text{(V.B.44a)}$$

$$Y_n = (1 \vdots 0 \vdots 0) \begin{pmatrix} P_n \\ V_n \\ A_n \end{pmatrix} + V_n, \quad n = 0, 1, \ldots, \quad \text{(V.B.44b)}$$

which leads to the estimator recursions

$$\begin{pmatrix} \hat{P}_{t+1|t} \\ \hat{V}_{t+1|t} \\ \hat{A}_{t+1|t} \end{pmatrix} = \begin{pmatrix} \hat{P}_{t|t} + T_s \hat{V}_{t|t} \\ \hat{V}_{t|t} + T_s \hat{A}_{t|t} \\ \rho \hat{A}_{t|t} \end{pmatrix} \quad \text{(V.B.45a)}$$

and

$$\begin{pmatrix} \hat{P}_{t|t} \\ \hat{V}_{t|t} \\ \hat{A}_{t|t} \end{pmatrix} = \begin{pmatrix} \hat{P}_{t|t-1} \\ \hat{V}_{t|t-1} \\ \hat{A}_{t|t-1} \end{pmatrix} + \begin{pmatrix} K_{t,1} \\ K_{t,2} \\ K_{t,3} \end{pmatrix} (Y_t - \hat{P}_{t|t-1}), \quad \text{(V.B.45b)}$$

where the gains are given by

$$\begin{pmatrix} K_{t,1} \\ K_{t,2} \\ K_{t,3} \end{pmatrix} = \begin{pmatrix} \Sigma_{t|t-1}(1, 1)/(\Sigma_{t|t-1}(1, 1) + r) \\ \Sigma_{t|t-1}(2, 1)/(\Sigma_{t|t-1}(1, 1) + r) \\ \Sigma_{t|t-1}(3, 1)/(\Sigma_{t|t-1}(1, 1) + r) \end{pmatrix}. \quad \text{(V.B.46)}$$

Note that we now must track the acceleration in addition to position and velocity. As in the lower-order model of Example V.B.3, the gain vector in (V.B.45b) is sometimes replaced in practice with a constant vector, usually denoted by

$$\begin{pmatrix} \alpha \\ \beta/T_s \\ \gamma/T_s^2 \end{pmatrix},$$

in order to reduce computational requirements. The result is known as an α—β—γ *tracker*, and the three parameters α, β, and γ are chosen to given desired performance characteristics.

The example above illustrates how dependence in the input sequence can be handled in the Kalman-Bucy filtering model. For a more detailed discussion of the issue of dependence, the reader is referred to Anderson and Moore (1979). The other principal assumptions in the Kalman model are the linearity of the state and measurement equations and the Gaussianity of the statistics. The latter assumption can be dropped if one is interested in optimizing over all linear filters rather than over all estimators as we have done here. Note that the Kalman-Bucy filter is specified by the second-order statistics (mean and covariances) of the random quantities in the model, and it is in fact the optimum (MMSE) estimator among all linear filters for any initial condition, input and noise sequences with these given second-order statistics (whether they are Gaussian or not). This issue is discussed in Section V.C. The assumption of linearity in the state and observation equations is more difficult to relax than that of Gaussianity. Without this linearity the MMSE state estimation problem becomes quite difficult analytically. Nevertheless, there are several useful techniques for dealing with state estimation in nonlinear systems. Some of these are discussed in Section VII.C in the context of continuous-time signal estimation.

V.C Linear Estimation

In Section V.B we considered optimum estimation in the linear stochastic system model with Gaussian statistics. As noted above, the Kalman-Bucy filter is optimum not only for this model but is also optimum among all linear estimators for the same model with non-Gaussian statistics provided that the second-order statistics of the model (i.e., means and covariances) remain unchanged. The latter result is a particular case of a general theory of optimum linear estimation in which only second-order statistics are needed to specify the optimum procedures. In this section we develop this idea further, and in the following section we apply this theory to a general class of problems known as Wiener-Kolmogorov filtering.

Suppose that we have two sequences of random variables $\{Y_n\}_{n=-\infty}^{\infty}$ and $\{X_n\}_{n=-\infty}^{\infty}$. We observe Y_n some set of times $a \leqslant n \leqslant b$ and we wish to estimate X_t from these observations for some particular time t. Of course, the optimum estimator (in the MMSE sense) is the conditional mean, $\hat{X}_t = E\{X_t | Y_a^b\}$, and the computation of this estimate has been discussed previously. However, if the number of observations $(b-a+1)$ is large, this computation can be quite cumbersome unless the problem exhibits special structure (as in the Kalman-Bucy model). Furthermore, the determination of the conditional mean generally requires knowledge of the joint distribution of the variables $X_t, Y_a, ..., Y_b$, knowledge which may be impractical (or impossible) to obtain in practice.

One way of circumventing the first of these problems is to constrain the estimators to be considered to be of some computationally convenient form, and then to minimize the MSE over this constrained class. One such constraint that is quite useful in this context is the *linear constraint*, in which we consider estimates $\hat{X}_t$ of the form

$$\hat{X}_t = \sum_{n=a}^{b} h_{t,n} Y_n + c_t, \qquad \text{(V.C.1)}$$

where $h_{t,a}, ..., h_{t,b}$, and c_t are scalars.† As we shall see below, this constraint also solves the second problem of having to specify the joint distribution of all variables, since only knowledge of *second-order statistics* will be needed to optimize over linear estimates. Before considering this optimization, we must first note some analytical properties of the sum (V.C.1).

For finite a and b, the meaning of the sum in (V.C.1) is clear. However, we will also be interested in cases in which $a=-\infty, b=+\infty$, or both. Although the meaning of (V.C.1) is clear from a practical viewpoint in such cases, for analytical purposes we must define precisely what we mean by these infinite sums of random variables. The most useful definition in this context is the mean-square sum, in which, for example, for $a=-\infty$ and b finite, the equation (V.C.1) means that

$$\lim_{m \to -\infty} E\left\{\left(\sum_{n=m}^{b} h_{t,n} Y_n + c_t - \hat{X}_t\right)^2\right\} = 0. \qquad \text{(V.C.2)}$$

The sum in (V.C.1) is defined similarly for $b=+\infty$ with a finite and for $a=-\infty, b=+\infty$. Because of the limiting

† Estimates of the form (V.C.1) are more properly termed *affine*. Because of the additive constant c_t, they are not actually linear. However, the term "linear" is fairly standard in this context, so we will use it here. It should be noted that if $X_t, Y_a, ..., Y_b$ are jointly Gaussian random variables, then $E\{X_t | Y_a^b\}$ is of the form (V.C.1), so optimization over linear estimates yields globally optimum estimators in this particular case.

definition of (V.C.2), the observation set for $a=-\infty$ and b finite should be interpreted as $a<t\leqslant b$ rather than $a\leqslant t\leqslant b$, with a similar interpretation for $b=+\infty$.

In order to proceed further with the linear estimation problem, we assume for the remainder of this section that $\{X_n\}_{n=-\infty}^{\infty}$ and $\{Y_n\}_{n=-\infty}^{\infty}$ are second-order sequences, i.e., that $E\{X_n^2\}<\infty$ and $E\{Y_n^2\}<\infty$ for all n. Also, we denote by H_a^b the set of all estimates of the form (V.C.1) based on Y_a^b. The following preliminary results concerning H_a^b will be used later.

Proposition V.C.1:

Suppose that $\hat{X}_t \in H_a^b$. Then

(i) $E\{(\hat{X}_t)^2\}<\infty$; and

(ii) if Z is a random variable satisfying $E\{Z^2\}<\infty$, then

$$E\{Z\hat{X}_t\} = \sum_{n=a}^{b} h_{t,n} E\{ZY_n\} + c_t E\{Z\}.$$

Proof: These two properties are obvious if a and b are both finite. In this case, property (i) follows from successive application of the inequality, $(x+y)^2\leqslant 4(x^2+y^2)$, and property (ii) is simply the linearity property of expectation. To prove these properties for the situation in which a, b, or both is infinite, we consider the specific case in which $a=-\infty$ and b is finite. (Proofs for the other two cases are identical to this one.)

To prove property (i) in this case we write, for $m<b$,

$$\hat{X}_t = \sum_{n=m}^{b} h_{t,n} Y_n + c_t + (\hat{X}_t - \sum_{n=m}^{b} h_{t,n} Y_n - c_t). \quad \text{(V.C.3)}$$

Again, using the inequality $(x+y)^2\leqslant 4(x^2+y^2)$ and taking expectations, we have

$$E\{(\hat{X}_t)^2\} \leqslant 4E\{(\sum_{n=m}^{b} h_{t,n} Y_n + c_t)^2\}$$
$$+ 4E\{(\hat{X}_t - \sum_{n=m}^{b} h_{t,n} Y_n - c_t)^2\}. \tag{V.C.4}$$

The first term on the right-hand side of (V.C.4) is finite by the validity of property (i) for finite a and b. The second term on the right-hand side of (V.C.4) converges to zero an m approaches $-\infty$. Thus there must be a value of m that makes this term finite, which implies that $E\{(\hat{X}_t)^2\} < \infty$.

To prove property (ii), we consider for $m < b$ the quantity

$$E\{Z\hat{X}_t\} - \sum_{n=m}^{b} h_{t,n} E\{XY_n\} - c_t E\{Z\} =$$
$$E\{Z(\hat{X}_t - \sum_{n=m}^{b} h_{t,n} Y_n - c_t)\}. \tag{V.C.5}$$

From the Schwarz inequality we have

$$|E\{Z(\hat{X}_t - \sum_{n=m}^{b} h_{t,n} Y_n - c_t)\}|^2$$
$$\leqslant E\{Z^2\} E\{(\hat{X}_t - \sum_{n=m}^{b} h_{t,n} Y_n - c_t)^2\}. \tag{V.C.6}$$

By assumption $E\{Z^2\} < \infty$ and by definition $E\{(\hat{X}_t - \sum_{n=m}^{b} h_{t,n} Y_n - c_t)^2\} \to 0$ as $m \to -\infty$. Thus (V.C.5) and (V.C.6) imply property (ii). This completes the proof of this proposition.

□

Having constrained ourselves to estimators of the form (V.C.1), we would like to find the best such estimate in the minimum-mean-squared-error sense; i.e., we would like to solve the problem

$$\min_{\hat{X}_t \in H_a^b} E\{(\hat{X}_t - X_t)^2\}. \tag{V.C.7}$$

The solution to this problem can be characterized by the following.

Proposition V.C.2: The Orthogonality Principle
$\hat{X}_t \in H_a^b$ solves (V.C.7) if and only if

$$E\{(\hat{X}_t - X_t)Z\} = 0 \text{ for all } Z \in H_a^b. \tag{V.C.8}$$

Proof: First suppose that $\hat{X}_t$ satisfies (V.C.8), and let $\tilde{X}_t$ be any other estimate in H_a^b. Then the MSE associated with $\tilde{X}_t$ is given by

$$\begin{aligned} E\{(X_t - \tilde{X}_t)^2\} &= E\{(X_t - \hat{X}_t + \hat{X}_t - \tilde{X}_t)^2\} \\ &= E\{(X_t - \hat{X}_t)^2\} \\ &+ 2E\{(X_t - \hat{X}_t)(\hat{X}_t - \tilde{X}_t)\} \\ &+ E\{(\hat{X}_t - \tilde{X}_t)^2\}. \end{aligned} \tag{V.C.9}$$

It is easy to see that $\hat{X}_t \in H_a^b$ and $\tilde{X}_t \in H_a^b$ imply that $(\hat{X}_t - \tilde{X}_t) \in H_a^b$, and thus the second term on the right-hand side of (V.C.9) is zero. This gives

$$\begin{aligned} E\{(X_t - \tilde{X}_t)^2\} &= E\{(X_t - \hat{X}_t)^2\} + E\{(\hat{X}_t - \tilde{X}_t)^2\} \\ &\geqslant E\{(X_t - \hat{X}_t)^2\}. \end{aligned} \tag{V.C.10}$$

Since $\tilde{X}_t$ was chosen arbitrarily, (V.C.10) proves the sufficiency of (V.C.8) for $\hat{X}_t$ to solve (V.C.7).

To prove the necessity of (V.C.8), suppose that $\tilde{X}_t \in H_a^b$ and that there is a $Z \in H_a^b$ such that $E\{(X_t - \hat{X}_t)Z\} \neq 0$. Define a new estimator $\hat{X}_t$ by

$$\hat{X}_t = \tilde{X}_t + \frac{E\{(X_t - \tilde{X}_t)Z\}}{E\{Z^2\}}. \qquad \text{(V.C.11)}$$

(Note that the condition $E\{(X_t - \tilde{X}_t)Z\} \neq 0$ implies that $E\{Z^2\} > 0$.) A straightforward computation gives that

$$E\{(X_t - \hat{X}_t)\}^2 = E\{(X_t - \tilde{X}_t)^2\} - \frac{|E\{(X_t - \tilde{X}_t)Z\}|^2}{E\{Z^2\}} \qquad \text{(V.C.12)}$$

$$< E\{(X_t - \tilde{X}_t)^2\}.$$

Thus $\hat{X}_t$ is a better estimator than $\tilde{X}_t$, so $\tilde{X}_t$ cannot solve (V.C.7). This proves the necessity of (V.C.8) and completes the proof of this proposition.

□

Proposition V.C.2 says that $\hat{X}_t$ is a MMSE linear estimator of X_t given Y_a^b if and only if the estimation error, $X_t - \hat{X}_t$, is orthogonal to every linear function of the observations Y_a^b. This result is known as the *orthogonality principle.*† This result is a special case of a more general result in analysis known as the *projection theorem*, which has the following familiar form in the particular case of a finite-dimensional vector space.

Suppose that $\underline{x}$ and $\underline{y}$ are two vectors of the same dimension, and suppose that we would like to approximate $\underline{x}$ by a constant, say α, times $\underline{y}$ such that the length of the error vector $\underline{x} - \alpha\underline{y}$ is as small as possible. It is easy to see that α minimizes this length if and only if the error vector is perpendicular (i.e., orthogonal) to the line that is aligned along $\underline{y}$ (see Fig. V.C.1) and hence to every constant multiple of $\underline{y}$. The resulting approximation is the *projection of* $\underline{x}$ *in the* $\underline{y}$ *direction.*

---

† It is interesting to note that the conditional-mean estimator $\hat{X}_t = E\{X_t | Y_a^b\}$ uniquely satisfies the analogous condition

$$E\{(X_t - \hat{X}_t)Z\} = 0 \text{ for all } Z \in G_a^b,$$

where G_a^b denotes the set of all real-valued functions g of Y_a^b satisfying $E\{g^2(Y_a^b)\} < \infty$.

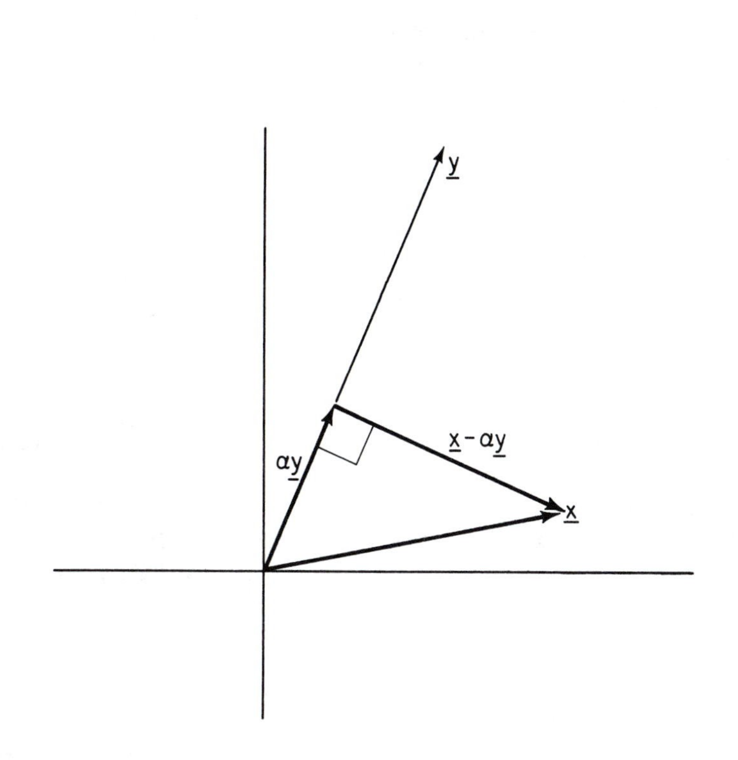

Fig. V.C.1: *Illustration of the orthogonality principle.*

The analogy between the problem and that of linear MMSE estimation is straightforward. The vector $\underline{y}$ is analogous to our observations Y_a^b and the line aligned along $\underline{y}$ is analogous to the set of all linear estimates H_a^b. The vector $\underline{x}$ corresponds to our quantity X_t to be estimated, and the length of the error vector $\|\underline{x}-\alpha\underline{y}\|^2$ is analogous to the MSE, $E\{(X_t-\hat{X}_t)^2\}$. Thus we can think of the linear MMSE estimate as being the *projection* of X_t onto the data Y_a^b.

The result of Proposition V.C.2 characterizes solutions to (V.C.7). A more convenient form of this result for finding such solutions is given by the following result.

Proposition V.C.3: An Alternative Orthogonality Condition

$\hat{X}_t$ solves (V.C.7) if and only if

$$E\{\hat{X}_t\} = E\{X_t\} \tag{V.C.13}$$

and

$$E\{(X_t - \hat{X}_t)Y_l\} = 0 \text{ for all } a \leqslant l \leqslant b. \tag{V.C.14}$$

Proof: The necessity of (V.C.13) and (V.C.14) follows from the application the orthogonality condition (V.C.8) to the particular elements of $H_a^b, Z=1$ and $Z=Y_l$, respectively. The sufficiency follows from property (ii) of Proposition V.C.1. In particular, if we assume that $\hat{X}_t$ satisfies (V.C.13) and (V.C.14), then with $Z=\sum_{n=a}^{b} h_{t,n} Y_n + c_t$, we have

$$E\{(X_t - \hat{X}_t)Z\} = \sum_{n=a}^{b} h_{t,n} E\{(X_t - \hat{X}_t)Y_n\}$$
$$+ c_t E\{X_t - \hat{X}_t\} = 0, \tag{V.C.15}$$

which gives (V.C.8). This completes the proof.

□

Using Proposition V.C.2, we can obtain equations specifying the coefficients of an optimum estimator of the form (V.C.1). In particular, on substituting (V.C.1) into (V.C.13), we have

$$E\{\sum_{n=a}^{b} h_{t,n} Y_n + c_t\} = E\{X_t\},$$

from which we have [using property (ii) of Proposition V.C.1 with $Z=1$]

$$c_t = E\{X_t\} - \sum_{n=a}^{b} h_{t,n} E\{Y_n\}. \qquad \text{(V.C.16)}$$

From (V.C.1) and (V.C.14), we have the relationship

$$E\{(X_t - \sum_{n=a}^{b} h_{t,n} Y_n - c_t) Y_l\} = 0, \quad a \leqslant l \leqslant b. \qquad \text{(V.C.17)}$$

Substituting (V.C.16) into (V.C.17), we get, successively,

$$E\{[(X_t - E\{X_t\}) - \sum_{n=a}^{b} h_{t,n} (Y_n - E\{Y_n\})] Y_l\} = 0, \; a \leqslant l \leqslant b,$$

$$E\{(X_t - E\{X_t\}) Y_l\} = \sum_{n=a}^{b} h_{t,n} E\{(Y_n - E\{Y_n\}) Y_l\}, \; a \leqslant l \leqslant b,$$

$$Cov\,(X_t, Y_l) = \sum_{n=a}^{b} h_{t,n}\; Cov\,(Y_n, Y_l), \quad a \leqslant l \leqslant b,$$

and finally

$$C_{XY}(t, l) = \sum_{n=a}^{b} h_{t,n} C_Y(n, l), \quad a \leqslant l \leqslant b, \qquad \text{(V.C.18)}$$

where $C_{XY}(t, l) \triangleq Cov\,(X_t, Y_l)$ is the *cross-covariance function* of the sequences $\{X_n\}_{n=-\infty}^{\infty}$ and $\{Y_n\}_{n=-\infty}^{\infty}$, and where $C_Y(n, l) \triangleq Cov(Y_n, Y_l)$ is the *autocovariance function* of the sequence $\{Y_n\}_{n=-\infty}^{\infty}$.

Equations (V.C.16) and (V.C.18) give equations that are necessary and sufficient for a set of coefficients $\{h_{t,n}\}_{n=a}^{b}$ and c_t to yield an optimum linear estimator of X_t from Y_a^b. Note that these equations involve only the means, covariances, and cross-covariances (i.e., the *second-order statistics*) of Y_a^b and X_t. This provides a significant practical advantage over the conditional-mean estimator, $E\{X_t | Y_a^b\}$ which in general requires the joint distribution of Y_a^b and X_t, since second-order statistics are much easier to model analytically or to estimate accurately from observed data than are multivariate distribution functions. If $\{h_{t,n}\}_{n=a}^{b}$ can be found to solve (V.C.18), the optimum choice of c_t is immediate from (V.C.16). Examination of (V.C.16) reveals that the role of c_t is to adjust the mean of $\hat{X}_t$ to equal that of X_t, so that an optimum linear estimate will always be of the form

$$\hat{X}_t = E\{X_t\} + \sum_{n=a}^{b} h_{t,n}(Y_n - E\{Y_n\}). \qquad \text{(V.C.19)}$$

Thus for the purpose of discussion we can, without loss of generality, assume that the means of $\{X_n\}_{n=-\infty}^{\infty}$ and $\{Y_n\}_{n=-\infty}^{\infty}$ are zero, which we henceforth do. With this assumption we always have $c_t = 0$ and (V.C.16) is unnecessary.

Equation (V.C.18) is thus the key equation determining the optimum linear estimator. This equation is known as the *Wiener-Hopf equation*. For finite a and b, this equation is quite easy to solve in principle. In particular, we note that (V.C.18) is a set of $(b-a+1)$ linear equations in $(b-a+1)$ unknowns. This can be rewritten in matrix form as

$$\underline{\sigma}_{XY}(t) = \Sigma_Y \underline{h}_t, \qquad \text{(V.C.20)}$$

where $\underline{\sigma}_{XY}(t) \triangleq [C_{XY}(t,a), \ldots, C_{XY}(t,b)]^T$, $\underline{h}_t \triangleq (h_{t,a}, \ldots, h_{t,b})^T$, and Σ_Y is the covariance matrix of the vector $(Y_a, \ldots, Y_b)^T$. Assuming that Σ_Y is positive definite,[†] we see from (V.C.20) that the optimum estimator coefficients are given by

$$\underline{h}_t = \Sigma_Y^{-1} \underline{\sigma}_{XY}(t). \qquad \text{(V.C.21)}$$

Thus for finite a and b, the MMSE estimation problem is, in principle, solved. In practice, however, the determination of these coefficients sometimes presents computational difficulties because of the inversion of the matrix Σ_Y. In

† Since Σ_Y is a covariance matrix it must be at least nonnegative definite. If it is not strictly positive definite, then this implies that there are redundant observations as noted in Section III.B. Even in this case however, (V.C.20) has a solution, although not a unique one.

general, inversion of a $k \times k$ matrix requires a number of basic computational operations of the order of the k^3. In our case k equals the number of observations which, for many signal estimation applications, grows linearly with time. So, in general, the computation of optimum coefficients from (V.C.21) cannot be accomplished in real time. For this reason the study of linear signal estimation is dominated by the investigation of particular models which allow for more efficient computation of these coefficients. One such model is the Kalman-Bucy model of Section V.B, which we discuss further below. Two other important models of this type are the Levinson model, which yields an efficient computational algorithm for the optimum filter coefficients, and the Wiener-Kolmogorov model, which essentially overcomes this problem by allowing a to be $-\infty$. The Levinson model is discussed in the following example, and the Wiener-Kolmogorov model is discussed in Section V.D.

Example V.C.1: Levinson Filtering

Levinson filtering is concerned with one-step prediction of a random sequence whose second-order statistics are stationary in time. In particular, in this model the autocovariance function of our observation sequence $\{Y_n\}_{n=-\infty}^{\infty}$ is assumed to satisfy the condition

$$C_Y(n, l) = C_Y(n-l, 0) \qquad \text{(V.C.22)}$$

for all integers n and l. Such a sequence is said to be *covariance stationary* or *wide-sense stationary* (w.s.s.), and for convenience we usually write the autocovariance function of a w.s.s. sequence as a function of single variable, the time difference, by suppressing the 0 in the second argument on the right-hand side of (V.C.22); i.e.,

$$C_Y(n, l) \equiv C_Y(n-l). \qquad \text{(V.C.23)}$$

Note that, since $Cov(Y_n, Y_l) = Cov(Y_l, Y_n)$, the function C_Y is symmetric: $C_Y(n-l) = C_Y(l-n)$.

In the Levinson filtering problem we observe Y_n for $0 \leqslant n \leqslant t$ and we wish to estimate Y_{t+1}. In our previous notation we have $a=0, b=t$, and $X_t = Y_{t+1}$. The cross-covariance function of $\{X_n\}_{n=-\infty}^{\infty} \equiv \{Y_{n+1}\}_{n=-\infty}^{\infty}$ and $\{Y_n\}_{n=-\infty}^{\infty}$ is thus

$$
\begin{aligned}
C_{XY}(t, l) &= Cov\ (X_t, Y_l) \\
&= Cov\ (Y_{t+1}, Y_l) \qquad \text{(V.C.24)} \\
&= C_Y(t+1-l),
\end{aligned}
$$

and, of course, the $n-l^{\underline{th}}$ element of $\mathbf{\Sigma}_Y$ is $C_Y(n-l)$. Thus the Wiener-Hopf equation (V.C.20) becomes

$$
\begin{pmatrix} C_Y(t+1) \\ C_Y(t) \\ \cdot \\ \cdot \\ \cdot \\ \cdot \\ C_Y(1) \end{pmatrix} = \begin{pmatrix} C_Y(0) & C_Y(1) & \cdots & C_Y(t) \\ C_Y(1) & C_Y(0) & \cdots & C_Y(t-1) \\ \cdot & \cdot & & \cdot \\ \cdot & \cdot & & \cdot \\ \cdot & \cdot & & \cdot \\ & & & C_Y(1) \\ C_Y(t) & \cdots & C_Y(1) & C_Y(0) \end{pmatrix} \begin{pmatrix} h_{t,0} \\ h_{t,1} \\ \cdot \\ \cdot \\ \cdot \\ \cdot \\ h_{t,t} \end{pmatrix}, \qquad \text{(V.C.25)}
$$

a set of equations sometimes known as the *Yule-Walker equations.*

Because $\{Y_n\}_{n=-\infty}^{\infty}$ is w.s.s., the matrix $\mathbf{\Sigma}_Y$ on the right-hand side of (V.C.25) is a *Toeplitz* matrix, which means that its entries are constant along the diagonals [since $C_Y(n, l) = C_Y(n-l)$]. Unlike general covariance matrices, $k \times k$ Toeplitz matrices can be inverted in a number of operations that is of the order of k^2. [A well-known algorithm for doing this is due to Trench (1964).] Thus for any linear MMSE problem in which the observations are w.s.s., the complexity of computing the estimator coefficients is reduced by a factor equal to the number of observations. However, in the Levinson problem there is additional structure that allows for a further simplification in computing the estimator coefficients. In particular, the vector $\underline{\sigma}_{XY}(t)$ on the left of (V.C.25) is like the first row of $\mathbf{\Sigma}_Y$ shifted by one time unit.

This structure allows the coefficients to be computed recursively in t.

It is conventional in this problem to rewrite the predictor $\hat{Y}_{t+1}=\sum_{n=0}^{t} h_{t,n} Y_n$ as $\hat{Y}_{t+1}=-\sum_{n=0}^{t} a_{t+1,t+1-n} Y_n$. The coefficients $a_{t,1}, \ldots, a_{t,t}$ can then be updated recursively (in t) through the following algorithm, known as the *Levinson algorithm*:

$$a_{t+1,k} = a_{t,k} - k_t a_{t,t+1-k}, \quad k = 1, \ldots, t, \quad \text{(V.C.26)}$$

and

$$a_{t+1,t+1} = -k_t, \quad \text{(V.C.27)}$$

where k_t is generated recursively with $\epsilon_t \triangleq E\{(Y_t - \hat{Y}_t)^2\}$ via

$$\epsilon_{t+1} = (1-k_t^2)\epsilon_t \quad \text{(V.C.28)}$$

and

$$k_t = \frac{C_Y(t+1) + \sum_{k=1}^{t} a_{t,k} C_Y(t+1-k)}{\epsilon_t}. \quad \text{(V.C.29)}$$

This algorithm is initialized by $k_0=-C_Y(1)/C_Y(0)$ and $\epsilon_0=C_Y(0)$. Note that this algorithm computes the MMSE, ϵ_t, as a by-product. It also computes the coefficients $k_1, k_2, \ldots$, which are known as *partial correlation* (PARCOR) *coefficients* or *reflection coefficients*. The latter coefficients are useful for implementing the one-step predictor using a lattice filter. A derivation of the Levinson algorithm is found by Honig and Messerschmidt (1984), together with a discussion of the implementation and several applications of one-step predictors.

The linear estimation problem can be extended straightforwardly to the case in which the observation sequence is a sequence of vectors (say k-dimensional) and the quantity to be estimated is also a vector (say, m-dimensional). In this case, we consider estimates of the form

$$\hat{\underline{X}}_t = \sum_{n=a}^{b} \mathbf{H}_{t,n} \underline{Y}_n + \underline{c}_t, \tag{V.C.30}$$

where $\{\mathbf{H}_{t,n}\}_{n=a}^{b}$ is a sequence of $m \times k$ matrices and $\underline{c}_t \in \mathbb{R}^m$. With a or b infinite, (V.C.30) is defined in the mean-norm sense; e.g., with $a = -\infty$ and b finite,

$$\lim_{j \to -\infty} E\left\{ \left\| \sum_{n=j}^{b} \mathbf{H}_{t,n} \underline{Y}_n + \underline{c}_t - \hat{\underline{X}}_t \right\|^2 \right\} = 0,$$

where $\|\underline{x}\|^2 \triangleq \underline{x}^T \underline{x}$. If we wish to choose an estimate to solve

$$\min_{\hat{\underline{X}} \in \underline{H}_a^b} E\{\|\underline{X}_t - \hat{\underline{X}}_t\|^2\},$$

where $\underline{H}_a^b$ is the set of all estimators of the form (V.C.30), it follows similarly to Proposition V.C.2 that $\hat{\underline{X}}_t$ is optimum if and only if

$$E\{(\underline{X}_t - \hat{\underline{X}}_t)^T \underline{Z}\} = 0 \text{ for all } \underline{Z} \in \underline{H}_a^b. \tag{V.C.31}$$

Equation (V.C.31) can be transformed into the equivalent conditions

$$E\{\hat{\underline{X}}_t\} = E\{\underline{X}_t\} \tag{V.C.32a}$$

and

$$E\{(\underline{X}_t - \hat{\underline{X}}_t)\underline{Y}_l^T\} = \mathbf{O}, \quad a \leqslant l \leqslant b, \tag{V.C.32b}$$

where $\mathbf{O}$ denotes the matrix of all zeroes. These equations in turn give an equation for the optimum $\underline{c}_t$ and a vector Wiener-Hopf equation:

$$\mathbf{C}_{XY}(t,l) = \sum_{n=a}^{b} \mathbf{H}_{t,n}\mathbf{C}_{Y}(n,l),\ a \leqslant l \leqslant b, \quad \text{(V.C.33)}$$

where $\mathbf{C}_{XY}(t,l) \triangleq Cov(\underline{X}_t, \underline{Y}_l)$ is the (matrix) cross-covariance function of $\{\underline{X}_n\}_{n=-\infty}^{\infty}$ and $\{\underline{Y}_n\}_{n=-\infty}^{\infty}$ and similarly $\mathbf{C}_Y(n,l) \triangleq Cov(\underline{Y}_n, \underline{Y}_l)$. Note that $\mathbf{C}_{XY}(t,l)$ and $\mathbf{C}_Y(n,l)$ are $m \times k$ and $k \times k$ matrices, respectively.

For finite a and b, the vector Wiener-Hopf equation (V.C.33) gives a set of $(b-a+1)\times m \times k$ linear equations in the same number of unknowns. It can thus be solved by matrix inversion subject to positive definiteness of the covariance matrix of the $(b-a+1)$ k-dimensional vector $(\underline{Y}_a^T, \underline{Y}_{a+1}^T, \ldots, \underline{Y}_b^T)^T$. In fact, the minimization of the mean norm error $E\{\|\underline{X}_t - \hat{\underline{X}}_t\|^2\}$ is equivalent to minimizing the mean-square error on each component of $\underline{X}_t$. So the vector Wiener-Hopf equation is essentially a set of m scalar Wiener-Hopf equations, each with $(b-a+1)k$ observations. Unfortunately, this structure does not simplify the solution since it is the observation dimension that affects the computational burden most. As in the scalar case, computational issues are often dominant in the study of these problems. The Levinson problem can be formulated in the vector case as well as in the scalar case, with an efficient solution algorithm similar to that of Example V.C.1. Moreover, in addition to its role as a global MMSE estimator in the linear-Gaussian model of Section V.B, the Kalman-Bucy filter can also be interpreted as a linear MMSE estimator in a less restrictive model. This result is summarized in the following example.

Example V.C.2: The Kalman-Bucy Filter as a Linear MMSE Estimator

Consider the linear stochastic system model

$$\underline{X}_{n+1} = \mathbf{F}_n \underline{X}_n + \mathbf{G}_n \underline{U}_n, \quad n = 0, 1, \ldots \quad \text{(V.C.34a)}$$

$$\underline{Y}_n = \mathbf{H}_n \underline{X}_n + \underline{V}_n, \quad n = 0, 1, \ldots, \quad \text{(V.C.34b)}$$

where, for each $n \geqslant 0$, $\underline{X}_n$, $\underline{U}_n$, $\underline{Y}_n$, and $\underline{V}_n$ are random vectors of dimension m, s, k, and k, respectively, and $\mathbf{F}_n$, $\mathbf{G}_n$, and

$\mathbf{H}_n$ are matrices of appropriate dimensions. We assume that $\{\underline{U}_n\}_{n=-\infty}^{\infty}$ and $\{\underline{V}_n\}_{n=-\infty}^{\infty}$ are uncorrelated sequences of zero-mean uncorrelated random vectors [i.e., $Cov(\underline{V}_n, \underline{U}_l)=\mathbf{O}$ for all n and l and $Cov(\underline{U}_n, \underline{U}_l)=Cov(\underline{V}_n, \underline{V}_l)=\mathbf{O}$ for all $n \neq l$, where $\mathbf{O}$ denotes a matrix of all zeros], and that the initial condition $\underline{X}_0$ is uncorrelated with both $\{\underline{U}_n\}_{n=-\infty}^{\infty}$ and $\{\underline{V}_n\}_{n=-\infty}^{\infty}$. We also assume that $\underline{U}_n$ and $\underline{V}_n$ have known covariance matrices $\mathbf{Q}_n$ and $\mathbf{R}_n$, respectively, for each n, and that $\underline{X}_0$ has known mean $\underline{m}_0$ and covariance matrix $\mathbf{\Sigma}_0$. Apart from these assumptions, the statistics of the various random quantities are arbitrary (e.g, no Gaussian assumption is made here).

Within the assumptions above it can be shown that the Kalman-Bucy filtering recursions of Proposition V.B.1 give the linear minimum-mean-norm-error estimators of $\underline{X}_t$ and $\underline{X}_{t+1}$ from the measurements $\underline{Y}_0^t$. Although we will not develop this result in detail here,† the application of the orthogonality principle can be illustrated in deriving the estimator time update, $\hat{\underline{X}}_{t+1|t} = \mathbf{F}_t \hat{\underline{X}}_{t|t}$. In particular, suppose that $\hat{\underline{X}}_{t|t}$ is the best linear estimator of $\underline{X}_t$ given $\underline{Y}_0^t$ and consider the quantity

$$E\{(\underline{X}_{t+1} - \mathbf{F}_t \hat{\underline{X}}_{t|t})^T \underline{Z}\} \tag{V.C.35}$$

for $\underline{Z} \in \underline{H}_0^t$. Using the state equation (V.C.34a), (V.C.35) becomes

$$\begin{aligned} &E\{(\mathbf{F}_t \underline{X}_t + \mathbf{G}_t \underline{U}_t - \mathbf{F}_t \hat{\underline{X}}_{t|t})^T \underline{Z}\} \\ &= E\{(\underline{X}_t - \hat{\underline{X}}_{t|t})^T \mathbf{F}_t^T \underline{Z}\} + E\{\underline{U}^T \underline{Z}\} \mathbf{G}_t^T. \end{aligned} \tag{V.C.36}$$

Since $\hat{\underline{X}}_{t|t}$ is assumed to be the best linear estimator of $\underline{X}_t$ from $\underline{Y}_0^t$ and since $\mathbf{F}_t^T \underline{Z} \in \underline{H}_0^t$ whenever $\underline{Z} \in \underline{H}_0^t$, the first term on the right-hand side of (V.C.36) is zero for any $\underline{Z} \in \underline{H}_0^t$. With regard to the second term on the right of (V.C.36), we

† Actually, this result can be inferred from the optimality of the Kalman-Bucy filter in the Gaussian case and its linearity. In particular, the model above includes the Gaussian model as a special case. Thus if some other structure were the best linear estimator for this model, then the Kalman-Bucy filter could not be globally optimum for the particular case of Gaussian statistics.

note that $\underline{Z}$ is a linear transformation of $\underline{Y}_0^t$, which in turn is a linear transformation of $\underline{X}_0, \underline{U}_0^{t-1}$, and $\underline{V}_0^t$, all of which are uncorrelated with $\underline{U}_t$. Thus $\underline{U}_t$ and $\underline{Z}$ are uncorrelated and we have that $E\{(\underline{X}_{t+1} - \mathbf{F}_t \hat{\underline{X}}_{t|t})^T \underline{Z}\} = 0$ for all $\underline{Z} \in \underline{H}_0^t$, implying from the orthogonality principle that $\mathbf{F}_t \hat{\underline{X}}_{t|t}$ is the best linear estimator of $\underline{X}_{t+1}$ given $\underline{Y}_0^t$. The proof of the covariance time update is almost identical to that given in Proposition V.B.1 for the Gaussian case.

V.D Wiener-Kolmogorov Filtering

In Section V.C we derived the Wiener-Hopf equation, which specifies the coefficients for optimum linear estimation of one random variable, X_t, from observation of a set of other random variables, $Y_a, \ldots, Y_b$. In most signal estimation applications the number of observations $(b-a+1)$ grows linearly with t, so further assumptions are usually needed in order to compute coefficients of the corresponding estimator efficiently. Two such sets of assumptions are those made by the Levinson and Kalman-Bucy filtering models. Another set of simplifying assumptions, known as the Wiener-Kolmogorov model, leads to the solution of the optimum linear estimation problem for a wide class of signal estimation applications. In this section we develop the latter model in some detail.

As in the Levinson problem, we assume that the (scalar) observation sequence is wide-sense stationary; i.e., $C_Y(n,l) \triangleq Cov(Y_n, Y_l) = C_Y(n-l, 0) \equiv C_Y(n-l)$ for all integers n and l. We also assume that the observation sequence and the (scalar) sequence $\{X_n\}_{n=-\infty}^{\infty}$ are *jointly wide-sense stationary*; i.e., we assume that $C_{XY}(t,n) \triangleq Cov(X_t, Y_n) = C_{XY}(t-n, 0) \equiv C_{XY}(t-n)$ for all integers t and n. (We continue to assume, without loss of generality, that all X_n's and Y_n's have zero means.) We also assume that the number of observations is infinite, and we will consider two such cases: the so-called *noncausal Wiener-Kolmogorov filtering problem*, in which we take $a=-\infty$ and $b=+\infty$; and the *causal Wiener-Kolmogorov filtering problem*, in which $a=-\infty$ and $b=t$. We treat the noncausal case first because it solution is simpler.

V.D.1 Noncausal Wiener-Kolmogorov Filtering

The noncausal Wiener-Kolmogorov problem is so called because we are estimating at time t based on observations for all times, $-\infty < t < \infty$. Thus the estimate

$$\hat{X}_t = \sum_{n=-\infty}^{\infty} h_{t,n} Y_n, \qquad \text{(V.D.1)}$$

if thought of as a linear filtering operation on the sequence $\{Y_n\}_{n=-\infty}^{\infty}$, is not necessarily causal; that is, the impulse response $\{h_{t,n}\}_{n=-\infty}^{\infty}$ may not satisfy $h_{t,n}=0$ for $t>n$. This implies that the estimate at the "present" time t may depend on observations at future times $n>t$. Obviously, for real-time estimation problems one should restrict attention to causal filters; however, for applications in which the data have been stored or in which the index t is a spatial parameter rather than a time parameter (as in image or array processing), this type of causality is not an issue.

The Wiener-Hopf equation (V.C.18) for this problem is

$$C_{XY}(t,l) = \sum_{n=-\infty}^{\infty} h_{t,n} C_Y(n,l), \; -\infty < l < \infty, \qquad \text{(V.D.2)}$$

which, from the stationarity assumptions, can be written as

$$C_{XY}(t-l) = \sum_{n=-\infty}^{\infty} h_{t,n} C_Y(n-l), \; -\infty < l < \infty. \qquad \text{(V.D.3)}$$

To put (V.D.3) in a more tractable form, let us define a new variable, $\tau = t-l$, from which we have

$$C_{XY}(\tau) = \sum_{n=-\infty}^{\infty} h_{t,n} C_Y(n+\tau-t), \; -\infty < \tau < \infty. \qquad \text{(V.D.4)}$$

Now, changing variables in the sum with the substitution $\alpha = t-n$, the Wiener-Hopf equation becomes

$$C_{XY}(\tau) = \sum_{\alpha=-\infty}^{\infty} h_{t,t-\alpha} C_Y(\alpha-\tau), \; -\infty<\tau<\infty. \qquad \text{(V.D.5)}$$

Note that the variable t appears in (V.D.5) only in the coefficient sequence $\{h_{t,t-\alpha}\}_{\alpha=-\infty}^{\infty}$. This implies that if the Wiener-Hopf equation has a solution in this case, we can choose that solution independently of t. That is, an optimum $\{h_{t,n}\}_{n=-\infty}^{\infty}$ can be chosen such that $h_{t,t-\alpha}$ depends only on α, or equivalently, that $h_{t,t-\alpha}=h_{\alpha,0}$ for all integers t and α. Thus if a solution exists, it can be chosen to be *time-invariant* (or *shift-invariant*) with coefficient sequence $h_{t,n}=h_{t-n,0} \triangleq h_{t-n}$, where for convenience we suppress the second index. With this observation, and noting that $C_Y(\alpha-\tau)=C_Y(\tau-\alpha)$, the Wiener-Hopf equation becomes

$$C_{XY}(\tau) = \sum_{\alpha=-\infty}^{\infty} h_{\alpha} C_Y(\tau-\alpha), \; -\infty<\tau<\infty. \qquad \text{(V.D.6)}$$

The right-hand side of (V.D.6) is recognized as the discrete-time *convolution* of the sequences $\{h_n\}_{n=-\infty}^{\infty}$ and $\{C_Y(n)\}_{n=-\infty}^{\infty}$. Thus (V.D.6) is a convolution equation, which can be converted to a simple algebraic equation by converting to the frequency domain. In particular, on assuming that the following discrete-time Fourier transforms exist:

$$H(\omega) \triangleq \sum_{n=-\infty}^{\infty} h_n e^{-i\omega n}, \; -\pi\leqslant\omega\leqslant\pi, \qquad \text{(V.D.7)}$$

$$\phi_{XY}(\omega) \triangleq \sum_{n=-\infty}^{\infty} C_{XY}(n) e^{-i\omega n}, \; -\pi\leqslant\omega\leqslant\pi, \qquad \text{(V.D.8)}$$

and

$$\phi_Y(\omega) \triangleq \sum_{n=-\infty}^{\infty} C_Y(n) e^{-i\omega n}, \; -\pi\leqslant\omega\leqslant\pi, \qquad \text{(V.D.9)}$$

the Wiener-Hopf equation becomes

$$\phi_{XY}(\omega) = H(\omega)\phi_Y(\omega), \ -\pi \leqslant \omega \leqslant \pi. \qquad \text{(V.D.10)}$$

Note that H is the *transfer function* of the filter $\{h_n\}_{n=-\infty}^{\infty}$, ϕ_Y is the *power spectral density* (or spectrum) of the sequence $\{Y_n\}_{n=-\infty}^{\infty}$, and ϕ_{XY} is the *cross power spectral density* of the sequences $\{X_n\}_{n=-\infty}^{\infty}$ and $\{Y_n\}_{n=-\infty}^{\infty}$.

In the form (V.D.10), the Wiener-Hopf equation is easily solved for the transfer function of the optimum estimator, i.e.,

$$H(\omega) = \frac{\phi_{XY}(\omega)}{\phi_Y(\omega)}, \ -\pi \leqslant \omega \leqslant \pi, \qquad \text{(V.D.11)}$$

from which the (time-invariant) filter coefficients become,[†]

$$h_n = \frac{1}{2\pi}\int_{-\pi}^{\pi} \frac{\phi_{XY}(\omega)}{\phi_Y(\omega)} e^{i\omega n} d\omega, \ n \in \mathbb{Z}. \qquad \text{(V.D.12)}$$

Within the assumptions made above, (V.D.11) or (V.D.12) specifies the optimum linear estimator in the non-causal Wiener-Kolmogorov problem. Before giving a specific example to illustrate this result, it is of interest to consider the value of the mean squared error incurred by using this filter. In particular, we would like an expression for the minimum value of the MSE,

$$MMSE \triangleq \min_{\hat{X}_t \in H_{-\infty}^{\infty}} E\{(X_t - \hat{X}_t)^2\}.$$

We have that

$$MMSE = E\{(X_t - \hat{X}_t)^2\},$$

where $\hat{X}_t$ is the optimum estimate. We can write

---

† Here, and elsewhere in this book, $\mathbb{Z}$ denotes the set of all integers.

$$MMSE = E\{(X_t - \hat{X}_t)^2\}$$

$$= E\{(X_t - \hat{X}_t)X_t\} - E\{(X_t - \hat{X}_t)\hat{X}_t\} \quad \text{(V.D.13)}$$

$$= E\{(X_t - \hat{X}_t)X_t\} = E\{X_t^2\} - E\{\hat{X}_t X_t\},$$

where the disappearance of the term $E\{(X_t - \hat{X}_t)\hat{X}_t\}$ is due to the orthogonality principle.

Consider first the second term on the right-hand side of (V.D.13). We have

$$E\{\hat{X}_t X_t\} = E\{(\sum_{n=-\infty}^{\infty} h_{t-n} Y_n)X_t\} = \sum_{n=-\infty}^{\infty} h_{t-n} E\{Y_n X_t\}$$

$$= \sum_{n=-\infty}^{\infty} h_{t-n} C_{XY}(t-n) = \sum_{\alpha=-\infty}^{\infty} h_\alpha C_{XY}(\alpha), \quad \text{(V.D.14)}$$

where we have made the substitution $\alpha = t - n$. Noting that the right-hand side of (V.D.14) is the zeroth term of the convolution of $\{h_n\}_{n=-\infty}^{\infty}$ and $\{C_{XY}(-n)\}_{n=-\infty}^{\infty}$, we have that

$$E\{\hat{X}_t X_t\} = \frac{1}{2\pi}\int_{-\pi}^{\pi} H(\omega)\overline{\Phi}_{XY}(\omega)d\omega, \quad \text{(V.D.15)}$$

when $\overline{\Phi}_{XY}(\omega)$ is the discrete-time Fourier transform of the sequence $\{C_{XY}(-n)\}_{n=-\infty}^{\infty}$; i.e.,

$$\overline{\Phi}_{XY}(\omega) = \sum_{n=-\infty}^{\infty} C_{XY}(-n)e^{-i\omega n}, \; -\pi \leqslant \omega \leqslant \pi. \quad \text{(V.D.16)}$$

Setting $\alpha = -n$ in (V.D.16), we have

$$\overline{\Phi}_{XY}(\omega) = \sum_{\alpha=-\infty}^{\infty} C_{XY}(\alpha)e^{i\omega\alpha} = \phi_{XY}^*(\omega), \quad \text{(V.D.17)}$$

where the superscript * denotes complex conjugation. From

(V.D.11) and (V.D.15), we thus have

$$E\{\hat{X}_t X_t\} = \frac{1}{2\pi}\int_{-\pi}^{\pi} \frac{\phi_{XY}(\omega)}{\phi_Y(\omega)}\phi_{XY}^*(\omega)d\,\omega$$
$$= \frac{1}{2\pi}\int_{-\pi}^{\pi} \frac{|\phi_{XY}(\omega)|^2}{\phi_Y(\omega)}d\,\omega. \tag{V.D.18}$$

Assuming that the sequence $\{X_n\}_{n=-\infty}^{\infty}$ is w.s.s. with power spectrum ϕ_X, we can write the first term on the right-hand side of (V.D.13) as

$$E\{X_t^2\} = C_X(0) = \frac{1}{2\pi}\int_{-\pi}^{\pi} \phi_X(\omega)d\,\omega,$$

from which we have

$$MMSE = \frac{1}{2\pi}\int_{-\pi}^{\pi} \left[\phi_X(\omega) - \frac{|\phi_{XY}(\omega)|^2}{\phi_Y(\omega)}\right]d\,\omega. \tag{V.D.19}$$

Equation (V.D.19) can be rewritten as

$$MMSE = \frac{1}{2\pi}\int_{-\pi}^{\pi} \left[1 - \frac{|\phi_{XY}(\omega)|^2}{\phi_X(\omega)\phi_Y(\omega)}\right]\phi_X(\omega)d\,\omega, \tag{V.D.20}$$

so that the performance of the optimum filter is the integral of the function $[1-|\phi_{XY}|^2/\phi_X\phi_Y]$ weighted by ϕ_X. A property of the cross-spectrum is that $|\phi_{XY}(\omega)|^2 \leqslant \phi_X(\omega)\phi_Y(\omega)$, $-\pi \leqslant \omega \leqslant \pi$, with equality for all $\omega \epsilon [-\pi, \pi]$ if and only if the sequence $\{X_n\}_{n=-\infty}^{\infty}$ and $\{Y_n\}_{n=-\infty}^{\infty}$ are perfectly correlated (i.e., $X_t \epsilon H_{-\infty}^{\infty}$ for all $t \epsilon \mathbb{Z}$). Thus (V.D.20) shows that the MMSE ranges from $E\{X_t^2\}$ to zero as the relationship between the sequences $\{X_n\}_{n=-\infty}^{\infty}$ and $\{Y_n\}_{n=-\infty}^{\infty}$ ranges from complete uncorrelatedness [$\phi_{XY}(\omega)=0$, $-\pi \leqslant \omega \leqslant \pi$] to perfect correlation. Note that in the first of these two extremes we have $\hat{X}_t = E\{X_t\} \equiv 0$, and in the latter one we have $\hat{X}_t = X_t$. The interesting cases, of course, are between these two extremes, in which case $0 < MMSE < E\{X_t^2\}$.

To illustrate noncausal Wiener-Kolmogorov filtering, we consider the following example.

Example V.D.1: Signal Estimation in Additive Noise

Consider the observation model

$$Y_n = S_n + N_n, \; n \in \mathbb{Z}, \tag{V.D.21}$$

where $\{S_n\}_{n=-\infty}^{\infty}$ and $\{N_n\}_{n=-\infty}^{\infty}$ are uncorrelated, zero-mean, w.s.s sequences representing signal and noise, respectively. Suppose that the quantity we wish to estimate at time t is the signal at time $t+\lambda$ for some integer λ; i.e.,

$$X_t = S_{t+\lambda}. \tag{V.D.22}$$

This problem represents filtering ($\lambda=0$), prediction ($\lambda>0$), or smoothing ($\lambda<0$) of the signal. Denoting the power spectra of signal and noise by ϕ_S and ϕ_N, respectively, it is straightforward to show that

$$\phi_Y(\omega) = \phi_S(\omega) + \phi_N(\omega), \; -\pi \leqslant \omega \leqslant \pi, \tag{V.D.23}$$

$$\phi_{XY}(\omega) = e^{i\omega\lambda}\phi_S(\omega), \; -\pi \leqslant \omega \leqslant \pi, \tag{V.D.24}$$

and

$$\phi_X(\omega) = \phi_S(\omega), \; -\pi \leqslant \omega \leqslant \pi. \tag{V.D.25}$$

From (V.D.11), (V.D.23), and (V.D.24) we see that the transfer function of the optimum noncausal filter is

$$H(\omega) = \frac{e^{i\omega\lambda}\phi_S(\omega)}{\phi_S(\omega) + \phi_N(\omega)}, \; -\pi \leqslant \omega \leqslant \pi. \qquad \text{(V.D.26)}$$

The interpretation of this filter is straightforward. The term $e^{i\omega\lambda}$ is a (unit-magnitude) phase term which corresponds to a shift of λ time units in the time domain. Thus this term merely time-shifts the data sequence to account for the fact that we wish to estimate the signal at time $t+\lambda$. Note that this shifting is causal if $\lambda \leqslant 0$ and is noncausal if $\lambda > 0$. The remaining term in the filter transfer function is its magnitude, $\phi_S/(\phi_S+\phi_N)$, which represents the gain of the filter. We can rewrite this term as

$$|H(\omega)| = \frac{\phi_S(\omega)/\phi_N(\omega)}{\phi_S(\omega)/\phi_N(\omega)+1}, \; -\pi \leqslant \omega \leqslant \pi. \qquad \text{(V.D.27)}$$

Note that this term ranges from zero to unity as the ratio $\phi_S(\omega)/\phi_N(\omega)$ ranges from zero to infinity. (Note that power spectra are real nonnegative functions.) In particular, if $\phi_S(\omega)/\phi_N(\omega) << 1$, then $|H(\omega)| \cong 0$, and if $\phi_S(\omega)/\phi_N(\omega) >> 1$, then $|H(\omega)| \cong 1$. The quantity $\phi_S(\omega)/\phi_N(\omega)$ can be interpreted as a measure of the signal-to-noise power ratio at frequency ω. Thus we see that if the noise is dominant at a given frequency, then the filter gain at that frequency is essentially zero, and if the signal is dominant, the gain is essentially unity. Between these extremes the gain is chosen to balance the effect of distorting the signal (caused by less than unity gain) and the effect of allowing noise to pass through the filter (caused by greater than zero gain).

From (V.D.20) and (V.D.23) through (V.D.25) the performance of the noncausal Wiener-Kolmogorov filter is given in this case by

$$MMSE = \frac{1}{2\pi}\int_{-\pi}^{\pi} \frac{\phi_S(\omega)\phi_N(\omega)}{\phi_S(\omega) + \phi_N(\omega)} d\omega. \qquad \text{(V.D.28)}$$

(Note that the shift λ is irrelevant to performance, as it should be for this noncausal case.) Since $\phi_S(\omega)/[\phi_S(\omega)+\phi_N(\omega)] \leqslant 1$ and $\phi_N(\omega)/[\phi_S(\omega)+\phi_N(\omega)] \leqslant 1$,

(V.D.28) implies that the MMSE is never larger than the minimum of the average signal power $[(1/2\pi)\int_{-\pi}^{\pi}\phi_S(\omega)d\,\omega]$ and the average noise power $[(1/2\pi)\int_{-\pi}^{\pi}\phi_N(\omega)d\,\omega]$; i.e.,

$$MMSE \leqslant \min\left\{\frac{1}{2\pi}\int_{-\pi}^{\pi}\phi_S(\omega)d\,\omega,\ \frac{1}{2\pi}\int_{-\pi}^{\pi}\phi_N(\omega)d\,\omega\right\}. \quad \text{(V.D.29)}$$

It achieves the first of these quantities in the limit as the ratio $\phi_S(\omega)/\phi_N(\omega)$ approaches zero uniformly in $[-\pi, \pi]$, in which case the optimum filter becomes a no-pass $[H(\omega) \cong 0]$ filter; and it achieve the second of these quantities in the limit as $\phi_S(\omega)/\phi_N(\omega)$ increases without bound uniformly in $[-\pi, \pi]$, in which case the optimum filter becomes an all-pass $[H(\omega) \cong 1]$ filter. Equation (V.D.28) also indicates that the MMSE is zero if and only if $\phi_S(\omega)\phi_N(\omega)=0$ for almost all $\omega \epsilon [-\pi, \pi]$, a condition that holds, for example, when the signal and noise occupy different parts of the frequency band.

V.D.2 Causal Wiener-Kolmogorov Filtering

As noted previously, noncausal linear estimators are not suitable for applications in which real-time estimates are desired. Since many applications do require real-time estimates, it is thus of interest to consider the causal Wiener-Kolmogorov problem, in which we wish to estimate X_t based on observation of the sequence $\{Y_n\}_{n=-\infty}^{\infty}$ only up to time t. This corresponds to the case $a=-\infty$ and $b=t$ in the notation employed previously.

To develop the solution to this problem, we first note that the set $H_{-\infty}^{t}$ is a subset of $H_{-\infty}^{\infty}$. Thus if the solution to the noncausal Wiener-Kolmogorov problem happens to be causal, it also solves the causal Wiener-Kolmogorov problem. Unfortunately, except under very special circumstances, the solution to the noncausal problem will in fact be strictly noncausal. However, there is a very definite relationship between the solutions to the causal and noncausal Wiener-Kolmogorov problems. In particular, let $\hat{X}_t$ and $\tilde{X}_t$ denote, respectively, the causal and noncausal Wiener-Kolmogorov estimates for a given model. Since $(X_t-\hat{X}_t)=(\tilde{X}_t-\hat{X}_t)+(X_t-\tilde{X}_t)$, we have for any $Z \epsilon H_{-\infty}^{t}$ that

$$0 = E\{(X_t - \hat{X}_t)Z\}$$
$$= E\{(\tilde{X}_t - \hat{X}_t)Z\} + E\{(X_t - \tilde{X}_t)Z\}. \tag{V.D.30}$$

Note that the term $E\{(X_t - \tilde{X}_t)Z\}$ is zero due to the orthogonality principle applied to $\hat{X}_t$ and to the fact that $Z \in H_{-\infty}^{\infty}$ (since $H_{-\infty}^{t} \subset H_{-\infty}^{\infty}$). Thus we have from (V.D.30) that

$$E\{(\tilde{X}_t - \hat{X}_t)Z\} = 0 \text{ for all } Z \in H_{-\infty}^{t}. \tag{V.D.31}$$

Equation (V.D.31) and the orthogonality principle imply that $\hat{X}_t$ is the MMSE estimate of $\tilde{X}_t$ among all estimates in $H_{-\infty}^{t}$. In other words, $\hat{X}_t$, which is the projection of X_t onto $H_{-\infty}^{t}$, can be obtained by first projecting X_t onto $H_{-\infty}^{\infty}$ to get $\tilde{X}_t$ and then projecting $\tilde{X}_t$ onto $H_{-\infty}^{t}$. A geometric analogy to this fact can be seen by considering the space $\mathbb{R}^3$ with the standard orthogonal axes labeled x, y, and z. To find the x-projection of a vector in $\mathbb{R}^3$, we can first project the vector onto the x-y plane (or the x-z plane) and then project the result onto the x-axis. This works because the x-axis is a subset of the $x-y$ (or $x-z$) plane.

In view of the above, a good starting point in seeking a causal MMSE estimator is to consider first the noncausal MMSE estimator

$$\tilde{X}_t = \sum_{n=-\infty}^{\infty} \tilde{h}_{t-n} Y_n \tag{V.D.32}$$

with $\{\tilde{h}_n\}_{n=-\infty}^{\infty}$ given by (V.D.12). Since we would like to project $\tilde{X}_t$ onto the set of linear estimates generated by $\{Y_n\}_{n=-\infty}^{t}$, we might be tempted to try a simple truncation of (V.D.32); i.e., we might consider the estimate

$$\overline{X}_t \triangleq \sum_{n=-\infty}^{t} \tilde{h}_{t-n} Y_n. \tag{V.D.33}$$

If $\overline{X}_t$ is the projection of $\tilde{X}_t$ onto $H_{-\infty}^{t}$, then the error, $\tilde{X}_t - \overline{X}_t = \sum_{n=t+1}^{\infty} \tilde{h}_{t-n} Y_m$, must be orthogonal to Y_m for all

$m \leqslant t$. However, since the sequences $\{Y_n\}_{n=t+1}^{\infty}$ and $\{Y_n\}_{n=-\infty}^{t}$ are usually correlated, and since the coefficient sequence $\{\tilde{h}_n\}_{n=-\infty}^{\infty}$ is chosen to satisfy a different orthogonality condition, it is unlikely that $(\tilde{X}_t - \overline{X}_t)$ will be orthogonal to $H_{-\infty}^t$ for the general case. However, one situation in which $\overline{X}_t$ of (V.D.33) will satisfy the required orthogonality condition is when $\{Y_n\}_{n=-\infty}^{\infty}$ is a sequence of uncorrelated random variables. We then have

$$E\{(\tilde{X}_t - \overline{X}_t)Y_m\} = \sum_{n=t+1}^{\infty} \tilde{h}_{t-n} E\{Y_n Y_m\}$$

$$= \sigma^2 \sum_{n=t+1}^{\infty} \tilde{h}_{t-n} \delta_{n,m} = 0, \; m \leqslant t, \tag{V.D.34}$$

where $\delta_{n,m}$ is the Kronecker delta ($\delta_{n,m}=1$ if $n=m$ and $\delta_{n,m}=0$ if $n \neq m$) and where $\sigma^2 = E\{Y_n^2\}$. (Recall that $\{Y_n\}_{n=-\infty}^{\infty}$ is assumed to be w.s.s. and zero-mean.)

Thus from the above we see that if we could first convert $\{Y_n\}_{n=-\infty}^{\infty}$ into an equivalent w.s.s. sequence $\{Z_n\}_{n=-\infty}^{\infty}$ of *uncorrelated* random variables by a causal linear operation, then the causal Wiener-Kolmogorov estimator would be given by simple truncation [as in (V.D.33)] of the optimum non-causal estimator of X_t based on $\{Z_n\}_{n=-\infty}^{\infty}$. Since such a sequence $\{Z_n\}_{n=-\infty}^{\infty}$ would have a constant, say unity spectrum, this latter estimator would be given by

$$\hat{X}_t = \sum_{n=-\infty}^{t} \hat{h}_{t-n} Z_n, \tag{V.D.35}$$

where

$$\hat{h}_n = \frac{1}{2\pi} \int_{-\pi}^{\pi} \phi_{XZ}(\omega) e^{i\omega n} d\omega = C_{XZ}(n), \; n \geqslant 0, \tag{V.D.36}$$

with ϕ_{XZ} and C_{XZ} the cross spectrum and cross covariance, respectively, between $\{X_n\}_{n=-\infty}^{\infty}$ and $\{Z_n\}_{n=-\infty}^{\infty}$.

The idea of causally converting $\{Y_n\}_{n=-\infty}^{\infty}$ into an equivalent white[†] sequence $\{Z_n\}_{n=-\infty}^{\infty}$ is not an unrealistic one in view of similar ideas that have arisen in the Gaussian detection and estimation problems of Sections III.B and V.B. In view of these earlier analyses, let us suppose that $\hat{Y}_{t|t-1}$ is the best linear prediction of Y_t from $\{Y_n\}_{n=-\infty}^{t-1}$; i.e., suppose that $\hat{Y}_{t|t-1}$ minimizes $E\{(Y_t - Z)^2\}$ over all $Z \in H_{-\infty}^{t-1}$. Also, let σ_t^2 denote the mean-squared-error in this prediction; i.e.,

$$\sigma_t^2 = E\{(Y_t - \hat{Y}_{t|t-1})^2\}. \tag{V.D.37}$$

Now define a sequence $\{Z_n\}_{n=-\infty}^{\infty}$ by

$$Z_n = \frac{Y_n - \hat{Y}_{n|n-1}}{\sigma_n}, \; n \in \mathbb{Z}, \tag{V.D.38}$$

and note that $Z_n \in H_{-\infty}^{n}$. We have $E\{Z_n^2\}=1, E\{Z_n\}=0$, and $Cov(Z_n, Z_m) = E\{Z_n Z_m\}$. With $m < n$ we have

$$E\{Z_n Z_m\} = (1/\sigma_n) E\{(Y_n - \hat{Y}_{n|n-1}) Z_m\} = 0,$$

by the orthogonality principle since $Z_m \in H_{-\infty}^{n-1}$. Similarly, $E\{Z_n Z_m\}=0$ for $m > n$. Thus, this $\{Z_n\}_{n=-\infty}^{\infty}$ is a white sequence obtained by causal linear transformation of $\{Y_n\}_{n=-\infty}^{\infty}$.

Now if we could show that $\{Z_n\}_{n=-\infty}^{\infty}$ is equivalent to $\{Y_n\}_{n=-\infty}^{\infty}$ for the purposes of linear MMSE estimation, then the causal Wiener-Kolmogorov estimation problem is effectively reduced to the problem of one-step linear prediction. This follows since one-step linear prediction can be used to prewhiten the observations via (V.D.38), and then the causal estimator of X_t from the prewhitened data is given straightforwardly by (V.D.35) and (V.D.36). Such an equivalence between $\{Z_n\}_{n=-\infty}^{\infty}$ and $\{Y_n\}_{n=-\infty}^{\infty}$ can in fact be established within a mild condition on the spectrum of

---

† Here we use the term white to denote a w.s.s. sequence of zero-mean uncorrelated random variables. Such a sequence will have a spectrum that is constant for $\omega \in [-\pi, \pi]$. Thus by analogy with white light, which is light containing equal levels of all visible wavelengths, such a sequence is termed *white*.

$\{Y_n\}_{n=-\infty}^{\infty}$. To show this, we now turn to the analysis of the specific problem of linear prediction.

Linear Prediction

Consider the specific causal Wiener-Kolmogorov problem in which $X_t = Y_{t+\lambda}$, where λ is a positive integer. Note that for $\lambda=1$, this is similar to the Levinson problem except that our observations now extend back in time to $-\infty$. To seek a solution to this problem, we first give the following result, which is central to the theory of linear prediction.

Proposition V.D.1: The Spectral Factorization Theorem

Suppose that $\{Y_n\}_{n=-\infty}^{\infty}$ has a spectrum satisfying the so-called *Paley-Wiener condition*, given by

$$c_0 \triangleq \frac{1}{2\pi}\int_{-\pi}^{\pi} \log \phi_Y(\omega) d\omega > -\infty. \qquad \text{(V.D.39)}$$

Then ϕ_Y can be written as $\phi_Y(\omega) = \phi_Y^+(\omega)\phi_Y^-(\omega)$, $-\pi \leqslant \omega \leqslant \pi$, where ϕ_Y^+ and ϕ_Y^- are two functions satisfying $|\phi_Y^+(\omega)|^2 = |\phi_Y^-(\omega)|^2 = \phi_Y(\omega)$,

$$\frac{1}{2\pi}\int_{-\pi}^{\pi} \phi_Y^+(\omega) e^{in\omega} d\omega = 0 \text{ for all } n < 0 \qquad \text{(V.D.40a)}$$

and

$$\frac{1}{2\pi}\int_{-\pi}^{\pi} \phi_Y^-(\omega) e^{in\omega} d\omega = 0 \text{ for all } n > 0. \qquad \text{(V.D.40b)}$$

Moreover, (V.D.40a) [resp. (V.D.40b)] is also satisfied when ϕ_Y^+ [resp. ϕ_Y^-] is replaced by $1/\phi_Y^+$ [resp. $1/\phi_Y^-$].†

† Note that (V.D.39) implies that ϕ_Y is nonzero. Since $|\phi_Y^+|^2 = |\phi_Y^-|^2 = \phi_Y$, we see that ϕ_Y^+ and ϕ_Y^- must also be nonzero, so $1/\phi_Y^+$ and $1/\phi_Y^-$ are well defined. With $1/\phi_Y^+$ satisfying (V.D.40a), ϕ_Y^+ is said to be of *minimum phase*.

Proof: A complete proof of the result can be found, for example, in Ash and Gardner (1975). Here we will outline the key ideas in this argument.

First we note that since $\log(x)$ is a concave function of x, Jensen's inequality implies that

$$c_0 \leqslant \log\left(\frac{1}{2\pi}\int_{-\pi}^{\pi}\phi_Y(\omega)d\omega\right) = \log(E\{Y^2\}) < \infty.$$

Thus the condition (V.D.39) is equivalent to the condition $|c_0| < \infty$. This allows us to write $\log \phi_Y$ as a discrete-time Fourier transform

$$\log \phi_Y(\omega) = \sum_{n=-\infty}^{\infty} c_n e^{-i\omega n}, \; -\pi < \omega \leqslant \pi, \quad \text{(V.D.41a)}$$

where

$$c_n \triangleq \frac{1}{2\pi}\int_{-\pi}^{\pi} e^{in\omega} \log \phi_Y(\omega), \; n \in \mathbb{Z}. \quad \text{(V.D.41b)}$$

From (V.D.41a) we can write

$$\begin{aligned}\phi_Y(\omega) &= \exp\left\{\sum_{n=-\infty}^{\infty} c_n e^{-i\omega n}\right\} \\ &= \phi_Y^+(\omega)\phi_Y^-(\omega), \; -\pi \leqslant \omega \leqslant \pi,\end{aligned} \quad \text{(V.D.42)}$$

where

$$\phi_Y^+(\omega) \triangleq \exp\left\{\frac{c_0}{2} + \sum_{n=1}^{\infty} c_n e^{-i\omega n}\right\} \quad \text{(V.D.43)}$$

and

$$\phi_Y^-(\omega) \triangleq \exp\left\{\frac{c_0}{2} + \sum_{n=-\infty}^{-1} c_n e^{-i\omega n}\right\}. \qquad \text{(V.D.44)}$$

We would now like to show that the functions ϕ_Y^+ and ϕ_Y^- have the desired properties. To do so, we must first note that all power spectral densities are even-symmetric functions [i.e., $\phi_Y(-\omega)=\phi_Y(\omega)$], a fact that follows straightforwardly from the even symmetry of C_Y. The even symmetry of $\phi_Y(\omega)$ implies that $\log \phi_Y(\omega)$ is also even symmetric, which in turn implies that the coefficients $\{c_n\}_{n=-\infty}^{\infty}$ of (V.D.41b) are real and even symmetric (i.e., $c_{-n}=c_n=c_n^*$). Thus we can write

$$\begin{aligned}\phi_Y^-(\omega) &= \exp\left\{\frac{c_0}{2} + \sum_{n=1}^{\infty} c_{-n} e^{i\omega n}\right\} \\ &= \exp\left\{\frac{c_0}{2} + \sum_{n=1}^{\infty} c_n e^{i\omega n}\right\} = [\phi^+(\omega)]^*.\end{aligned} \qquad \text{(V.D.45)}$$

Equation (V.D.45) implies that $|\phi_Y^-(\omega)|^2=|\phi_Y^+(\omega)|^2= \phi_Y^+(\omega)[\phi_Y^+(\omega)]^*=\phi_Y^+(\omega)\phi_Y^-(\omega)=\phi_Y(\omega)$.

To verify (V.D.40a) we first note that e^z has the power series expansion

$$e^z = \sum_{k=0}^{\infty} \frac{z^k}{k!}. \qquad \text{(V.D.46)}$$

So we can write

$$\phi_Y^+(\omega) = e^{c_0/2}\left[\sum_{k=0}^{\infty}\left(\sum_{n=1}^{\infty} c_n e^{-i\omega n}\right)^k / k!\right]. \qquad \text{(V.D.47)}$$

Inspection of (V.D.47) reveals that its right-hand side contains only nonnegative powers of $e^{-i\omega}$. This implies that the Fourier components of $\phi_Y^+(\omega)$ with negative indices are all zero. This is (V.D.40a). Equation (V.D.40b) and the analogous conditions for $1/\phi_Y^+$ and $1/\phi_Y^-$ follow by similar arguments.

□

It is interesting to note that the spectral factorization of ϕ_Y into a product of causal and anticausal parts is analogous to the Cholesky decomposition of a covariance matrix into lower and upper triangular factors.

We can use the spectral decomposition of Proposition V.D.1 to find linear MMSE predictors. To do so we henceforth assume that ϕ_Y satisfies (V.D.39). Consider the time-invariant linear filter with transfer function $H(\omega)=1/\phi_Y^+(\omega)$. Note that this is a causal filter by way of condition (V.D.40a) applied to $1/\phi_Y^+$. Suppose that we apply the sequences $\{Y_n\}_{n=-\infty}^{\infty}$ to this filter and let $\{W_n\}_{n=-\infty}^{\infty}$ denote the output sequence. A well-known result in the analysis of second-order random sequences is that the output of a time-invariant linear filter driven by a w.s.s. process is also w.s.s. and that the spectrum of the output process is given by $|H(\omega)|^2\phi(\omega)$, where H is the filter transfer function and ϕ is the input spectrum [see, e.g., Wong (1983)]. Thus $\{W_n\}_{n=-\infty}^{\infty}$ is a w.s.s. sequence and its spectrum is

$$\phi_W(\omega) = \left|\frac{1}{\phi_Y^+(\omega)}\right|^2 \phi_Y(\omega)$$

$$= \frac{\phi_Y(\omega)}{|\phi_Y^+(\omega)|^2} = 1, \ -\pi \leqslant \omega \leqslant \pi. \tag{V.D.48}$$

Since a constant spectrum corresponds to a white sequence, we see that the filter $1/\phi_Y^+(\omega)$ is a whitening filter for $\{Y_n\}_{n=-\infty}^{\infty}$. Moreover, $\{Y_n\}_{n=-\infty}^{\infty}$ is obtained causally from $\{W_n\}_{n=-\infty}^{\infty}$ by applying the latter sequence to the filter with transfer function $\phi_Y^+(\omega)$. Thus we can write

$$Y_t = \sum_{n=-\infty}^{t} f_{t-n} W_n \, , t \in \mathbb{Z} \, , \tag{V.D.49}$$

where

$$f_n = \frac{1}{2\pi}\int_{-\pi}^{\pi} \phi_Y^+(\omega) e^{i\omega n} \, d\omega, \; n \geq 0. \qquad \text{(V.D.50)}$$

Equation (V.D.49) gives a representation for $\{Y_n\}_{n=-\infty}^{\infty}$ as the output of a time-invariant linear filter driven by a white sequence. This representation can be used to derive the optimum linear predictor of the sequence $\{Y_n\}_{n=-\infty}^{\infty}$. In particular, we note that for $\lambda>0$,

$$\begin{aligned} Y_{t+\lambda} &= \sum_{n=-\infty}^{t+\lambda} f_{t+\lambda-n} W_n \\ &= \sum_{n=t+1}^{t+\lambda} f_{t+\lambda-n} W_n + \sum_{n=-\infty}^{t} f_{t+\lambda-n} W_n . \end{aligned} \qquad \text{(V.D.51)}$$

Since $\{W_n\}_{n=-\infty}^{\infty}$ is white, the variables $W_{t+1}, \ldots, W_{t+\lambda}$ are orthogonal to $\{W_n\}_{n-\infty}^{t}$, and the representation (V.D.49) thus implies that they are orthogonal to $H_{-\infty}^{t}$. On rearranging (V.D.51) as

$$Y_{t+\lambda} - \sum_{n=-\infty}^{t} f_{t+\lambda-n} W_n = \sum_{n=t+1}^{t+\lambda} f_{t+\lambda-n} W_n ,$$

we then have that $Y_{t+\lambda} - \sum_{n=-\infty}^{t} f_{t+\lambda-n} W_n$ is orthogonal to $H_{-\infty}^{t}$. Since $\sum_{n=-\infty}^{t} f_{t+\lambda-n} W_n \in H_{-\infty}^{t}$, the orthogonality principle implies that the linear prediction of $Y_{t+\lambda}$ from $\{Y_n\}_{n=-\infty}^{t}$ is given by

$$\hat{Y}_{t+\lambda} = \sum_{n=-\infty}^{t} f_{t+\lambda-n} W_n . \qquad \text{(V.D.52)}$$

The linear prediction filter of (V.D.52) can be thought of as a series connection of two time-invariant linear filters, as depicted in Fig. V.D.1. The first of these two filters is the whitening filter with transfer function $1/\phi_Y^+(\omega)$. The second filter has impulse response

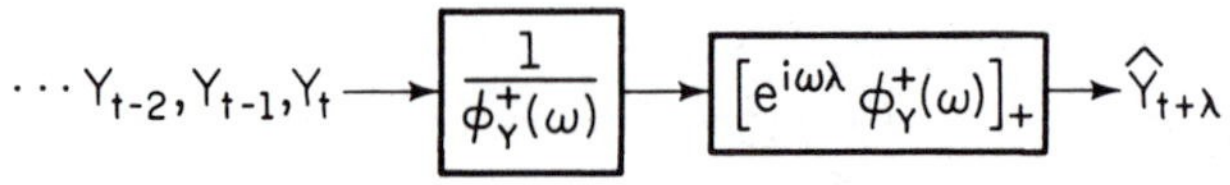

$$\cdots Y_{t-2}, Y_{t-1}, Y_t \longrightarrow \boxed{\frac{1}{\phi_Y^+(\omega)}} \longrightarrow \boxed{[e^{i\omega\lambda}\phi_Y^+(\omega)]_+} \longrightarrow \hat{Y}_{t+\lambda}$$

Fig. V.D.1: Representation of the optimum pure-prediction filter

$$\begin{cases} f_{n+\lambda} \text{ for } n \geqslant 0 \\ \\ 0 \qquad \text{for } n < 0. \end{cases} \tag{V.D.53}$$

If we define the operation $[H(\omega)]_+$ by

$$[H(\omega)]_+ = \sum_{n=0}^{\infty} h_n e^{-i\omega n}, \tag{V.D.54}$$

where $h_n = (1/2\pi)\int_{-\pi}^{\pi} H(\omega) e^{i\omega n} d\omega$, the filter of (V.D.53) has transfer function $[e^{i\omega\lambda}\phi_Y^+(\omega)]_+$. Thus the overall transfer function for optimum linear prediction λ steps into the future can be written as

$$\frac{1}{\phi_Y^+(\omega)}[e^{i\omega\lambda}\phi_Y^+(\omega)]_+ \triangleq H_\lambda(\omega). \tag{V.D.55}$$

The representation of (V.D.49) and (V.D.52) also allows us to write an expression for the mean-squared-error incurred in optimum linear prediction. In particular, from (V.D.51) we have that

$$MMSE = \min_{Z \in H_{-\infty}^{t}} E\{(Y_{t+\lambda} - Z)^2\} = E\{(Y_{t+\lambda} - \hat{Y}_{t+\lambda})^2\}$$

$$= E\left\{\left(\sum_{n=t+1}^{t+\lambda} f_{t+\lambda-n} W_n\right)^2\right\} = \sum_{n=t+1}^{t+\lambda} f_{t+\lambda-n}^2, \tag{V.D.56}$$

where the last inequality follows from the orthogonality of $W_{t+1}, W_{t+2}, \ldots,$ and $W_{t+\lambda}$. A simple change of variables in (V.D.56) gives

$$MMSE = \sum_{n=0}^{\lambda-1} f_n^2. \tag{V.D.57}$$

The coefficients $\{f_n\}_{n=0}^{\lambda-1}$ can be obtained from (V.D.47) since f_n is the $n^{\underline{th}}$ Fourier coefficient of ϕ_Y^+, although this is somewhat tedious for large λ. The most interesting case is one-step prediction (λ=1). It is easily seen from (V.D.47) that $a_0 = e^{c_0/2}$; thus the value of MMSE for one-step prediction is

$$MMSE = \exp\left\{\frac{1}{2\pi}\int_{-\pi}^{\pi} \log \phi_Y(\omega) d\omega\right\}, \tag{V.D.58}$$

a result known as the *Kolmogorov-Szegö-Krein formula.*

Equation (V.D.51) and the orthogonality of $\{W_n\}_{n=-\infty}^{\infty}$ imply that†

$$\sum_{n=0}^{\infty} f_n^2 = E\{(Y_{t+\lambda})^2\} < \infty. \tag{V.D.59}$$

Comparing (V.D.57) and (V.D.59), we see that as we try to

† Equivalently, Parceval's formula gives

$$\sum_{n=0}^{\infty} f_n^2 = \frac{1}{2\pi}\int_{-\pi}^{\pi} |\phi_Y^+(\omega)|^2 d\omega = \frac{1}{2\pi}\int_{-\pi}^{\pi} \phi_Y(\omega) d\omega = E\{Y_t^2\},\ t \in \mathbb{Z}.$$

predict further into the future (i.e., as λ increases), the minimum mean squared prediction error approaches the mean squared value of the quantity we are trying to predict. An equivalent interpretation is that if we fix the time at which we are trying to estimate $\{Y_n\}_{n=-\infty}^{\infty}$ (i.e., fix $t+\lambda$) and let t approach $-\infty$, then in the limit as $t\rightarrow-\infty$ the observations are of no use in predicting $Y_{t+\lambda}$. In other words, for fixed n, no part of Y_n can be determined from the infinite past. A sequence with this property is said to be *purely nondeterministic*, and a sequence has this property if and only if it has the representation $\sum_{n=-\infty}^{t} f_{t-n} W_n$, with $\{W_n\}_{n=-\infty}^{\infty}$ white and $\sum_{n=0}^{\infty} f_n^2<\infty$. Proposition V.D.1 and the analysis following it show that the Paley-Wiener condition is sufficient for the existence of this type of representation. It can be shown that this condition is also a necessary condition for the existence of such a representation [see Ash and Gardner (1975) for further discussion of this notion.]

Having determined a mechanism for causally prewhitening a covariance stationary sequence of observations, the solution to the general causal Wiener-Kolmogorov problem follows almost immediately. In particular, assuming that $\{Y_n\}_{n=-\infty}^{\infty}$ satisfies the Paley-Wiener condition, it can be whitened (into, say, $\{Z_n\}_{n=-\infty}^{\infty}$) by passing it through the filter $1/\phi_Y^+$. Then we need only find the cross spectrum ϕ_{XZ} and (V.D.35) and (V.D.36) give us the optimum filter to follow the prewhitener as $[\phi_{XZ}(\omega)]_+$, where the notation $[\cdot]_+$ is as defined in (V.D.54). It is a straightforward exercise to show that if $\{Y_n\}_{n=-\infty}^{\infty}$ is passed through a filter with transfer function H to get the sequence $\{Z_n\}_{n=-\infty}^{\infty}$, then $\phi_{XZ}=\phi_{XY}H^*$. Thus since in our case $H=1/\phi_Y^+$ we have

$$\phi_{XZ}(\omega)=\frac{\phi_{XY}(\omega)}{[\phi_Y^+(\omega)]^*}=\frac{\phi_{XY}(\omega)}{\phi_Y^-(\omega)}. \qquad \text{(V.D.60)}$$

We then have that the causal Wiener-Kolmogorov filter for estimating X_t from $\{Y_n\}_{n=-\infty}^{t}$ has transfer function

$$\frac{1}{\phi_Y^+(\omega)}\left[\frac{\phi_{XY}(\omega)}{\phi_Y^-(\omega)}\right]_+. \qquad \text{(V.D.61)}$$

It is interesting to note that the noncausal Wiener-Kolmogorov filter has transfer function [see (V.D.11)]

$$\frac{\phi_{XY}(\omega)}{\phi_Y(\omega)} = \frac{1}{\phi_Y^+(\omega)} \left[\frac{\phi_{XY}(\omega)}{\phi_Y^-(\omega)} \right]. \tag{V.D.62}$$

Thus both filters can be represented as a series connection of a causal prewhitener with a second filter. In the noncausal filter this second filter is ϕ_{XY}/ϕ_Y^- and in the causal case it is the (additive) causal part of ϕ_{XY}/ϕ_Y^-.

Factorization of Rational Spectra

Note that the key step in designing a causal Wiener-Kolmogorov filter is the factorization of the observation spectrum ϕ_Y. Since ϕ_Y^+ and ϕ_Y^- can be written in terms of the Fourier coefficients $\{c_n\}_{n=-\infty}^{\infty}$ of $\log \phi_Y$ [see (V.D.43) and (V.D.44)] this factorization can be performed numerically by computing the c_n's. (The logarithm of the spectrum is often termed the *cepstrum*.) For a large class of spectra of interest in practice, however, spectral factorization can be viewed as the factorization of complex polynomials. We discuss this issue briefly in the following paragraphs.

The power spectrum of a random sequence $\{Y_n\}_{n=-\infty}^{\infty}$ is said to be *rational* if it can be written as the ratio of two real trigonometric polynomials; i.e., ϕ is rational if we can write

$$\phi_Y(\omega) = \frac{n_0 + 2\sum_{k=1}^{p} n_k \cos k\omega}{d_0 + 2\sum_{k=1}^{m} d_k \cos k\omega}, \tag{V.D.63}$$

where m and p are positive integers and $n_0, \ldots, n_p$, $d_0, \ldots, d_m$ are real numbers. Many random sequences arising in practice have this type of spectrum, and most covariance stationary sequences have spectra that can be approximated arbitrarily closely by rational spectra with large enough choice of the orders m and p. Since power spectra must be even symmetric about ω=0, the polynomials in (V.D.62) contain only cosine terms and no sine terms.

Since $2 \cos k\omega = e^{ik\omega} + e^{-ik\omega}$, the spectrum of (V.D.63) can be written as

$$\phi_Y(\omega) = \frac{N(e^{i\omega})}{D(e^{i\omega})}, \tag{V.D.64}$$

where N and D are polynomials of a complex variable z defined by

$$N(z) = \sum_{k=-p}^{p} n_{|k|} z^{-k} \tag{V.D.65a}$$

and

$$D(z) = \sum_{k=-m}^{m} d_{|k|} z^{-k}. \tag{V.D.65b}$$

Note that $z^p N(z)$ is a $(2p)^{\underline{th}}$ order polynomial, so it has $2p$ roots $z_1, z_2, ..., z_{2p}$, and can be written as $n_p \Pi_{k=1}^{2p}(z - z_k)$. Thus we can write

$$N(z) = n_p z^{-p} \prod_{k=1}^{2p} (z - z_k). \tag{V.D.66}$$

Since $N(z) = N(1/z)$, the roots $z_1, ..., z_{2p}$ must be in reciprocal pairs; i.e., for each root z_k there is another root equal to $1/z_k$. Assuming for notational convenience that $z_1, ..., z_{2p}$ are ordered such that $|z_1| \geqslant |z_2| \geqslant ... \geqslant |z_{2p}|$, it is straightforward to write

$$N(z) = B(z) B(1/z), \tag{V.D.67}$$

where

$$B(z) = [(-1)^p n_p / z_1 z_2 ... z_p]^{1/2} \prod_{k=1}^{p} (z^{-1} - z_k). \tag{V.D.68}$$

Since $|z_1| \geqslant |z_2| \geqslant \dots \geqslant |z_{2p}|$, we must have $|z_{2p}| = 1/|z_1|$, $|z_{2p-1}| = 1/|z_2|, \dots, |z_{p+1}| = 1/|z_p|$, from which it follows that $|z_1| \geqslant |z_2| \geqslant \dots \geqslant |z_p| \geqslant 1$. Note that $B(z)$ can be expanded into the form

$$B(z) = \sum_{k=0}^{p} b_k z^{-k}. \tag{V.D.69}$$

Similarly, the polynomial $D(z)$ can be written as

$$D(z) = A(z)A(1/z) \tag{V.D.70}$$

with $A(z)$ in the form

$$\begin{aligned} A(z) &= [d_m / p_1 p_2 \cdots p_m]^{1/2} \prod_{k=1}^{m} (z^{-1} - p_k) \\ &= \sum_{k=0}^{m} a_k z^{-k}, \end{aligned} \tag{V.D.71}$$

where $|p_1| \geqslant |p_2| \geqslant \dots \geqslant |p_m| \geqslant 1$.

We see from the above that the rational spectrum $\phi_Y(\omega)$ can be written as

$$\phi_Y(\omega) = \frac{B(e^{i\omega})B(e^{-i\omega})}{A(e^{i\omega})A(e^{-i\omega})}. \tag{V.D.72}$$

We assume henceforth that none of the roots of $B(z)$ or $A(z)$ is on the unit circle $|z| = 1$ (i.e., we assume that $|z_p| > 1$ and $|p_m| > 1$). [This ensures that $\phi_Y(\omega)$ is bounded from above and is bounded away from zero from below, which in turn implies that it satisfies the Paley-Wiener condition.] It is not hard to show [see, e.g., Oppenheim and Schafer (1975)] that both $B(e^{i\omega})/A(e^{i\omega})$ and $A(e^{i\omega})/B(e^{i\omega})$ are causal stable transfer functions, and that both $B(e^{-i\omega})/A(e^{-i\omega})$ and $A(e^{-i\omega})/B(e^{-i\omega})$ (V.D.28) are anticausal stable transfer functions. It follows from this and (V.D.72) that the spectral factors of ϕ_Y are

$$\phi_Y^+(\omega) = B(e^{i\omega})/A(e^{i\omega}) \qquad \text{(V.D.73a)}$$

and

$$\phi_Y^-(\omega) = [\phi_Y^+(\omega)]^* = B(e^{-i\omega})/A(e^{-i\omega}). \qquad \text{(V.D.73b)}$$

The whitening filter for $\{Y_n\}_{n=-\infty}^{\infty}$ is now given by

$$\frac{1}{\phi_Y^+(\omega)} = \frac{A(e^{i\omega})}{B(e^{i\omega})}. \qquad \text{(V.D.74)}$$

Equivalently, with $\{Z_n\}_{n=-\infty}^{\infty}$ representing the whitened sequence, we can say that the output of the filter $A(e^{i\omega})$ when applied to $\{Y_n\}_{n=-\infty}^{\infty}$ equals the output of the filter $B(e^{i\omega})$ when applied to $\{Z_n\}_{n=-\infty}^{\infty}$. From (V.D.69) and (V.D.71) the impulse responses of $A(e^{i\omega})$ and $B(e^{i\omega})$ are, respectively,

$$\begin{cases} a_n & \text{if } 0 \leqslant n \leqslant m \\ 0 & \text{otherwise} \end{cases} \qquad \text{(V.D.75)}$$

and

$$\begin{cases} b_n & \text{if } 0 \leqslant n \leqslant p \\ 0 & \text{otherwise.} \end{cases} \qquad \text{(V.D.76)}$$

This implies that $\{Y_n\}_{n=-\infty}^{\infty}$ and $\{Z_n\}_{n=-\infty}^{\infty}$ are related by

$$\sum_{k=0}^{m} a_k Y_{n-k} = \sum_{k=0}^{p} b_k Z_{n-k}, \; n \in \mathbb{Z}. \qquad \text{(V.D.77)}$$

Thus the sequence $\{Z_n\}_{n=-\infty}^{\infty}$ satisfies the recursion

$$b_0 Z_n = -\sum_{k=1}^{p} b_n Z_{n-k} + \sum_{k=0}^{m} a_k Y_{n-k} . \qquad \text{(V.D.78)}$$

This recursion represents a finite-dimensional linear digital filter, as illustrated in Fig. V.D.2. (In the figure z^{-1} represents a delay of one time unit.)

Note that (V.D.77) also implies that $\{Y_n\}_{n=-\infty}^{\infty}$ is generated from $\{Z_n\}_{n=-\infty}^{\infty}$ by the recursion

$$a_0 Y_n = -\sum_{k=1}^{m} a_k Y_{n-k} + \sum_{k=0}^{p} b_k Z_{n-k} , \, n \in \mathbb{Z} . \qquad \text{(V.D.79)}$$

A sequence generated in this fashion from a white sequence is said to be an *autoregressive/moving-average sequence with autoregressive order m and moving-average order p*, or an ARMA(m, p) sequence. An ARMA (m, 0) sequence is called an *autoregressive sequence of order m* [AR(m)] and an ARMA(0, p) sequence is called a *moving average of order p* [MA(p)]. [With m=0, the first sum in (V.D.79) is taken to be

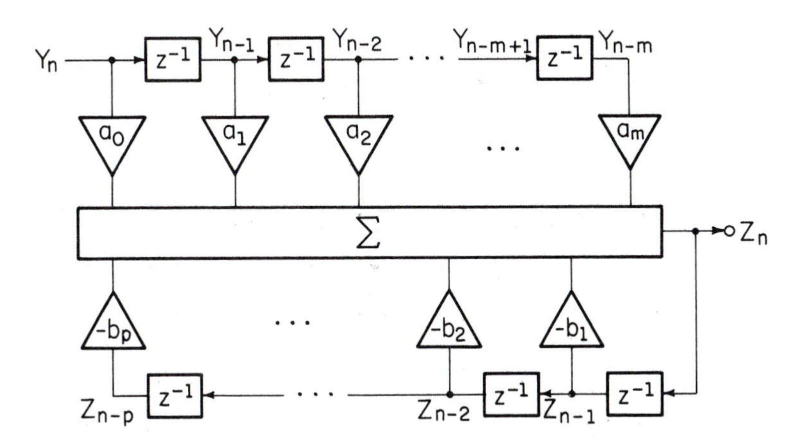

Fig. V.D.2: Whitening filter for a sequence with a rational spectrum.

zero.] ARMA models are closely related to the state-space models arising in Kalman-Bucy filtering, and some of their properties relevant to filtering can be found in Anderson and Moore (1979).

Example V.D.2: Pure Prediction of a Wide-Sense Markov Sequence

A simple but useful model for the correlation structure of covariance stationary random sequences is the so-called *wide-sense Markov model*:

$$C_Y(n) = P\, r^{|n|}, \quad n \in \mathbb{Z}, \tag{V.D.80}$$

where $|r|<1$ and $P>0$. The power spectrum corresponding to (V.D.80) is given by [see, e.g., Thomas (1971)]

$$\phi_Y(\omega) = \frac{P(1-r^2)}{1-2r\cos\omega+r^2}. \tag{V.D.81}$$

Note that (V.D.81) is a rational spectrum, and using $2\cos\omega = e^{i\omega}+e^{-i\omega}$, we have

$$\phi_Y(\omega) = \frac{P(1-r^2)}{1-re^{i\omega}-re^{-i\omega}+r^2}$$

$$= \frac{P(1-r^2)}{(1-re^{-i\omega})(1-re^{i\omega})} \tag{V.D.82}$$

$$= \frac{1}{A(e^{i\omega})A(e^{-i\omega})},$$

where

$$A(z) = a_0 + a_1 z^{-1}, \tag{V.D.83}$$

with $a_0=[P(1-r^2)]^{-1/2}$ and $a_1=-r\,[P(1-r^2)]^{-1/2}$.

Suppose that we wish to predict $\{Y_n\}_{n=-\infty}^{\infty}$ λ steps into the future. The transfer function of the optimum prediction is given by (V.D.55), which in this case becomes

$$\hat{H}_\lambda(\omega) = A(e^{i\omega}) \left[\frac{e^{i\omega\lambda}}{A(e^{i\omega})} \right]_+ . \qquad \text{(V.D.84)}$$

On using the geometric series, $\sum_{k=0}^{\infty} x^k = 1/(1-x)$ for $|x|<1$, we have

$$\frac{1}{A(z)} = \frac{1}{a_0(1-rz^{-1})} = \frac{1}{a_0} \sum_{n=0}^{\infty} r_n z^{-1}, \qquad \text{(V.D.85)}$$

which converges for $|z|=1$ since $|r|<1$. So $1/A(e^{i\omega}) = (1/a_0)\sum_{n=0}^{\infty} r^n e^{-i\omega n}$ and we have the following steps:

$$\left[\frac{e^{i\omega\lambda}}{A(e^{i\omega})} \right]_+ = \left[\frac{1}{a_0} \sum_{n=0}^{\infty} r^n e^{-i\omega(n-\lambda)} \right]_+$$

$$= \frac{1}{a_0} \sum_{n=\lambda}^{\infty} r^n e^{-i\omega(n-\lambda)} \qquad \text{(V.D.86)}$$

$$= \frac{1}{a_0} \sum_{l=0}^{\infty} r^{l+\lambda} e^{-i\omega l} = \frac{r^\lambda}{A(e^{i\omega})}.$$

Considering (V.D.84) and (V.D.86), we have that $H_\lambda(\omega) = r^\lambda$; that is, in this case the optimum predictor is a pure gain. The impulse response of the predictor is thus $h_0 = r^\lambda$ and $h_n = 0$, $n \neq 0$, so we have simply

$$\hat{Y}_{t+\lambda} = r^\lambda Y_t . \qquad \text{(V.D.87)}$$

The mean squared prediction error is easily computed from (V.D.87) and (V.D.80) as

$$E\{(Y_{t+\lambda}-\hat{Y}_{t+\lambda})^2\} = E\{Y_{t+\lambda}^2\} - E\{Y_{t+\lambda}\hat{Y}_{t+\lambda}\}$$

$$= E\{Y_{t+\lambda}^2\} - r^{\lambda}E\{Y_{t+\lambda}Y_t\}$$

$$= C_Y(0) - r^{\lambda}C_Y(\lambda) \qquad \text{(V.D.88)}$$

$$= P(1-r^{2\lambda}).$$

Since $|r|<1$, the prediction error increases monotonically from $(1-r^2)P$ to P as λ increases from 1 to ∞.

Example V.D.3: Pure Prediction of AR(m) Sequences

In view of (V.D.83), a wide-sense Markov sequence is an AR (1) sequence. In particular, $\{Y_n\}_{n=-\infty}^{\infty}$ is generated by

$$Y_{t+1} = rY_t + [P(1-r^2)]^{1/2} Z_{t+1},\ t \in \mathbb{Z}, \qquad \text{(V.D.89)}$$

where $\{Z_n\}_{n=-\infty}^{\infty}$ is white. Since Z_{t+1} is orthogonal to $\{Z_n\}_{n=-\infty}^{t}$ and hence to $\{Y_n\}_{n=-\infty}^{t}$, we see from (V.D.89) that $(Y_{t+1}-rY_t)$ is orthogonal to $\{Y_n\}_{n=-\infty}^{t}$ and thus the orthogonality principle implies that rY_t is the MMSE linear estimate of Y_{t+1} from $\{Y_n\}_{n=-\infty}^{t}$. This is (V.D.87) for $\lambda=1$.

Similarly, for any autoregressive sequence

$$Y_{t+1} = -\sum_{k=1}^{m} a_k Y_{t+1-k} + b_0 Z_{t+1},\ t \in \mathbb{Z} \qquad \text{(V.D.90)}$$

(without loss of generality we take $a_0=1$), the quantity $Y_{t+1} + \sum_{k=1}^{m} a_k Y_{t+1-k} \equiv b_0 Z_{t+1}$ is orthogonal to $\{Y_n\}_{n=-\infty}^{t}$. So the optimum one-step predictor is

$$\hat{Y}_{t+1} = -\sum_{k=1}^{m} a_k Y_{t+1-k}. \qquad \text{(V.D.91)}$$

The minimum mean squared prediction error is simply

$$MMSE = E\{(Y_{t+1} + \sum_{k=1}^{m} a_k Y_{t+1-k})^2\}$$

$$= E\{b_0^2 Z_{t+1}^2\} = b_0^2 E\{Z_{t+1}^2\} = b_0^2 . \qquad \text{(V.D.92)}$$

For the $AR(1)$ case $b_0^2 = P(1-r^2)$, which agrees with (V.D.88). In general, the Kolmogorov-Szegö-Krein formula (V.D.58) gives

$$b_0^2 = \exp\{\frac{1}{2\pi}\int_{-\pi}^{\pi} \log \phi_Y(\omega) d\omega\}. \qquad \text{(V.D.93)}$$

Example V.D.4: Filtering, Prediction, and Smoothing of Wide-Sense Markov Sequences in White Noise

Consider the observation model

$$Y_n = S_n + N_n, \ n \in \mathbb{Z}, \qquad \text{(V.D.94)}$$

where $\{S_n\}_{n=-\infty}^{\infty}$ and $\{N_n\}_{n=-\infty}^{\infty}$ are zero-mean orthogonal wide-sense stationary sequences. Assume that $\{N_n\}_{n=-\infty}^{\infty}$ is white with $E\{N_n^2\} = v_N^2$ and that $\{S_n\}_{n=-\infty}^{\infty}$ is wide-sense Markov with $C_S(n) = Pr^{|n|}, n \in \mathbb{Z}$. Referring to Example V.D.2, and using the orthogonality of $\{S_n\}_{n=-\infty}^{\infty}$ and $\{N_n\}_{n=-\infty}^{\infty}$, the spectrum of the observation is given by

$$\phi_Y(\omega) = \phi_S(\omega) + \phi_N(\omega) = \frac{P(1-r^2)}{(1-re^{-i\omega})(1-re^{i\omega})} + v_N^2$$

$$= \frac{P(1-r^2) + v_N^2(1-re^{-i\omega})(1-re^{i\omega})}{(1-re^{-i\omega})(1-re^{i\omega})},$$

which is a rational spectrum.

The denominator polynomial in ϕ_Y is already factored as $A(z)A(1/z)$ with $A(z) = 1 - rz^{-1}$. The numerator polynomial is $N(z) = n_1 z + n_0 + n_1 z^{-1}$ with $n_0 = P(1-r^2) + v_N^2(1+r^2)$

and $n_1 = -v_N^2 r$. Using the quadratic formula we can write $N(z)$ as

$$N(z) = n_1 z^{-1}(z - z_1)(z - 1/z_1),$$

where

$$z_1 = -[(n_0^2 - 4n_1^2)^{1/2} + n_0]/2n_1.$$

Note that $|z_1| > 1$, and thus $N(z) = B(z)B(1/z)$, where

$$B(z) = \sqrt{-n_1/z_1}(z^{-1} - z_1) = b_0 + b_1 z^{-1}, \quad \text{(V.D.96)}$$

with $b_0 = -z_1\sqrt{-n_1/z_1}$ and $b_1 = \sqrt{-n_1/z_1}$. The whitening filter in this case thus becomes

$$\frac{1}{\phi_Y^+(\omega)} = \frac{A(e^{i\omega})}{B(e^{i\omega})} = \frac{1 - re^{-i\omega}}{b_0 + b_1 e^{-i\omega}}. \quad \text{(V.D.97)}$$

As in Example V.D.1, suppose that we are interested in estimating the signal sequence $\{S_n\}_{n=-\infty}^{\infty}$ at time $t+\lambda$. Then $X_t = S_{t+\lambda}$ and the required cross spectrum is given [see (V.D.24)] by

$$\phi_{XY}(\omega) = e^{i\omega\lambda}\phi_S(\omega) = \frac{P(1-r^2)e^{i\omega\lambda}}{A(e^{i\omega})A(e^{-i\omega})}. \quad \text{(V.D.98)}$$

Applying (V.D.97) and (V.D.98) to (V.D.61), the transfer function of the optimum filter is given by

$$H(\omega) = \frac{A(e^{i\omega})}{B(e^{i\omega})}\left[\frac{P(1-r^2)e^{i\omega\lambda}}{A(e^{i\omega})B(e^{-i\omega})}\right]_+$$

$$\text{(V.D.99)}$$

$$= \left\{\frac{1 - re^{-i\omega}}{b_0 + b_1 e^{-i\omega}}\right\}\left[\frac{P(1-r^2)e^{-i\omega\lambda}}{(1 - re^{-i\omega})(b_0 + b_1 e^{i\omega})}\right]_+.$$

To simplify (V.D.99), consider the function of complex variable z given by

$$\hat{H}(z) = \left(\frac{P(1-r^2)}{1-rz^{-1}}\right)\left(\frac{1}{b_0 + b_1 z}\right). \qquad \text{(V.D.100)}$$

Using a partial fraction expansion, we can write

$$\hat{H}(z) = \frac{k'}{1-rz^{-1}} + \frac{k'}{1-z/z_1}, \qquad \text{(V.D.101)}$$

where $z_1 = -b_0/b_1$ and $k' = P(1-r^2)/(b_0 + b_1 r)$. Using the geometric series, $\hat{H}$ becomes

$$\begin{aligned} \hat{H}(z) &= k' \sum_{n=0}^{\infty} r^n z^{-1} + k' \sum_{n=0}^{\infty} z_1^{-1} z^n \\ &= k' \sum_{n=0}^{\infty} r^n z^{-1} + k' \sum_{n=-\infty}^{0} z_1^n z^{-1}. \end{aligned} \qquad \text{(V.D.102)}$$

The impulse response of $\hat{H}$ is then

$$\hat{h}_n = \begin{cases} k' z_1^n \text{ if } n < 0 \\ 2k' \text{ if } n = 0 \\ k' r^n \text{ if } n > 0. \end{cases} \qquad \text{(V.D.103)}$$

The impulse response of $e^{i\omega\lambda}\hat{H}(\omega)$ thus becomes

$$\hat{h}_{n+\lambda} = \begin{cases} k' z_1^{\lambda} z_1^{n} & \text{if } n < -\lambda \\ 2k' & \text{if } n = -\lambda \\ k' r^{\lambda} r^{n} & \text{if } n > -\lambda. \end{cases} \qquad \text{(V.D.104)}$$

The filter $\hat{h}_{n+\lambda}$ is illustrated in Fig. V.D.3. In order to get $\tilde{H}(\omega) \triangleq [e^{i\omega\lambda}\hat{H}(\omega)]_+$, we must truncate $\hat{h}_{n+\lambda}$ to be causal. From (V.D.104) we have that for $\lambda > 0$, the truncated impulse response is

$$\tilde{h}_n = \begin{cases} 0 & \text{if} \quad n < 0 \\ k' r^{\lambda} r^{n} & \text{if} \quad n \geqslant 0, \end{cases} \qquad \text{(V.D.105)}$$

and for $\lambda \leqslant 0$ it becomes

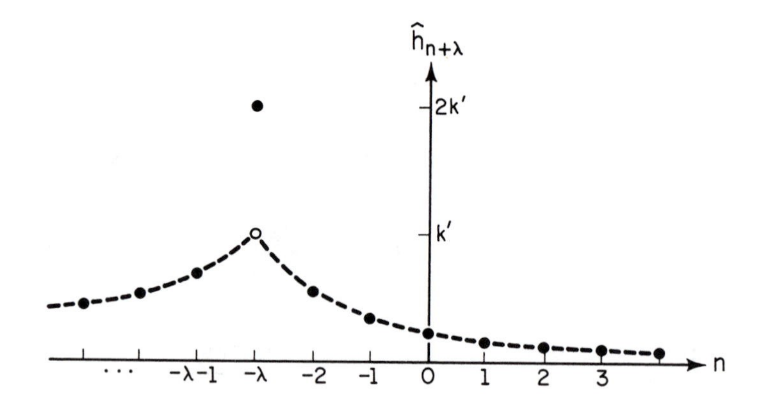

Fig. V.D.3: Impulse response of the filter $e^{i\omega\lambda}\hat{H}(\omega)$.

$$\tilde{h}_n = \begin{cases} 0 \text{ if } n < 0 \\ k' z_1^{\lambda} z_1^{n} \text{ if } 0 \leq n < -\lambda \\ 2k' \text{ if } n = -\lambda \\ k' r^{\lambda} r^{n} \text{ if } n > -\lambda. \end{cases} \tag{V.D.106}$$

The two cases $\lambda < 0$ and $\lambda > 0$ are illustrated in Figs. V.D.4 and V.D.5, respectively.

To carry the analysis further, let us consider the case of prediction ($\lambda > 0$). From (V.D.105) we have that

$$\begin{aligned} \tilde{H}(e^{i\omega}) &= \sum_{n=0}^{\infty} \tilde{h}_n e^{-in\omega} \\ &= \sum_{n=0}^{\infty} k' r^{\lambda} r^{n} e^{-in\omega} = \frac{k' r^{\lambda}}{1 - re^{-in\omega}}. \end{aligned} \tag{V.D.107}$$

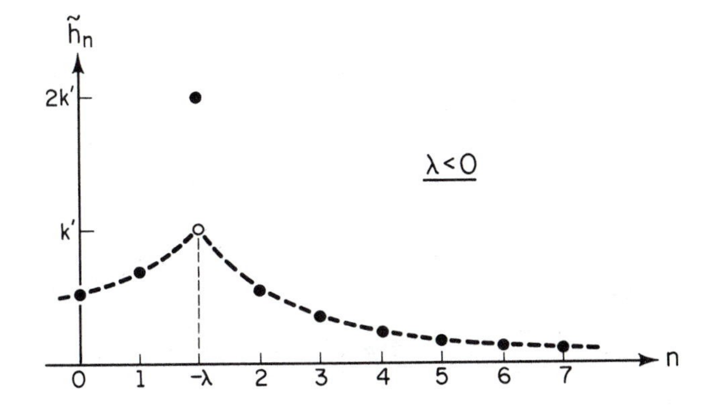

Fig. V.D.4: Impulse response of the filter $[e^{i\omega\lambda}\hat{H}(\omega)]_+$ *for* $\lambda < 0$ *(smoothing).*

Combining (V.D.99) and (V.D.107), we see that the optimum prediction filter has transfer function

$$H(\omega) = \frac{k'r^{\lambda}}{b_0 + b_1 e^{-i\omega}} = \frac{k'r^{\lambda}/b_0}{1 - e^{-i\omega}/z_1}. \qquad \text{(V.D.108)}$$

The impulse response of the optimum predictor is thus

$$h_n = \begin{cases} 0 \text{ if } n < 0, \\ \dfrac{k'r^{\lambda}}{b_0} z_1^{-n} \text{ if } n \geqslant 0. \end{cases} \qquad \text{(V.D.109)}$$

Alternatively, this optimum predictor can be implemented recursively by

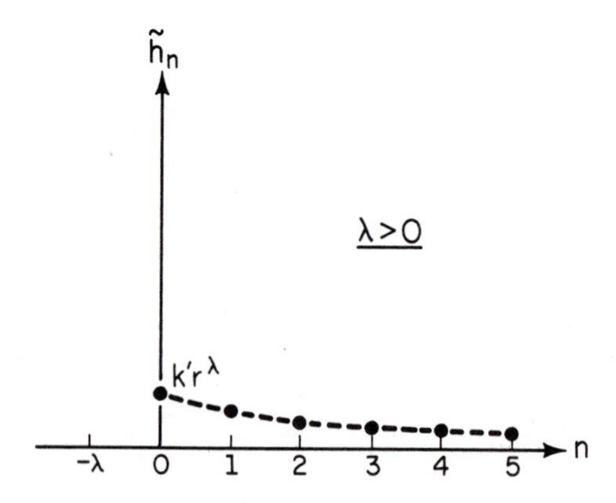

Fig. V.D.5: Impulse response of the filter $[e^{i\omega\lambda}\hat{H}(\omega)]_+$ *for* $\lambda > 0$ *(prediction).*

$$\hat{S}_{t+\lambda|t} = \frac{1}{z_1}\hat{S}_{t-1+\lambda|t-1} + \frac{k'r^{\lambda}}{b_0}Y_t, \; t \in \mathbb{Z} \quad \text{(V.D.110)}$$

where $\hat{S}_{t+\lambda|t}$ and $\hat{S}_{t-1+\lambda|t-1}$ denote the optimum predictor of $S_{t+\lambda}$ from $\{Y_n\}_{n=-\infty}^{t}$ and $S_{t-1+\lambda}$ from $\{Y_n\}_{n=-\infty}^{t-1}$, respectively. Note that when $v_N^2=0$ (i.e., when there is no noise), $\hat{S}_{t+\lambda|t}$ reduces straightforwardly to the pure predictor derived in Example V.D.2.

It is of interest to consider the case of one-step prediction ($\lambda=1$) further. Straightforward algebra yields that $z_1^{-1}=(r-k'r/b_0)$, so that (V.D.110) can be rewritten as

$$\hat{S}_{t+1|t} = r\hat{S}_{t|t-1} + \frac{k'r}{b_0}(Y_t - \hat{S}_{t|t-1}). \quad \text{(V.D.111)}$$

This form is reminiscent of the Kalman-Bucy prediction filter of Section V.B, which updates the one-step predictor in a state-space model in this same fashion. In fact, since $\{S_n\}_{n=-\infty}^{\infty}$ in this case is an AR (1) sequence, it can be generated via [see (V.D.89)]

$$S_{n+1} = rS_n + [P(1-r^2)]^{1/2}W_n, \; n \in \mathbb{Z}. \quad \text{(V.D.112)}$$

where $\{W_n\}_{n=-\infty}^{\infty}$ is white with unit variance ($\{W_n\}_{n=-\infty}^{\infty}$ is the prewhitened signal). The observation model is

$$Y_n = S_n + N_n, \; n \in \mathbb{Z}. \quad \text{(V.D.113)}$$

where $\{N_n\}_{n=-\infty}^{\infty}$ is a white sequence with variance v_N^2. Since $\{N_n\}_{n=-\infty}^{\infty}$ and $\{S_n\}_{n=-\infty}^{\infty}$ are orthogonal, so are $\{N_n\}_{n=-\infty}^{\infty}$ and $\{W_n\}_{n=-\infty}^{\infty}$. Thus (V.D.112) and (V.D.113) is a scalar time-invariant Kalman-Bucy model with white orthogonal noises. Thus from Example V.C.2 we know that the Kalman-Bucy filter provides the linear MMSE estimates of S_t and S_{t+1} given $\{Y_n; n \leqslant t\}$. The basic difference between this case and that treated in Section V.B is that (V.D.112) and (V.D.113) is a stationary or steady-state model. Its Kalman-Bucy prediction filter is thus the steady-state version derived in Example V.B.2, which is identical to (V.D.111) with the appropriate identification of equivalent parameters. In particular, the parameter set (a, b, c, q, r) in the Kalman-Bucy model of Example V.B.2 corresponds to the parameter set $(r, \sqrt{P(1-r^2)}, 1, 1, \nu_N^2)$ here.

Thus in the scalar time-invariant case, we can think of the Wiener-Kolmogorov filter as a steady-state version of the Kalman-Bucy filter, or, conversely, we can think of the Kalman-Bucy filter as a version of the Wiener-Kolmogorov filter that includes transient behavior. A similar identification can be made between other stable time-invariant Kalman-Bucy models and Wiener-Kolmogorov filters for signals with rational spectra observed in white noise [see, e.g., Anderson and Moore (1979) for further discussion of this issue.] Note, however, that Wiener-Kolmogorov filtering applies to more general spectral models for signals and noise, and that Kalman-Bucy filtering also applies to time-varying and unstable state-space models.

V.E Exercises

1. Show directly (i.e., without using the facts that $\hat{\underline{X}}_{t|t-1} = E\{\underline{X}_t | \underline{Y}_0^{t-1}\}$ and $\hat{\underline{X}}_{t|t} = E\{\underline{X}_t | \underline{Y}_0^t\}$) that the filtering and prediction errors generated by the Kalman filter are orthogonal to the data. I.e., show that

$$E\{(\underline{X}_t - \hat{\underline{X}}_{t|t})\underline{Y}_k^T\} = \mathbf{O}, \quad 0 \leqslant k \leqslant t$$

and

$$E\{(\underline{X}_t - \hat{\underline{X}}_{t|t-1})\underline{Y}_k^T\} = \mathbf{O}, \quad 0 \leqslant k \leqslant t-1.$$

where $\mathbf{O}$ denotes a matrix of all zeros.

2. Suppose the state equation in the Kalman-Bucy model is modified as follows:

$$\underline{X}_{k+1} = F_k \underline{X}_k + G_k \underline{U}_k + \Gamma_k \underline{s}_k$$

where $\{\underline{s}_k\}_{k=0}^{\infty}$ is a known sequence of vectors and $\{\Gamma_k\}_{k=0}^{\infty}$ is a known sequence of matrices of appropriate dimension. (Note, e.g., that $\{\underline{s}_k\}_{k=0}^{\infty}$ could be a sequence of controls.) Find the appropriate modification of the

Kalman-Bucy recursions.

3. Repeat Exercise 2 for the situation in which each $\underline{s}_k$ is allowed to be a function of the past measurement; i.e., $\underline{s}_k$ can be a function of $\underline{Y}_0^k$. (So, for example, $\{\underline{s}_k\}_{k=0}^{\infty}$ could be a sequence of feedback controls.)

4. Suppose we return to the original Kalman-Bucy model, but allow for correlation between the state and measurement noises. I.e., assume everything as before except

$$Cov(\underline{U}_k, \underline{V}_l) = \begin{cases} \mathbf{C}_k, \; k = l \\ \\ \mathbf{O} \;, \; k \neq l \end{cases}$$

where $\mathbf{C}_k$ is a matrix of appropriate dimension. Show that the Kalman predictor is given by

$$\underline{\hat{X}}_{t+1|t} = \mathbf{F}_t \underline{\hat{X}}_{t|t-1} + \mathbf{K}_t (\underline{Y}_t - \mathbf{H}_t \underline{\hat{X}}_{t|t-1})$$

with

$$\underline{\hat{X}}_{0|-1} = \underline{m}_0,$$

where

$$\mathbf{K}_t = (\mathbf{F}_t \mathbf{\Sigma}_{t|t-1} \mathbf{H}_t^T + \mathbf{G}_t \mathbf{C}^t)(\mathbf{H}_t \mathbf{\Sigma}_{t|t-1} \mathbf{H}_t^T + \mathbf{R}_t)^{-1}$$

and

$$\mathbf{\Sigma}_{t+1|t} = \mathbf{F}_t \mathbf{\Sigma}_{t|t-1} \mathbf{F}_t^T$$
$$- \mathbf{K}_t (\mathbf{F}_t \mathbf{\Sigma}_{t|t-1} \mathbf{H}_t^T + \mathbf{G}_t \mathbf{C}_t) + \mathbf{G}_t \mathbf{Q}_t \mathbf{G}_t^T$$

with

$$\Sigma_{0|-1} = Cov(\underline{X}_0).$$

5. Consider the model (X_k's are scalars)

$$X_{k+1} = \tfrac{1}{2}X_k + U_k, \quad k = 0, 1, \ldots$$

$$Y_k = \Theta X_k + V_k, \quad k = 0, 1, \ldots$$

where $U_0, U_1, \ldots$ are i.i.d. $N(0, q)$ random variables, $V_0, V_1, \ldots$ are i.i.d. $N(0, r)$ random variables, X_0 is a $N(0, \Sigma_0)$ random variable, and all U_k's, V_k's, and X_0 are independent of one another.
(a) Suppose $\Theta \equiv 1$ and $r = q = 1$. Find the initial state variance Σ_0 such that the optimal prediction filter (i.e., $\hat{X}_{t+1|t}$) is time-invariant. Write the recursion for $\hat{X}_{t+1|t}$ in this case. What is the mean squared prediction error for this case?
(b) Suppose $r = q = 1$ and Σ_0 is the answer from part (a). What is the structure of the optimum decision rules for deciding $\Theta = 0$ versus $\Theta = 1$ based on observations for $k = 0, 1, \ldots, n$?

6. Consider the standard Kalman-Bucy model with states $\underline{X}_k$ and observations $\underline{Y}_k$. Suppose $0 \leqslant j \leqslant t$ and we wish to estimate $\underline{X}_j$ from $\underline{Y}_0^t$. Consider the estimator defined recursively (in t) by

$$\hat{\underline{X}}_{j|t} = \hat{\underline{X}}_{j|t-1} + \mathbf{K}_t^a(\underline{Y}_t - \mathbf{H}_t \hat{\underline{X}}_{t|t-1})$$

where

$$\mathbf{K}_t^a = \mathbf{\Sigma}_{t|t-1}^a \mathbf{H}_t^T [\mathbf{H}_t \mathbf{\Sigma}_{t|t-1}^a \mathbf{H}_t^T + \mathbf{R}_t]^{-1}$$

and

$$\mathbf{\Sigma}_{t+1|t}^a = \mathbf{\Sigma}_{t|t-1}^a [\mathbf{F}_t - \mathbf{K}_t \mathbf{H}_t]^T$$

with

$$\mathbf{\Sigma}_{j|j-1}^a = \mathbf{\Sigma}_{j|j-1}$$

where $\mathbf{H}_t, \hat{\underline{X}}_{t|t-1}, \mathbf{\Sigma}_{t|t-1}, \mathbf{R}_t, \mathbf{F}_t$, and $\mathbf{K}_t$ are as in the one-step prediction problem.

(a) Show that $\mathbf{\Sigma}_{t|t-1}^{q} = E\{(\underline{X}_j - \hat{\underline{X}}_{j|t-1})\underline{X}_t^T\}$.

(b) Show that $\hat{\underline{X}}_{j|t} = E\{\underline{X}_j | \underline{Y}_0^t\}$.

7. Consider the observation model:

$$Y_k = N_k + \Theta s_k, \quad k = 1, 2, \ldots$$

where $N_1, N_2, \ldots$ are i.i.d. $N(0, \sigma^2)$ random variables, $\Theta \sim N(\mu, v^2)$ is independent of $N_1, N_2, \ldots$, and $s_1, s_2, \ldots$ is a known sequence. Let $\hat{\theta}_n$ denote the MMSE estimate of Θ given $Y_1, \ldots, Y_n$. Find recursions for $\hat{\theta}_n$ and for the minimum mean squared error, $E\{(\hat{\theta}_n - \Theta)^2\}$, by recasting this problem as a Kalman filtering problem.

8. Verify that the Levinson algorithm given in (V.C.26) - (V.C.29) solves the Yule-Walker equations (V.C.25).

9. Consider the model of Example V.D.1 with λ=0. Consider an estimate of S_t given by

$$\hat{S}_t = \sum_{n=-\infty}^{\infty} h_{t-n} \lambda_n .$$

Show that

$$E\{(\hat{S}_{t+\lambda} - S_{t+\lambda})^2\} = \frac{1}{2\pi}\int_{-\pi}^{\pi} |1 - H(\omega)|^2 \phi_S(\omega) d\omega$$

$$+ \frac{1}{2\pi}\int_{-\pi}^{\pi} |H(\omega)|^2 \phi_N(\omega) d\omega.$$

where H is the transfer function corresponding to $\{h_n\}_{n=-\infty}^{\infty}$. Interpret the two terms in this error formula.

10. Show directly that the filter with transfer function

$$H(\omega) = \frac{\phi_S(\omega)}{\phi_S(\omega) + \phi_N(\omega)}$$

minimizes the error expression given in Exercise 9.

VI SIGNAL DETECTION IN CONTINUOUS TIME

VI.A Introduction

In the preceding chapters we have presented the basic principles of signal detection and estimation, assuming throughout that our observation set is either a set of vectors or is a discrete set. Throughout this analysis a key role was played by a family of densities $\{p_\theta; \theta \epsilon \Lambda\}$ on the observation space, either through the likelihood ratio in hypothesis testing, through the computation of an *a posteriori* parameter distribution in Bayesian estimation, or through the study of MVUEs and MLEs in nonBayesian parameter estimation. This necessity of specifying a family of densities on the observation space is the primary reason for restricting our observation sets in the way that we have done. In particular, as we have seen, all the problems considered thus far have been treated using the ordinary probability calculus of probability density functions and probability mass functions.†

Although the observation sets treated thus far are of considerable interest in practice, there are many applications in which our observations are best modeled as a continuous-time random process. That is, our overall observation Y is a collection of random variables $\{Y_t; t \epsilon [0, T]\}$ indexed by a continuous parameter t, where for convenience we have chosen our observation interval to be $[0, T]$ for some $T > 0$. In this chapter and the following one, we consider signal detection and estimation problems with this type of observations. Signal detection is treated in this chapter, with signal estimation

† An exception is the linear estimation problem treated in Section V.D. Since we needed only a second-order statistical description for this problem, we were in fact able to extend our observation set to include sets of infinite sequences.

being treated in Chapter VII.

In continuous-time problems, the observation set Γ becomes a set each of whose elements is a continuous-time waveform. Such a set is called a *function space*. In order to model signal detection and estimation problems in this setting, we need to construct families of densities on such sets. Since a density is a function that can be integrated (or summed) to give probabilities, the notion of a density in continuous time requires a method of integration on function spaces. This type of integration requires analytical techniques beyond those of ordinary calculus, and thus before treating signal detection and estimation problems in continuous time we must first develop some analytical tools for dealing with such problems.

In Section VI.B we discuss very briefly the theory of integration in abstract spaces. The purpose of this treatment is not to provide the reader with the details of this theory but rather to indicate how the notion of a density can be extended to function spaces, and in turn how the necessary modeling for continuous time problems can be accomplished. We also consider in Section VI.B a representation for continuous time random processes (the Karhunen-Loéve expansion), which allows the reduction of such processes to equivalent discrete-time processes. This representation is the key to the solution of the signal detection problems presented in the remainder of the chapter. In Sections VI.C and VI.D we turn to specific problems of signal detection with continuous time observations. The theory of such problems is no different from that of the previous chapters once the appropriate families of densities have been specified. Thus these sections are concerned primarily with specific methods for finding appropriate density classes for models of interest in applications, although the systems and performance aspects of the resulting procedures are also discussed. Section VI.C is concerned with the problem of detecting deterministic (i.e., coherent) signs in Gaussian noise, and Section VI.D with the detection of stochastic signals in Gaussian noise.

VI.B Mathematical Preliminaries

VI.B.1 Densities in Function Spaces

As noted in the introduction, in order to consider signal detection and estimation problems with continuous-time observations, we can no longer restrict attention to observation sets Γ that are either Euclidean or discrete. Thus for now we will allow Γ to be an arbitrary set and the corresponding event class G to be an arbitrary σ-algebra of subsets of Γ. In this section we discuss the notion of a density on such observation sets and relate this notion to those defined for the specific observation spaces considered in the preceding chapters. Our purpose here is to give the reader a basic idea of what is meant by such densities, not to provide a complete development of this theory, which is beyond the scope of this book.

To specify densities on function spaces we need to consider integration in such spaces. To define integrals on an arbitrary space we begin with the notion of a *measure*, defined as follows.

Definition VI.B.1: Measure

A function μ: $G \to [0, \infty]$ is a *measure* on (Γ, G) if it satisfies the following two properties.

(i) $\mu(\phi)=0$, where ϕ denotes the null set; and

(ii) if $G_1, G_2, \ldots$ is a sequence of disjoint sets (i.e., $G_i G_j = \phi, i \neq j$) in G then $\mu(\bigcup_{i=1}^{\infty} G_i) = \sum_{i=1}^{\infty} \mu(G_i)$.

Perhaps the most familiar example of a measure is a *probability measure* (or *probability distribution*) which satisfies the additional condition $\mu(\Gamma)=1$. [This condition and properties (i) and (ii) of the definition are often called the *axioms of probability*.] Integration with respect to probability measures is a familiar operation. In particular, suppose that μ is a probability measure on (Γ, G) and X is a measurable[†] function from (Γ, G) to $(\mathbb{R}, B)$. Then X induces a probability distribution,

† A function X from (Γ, G) to $(\mathbb{R}, B)$ is *measurable* if for every set $A \in B$, $\{y \in \Gamma | X(y) \in A\}$ is in G. This condition implies that events on the real line described in terms of X [e.g., "$X(y)$ is in A"] are mapped back into events in the original observation space. This condition allows us to define a probability distribution induced on $(\mathbb{R}, B)$ by X as described above.

P_X, on $(\mathbb{R}, B)$ via the definition

$$P_X(A) = \mu(X^{-1}(A)), \quad A \in B, \tag{VI.B.1}$$

where $X^{-1}(A) \triangleq \{y \in \Gamma | X(y) \in A\}$.

X is thus a random variable and its expectation $E\{X\}$ can be defined in the usual way. This quantity is usually thought of as the averaging of a real variable x weighted by the distribution P_X; however, it can also be thought of as an averaging of $X(y)$ weighted by the probability measure μ on (Γ, G). To denote the latter interpretation, we write

$$E\{X\} = \int_\Gamma X(y)\mu(dy) \equiv \int X d\mu, \tag{VI.B.2}$$

and in this context $E\{X\}$ is termed the *integral of* X *with respect to* μ. If X_1 and X_2 are two functions on (Γ, G) with $E\{X_1\}$ and $E\{X_2\}$ well defined, and α and β are scalars, then we know from the properties of expectation that $E\{\alpha X_1 + \beta X_2\} = \alpha E\{X_1\} + \beta E\{X_2\}$. In the notation above, this implies that

$$\int (\alpha X_1 + \beta X_2) d\mu = \alpha \int X_1 d\mu + \beta \int X_2 d\mu. \tag{VI.B.3}$$

Thus the quantity defined in (VI.B.2) is linear as a function of X, which is the principal requirement placed on an integral.

For any event $F \in G$, we can define a particular measurable function from Γ to $\mathbb{R}$, by

$$I_F(y) = \begin{cases} 1 \text{ if } y \in F \\ 0 \text{ if } y \in F^c. \end{cases} \tag{VI.B.4}$$

As mentioned in Chapter III, this function is called the *indicator function* of F. Since I_F defines a discrete random variable taking on the value 1 with probability $\mu(F)$ and 0 with probability $\mu(F^c)$, we see that

$$E\{I_F\} = 1\mu(F)+0\mu(F^c) = \mu(F), \qquad \text{(VI.B.5)}$$

or, equivalently,

$$\int I_F \, d\mu = \mu(F). \qquad \text{(VI.B.6)}$$

Thus the measure μ of any set F can be obtained from the integral $\int X d\mu$ by proper choice of X, and since $\int X d\mu$ is defined from μ, the integral $\int X d\mu$ and the measure μ are really equivalent notions. The quantity $\int I_F \, d\mu$ is sometimes written as $\int_F d\mu$ or $\int_F \mu(dy)$, and for a function X and a set $F \in G$ we write

$$\int_\Gamma I_F(y) X(y) \mu(dy) = \int_F X(y) \mu(dy). \qquad \text{(VI.B.7)}$$

From the discussion above we see that integration of a function X with respect to a probability measure μ is nothing more than the familiar expectation of X.† This notion of integration with respect to a measure can be extended easily to types of measures other than probability measures. In particular, we say that a measure μ is *finite* if $\mu(\Gamma)<\infty$, and we say that μ is σ*-finite* if Γ can be written as the union of disjoint events $\Gamma_1, \Gamma_2, \ldots$, each of which satisfies $\mu(\Gamma_i)<\infty$.

The properties of finite measures are essentially identical to those of probability measures, aside from the normalization $\mu(\Gamma)=1$. In particular, for any finite measure μ we can define a probability measure μ' by

$$\mu'(F) = \mu(F)/\mu(\Gamma), \quad F \in G. \qquad \text{(VI.B.8)}$$

Integration of a function X with respect to μ is then defined straightforwardly by

† It should be noted that $\int X d\mu$ can be defined first, and then $E\{X\}$ can be defined as $\int X d\mu$ [see, e.g., Billingsley (1979)]. This approach is more customary than the above; however, our purpose here is to define the unfamiliar $\int X d\mu$ in terms of the familiar $E\{X\}$.

$$\int X d\mu = \mu(\Gamma) \int X d\mu'. \qquad \text{(VI.B.9)}$$

Similarly, if μ is a σ-finite measure, we can decompose it into a sum of finite measures via

$$\mu(F) = \sum_{i=1}^{\infty} \mu_i(F), \quad F \in G, \qquad \text{(VI.B.10)}$$

where, for each i,

$$\mu_i(F) = \mu(F \cap \Gamma_i), \quad F \in G. \qquad \text{(VI.B.11)}$$

It follows easily from property (ii) in the definition of a measure that $\mu(F \cap \Gamma_i) \leqslant \mu(\Gamma_i) < \infty$. Thus for a σ-finite measure we can define

$$\int X d\mu = \sum_{i=1}^{\infty} \int X d\mu_i, \qquad \text{(VI.B.12)}$$

where $\int X d\mu_i$ is defined as in (VI.B.9).†

Having defined integration as described above (such integrals are called *Lebesgue-Stieltjes integrals*), we can now provide the notion of a probability density needed for our purposes. First we give the following definition.

Definition VI.B.2: Absolute Continuity of Measures

Suppose that μ_0 and μ_1 are two measures on (Γ, G). We say that μ_1 *is absolutely continuous with respect to* μ_0 (or that μ_0 *dominates* μ_1) if the condition $\mu_0(F)=0$ implies that $\mu_1(F)=0$. We use the notation $\mu_1 << \mu_0$ to denote this condition.

† In defining (VI.B.12) we must worry about the convergence of the sum. This sum is defined as $+\infty$ if $\sum_{i=1}^{\infty} \int X^+ d\mu_i = \infty$ and $\sum_{i=1}^{\infty} \int X^- d\mu_i < \infty$ where $X^+(y) = \max\{X(y), 0\}$ and $X^-(y) = \max\{-X(y), 0\}$. Similarly, $\int X d\mu = -\infty$ if $\sum_{i=1}^{\infty} \int X^+ d\mu_i < \infty$ and $\sum_{i=1}^{\infty} \int X^- d\mu_i = \infty$. $\int X d\mu$ is left undefined if both $\sum_{i=1}^{\infty} \int X^+ d\mu_i$ and $\sum_{i=1}^{\infty} \int X^- d\mu_i$ are infinite.

With this definition we can state the following key result.

Proposition VI.B.1: The Radon-Nikodym Theorem

Suppose that μ_0 and μ_1 are σ-finite measures on (Γ, G) and $\mu_1 << \mu_0$. Then there exists a measurable function f : $\Gamma \to \mathbb{R}$ such that

$$\mu_1(F) = \int_F f \, d\mu_0 \text{ for all } F \in G . \qquad \text{(VI.B.13)}$$

Moreover, f is uniquely defined except possibly on a set G_0 with $\mu_0(G_0)=0$.

The proof of this result can be found in most books on integration and measure [e.g., Billingsley (1979)], and is omitted here. Note that the theorem states that the μ_1-measure of any event F can be obtained from μ_0 by integration of the function f over F. This function f is called the *Radon-Nikodym derivative* of μ_1 with respect to μ_0 and is written as $f = d\mu_1/d\mu_0$. In the observation sets considered in earlier chapters, Radon-Nikodym derivatives take on familiar forms, as is shown in the following two examples.

Example VI.B.1: Radon-Nikodym Derivatives on $(\mathbb{R}^n, B^n)$

Suppose that $(\Gamma, G)=(\mathbb{R}, B)$. It can be shown that a unique measure μ can be defined on $(\mathbb{R}, B)$ such that the μ-measure of any interval is its length [i.e., $\mu([a,b])=|b-a|$]. This measure is known as *Lebesgue measure*, and it is σ-finite since we can decompose $\mathbb{R}$ into the union of countably many intervals, each having finite length. If g is a function with an ordinary Riemann integral $\int_{-\infty}^{\infty} g(y)dy$, it turns out that $\int_{\mathbb{R}} g d\mu = \int_{-\infty}^{\infty} g(y)dy$. That is, integration with respect to Lebesgue measure is essentially the same as the integration of ordinary calculus (although a function can be Lebesgue integrable without being Riemann integrable).

Now suppose that P is a probability measure on $(\mathbb{R}, B)$ corresponding to a continuous random variable, and let p denote the corresponding probability density function (pdf).

Then we know that for any $A \in B$, $P(A) = \int_A p(y)dy$, and since $\int_A p(y)dy = \int_A p d\mu \equiv \int_A p(y)\mu(dy)$, we can see that $p = dP/d\mu$. Thus probability density functions of continuous random variables are Radon-Nikodym derivatives of the corresponding probability measures with respect to Lebesgue measure.

Lebesgue measure is the "natural" measure on $\mathbb{R}$ because it corresponds to length, which is the most natural number to associate with any interval. Lebesgue measures can be defined similarly in $\mathbb{R}^2$ and $\mathbb{R}^3$ by using area and volume, respectively, as natural measures of size. This idea can be further extended to $\mathbb{R}^n$ for $n > 3$ by using the usual notion of "volume" in $\mathbb{R}^3$. In each of these cases the pdf's of continuous random vectors are the Radon-Nikodym derivatives of the corresponding probability measures with respect to Lebesgue measure.

Example VI.B.2: Radon-Nikodym Derivatives on Discrete Sets

Suppose that Γ is a discrete set $\{\gamma_1, \gamma_2, ...\}$ and let $G = 2^\Gamma$. Define a measure μ on (Γ, G) by

$$\mu(F) = \textit{the number of elements in } F. \quad \text{(VI.B.14)}$$

Thus μ counts the number of elements in an event F and assigns that number to F. This measure is called *counting measure*. It is σ-finite since Γ is the union of the sets $\{\gamma_i\}$, each of which has a single element. It is easily seen that for any function $g : \Gamma \to \mathbb{R}$,

$$\int_\Gamma g(y)\mu(dy) = \sum_{i=1}^{\infty} g(\gamma_i). \quad \text{(VI.B.15)}$$

That is, integration with respect to counting measure on Γ is simply summation.

Suppose that P is a probability measure on Γ with probability mass function p. Then, for any $F \in G$,

$$P(F) = \sum_{\gamma_i \in F} p(\gamma_i) = \int_F p(y)\mu(dy). \qquad \text{(VI.B.16)}$$

Thus $p = dP/d\mu$, so we see that probability mass functions are Radon-Nikodym derivatives with respect to counting measure.

Examples VI.B.1 and VI.B.2 illustrate that the densities we have used in the preceding chapters are special cases of the Radon-Nikodym derivative. We also see that the notation introduced to treat both continuous and discrete cases simultaneously [i.e., $E\{g(Y)\} = \int_\Gamma g(y)p(y)\mu(dy)$] has a more general interpretation as integration with respect to a measure μ. The extension of the results of preceding chapters to more general observation spaces can thus be accomplished in the following way.

For our purposes the observation model of interest always consists of a family $\{P_\theta; \theta \in \Lambda\}$ of distributions (i.e., probability measures) on (Γ, G). If there is a σ-finite measure μ on (Γ, G) such that $P_\theta << \mu$ for all $\theta \in \Lambda$, then we can generate a family of densities $\{p_\theta; \theta \in \Lambda\}$ where $p_\theta = dP_\theta / d\mu$, $\theta \in \Lambda$. With this more general interpretation of p_θ, all of the general results we developed for hypothesis testing and parameter estimation hold for more general observation spaces, including the general results in Chapters II and IV for hypotheses testing and parameter estimation, respectively, and the development of the Chernoff bounds in Section III.C. Moreover, many other results developed specifically for sequences of i.i.d. real observations can be extended directly to sequences of i.i.d. observations from general observation spaces, with the role of the marginal density being filled by a Radon-Nikodym derivative.† These include many of the discrete-time detection results of Chapter III, as well as asymptotic analysis of maximum-likelihood estimates discussed in Section IV.D.

Thus the notions of Lebesgue-Stieltjes integration and Radon-Nikodym differentiation described above provide the analytical basis for dealing with signal detection and estimation in continuous time. As background for considering the

† Problems of this type arise in applications such as array processing, in which we have a set of n observations, each of which is the waveform appearing at the output of a sensor in an array.

signal detection problem, we consider some additional properties for the particular situation of binary parameter sets $\Lambda=\{0, 1\}$.

Consider two probability measures P_0 and P_1 on (Γ, G). There always exists a σ-finite measure μ that dominates both P_0 and P_1; for example, the finite measure $\mu=P_0+P_1$ is easily seen to dominate both P_0 and P_1. Thus, without loss of generality, we assume the existence of a measure μ for which we can define densities $p_j \triangleq dP_j/d\mu$, $j=0, 1$. The Bayes, minimax, and Neyman-Pearson optimum tests of P_0 versus P_1 are all based on comparing the likelihood ratios $L=p_1/p_0$ to a threshold τ, announcing H_1 if $L(y)$ exceeds τ, announcing H_0 if $L(y)$ falls below τ, and possibly randomizing if $L(y)=\tau$. If it is also the case that $P_1 \ll P_0$, then P_1 is also differentiable with respect to P_0. It can be shown that for any μ dominating both P_0 and P_1, we have

$$\frac{dP_1}{dP_0} = \frac{dP_1/d\mu}{dP_0/d\mu} = \frac{p_1}{p_0}. \tag{VI.B.17}$$

Thus when $P_1 \ll P_0$, the likelihood ratio is simply the Radon-Nikodym derivative of P_1 with respect to P_0.

If $P_1 \ll P_0$, then the Radon-Nikodym theorem implies that we can write

$$P_1(F) = \int_F L dP_0, \text{ for all } F \in G. \tag{VI.B.18}$$

More generally, it is straightforward to show that for any pair of probability measures P_0 and P_1, there is always a measurable function $f : \Gamma \to \mathbb{R}$ and a set $H \in G$ with $P_0(H)=0$ such that

$$P_1(F) = \int_F f\, dP_0 + P_1(F \cap H) \text{ for all } F \in G. \tag{VI.B.19}$$

In fact, H can be chosen as the set

$$H = \{y \in \Gamma | p_1(y) > 0 \text{ and } p_0(y) = 0\}, \quad \text{(VI.B.20)}$$

and $f(y)$ can be chosen as $L(y)$ for $y \in H^c$ and arbitrarily for $y \in H$. [Since $P_0(H)=0$, the value of f on H is irrelevant in the integral of (VI.B.19).] Thus the optimum tests always choose H_1 when $y \in H$ and they choose according to the comparison of $f(y)$ with τ when $y \in H^c$.

In the representation of (VI.B.16), if $P_1(H)=0$, the second term on the right-hand side is always zero, $P_1 << P_0$, and $f = dP_1/dP_0$. On the other hand, if $P_1(H)=1$, then $P_1(F \cap H) = P_1(F)$ for all $F \in G$ and the *first* term on the right-hand side is always zero. In this case the event H occurs with probability 1 under P_1 and the event H^c occurs with probability 1 under P_0. This implies that we can distinguish between H_0 and H_1 with zero error probability by choosing H_1 when $y \in H$ and H_0 when $y \in H^c$. When this condition occurs, P_0 and P_1 are said to be *singular* (denoted by $P_0 \perp P_1$), and when P_0 and P_1 represent a signal detection model, this is called *singular detection*.

Singularity between measures essentially means that the support sets of the densities p_0 and p_1 are disjoint. For $\Gamma = \mathbb{R}^n$ or Γ discrete, this rarely occurs for practical models, so it is of little interest in applications. However, for continuous-time observations, singularity is a more subtle issue and it can occur for models that may seem otherwise reasonable. As an example in which singularity is obvious, suppose that we wish to detect the presence or absence of the signal waveform shown in Fig. VI.B.1 in additive noise. Under some mild assumptions, if the noise process has finite bandwidth the possible noise waveforms will all be continuous functions. Obviously, the presence of the signal will cause a discontinuity in the observations at $t = T/2$, whereas the observed waveform is continuous at $t = T/2$ if the signal is absent. Thus by observing the continuity or lack of continuity of the observed waveform at $t = T/2$, we can tell perfectly well whether or not the signal is present. (Note that this particular problem could not arise in discrete time since the notion of continuity in time is not present in that case.) That this particular model is singular is fairly obvious; however, singularity can occur in much less obvious ways. Essentially, singularity occurs in problems of detecting signals in additive noise when the signal

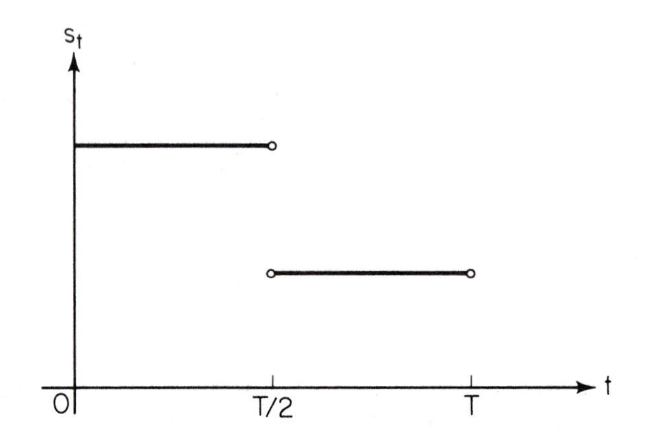

Fig. VI.B.1: A signal yielding singular detection in continuous additive noise.

can change faster than the noise can. Although singularity cannot occur in physical signal detection situations, detection models must be investigated for possible singularity because such singularity points to a lack of realism in the model. The problem of singular detection has been studied quite deeply, and we touch briefly on it in the following sections.

The conditions $P_1 << P_0$ and $P_1 \perp P_0$ are the two extremes of the representation (VI.B.19). Of course, we would have an intermediate case if $0 < P_1(H) < 1$. However, for most useful continuous-time signal detection models, we usually have one of the two extremes. [In fact, there are fairly general classes of continuous-time models in which we either have singularity or we have the condition that $P_1 << P_0$ and $P_0 << P_1$, a condition known as *equivalence* of P_0 and P_1 (denoted by $P_0 \equiv P_1$).] Thus the solution to continuous-time detection problems usually breaks down into these steps:

1. Determine whether or not the problem is singular.
2. If the problem is not singular, find the likelihood ratio.
3. Find the error probabilities.

Before considering these problems directly, we first present a representation for continuous-time processes which greatly facilitates their solution.

VI.B.2 Grenander's Theorem and the Karhunen-Loéve Expansion

Although the ideas described in Section VI.B.1 provide suitable notions of probability densities and likelihood ratios on function spaces, there is no indication from this construction as to how one might find representations for these quantities that can be interpreted for the purposes of implementing a detection or estimation system. In this and the following section we consider methods for finding such representations.

One way of computing likelihood ratios for continuous-time observation models is first to reduce the continuous-time observation process $\{Y_t; t \in [0,T]\}$ to an equivalent observation sequence, say $Z_1, Z_2, \ldots$. Supposing that this can be done, we might look for a likelihood ratio based on our observations by first truncating the infinite sequence $\{Z_k\}_{k=1}^{\infty}$ to the finite sequence $Z_1, \ldots, Z_n$. The likelihood ratio for $Z_1, Z_2, \ldots, Z_n$ can be determined by the methods described in Chapter III; then by taking limits as $n \to \infty$, we might hope for convergence to the likelihood ratio based on $\{Z_k\}_{k=1}^{\infty}$. In fact, such convergence is assured under fairly mild conditions, as is indicated by the following result.

Proposition VI.B.2: Grenander's Theorem

Suppose that $(\Gamma, G)=(\mathbb{R}^{\infty}, B^{\infty})^{\dagger}$ and let P_0 and P_1 be two probability measures on (Γ, G). Suppose further that for each positive integer n, Y_1^n is a continuous random vector under both P_0 and P_1, with density $p_j^n(Y_1^n)$ under P_j. For each n, define a function f_n on Γ by

$$f_n(y) = \frac{p_1^n(y_1^n)}{p_0^n(y_1^n)}, \quad y \in \Gamma. \qquad \text{(VI.B.21)}$$

Then

† As in Section III.D, $\mathbb{R}^{\infty}$ denotes the set of all one-sided sequences of real numbers; i.e., $\mathbb{R}^{\infty}=\{y \mid y=\{y_k\}_{k=1}^{\infty}$ and $y_k \in \mathbb{R}, k \geq 1\}$. B^{∞} denotes the σ-algebra of Borel sets in $\mathbb{R}^{\infty}$, which is the smallest σ-algebra containing all sets of the form $\{y=\{y_k\}_{k=1}^{\infty} \mid (y_1, \ldots, y_n) \in A\}, A \in B^n$, for $n=1,2,\ldots$.

$$f_n(Y) \to f(Y) \text{ in probability under } P_0,$$

$$f_n(Y) \to f(Y) \text{ in probability under } P_1 \text{ on } H^c,$$

and

$$f_n(Y) \to \infty \text{ in probability under } P_1 \text{ on } H,$$

where f and H are, respectively, the function and the event appearing in the representation (VI.B.19); i.e.,

$$P_1(F) = \int_F f \, dP_0 + P_1(F \cap H), \quad F \in G. \quad \text{(VI.B.22)}$$

The proof of this theorem can be found in Grenander (1981) and will be omitted here. This theorem says that we can compute likelihood ratios on spaces of infinite sequences by first truncating the sequence and then looking for the limit in probability of the likelihood ratios for the truncated sequence. The next step is to consider the problem of representing a continuous-time observation process in terms of an equivalent observation sequence. Before doing so, however, we first give a few necessary definitions.

A random process $\{Y_t; t \in [0,T]\}$ is said to be a *second-order* process if $E\{Y_t^2\} < \infty$ for all $t \in [0,T]$. For a second-order process, the *autocovariance function* is defined as

$$C_Y(t,u) = Cov(Y_t, Y_u), \quad t,u \in [0,T]. \quad \text{(VI.B.23)}$$

For a real-valued function h on $[0,T]$, the *mean-square integral* $\int_0^T h(t) Y_t \, dt$ is defined as the mean-square limit as $n \to \infty$ and $\Delta_n \triangleq \max_{1 \leq i \leq n} |t_i^{(n)} - t_{i-1}^{(n)}| \to 0$, of the sequence of sums†

---

† That is

$$E\{[\int_0^T h(t) Y_t \, dt - \sum_{i=1}^n h(\xi_i^{(n)}) Y_{\xi_i^{(n)}} (t_i^{(n)} - t_{i-1}^{(n)})]^2\} \to 0.$$

The limit of the sequence (VII.C.4) exists if and only if

$$\int_0^T \int_0^T h(t) h(u) R_Y(t,u) \, dt \, du < \infty,$$

where $R_Y(t,u) \triangleq E\{Y_t Y_u\} \equiv C_Y(t,u) + E\{Y_t\} E\{Y_u\}$. [See, e.g., Parzen (1962)].

$$\sum_{i=0}^{n} h(\xi_i^{(n)}) Y_{\xi_i^{(n)}} [t_i^{(n)} - t_{i-1}^{(n)}], \tag{VI.B.24}$$

where, for each positive integer n, $0 = t_0^{(n)} < t_1^{(n)} < \ldots < t_n^{(n)} = T$ forms a partition of $[0, T]$ and $\xi_i^n \epsilon [t_{i-1}^{(n)}, t_i^{(n)}]$ for $i = 1, \ldots, n$. Note that aside from the fact that the $Y_{\xi_i^{(n)}}$'s are random, (VI.B.24) is the ordinary Riemann sum that yields $\int_0^T h(t) Y_t dt$ in the limit. Because of the randomness of the summands, it is necessary to define the limit in a stochastic sense, and the mean-square sense is the most useful for our purposes. We will also have occasion to use the *mean-square Stieltjes integral* $\int_0^T h(t) dY_t$, defined as the mean-square limit of the sequence of sums

$$\sum_{i=1}^{n} h(\xi_i^{(n)}) (Y_{t_i^{(n)}} - Y_{t_{i-1}^{(n)}}), \tag{VI.B.25}$$

where the $t_i^{(n)}$'s and $\xi_i^{(n)}$'s are defined as in (VI.B.24).

With the definitions above we can now give the following result.

Proposition VI.B.3: Mercer's Theorem and the Karhunen- Loève Expansion

Suppose that $\{Y_t; t \epsilon [0, T]\}$ is a zero-mean second-order random process with autocovariance function $C_Y(t, u)$ that is continuous on the square $[0, T] \times [0, T] \equiv [0, T]^2$. Then C_Y can be expanded in the uniformly and absolutely convergent series

$$C_Y(t, u) = \sum_{k=1}^{\infty} \lambda_k \psi_k(t) \psi_k(u), \; (t, u) \epsilon [0, T]^2, \tag{VI.B.26}$$

where $\{\lambda_k\}_{k=1}^{\infty}$ and $\{\psi_k\}_{k=1}^{\infty}$ are the *eigenvalues* and corresponding *orthonormal eigenfunctions* of C_Y; i.e., $\{\lambda_n\}_{n=1}^{\infty}$ and $\{\psi_n\}_{n=1}^{\infty}$ are solutions to the integral equation

$$\lambda\psi(t) = \int_0^T C_Y(t, u)\psi(u)du, \quad 0 \le t \le T, \quad \text{(VI.B.27)}$$

with $\int_0^T \psi_n(t)\psi_m(t)dt = 0$ if $n \ne m$ and $\int_0^T \psi_n^2(t)dt = 1$.

Furthermore, $\{Y_t; t \in [0, T]\}$ can be represented by the following mean-square convergent series:

$$Y_t = \sum_{k=1}^{\infty} Z_k \psi_k(t), \quad 0 \le t \le T, \quad \text{(VI.B.28)}$$

where

$$Z_k = \int_0^T \psi_k(t) Y_t dt, \quad k = 1, 2, \ldots . \quad \text{(VI.B.29)}$$

The validity of the representation (VI.B.26) is known as *Mercer's theorem*, and the expansion of (VI.B.28) is known as the *Karhunen-Loève expansion*. Neither of these results will be proved here. Mercer's theorem is a classical theorem in analysis and a proof can be found, for example, in Lovitt (1950). The validity of the Karhunen-Loéve expansion is a consequence of Mercer's theorem, and a proof of this result can be found in Thomas (1971). Note that Mercer's theorem gives a spectral decomposition of C_Y in terms of its eigenvalues and orthonormal eigenfunctions, just as the spectral decomposition of an $n \times n$ covariance matrix $\mathbf{\Sigma}$,

$$\mathbf{\Sigma} = \sum_{k=1}^{n} \lambda_k \underline{v}_k \underline{v}_k^T$$

is a representation in terms of the eigenvalues $\{\lambda_k\}_{k=1}^n$ and orthonormal eigenvectors $\{\underline{v}_k\}_{k=1}^n$ of the matrix as discussed in Section III.B. (Both of these representations are particular examples of a more general representation from operator theory, so they are completely analogous to one another.) Similarly, the Karhunen-Loéve expansion is analogous to the decomposition of a zero-mean random n-vector $\underline{Y}$ as

$$\underline{Y} = \sum_{k=1}^{n} Z_k \underline{v}_k , \qquad \text{(VI.B.30)}$$

where $Z_k = \underline{v}_k^T \underline{Y}$. This analogy is discussed further as we apply the Karhunen-Loéve expansion.

The Karhunen-Loéve expansion provides a separation of the randomness and the time-variation in the process $\{Y_t ; t \in [0, T]\}$. In particular, the randomness in $\{Y_t ; t \in [0, T]\}$ is summarized in the sequence $\{Z_k\}_{k=1}^{\infty}$ while the time variation in the process is embodied in the sequence of functions $\{\psi_k\}_{k=1}^{\infty}$. The expansion (VI.B.28) combines these two to represent the process. Since $\{Y_t ; t \in [0, T]\}$ is determined from $\{Z_k\}_{k=1}^{\infty}$ via (VI.B.28) and $\{Z_k\}_{k=1}^{\infty}$ is determined from $\{Y_t ; t \in [0, T]\}$ via (VI.B.29), the sequence $\{Z_k\}_{k=1}^{\infty}$ is an equivalent observation to the process $\{Y_t ; t \in [0, T]\}$. Thus the Karhunen-Loéve expansion provides a mechanism for reducing a continuous-time random process to an equivalent sequence.

It should be noted that the condition that C_Y be continuous on $[0, T]^2$ is very mild, and this condition is easily shown to be equivalent to the condition that $\{Y_t ; t \in [0, T]\}$ is *mean-square continuous*; i.e., that

$$\lim_{u \to t} E\{(Y_u - Y_t)^2\} = 0 \text{ for all } t \in [0, T] \qquad \text{(VI.B.31)}$$

[see, e.g., Parzen (1962)]. Also, one does not lose generality from the condition that $\{Y_t ; t \in [0, T]\}$ be a zero-mean process, since if $\{Y_t ; t \in [0, T]\}$ does not have zero mean, one can use the proposition to expand the process $\{Y_t - E\{Y_t\} ; t \in [0, T]\}$, which does have zero mean. Since the covariance, C_Y, is invariant to changes in the mean, this does not change the eigenfunctions used in the expansion. However, note that this does add $E\{Y_t\}$ to the right-hand side of (VI.B.28) and changes (VI.B.29) to

$$Z_k = \int_0^T \psi_k(t)(Y_t - E\{Y_t\})dt , \quad k = 1, 2, \ldots \qquad \text{(VI.B.32)}$$

At first glance it may appear that the Karhunen-Loéve expansion and Grenander's theorem provide an approach for solving any signal detection problem based on continuous-time observations. Unfortunately, this is far from being true because of two difficulties that arise in this approach. The first of these difficulties has to do with the fact that the Karhunen-Loéve expansion is dependent on the probability distribution of $\{Y_t; 0 \leq t \leq T\}$ through the autocovariance function C_Y. To generate the sequence $\{Z_k\}_{k=1}^{\infty}$ of (VI.B.29) we need to know the eigenfunctions of C_Y. However, in binary hypothesis tests we do not know *a priori* which of the two possible distributions is valid; thus unless the autocovariance functions of $\{Y_t; t \in [0, T]\}$ under the two hypotheses have the same eigenfunctions, we cannot generate the appropriate observation sequence $\{Z_k\}_{k=1}^{\infty}$.

The foregoing difficulty can be circumvented for a fairly broad class of problems, as we shall see in Section VI.C. However, a more serious difficulty arises in the application of Grenander's theorem. Suppose that the eigenfunctions of C_Y are the same under both hypotheses, so that we can generate the sequence $\{Z_k\}_{k=1}^{\infty}$ via

$$Z_k = \int_0^T \psi_k(t) Y_t \, dt, \quad k = 1, 2, \ldots . \tag{VI.B.33}$$

To apply Grenander's theorem we must have the probability density functions of Z_1^n under each hypothesis and for each n. With the exception of some particular cases, it is extremely difficult, if not impossible, to find even the marginal density of one of the random variables Z_k. Thus without further assumptions, one cannot proceed to find likelihood ratios in this way. An important exception to this difficulty is the situation in which the random process $\{Y_t; t \in [0, T]\}$ is Gaussian. Since many physical phenomena (including thermal noise in electrical circuits) can be modeled accurately as Gaussian processes (defined below), this particular case is of considerable interest in applications. We now consider this case in some detail.

Perhaps the simplest way to define a *Gaussian random process* is to say that $\{Y_t; t \in [0, T]\}$ is Gaussian if all random vectors formed by sampling the process are Gaussian random vectors; i.e., if all vectors of the form $(Y_{t_1}, Y_{t_2}, \ldots, Y_{t_n})^T$, where n is a positive integer and $t_i \in [0, T], i=1, \ldots, n$, have the multivariate Gaussian distribution. Analogously to Gaussian random vectors, Gaussian random processes have the property that linear transformations of them are also Gaussian. Since integration is a linear operation, the sequence $\{Z_n\}_{n=1}^{\infty}$ in the Karhunen-Loéve expansion is obtained by linear transformation of $\{Y_t; t \in [0, T]\}$. Thus for each n, Z_1^n forms a Gaussian random vector, so its density can be specified from the means and covariances of the random variables $Z_1, \ldots, Z_n$.

To compute these quantities we first note that the operations of expectation and mean-square integration can be interchanged. (This follows from an argument similar to that given in Proposition V.C.1.) Thus we have

$$\begin{aligned} E\{Z_k\} &= E\{\int_0^T \psi_k(t) Y_t \, dt\} \\ &= \int_0^T \psi_k(t) E\{Y_t\} dt = 0, \end{aligned} \tag{VI.B.34}$$

where we have used the zero-mean assumption on $\{Y_t; t \in [0, T]\}$. Similarly, we have

$$\begin{aligned} Cov\,(Z_k, Z_m) = E\{Z_k Z_m\} &= E\{\int_0^T \psi_k(t) Y_t \, dt \int_0^T \psi_m(u) Y_u \, du\} \\ &= E\{\int_0^T \int_0^T \psi_k(t) \psi_m(u) Y_t Y_u \, du\} \\ &= \int_0^T \int_0^T \psi_k(t) \psi_m(u) E\{Y_t Y_u\} dt du \\ &= \int_0^T \int_0^T \psi_k(t) \psi_m(u) C_Y(t, u) dt du . \end{aligned} \tag{VI.B.35}$$

Upon performing the integration with respect to u on the

right-hand term of (VI.B.35), we have, from (VI.B.27), that

$$Cov\,(Z_k\,,Z_m\,) = \lambda_m \int_0^T \psi_k\,(t\,)\psi_m\,(t\,)dt\,. \qquad \text{(VI.B.36)}$$

Applying the orthonormality of $\{\psi_k\}_{k=1}^{\infty}$ we then have

$$Cov\,(Z_k\,,Z_m\,) = \begin{cases} \lambda_k \text{ if } k = m \\ \\ 0 \text{ if } k \neq m\,. \end{cases} \qquad \text{(VI.B.37)}$$

We see from (VI.B.37) that Z_k and Z_m are uncorrelated when $k \neq m$ and, since they are Gaussian, they are also independent. Thus when $\{Y_t\,;t \in [0,T\,]\}$ is a zero-mean Gaussian process, the Karhunen-Loéve coefficients $\{Z_k\}_{k=1}^{\infty}$ form a sequence of independent random variables with $Z_k \sim N\,(0,\lambda_k\,)$.†

VI.C The Detection of Deterministic and Partly Determined Signals in Gaussian Noise

Armed with the analytical methods of the Section VI.B, we may turn to the specific problem of signal detection with continuous-time observations. Our basic approach will be first to reduce the continuous-time observation set to an equivalent discrete-time set via the Karhunen-Loéve expansion, and then to apply Grenander's theorem to find the likelihood ratio. Due to the considerations discussed in Section VI.B, we restrict our investigation of this problem primarily to models in which signals and noise are Gaussian random processes, although the detection of non-Gaussian signals is considered briefly in Section VI.D. In the present section we treat the detection of deterministic or parametrically determined signals in additive Gaussian noise, and the following section treats the problem of detecting stochastic signals in Gaussian noise.

---

† Note that the correlation structure of (VI.B.37) does not depend on the Gaussian assumption, and in fact always holds for the coefficients in the Karhunen-Loéve expansion. It is this property of uncorrelated coefficients that makes the Karhunen-Loéve expansion unique among many possible Fourier-type expansions of a random process.

VI.C.1 Coherent Detection

A large class of signal detection problems arising in practice can be modeled by the following hypothesis pair:

$$\begin{aligned} H_0&: Y_t = N_t + s_t^0, \quad 0 \leq t \leq T \\ &\text{versus} \\ H_1&: Y_t = N_t + s_t^1, \quad 0 \leq t \leq T, \end{aligned} \tag{VI.C.1}$$

where $\{s_t^0; t \in [0, T]\}$ and $\{s_t^1; t \in [0, T]\}$ are two signal waveforms that are completely known, and where $\{N_t; t \in [0, T]\}$ is a random process representing additive noise. This is the continuous-time version of the coherent detection problem discussed in Chapter III, and such problems arise in a number of applications, including digital communications and radar.

To study the problem of (VI.C.1), we can consider the equivalent model

$$\begin{aligned} H_0&: Y_t = N_t, \ 0 \leq t \leq T \\ &\text{versus} \\ H_1&: Y_t = N_t + s_t, \ 0 \leq t \leq T, \end{aligned} \tag{VI.C.2}$$

since we can subtract the signal $\{s_t^0; t \in [0, T]\}$ from the observations and let s_t be the difference signal, $s_t^1 - s_t^0$, in (VI.C.1) to get a model of the form (VI.C.2). We assume that the noise $\{N_t; t \in [0, T]\}$ is a zero mean Gaussian random process with a continuous autocovariance function $C_N(t, u), (t, u) \in [0, T]^2$. We further assume that $\{s_t; t \in [0, T]\}$ is continuous and has a representation of the form

$$s_t = \sum_{k=1}^{\infty} \hat{s}_k \psi_k(t), \quad 0 \leq t \leq T, \tag{VI.C.3}$$

where $\{\psi_k\}_{k=1}^{\infty}$ is the sequence of orthonormal eigenfunctions of C_N, and where $\hat{s}_k$ is the component of $\{s_t; 0 \leq t \leq T\}$ along ψ_k; i.e.,

$$\hat{s}_k = \int_0^T s_t \psi_k(t)dt, \quad k = 1, 2, \ldots . \qquad \text{(VI.C.4)}$$

The reason for the latter assumption will become clear below. The case in which this representation does not hold is discussed later.

Using (VI.C.3) and the Karhunen-Loéve expansion for $\{N_t; t \in [0, T]\}$, we can write the hypothesis pair of (VI.C.2) as

$$H_0: Y_t = \sum_{k=1}^{\infty} \hat{N}_k \psi_k(t), \, 0 \leqslant t \leqslant T$$

versus (VI.C.5)

$$H_1: Y_t = \sum_{k=1}^{\infty} (\hat{N}_k + \hat{s}_k)\psi_k(t), \, 0 \leqslant t \leqslant T,$$

where

$$\hat{N}_k = \int_0^T \psi_k(t) N_t \, dt, \, k = 1, 2, \ldots . \qquad \text{(VI.C.6)}$$

From (VI.C.2) and (VI.C.4) through (VI.C.6) we see that under *both* hypothesis we can represent the observation process as

$$Y_t = \sum_{k=1}^{\infty} Z_k \psi_k(t), \quad 0 \leqslant t \leqslant T, \qquad \text{(VI.C.7)}$$

where

$$Z_k = \int_0^T Y_t \psi_k(t)dt = \begin{cases} \hat{N}_k \text{ under } H_0 \\ \\ \hat{N}_k + \hat{s}_k \text{ under } H_1. \end{cases} \qquad \text{(VI.C.8)}$$

Thus the hypothesis pair (VI.C.2) is equivalent to

$$H_0: Z_k = \hat{N}_k, \; k=1,2,\ldots$$

versus (VI.C.9)

$$H_1: Z_k = \hat{N}_k + \hat{s}_k, \; k=1,2,\ldots .$$

From Section VI.B we know that $\{\hat{N}_k\}_{k=1}^{\infty}$ is a sequence of independent random variables with $\hat{N}_k \sim N(0, \lambda_k)$, where $\{\lambda_k\}_{k=1}^{\infty}$ is the sequence of eigenvalues of C_N. Since $\{\hat{s}_k\}_{k=1}^{\infty}$ is a known sequence, the problem (VI.C.9) truncated at n observation is a problem of discrete-time coherent detection in Gaussian noise as treated in Section III.B. In particular, the likelihood ratio based on the first n observation in (VI.C.9) is given by

$$f_n(Z) = \exp\left\{\sum_{k=1}^{n} \hat{s}_k Z_k / \lambda_k - \tfrac{1}{2} \sum_{k=1}^{n} \hat{s}_k^2 / \lambda_k\right\}, \quad \text{(VI.C.10)}$$

where Z denotes $\{Z_k\}_{k=1}^{\infty}$. According to Grenander's theorem, the sequence $\{f_n(Z)\}$ converges in probability under P_0 and P_1 to the likelihood ratio on the set where the likelihood ratio, L, is finite, and it diverges in probability under P_1 on the set where L is infinite. Investigation of the convergence of (VI.C.10) leads to the following result.

Proposition VI.C.1: Grenander's Dichotomy

Let P_0 and P_1 denote the probability measures described by the hypotheses H_0 and H_1, respectively, in (VI.C.9). Then we have the following:

(i) If $\sum_{k=1}^{\infty}(\hat{s}_k)^2/\lambda_k = \infty$, then $P_0 \perp P_1$; and

(ii) if $\sum_{k=1}^{\infty}(\hat{s}_k)^2/\lambda_k < \infty$, then $P_0 \equiv P_1$ and

$$\frac{dP_1}{dP_0}(Z) = \exp\left[\sum_{k=1}^{\infty} \hat{s}_k Z_k / \lambda_k - \tfrac{1}{2} \sum_{k=1}^{\infty} (\hat{s}_k)^2 / \lambda_k\right], \quad \text{(VI.C.11)}$$

where the sum in the exponent is a mean-square sum.

Proof: To show property (i) we assume that $\sum_{k=1}^{\infty}(\hat{s}_k)^2/\lambda_k = \infty$ and consider the probability

$$P_1(f_n(Z) > b) = P_1\left(\sum_{k=1}^{n} \hat{s}_k Z_k / \lambda_k - \tfrac{1}{2} d_n^2 > \log b\right) \qquad \text{(VI.C.12)}$$

for $b > 0$, where $d_n^2 \triangleq \sum_{k=1}^{n}(\hat{s}_k)^2/\lambda_k$. Under H_1, $Z_1, \ldots, Z_n$ are independent $N(\hat{s}_k, \lambda_k)$ random variables, which implies that

$$\sum_{k=1}^{n} \hat{s}_k Z_k / \lambda_k - \tfrac{1}{2} d_n^2 \sim N(\tfrac{1}{2} d_n^2, d_n^2). \qquad \text{(VI.C.13)}$$

From (VI.C.13) we have that

$$P_1(f_n(Z) > b) = 1 - \Phi\left(\frac{\log b}{d_n} - \tfrac{1}{2} d_n\right). \qquad \text{(VI.C.14)}$$

where d_n is the positive square root of d_n^2. Now if $\lim_{n\to\infty} d_n^2 = \infty$, then

$$\lim_{n\to\infty} P_1(f_n(Z) > b) = 1 - \Phi(-\infty) = 1, \qquad \text{(VI.C.15)}$$

for all $b > 0$.

The condition (VI.C.15) implies that $f_n(Z)$ diverges in P_1-probability and thus that H (the set on which $L = \infty$) has probability 1 under H_1. This implies that $P_0 \perp P_1$.

To show property (ii), we now assume that $\sum_{k=1}^{\infty}(\hat{s}_k)^2/\lambda_k < \infty$. We must show that this condition implies both $P_1 \ll P_0$ and $P_0 \ll P_1$. To show that $P_1 \ll P_0$, it is sufficient via Grenander's theorem to the show that $f_n(Z)$ converges in probability under H_1. Equivalently, we can show that $\log f_n(Z)$ converges in probability under H_1. We do this by demonstrating that $\log f_n(Z)$ converges in the mean-square sense under H_1, which then implies convergence in probability [see, e.g., Thomas (1986)].

We can write

$$\log f_n(Z) = \sum_{k=1}^{n} \hat{s}_k Z_k / \lambda_k - \tfrac{1}{2} d_n^2$$
$$= \sum_{k=1}^{n} \hat{s}_k X_k / \lambda_k + \tfrac{1}{2} d_n^2, \tag{VI.C.16}$$

where $X_k \triangleq Z_k - \hat{s}_k$. Since $\lim d_n^2 < \infty$, we need only investigate the convergence of the first term on the right-hand side of (VI.C.16). A necessary and sufficient condition for a sequence $\{W_k\}_{k=1}^{\infty}$ to converge in the mean-square sense is *Cauchy's criterion* [see, e.g, Parzen (1962)],

$$\lim_{n \to \infty} \sup_{m \geqslant n} E\{(W_m - W_n)^2\} = 0. \tag{VI.C.17}$$

On defining $W_n \triangleq \sum_{k=1}^{n} \hat{s}_k X_k / \lambda_k$, we have that

$$E_1\{(W_m - W_n)^2\} = E_1\left\{\left(\sum_{k=n+1}^{m} \hat{s}_k X_k / \lambda_k\right)^2\right\}$$
$$= \sum_{k=n+1}^{m} \hat{s}_k^2 / \lambda_k, \tag{VI.C.18}$$

where the second equality follows from the fact that under H_1, $W_1, \ldots, W_n$ are independent $N(0, \lambda_k)$ random variables. Now, since the summands in (VI.C.18) are nonnegative,

$$\sup_{m \geqslant n} E_1\{(W_m - W_n)^2\} = \sum_{k=n+1}^{\infty} (\hat{s}_k)^2 / \lambda_k,$$

which approaches zero as $n \to \infty$ if d_n^2 converges. Thus the condition $\lim_{n \to \infty} d_n^2 < \infty$ is sufficient for $\log f_n(Z)$ to converge in mean-square under H_1. This implies that

$\lim_{n \to \infty} d_n^2 < \infty$ is a sufficient condition for $P_1 << P_0$.

To show that $\lim d_n^2 < \infty$ is sufficient for $P_0 << P_1$, we can reverse the roles of P_0 and P_1, and investigate the convergence in probability of $1/f_n(Z)$ under H_0. This is the same as showing the convergence in probability of $-\log f_n(Z)$ under H_0, which is a task almost identical to that of the paragraph above. Thus we omit this part, and this completes the proof.

□

Proposition VI.C.1 not only indicates when $P_1 << P_0$, but it also gives the surprising result that the two measures in (VI.C.9) [and equivalently, those in (VI.C.2)] are either equivalent or singular, with there being no intermediate possibilities. The key parameter determining equivalence or singularity is the sum

$$d^2 \triangleq \sum_{k=1}^{\infty} (\hat{s}_k)^2 / \lambda_k . \tag{VI.C.19}$$

It is straightforward to show from Mercer's theorem that

$$\begin{aligned} \sum_{k=1}^{\infty} \lambda_k &= \int_0^T C_N(t,t)dt \\ &= \int_0^T E\{N_t^2\}dt < \infty, \end{aligned} \tag{VI.C.20}$$

where the finiteness of the integral follows from the continuity of R_N. Thus since $\lambda_k = Var(Z_k) \geq 0$, it follows that $\lim_{k \to \infty} \lambda_k = 0$, which implies that a necessary (but not sufficient) condition for d^2 to be finite is that $\sum_{k=1}^{\infty} (\hat{s}_k)^2 < \infty$. It follows from the representation (VI.C.3) that the signal energy in $[0, T]$ is given by

$$\int_0^T s_t^2 dt = \sum_{k=1}^{\infty} (\hat{s}_k)^2. \tag{VI.C.21}$$

Thus in order for the problem of (VII.C.19) to be nonsingular, the signal must have finite energy, a condition that is not surprising. However, finite energy of the signal is not enough to assure lack of singularity here since the finiteness of d^2 requires that $(\hat{s}_k)^2/\lambda_k$ be summable. This implies a subtle relationship between the signal and the eigenstructure of the noise autocovariance.

Assuming that $d^2 < \infty$, the optimum Bayes, minimax, and Neyman-Pearson detection strategies for H_0 versus H_1 are of the form

$$\delta(Y_0^T) = \begin{cases} 1 & > \\ \gamma \text{ if } \sum_{k=1}^{\infty} \hat{s}_k Z_k / \lambda_k & = \log \tau + \tfrac{1}{2} d^2 \triangleq \tau'. \\ 0 & < \end{cases} \qquad \text{(VI.C.22)}$$

This structure is shown in Fig. VI.C.1. Note that it consists of an infinite bank of continuous-time correlators (or matched filters), each corresponding to a different eigenfunction of the noise. As we shall see below, the optimum detector can be represented more conveniently for the purposes of implementation; however, the representation of Fig. VI.C.1 is useful for interpretation. In particular, the eigenfunctions $\{\psi_k\}$ can be thought of as a set of orthogonal "directions" in which there is noise power, with λ_k being the noise power in the direction ψ_k and $\hat{s}_k$ being the signal component in that direction. The $k^{\underline{th}}$ correlator, $Z_k = \int_0^T \psi_k(t) Y_t \, dt$, represents the component of the observation process in the ψ_k direction, and the orthogonality of the ψ_k's makes these outputs $\{Z_k\}_{k=1}^{\infty}$ independent under each hypothesis. The detector then correlates these independent outputs in the same way as that derived for the analogous discrete-time problem discussed in Section III.B.

Note that the response of this overall system of Fig. VI.C.1 to the input signal only is $\sum_{k=1}^{\infty} (\hat{s}_k)^2/\lambda_k = d^2$, while the response to input noise only is $\sum_{k=1}^{\infty} \hat{s}_k \hat{N}_k / \lambda_k$. The first of these quantities is a nonrandom scalar and the second quantity is easily seen to have zero mean and variance d^2. Thus the signal-to-noise ratio at the threshold comparison in Fig. VI.C.1 is

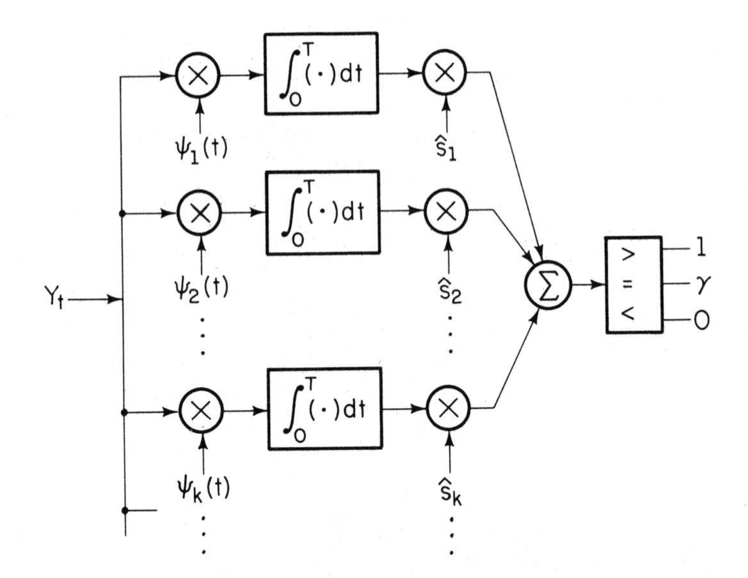

Fig. VI.C.1: Optimum detector structure for coherent detection in Gaussian noise.

$$SNR_0 = \frac{(\text{output signal amplitude})^2}{\text{output noise power}}$$

$$= d^2, \tag{VI.C.23}$$

and we see that the singularity condition can be interpreted as an infinite signal-to-noise-ratio case, where the signal-to-noise-ratio is defined at the output of the optimum detection system.

The test statistic, $\sum_{k=1}^{\infty} \hat{s}_k Z_k / \lambda_k$, is a mean-square sum of the independent Gaussian random variables $\{Z_k\}_{k=1}^{\infty}$, so it is also Gaussian. Under H_0, we have that $Z_k \sim N(0, \lambda_k)$ and it follows that

$$\sum_{k=1}^{\infty} \hat{s}_k Z_k / \lambda_k \sim N(0, d^2) \text{ under } H_0. \qquad \text{(VI.C.24)}$$

Under H_1, we have that $Z_k \sim N(\hat{s}_k, \lambda_k)$, and it follows similarly that

$$\sum_{k=1}^{\infty} \hat{s}_k Z_k / \lambda_k \sim N(d^2, d^2) \text{ under } H_0. \qquad \text{(VI.C.25)}$$

From (VI.C.24) and (VI.C.25) we can determine the performance of the optimum test (VI.C.22). (Note, for example, that the randomization is irrelevant.) In particular, the false-alarm probability is given by

$$P_0(\Gamma_1) = P_0\left(\sum_{k=1}^{\infty} \hat{s}_k Z_k / \lambda_k \geqslant \log \tau + d^2/2\right) \qquad \text{(VI.C.26a)}$$

$$= 1 - \Phi\left(\frac{\log \tau}{d} + \frac{d}{2}\right),$$

and the miss probability is given by

$$P_1(\Gamma_0) = P_1\left(\sum_{k=1}^{\infty} \hat{s}_k Z_k / \lambda_k \geqslant \log \tau + d^2/2\right) \qquad \text{(VI.C.26b)}$$

$$= \Phi\left(\frac{\log \tau}{d} - \frac{d}{2}\right).$$

Comparing these equations to the analogous ones from the problem of scalar location testing with Gaussian error (II.B.31), we see that the performance picture here is identical with that in the scalar problem aside from the new definition of d^2. Thus for example, for Bayes testing with uniform costs and equal priors the threshold τ equals 1 and the Bayes risk is

$$r(\delta_B) = \tfrac{1}{2} P_0(\Gamma_1) + \tfrac{1}{2} P_1(\Gamma_0) = 1 - \Phi(d/2), \qquad \text{(VI.C.27)}$$

as depicted in Fig. II.B.3. Similarly, for α-level Neyman-Pearson detection the threshold τ' is $\Phi^{-1}(1-\alpha)/d$ and the detection probability becomes

$$P_D = 1-\Phi[\Phi^{-1}(1-\alpha)-d\,]. \tag{VI.C.28}$$

The corresponding receiver operating characteristics are those depicted in Fig. II.D.4. Thus aside from the possibility of singularity, the problem of continuous-time detection of a coherent signal in Gaussian noise is exactly the same as the analogous scalar and vector problems once the detection statistic has been computed.

Before discussing an alternative, somewhat more practical, representation of the likelihood ratio for this problem, we first discuss briefly the situation in which the signal $\{s_t\,;t \in [0,T\,]\}$ does not have a representation of the form (VI.C.3). As noted above, the ψ_k's can be thought of as being a set of orthogonal directions in which there is noise power. Conversely, there is no noise power in any direction that is orthogonal to all of the ψ_k's. If the signal has finite energy but does not have a representation in the noise eigenfunctions, this implies that the signal has some energy in a direction orthogonal to all the ψ_k's. This in turn implies that there is signal energy in a direction where there is no noise energy, so by looking in this direction, perfect (i.e., singular) detection can be accomplished.

In particular, consider the function

$$f(t\,) \triangleq s_t - \sum_{k=1}^{\infty} \hat{s}_k \psi_k(t\,),\ \ 0 \leqslant t \leqslant T\,,$$

which is orthogonal to all the ψ_k's, since, for any positive integer m ,

$$\int_0^T f(t\,)\psi_m(t\,)dt = \int_0^T s_t\,\psi_m(t\,)dt - \sum_{k=1}^{\infty} \hat{s}_k \int_0^T \psi_m(t\,)\psi_m(t\,)dt$$

$$= \hat{s}_m - \hat{s}_m = 0.$$

If we consider the test statistic $\int_0^T f(t)Y_t\,dt$, its response to the signal is

$$\int_0^T f(t)s_t\,dt = \int_0^T s_t^2 dt - \sum_{k=1}^{\infty} (\hat{s}_k)^2,$$

which is positive if the representation $\sum \hat{s}_k \psi_k$ is not complete. Similarly its response to the noise is

$$\int_0^T f(t)N_t\,dt = \sum_{k=1}^{\infty} \hat{N}_k \int_0^T f(t)\psi_k(t)dt = 0.$$

Thus by examining the output of the single continuous-time correlator, $\int_0^T f(t)Y_t\,dt$, we can determine perfectly well whether or not the signal is present. Another way of looking at this is that $f(t)$ is an eigenfunction of C_N, say ψ_k, corresponding to an eigenvalue $\lambda_k = 0$, which, together with $(\hat{s}_k)^2 > 0$, makes $d^2 = \infty$ and thus gives singularity.

The representation for the likelihood ratio provided by Proposition VI.C.1 is very useful for analyzing performance and general structure in the hypothesis-testing problem of (VI.C.2). This approach also gives an idea of how one might approximately implement the detection system implied by this representation, by truncating the infinite filter bank of Fig. VI.C.1 at a point at which the lost signal-to-noise ratio, $\sum_{k=n+1}^{\infty}(\hat{s}_k)^2/\lambda_k$, is inconsequential. However, the likelihood ratio for this problem can be represented in another form that is suggestive of a more efficient implementation of the corresponding signal detection system. This representation is summarized in the following result.

Proposition VI.C.2: Pitcher's Theorem

Suppose that there is a function $H : [0, T] \rightarrow \mathbb{R}$ of bounded variation such that

$$s_t = \int_0^T C_N(t, u)dH(u), \quad 0 \leq t \leq T. \qquad \text{(VI.C.29)}$$

Then $P_0 \equiv P_1$,

$$\log \frac{dP_1}{dP_0}(Y_0^T) = \int_0^T Y_t \, dH(t) - \tfrac{1}{2} \int_0^T s_t \, dH(t) \qquad \text{(VI.C.30)}$$

and

$$d^2 = \int_0^T s_t \, dH(t). \qquad \text{(VI.C.31)}$$

Proof: Suppose that $\{s_t; t \in [0, T]\}$ has the representation (VI.C.29). By Mercer's theorem we can represent C_N as the uniformly and absolutely convergent series

$$C_N(t, u) = \sum_{k=1}^{\infty} \lambda_k \psi_k(t) \psi_k(u), \quad (t, u) \in [0, T]^2. \qquad \text{(VI.C.32)}$$

Inserting this into (VI.C.29) and interchanging order of integration and summation [which is permissible because of the uniform convergence of (VI.C.32)], we have

$$s_t = \sum_{k=1}^{\infty} \lambda_k \psi_k(t) \int_0^T \psi_k(u) dH(u), \quad 0 \leq t \leq T. \qquad \text{(VI.C.33)}$$

Comparing (VI.C.33) with the signal representation of (VI.C.3), we see that

$$\int_0^T \psi_k(u) dH(u) = \hat{s}_k / \lambda_k, \quad k = 1, 2, \ldots . \qquad \text{(VI.C.34)}$$

[Note that (VI.C.33) implies that the representation of (VI.C.3) is valid.] Since H is of bounded variation and C_N is continuous (and hence bounded on $[0, T]^2$), the integral

$$\int_0^T \int_0^T C_N(t,u)\,dH(u)\,dH(t) \qquad \text{(VI.C.35)}$$

is finite. Using Mercer's theorem and (VI.C.34), we have that this integral becomes

$$\sum_{k=1}^{\infty} \lambda_k \int_0^T \int_0^T \psi_k(t)\psi_k(u)\,dH(t)\,dH(u)$$

$$= \sum_{k=1}^{\infty} \lambda_k \int_0^T \psi_k(t)\,dH(t) \int_0^T \psi_k(u)\,dH(u) \qquad \text{(VI.C.36)}$$

$$= \sum_{k=1}^{\infty} \lambda_k \frac{\hat{s}_k}{\lambda_k} \frac{\hat{s}_k}{\lambda_k} = \sum_{k=1}^{\infty} (\hat{s}_k)^2/\lambda_k = d^2.$$

Thus $d^2 < \infty$ and Proposition VI.C.1 implies that $P_0 \equiv P_1$. Also, since $\int_0^T s_t\,dH(t)$ equals the integral of (VI.C.35), (VI.C.31) follows from (VI.C.36).

From (VI.C.7), we have that $Y_t = \sum_{k=1}^{\infty} Z_k \psi_n(t), 0 \leqslant t \leqslant T$. Again, using (VI.C.34), we then have

$$\int_0^T Y_t\,dH(t) = \sum_{k=1}^{\infty} Z_k \int_0^T \psi_k(t)\,dH(t)$$

$$\qquad \text{(VI.C.37)}$$

$$= \sum_{k=1}^{\infty} \hat{s}_k Z_k / \lambda_k .$$

Combining (VI.C.37) with (VI.C.11) and (VI.C.31) yields (VI.C.30). This completes the proof.

□

Pitcher's theorem suggests a simpler implementation of optimum detection systems for (VI.C.2) than is suggested by Grenander's theorem. In particular, the implementation of (VII.C.30) requires the computation of only the single integral,

$$\int_0^T Y_t \, dH(t),$$

rather than the sequence $\int_0^T \psi_k(t) Y_t \, dt$. Also, one needs to solve only one integral equation (VI.C.29) (*Pitcher's equation*) for H rather than to find all solutions to the eigenfunction equation $\lambda\psi(t) = \int_0^T C_N(t, u)\psi(u) du$, $0 \leq t \leq T$.

The relationship of this result to analogous discrete-time results is easily seen if we assume that $H(t)$ is differentiable with $h(t) = dH(t)/dt$. In this case, the integral equation (VI.C.29) becomes

$$s_t = \int_0^T C_N(t, u) h(u) du, \quad 0 \leq t \leq T, \qquad \text{(VI.C.38)}$$

and the optimum detection statistic becomes

$$\log \frac{dP_1}{dP_0}(Y_0^T) = \int_0^T Y_t h(t) dt - \tfrac{1}{2} \int_0^T s_t h(t) dt. \qquad \text{(VI.C.39)}$$

Now it is interesting to compare (VI.C.38) and (VI.C.39) with (III.B.25) and (III.B.23), respectively. In particular, on denoting $\underline{\tilde{s}}$ of (III.B.25) by $\underline{h}$, we see that (VI.C.38) is the continuous-time analog to the vector equation $\underline{s} = \Sigma_Y \underline{h}$, or

$$s_k = \sum_{j=1}^n (\Sigma_Y)_{kj} h_j, \quad 1 \leq k \leq n, \qquad \text{(VI.C.40)}$$

which specifies that the "pseudosignal" $\underline{h}$ in the optimum detector for discrete-time coherent detection in Gaussian noise. Similarly, (VI.C.39) is the analog to the corresponding discrete-time formula

$$\log \frac{p_1(\underline{Y})}{p_0(\underline{Y})} = \underline{h}^T \underline{Y} - \tfrac{1}{2} \underline{h}^T \underline{s} = \sum_{k=1}^n h_k Y_k - \tfrac{1}{2} \sum_{k=1}^n h_k s_k. \qquad \text{(VI.C.41)}$$

Thus we see that Pitcher's theorem provides the "direct"

representation of the likelihood ratio for continuous time, whereas Proposition VI.C.1 provides the "prewhitened" likelihood ratio formula.

The interpretation above relates Pitcher's theorem to the analogous discrete-time result. However, for continuous C_N, it is usually the case that $H(t)$ is not differentiable, so (VI.C.30) and (VI.C.31) must be given a slightly more general interpretation. We now give a specific example to illustrate the application of Pitcher's theorem.

Example VI.C.1: Coherent Detection in White Gaussian Noise

One of the most pervasive noise models in applications is that of Gaussian white noise. In discrete time, Gaussian white noise refers to a sequence of i.i.d. zero-mean Gaussian random variables. Such a sequence is wide-sense stationary with autocovariance function

$$\begin{aligned} C_N(k,l) &= C_N(k-l,0) \\ &\triangleq C_N(k-l) = \frac{N_0}{2}\delta_{k,l},\ k,l \in Z, \end{aligned} \tag{VI.C.42}$$

where $\delta_{k,l}$ is the Kronecker delta function ($\delta_{k,l}=0$ if $k \neq l$ and $\delta_{k,l}=1$ if $k=l$) and where $N_0/2$ is the variance of each element of the sequence. As noted in Chapter V, the term "white noise" comes from the fact that the power spectral density of the process,

$$\phi_N(\omega) \triangleq \sum_{k=-\infty}^{\infty} C_N(k)e^{-i\omega k} = \frac{N_0}{2},\ -\pi \leqslant \omega \leqslant \pi, \tag{VI.C.43}$$

is constant.

The analogous concept in continuous time would be a zero-mean Gaussian process with autocovariance function

$$C_N(t,u) = C_N(t-u,0) \triangleq C_N(t-u)$$
$$= \frac{N_0}{2}\delta(t-u),\ t,u \in \mathbb{R}, \qquad \text{(VI.C.44)}$$

where δ denotes the Dirac delta function.† The power spectral density in continuous time is the Fourier integral rather than the discrete-time Fourier transform of (VI.C.43). In particular, for (VI.C.44), we have

$$\phi_N(\omega) = \int_{-\infty}^{\infty} e^{-i\omega\tau}\frac{N_0}{2}\delta(\tau)d\tau = \frac{N_0}{2},\ -\infty<\omega<\infty. \qquad \text{(VI.C.45)}$$

Thus a process with the autocovariance function of (VI.C.44) has a flat spectrum, and $N_0/2$ is known as the *spectral height* of the noise.

Unfortunately, Gaussian white noise does not exist either as a physical phenomenon or as a mathematical random process in the ordinary sense. It is physically impossible for a phenomenon to have equal (nonzero) energy in all frequencies. The reasons that Gaussian white noise cannot exist as an ordinary random process are subtle [see, e.g., Skorohod (1974)]; however, it certainly does not exist as a second-order random process, since, from (VI.C.44), we have $E\{N_t^2\}=C_N(t,t)=(N_0/2)\delta(t-t)=\infty$. Nevertheless, the autocovariance structure assumed in (VI.C.44) is very convenient for a number of analytical purposes, so it is often used to model physical processes that are very wide-band relative to other processes in a given model. Basically, the analytical difficulties associated with white-noise models can be overcome as long as the white noise is the input to a linear system and it is the output of the system that we wish to analyze. In the context of signal detection, we can deal rigorously with white noise in the following way.

We wish to consider the detection problem

† The Dirac delta function is a so-called generalized function, the defining property of which is that $\int_{-\infty}^{\infty} f(x)\delta(x)dx = f(0)$ for any function f that is continuous at $x=0$. No real-valued function has this property required of δ, and $\delta(0)$ is usually interpreted as $+\infty$ with $\delta(x)=0$ for $x \neq 0$.

$$H_0: Y_t = N_t, \quad 0 \leqslant t \leqslant T$$

versus (VI.C.46)

$$H_1: Y_t = N_t + s_t, \quad 0 \leqslant t \leqslant T,$$

where $\{N_t ; t \in [0, T]\}$ represents white Gaussian noise and $\{s_t ; t \in [0, T]\}$ is a known signal. Regardless of other difficulties, since $\{N_t ; t \in [0, T]\}$ is not second order, we cannot apply the results of this section to find the likelihood ratio for (VI.C.46) directly. However, we can produce an equivalent problem that can be solved by integrating the observations in (VI.C.46) to get the following model:

$$H_0: X_t = W_t, \quad 0 \leqslant t \leqslant T$$

versus (VI.C.47)

$$H_1: X_t = W_t + m_t, \quad 0 \leqslant t \leqslant T,$$

where $X_t \triangleq \int_0^t Y_u \, du$, $W_t \triangleq \int_0^t N_u \, du$, and $m_t \triangleq \int_0^t s_u \, du$, $0 \leqslant t \leqslant T$. The process $\{W_t ; t \in [0, T]\}$ can easily be shown to be a Gaussian process with zero mean and autocovariance

$$C_W(t, u) = \frac{N_0}{2} \min\{t, u\}, \quad (t, u) \in [0, T]^2. \quad \text{(VI.C.48)}$$

A process with these characteristics is known as a *Wiener process*. Note that $C_W(t, u)$ is continuous on $[0, T]^2$ and thus we can apply the analysis above to find the likelihood ratio for this problem.

To do so, we solve Pitcher's equation (VI.C.29) for the hypothesis pair (VI.C.47), which is

$$m_t = \int_0^T C_W(t, u) dH(u), \quad 0 \leqslant t \leqslant T. \quad \text{(VI.C.49)}$$

Substituting for m_t and C_W, we have

$$\int_0^t s_u\,du = \frac{N_0}{2}\int_0^T \min\{t\,,u\}dH(u)$$

$$= \frac{N_0}{2}\int_0^t u\,dH(u) + \frac{N_0}{2}t\int_t^T dH(u) \qquad \text{(VI.C.50)}$$

$$= \frac{N_0}{2}\int_0^t u\,dH(u) + \frac{N_0}{2}t\,[H(T)-H(t)].$$

The first term on the right-hand side of (VI.C.50) can be integrated by parts to give

$$\int_0^t s_u\,du = \frac{N_0}{2}[tH(t) - \int_0^t H(u)du] + \frac{N_0}{2}t\,[H(T)-H(t)]$$

$$= \frac{N_0}{2}\left[\int_0^t [H(T)-H(u)]du\right], \quad 0\leqslant t\leqslant T. \qquad \text{(VI.C.51)}$$

Assuming that the signal is of bounded variation on $[0,T]$, the solution to Pitcher's equation is thus

$$H(t) = \begin{cases} H(T) - \frac{2}{N_0}s_t & \text{if } 0\leqslant t < T \\ H(T) & \text{if } t = T, \end{cases} \qquad \text{(VI.C.52)}$$

where $H(T)$ is arbitrary. This function s_t is illustrated in Fig. VI.C.2.

Assuming that that the signal is continuous at the end-point $t=T$, the log-likelihood ratio becomes

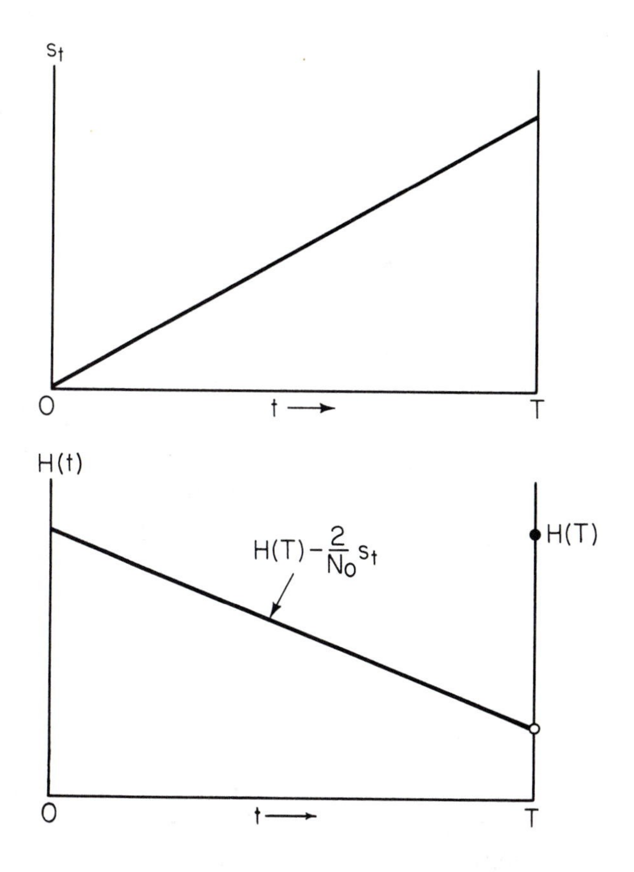

Fig. VI.C.2: Illustration of the solution to Pitcher's equation for Wiener noise.

$$\log \frac{dP_1}{dP_0}(X_0^T) = \frac{2}{N_0}\int_0^T (X_t - \tfrac{1}{2}\, m_t)ds_t$$

$$+ \frac{2}{N_0}(X_T - \tfrac{1}{2}\, m_T)s_T \qquad \text{(VI.C.53)}$$

since H takes a jump of $(2/N_0)s_T$ at $t=T$. Integrating by parts yields

$$\log \frac{dP_1}{dP_0}(X_0^T) = \frac{2}{N_0}\int_0^T s_t \, d\,(X_t - \tfrac{1}{2}\, m_t\,)$$

$$= \frac{2}{N_0}\int_0^T s_t \, dX_t - \frac{1}{N_0}\int_0^T s_t \, dm_t \qquad \text{(VI.C.54)}$$

$$= \frac{2}{N_0}\int_0^T s_t \, dX_t - \frac{1}{N_0}\int_0^T s_t^2 dt\ ,$$

since $s_t = dm_t/dt$.

The representation (VI.C.54) is known as the *Cameron-Martin formula*. As we shall discuss below, it actually can be shown to be valid for any signal with $\int_0^T s_t^2 dt \leqslant \infty$. The physical interpretation of (VI.C.54) is clearest if we think of the original observation Y_t as being the derivative of X_t. (Since the original model contained white noise, this is only a heuristic interpretation.) Then (VI.C.54) becomes

$$\log \frac{dP_1}{dP_0}(Y_0^T) = \frac{2}{N_0}\int_0^T s_t Y_t \, dt - \frac{1}{N_0}\int_0^T s_t^2 dt. \qquad \text{(VI.C.55)}$$

Recall that, in discrete time, the detection of a known signal $s_1, ..., s_n$ in observations $Y_1, ..., Y_n$ containing additive i.i.d. $N(0, N_0/2)$ noise is based on the log-likelihood ratio [see (III.B.9)]

$$\log \frac{p_1(Y_0^n)}{p_0(Y_0^n)} = \frac{2}{N_0}\sum_{k=1}^{n} s_k Y_k - \frac{1}{N_0}\sum_{k=1}^{n} s_k^2. \quad \text{(VI.C.56)}$$

The analogy between (VI.C.55) and (VI.C.56) is clear. In particular, (VI.C.55) is the continuous-time version of the correlator or matched filter, and the corresponding detection rule

$$\delta(Y_0^T) = \begin{cases} 1 & > \\ \gamma \text{ if } \dfrac{dP_1}{dP_0}(Y_0^T) & = \tau \\ 0 & < \end{cases} \quad \text{(VI.C.57)}$$

is depicted in its various forms in Fig. VI.C.3.

The performance of (VI.C.57) is based on the quantity d^2, given by Pitcher's theorem to be

$$d^2 = \int_0^T m_t \, dH(t) = \frac{2}{N_0}\int_0^T m_t \, ds_t + \frac{2}{N_0} m_T s_T \quad \text{(VI.C.58)}$$

$$= \frac{2}{N_0}\int_0^T s_t^2 dt.$$

Thus the performance here is determined by the total signal

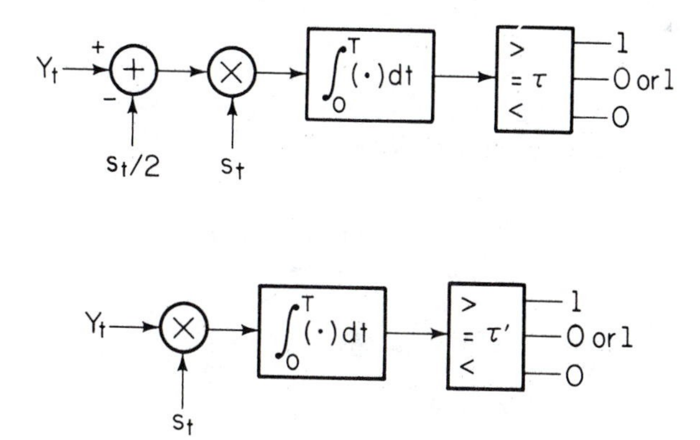

Fig. VI.C.3: Detector structure for coherent signals in white Gaussian noise.

energy divided by the spectral height of the noise. Note that singularity cannot occur here for a finite-energy signal.

The receiver structure of Fig. VI.C.3 could have been derived heuristically by applying Pitcher's theorem directly to the white noise model (VI.C.46). In particular, assuming that H is differentiable with $h(t)=dH(t)/dt$, we have the equation

$$\begin{aligned} s_t &= \int_0^T C_N(t,u)h(u)du \\ &= \frac{N_0}{2}\int_0^T \delta(t-u)h(u)du = \frac{N_0}{2}h(t), \end{aligned} \tag{VI.C.59}$$

from which $h(t)=(2/N_0)s_t$; and the likelihood ratio of (VII.C.47) becomes

$$\log \frac{dP_1}{dP_0}(Y_0^T) = \frac{2}{N_0}\left[\int_0^T s_t Y_t\,dt - \tfrac{1}{2}\int_0^T s_t^2 dt\right], \tag{VI.C.60}$$

which is identical to (VI.C.55). Of course, Pitcher's theorem cannot actually be applied in this way, but one can think of (VI.C.60) as being an approximation to the optimum receiver structure when the noise has a flat spectrum over a very wide (but finite) band of frequencies.

Solutions to Pitcher's equations can be found for a number of other covariance models, and some specific ones are given as exercises at the end of this chapter. Typically, the solution H is not differentiable, but it is often differentiable in the interior of the observation interval with possible discontinuities at the endpoints (as is the case for the Wiener process of Example VI.C.1). For this reason it is convenient to think of $H(t)$ as the integral of a function $h(t)$ that is real-valued in $(0,T)$ and that has singularities (i.e., Dirac delta functions) on the endpoints. A solution technique for Pitcher's equation using this interpretation has been proposed by Kailath (1966).

As a final comment on Pitcher's equation, it is sometimes assumed that the noise autocovariance function is of the form

$$C_N(t,u) = C_C(t,u) + \frac{N_0}{2}\delta(t-u), \quad (t,u)\in[0,T]^2, \quad \text{(VI.C.61)}$$

where C_C is a continuous autocovariance function and $N_0 > 0$; i.e., the noise is assumed to consist of a continuous part and an additive white part.† If, as in Example VI.C.1, we heuristically apply Pitcher's theorem to this case, it turns out that the solution H is usually differentiable and its derivative h thus satisfies the equation

$$s_t = \int_0^T C_C(t,u)h(u)du + \frac{N_0}{2}h(t), \quad 0 \leq t \leq T. \quad \text{(VI.C.62)}$$

Equation (VI.C.62) is a *Fredholm integral equation of the second kind* and solution techniques for this type of equation are well known (see, e.g., Lovitt (1950)). With $N_0=0$, (VI.C.62) is *Fredholm equation of the first kind*, and as noted above, solutions to this type of equation must often be interpreted as having singularities at $t=0$ and/or at $t=T$. The addition of the white noise for any $N_0>0$ suppresses these singularities. This approach can be made rigorous in a manner similar to that used in Example VI.C.1.

VI.C.2 Detection of Signals with Unknown Parameters

As, we discussed in Chapter III, it is common to have a signal detection problem in which the signal is of known form but contains some unknown parameters. For example, for sinusoidal signals, parameters such as phase, frequency, and amplitude are often unknown *a priori*.

To study this detection problem in continuous time, we will consider the following hypothesis pair:

† The white part is sometimes known as a *noise floor*.

$$H_0: Y_t = N_t, \quad 0 \leq t \leq T$$

versus (VI.C.63)

$$H_1: Y_t = s_t(\theta) + N_t, \quad 0 \leq t \leq T,$$

where $s_t(\theta)$ is a known function of the unknown parameter θ taking values in a set Λ, and where $\{N_t; t \in [0, T]\}$ represents Gaussian white noise with spectral height $N_0/2$. As in Example VI.C.1, we first convert this model to one that is better posed by integrating the observation to get the hypothesis pair†

$$H_0: X_t = W_t, \quad 0 \leq t \leq T$$

versus (VI.C.64)

$$H_1: X_t = W_t + \int_0^t s_u(\theta) du, \quad 0 \leq t \leq T,$$

where $X_t = \int_0^t Y_u \, du$ and where $\{W_t; t \in [0, T]\}$ is a Wiener process with autocovariance $C_W(t, u) = (N_0/2)\min\{t, u\}$. We assume that $\int_0^T s_t^2(\theta) dt < \infty$ for all $\theta \in \Lambda$. Thus for fixed θ, the likelihood ratio between H_0 and H_1 of (VI.C.64) is given by the Cameron-Martin formula (VI.C.54) as

$$L_\theta(X_0^T) \triangleq \frac{dP_\theta}{dP_0}(X_0^T)$$

$$= \exp\left\{ \frac{2}{N_0} \int_0^T s_t(\theta) dX_t - \frac{1}{N_0} \int_0^T s_t^2(\theta) dt \right\}. \quad \text{(VI.C.65)}$$

Assuming a parameter set $\Lambda \subset \mathbb{R}^m$, there are three different methods of interest for testing H_0 versus H_1, as we have discussed earlier. These are:

UMP testing, based on the criterion

† As, we shall see in Section VI.D, a wide class of Gaussian noise processes can be reduced linearly to the Wiener process, thus making (VI.C.64) a fairly general model for additive Gaussian noise.

$$\max_{\delta} P_D(\delta|\theta) \text{ for all } \theta \in \Lambda, \text{ subject to } P_F(\delta) \leqslant \alpha; \qquad \text{(VI.C.66)}$$

Simple testing (Bayes, minimax, and Neyman-Pearson), based on the assumption of a prior $w(\theta)$ on Λ under H_1; and

Maximum-likelihood testing, based on a comparison of the generalized likelihood ratio

$$\max_{\theta \in \Lambda} L_\theta(X_0^T) \qquad \text{(VI.C.67)}$$

to a threshold.

Recall than an α-level UMP test exists in this case if and only if the critical region $\Gamma_\theta = \{L_\theta(Y_0^T) > \tau_\theta\}$ can be chosen independently of θ such that $P_0(\Gamma_\theta) = \alpha$ for all $\theta \in \Lambda$. The simple tests are all solved by comparison of the averaged likelihood ratio

$$\int_\Lambda L_\theta(X_0^T) w(\theta) d\theta \qquad \text{(VI.C.68)}$$

to a threshold.

As with the discrete-time case, little further can be said about these problems without giving a more specific model for the signal. Also, once having determined the form (VI.C.65) for L_θ, there are not significant differences between the continuous- and discrete-time versions of this problem. Thus the following example, which is analogous to the discrete-time Example III.B.5, is sufficient to illustrate the approach.

Example VI.C.2: Detection of a Partly Determined Sinusoid in White Noise

Consider the signal model

$$s_t(\theta) = a_t \sin(\omega_c t + \theta), \quad 0 \leqslant t \leqslant T, \qquad \text{(VI.C.69)}$$

where ω_c is a known frequency satisfying the condition that $\omega_c T / \pi$ is an integer, $\{a_t; 0 \leqslant t \leqslant T\}$ is a known waveform satisfying

$$\frac{1}{T}\int_0^T a_t^2 dt \triangleq \overline{a^2} < \infty,$$

and θ is an unknown phase. For simplicity, we will assume that

$$\int_0^T a_t^2 \cos(2\omega_c t + \phi) dt = 0 \qquad \text{(VI.C.70)}$$

for all $0 \leqslant \phi \leqslant 2\pi$. This is true, for example, if a_t^2 is a constant or a raised cosine. Furthermore, (VI.C.70) holds approximately if a_t^2 is slowly varying relative to the sinusoid at frequency $2\omega_c$.

Given θ, the likelihood ratio from (VI.C.65) is given by

$$L_\theta(X_0^T) = \exp\left\{\frac{2}{N_0}\int_0^T a_t \sin(\omega_c t + \theta) dX_t \right.$$
$$\left. - \frac{1}{N_0}\int_0^T a_t^2 \sin^2(\omega_c t + \theta) dt \right\}. \qquad \text{(VI.C.71)}$$

On defining Y_c and Y_s by

$$Y_c = \int_0^T a_t \cos(\omega_c t) dX_t$$

and (VI.C.72)

$$Y_s = \int_0^t a_t \sin(\omega_c t) dX_t,$$

and applying the identity $\sin(a+b) = \cos a \sin b + \sin a \cos b$, the first term in the exponent of (VI.C.71) becomes

$$\frac{2}{N_0}(Y_c \sin\theta + Y_s \cos\theta).$$

Applying the identity $\sin^2(a) = \frac{1}{2} - \frac{1}{2}\cos(2a)$ the second term in this exponent becomes

$$\frac{\overline{a^2}T}{2N_0} - \frac{1}{2N_0}\int_0^T a_t^2 \cos(2\omega_c t + 2\theta)dt = \frac{\overline{a^2}T}{2N_0}, \qquad \text{(VI.C.73)}$$

where we have used the assumption (VI.C.70). We thus have

$$L_\theta(X_0^T) = e^{-\overline{a^2}T/2N_0} \times \exp\left\{\frac{2}{N_0}(Y_c \sin\theta + Y_s \cos\theta)\right\}. \qquad \text{(VI.C.74)}$$

On comparing (VI.C.74) with the likelihood ratio obtained in Example III.B.5, we see that the discrete-time and continuous-time conditional likelihood ratios for this problem are identical (with the identification $T=n$) except for the definition of the observables Y_c and Y_s. Thus for all forms of the likelihood ratio derived under the various assumptions made on $\{a_t\}$ and θ in Example III.B.5 and in the exercises in Chapter III are valid for this continuous-time case with the definition (VI.C.72) for Y_c and Y_s. Furthermore, the performance expressions derived in Example III.B.5 also hold in the continuous-time case. To see this, we note that given $\theta, \{X_t; 0 \leqslant t \leqslant T\}$ is a Gaussian random process under either hypothesis. Since Y_c and Y_s are linear transformations on X_0^T, they also are Gaussian (and jointly Gaussian) under either hypothesis (with θ fixed). It can be shown that,† for any two square-integrable functions f and g on $[0, T]$, we have

† This property is discussed in Section VI.D.

$$E\{\int_0^T f(t)dW_t\} = 0$$

and (VI.C.75)

$$E\{\int_0^T f(t)dW_t \int_0^T g(t)dW_t\} = \frac{N_0}{2}\int_0^T f(t)g(t)dt ,$$

when $\{W_t ; 0 \leqslant t \leqslant T\}$ is a zero-mean Wiener process with covariance $(N_0/2)\min\{t, u\}$. It follows straightforwardly from this that given θ,

$$E\{Y_c | H_j\} = j\frac{\overline{a^2}}{2}\sin\theta,$$

$$E\{Y_s | H_j\} = j\frac{\overline{a^2}}{2}\cos\theta, \tag{VI.C.76}$$

$$Var\ (Y_c | H_j) = Var\ (Y_s | H_j) = \tfrac{1}{4} N_0 T ,$$

and

$$Cov\ (Y_c , Y_s | H_j) = 0$$

for $j=0$ and $j=1$. Comparing (VI.C.76) with the corresponding quantities from Example III.B.5, we see that the statistics of Y_c and Y_s, when conditioned on θ, are identical to those of their discrete-time counterparts (again with the identification $T=n$). Thus since the likelihood-ratio expressions are also identical, the performance expressions of Example III.B.5 hold for the continuous-time problem as well.

We see from this example that as with the coherent detection problems, the continuous-time detection problem for signals with random parameters follows closely the analogous problem for discrete-time observations once the relevant likelihood ratios have been determined. We return to the model of (VI.C.64) in connection with the estimation of signal parameters in Chapter VII.

VI.D The Detection Of Random Signals In Gaussian Noise

In Section VI.C we have considered problems involving signals that either are completely known or are known within a set of unknown parameters. However, as noted in Chapter III, there are a number of applications in which signals of interest are best modeled as being purely random processes. This type of model is useful in applications, such as sonar and radio astronomy, in which turbulence of the propagation medium diffuses any parametric structure that the signals of interest may have had originally. In this section we consider the problem of detecting random signals in noise when the noise is Gaussian.

In particular, we consider the detection of a random signal $\{S_t\,;0\leqslant t\leqslant T\}$ in additive white Gaussian noise with spectral height $N_0/2$. As before in dealing with white noise, we integrate the observations first to obtain the model

$$H_0\colon X_t = W_t\,,\quad 0\leqslant t\leqslant T$$

versus

$$H_1\colon X_t = \int_0^t S_u\,du + W_t\,,\quad 0\leqslant t\leqslant T\,, \tag{VI.D.1}$$

where $\{W_t\,;0\leqslant t\leqslant T\}$ is a Wiener process and $\int_0^t S_u\,du$ denotes mean-square integration of $\{S_u\,;0\leqslant t\leqslant T\}$. As we shall see, the restriction to Wiener noise is not severe since "whitening" results can be used to convert many Gaussian noises linearly to the Wiener case. We consider principally the case in which $\{S_t\,;0\leqslant t\leqslant T\}$ is also a Gaussian random process; however, the situation in which $\{S_t\,;0\leqslant t\leqslant T\}$ is not Gaussian is also considered briefly.

VI.D.1 Preliminary Results on Wiener Processes

Before considering the detection problem of interest, we first give some properties of Wiener processes that will be useful in solving this problem. Throughout our discussion $\{W_t\,;0\leqslant t\leqslant T\}$ will denote a Wiener process with parameter $N_0/2$; i.e., $\{W_t\,;0\leqslant t\leqslant T\}$ is a zero-mean Gaussian process with $E\{W_t W_u\}=(N_0/2)\min\{t\,,u\}$. Some relevant properties of Wiener processes are summarized in the following.

Proposition VI.D.1: Properties of the Wiener Process

$\{W_t; 0 \leqslant t \leqslant T\}$ has the following properties.

(i) $W_0=0$ with probability 1.

(ii) $E\{(W_t-W_s)(W_u-W_v)\}=(N_0/2)l\{(u,v)\cap(t,s)\}$ for all $t>s\geqslant 0$ and $u>v\geqslant 0$, where $l\{(a,b)\}=|b-a|$.

(iii) $\int_0^T f_t\,dW_t$ exists as a m.s. integral if and only if f is square-integrable in $[0,T]$.

(iv) If $\int_0^T f_t\,dW_t$ exists, then $E\{\int_0^T f_t\,dW_t\}=0$.

(v) If f and g are square-integrable on $[0,T]$, then

$$E\{\int_0^T f_t\,dW_t \int_0^T g_t\,dW_t\} = \int_0^T f_t g_t\,dt\,.$$

(vi) If $\int_0^T f_t\,dW_t$ exists, then

$$E\{W_t \int_0^T f_u\,dW_u\} = (N_0/2)\int_0^t f_u\,du$$

for $0 \leqslant t \leqslant T$.

Proof: Properties (i) and (ii) are simple consequences of the autocovariance structure of W_0^T. Property (iii) follows from property (ii) and the Cauchy criterion for mean-square convergence [see (VI.C.17)]. Properties (iv) through (vi) are consequences of the second-order statistics of W_0^T and the properties allowing interchange of expectation and mean-square limits discussed in Chapter V. We will prove property (v), leaving proof of the others as an exercise.

To prove property (v) we first note that if $\{X_n\}_{n=1}^{\infty}$ and $\{Y_n\}_{n=1}^{\infty}$ are second-order sequences converging in mean-square to X and Y, respectively, then $E\{X_nY_n\}\to E\{XY\}$. Recall that $\int_0^T f_t\,dW_t$ is the mean-square limit of sums of the form $\sum_{i=1}^n f_{\xi_i^{(n)}}[W_{t_1^{(n)}}-W_{t_{i-1}^{(n)}}]$ where $0=t_0^{(n)}<t_0^{(n)}<\ldots<t_1^{(n)}=T$ and $\xi_i^{(n)}\epsilon[t_{t-1}^{(n)},t_i^{(n)}]$, and where the limit is taken as $\Delta_n \triangleq \max_{1\leqslant i\leqslant n}|t_i^{(n)}-t_{i-1}^{(n)}|$ approaches zero. Also, $\int_0^T g_t\,dW_t$ can be written as the same limit with $f_{\xi_i^{(n)}}$ replaced by $g_{\xi_i^{(n)}}$. Thus

$$E\{\int_0^T f_t\,dW_t \int_0^T g_t\,dW_t\}$$

$$= \lim_{\Delta_n \to 0} E\left[\sum_{i=1}^{n} f_{\xi_i^{(n)}}[W_{t_i^{(n)}} - W_{t_{i-1}^{(n)}}] \sum_{j=1}^{n} g_{\xi_j^{(n)}}[W_{t_j^{(n)}} - W_{t_{j-1}^{(n)}}]\right] \quad \text{(VI.D.2)}$$

$$= \lim_{\Delta_n \to 0} \sum_{i=1}^{n} \sum_{j=1}^{n} f_{\xi_j^{(n)}} g_{\xi_i^{(n)}} E\{[W_{t_i^{(n)}} - W_{t_{i-1}^{(n)}}][W_{t_j^{(n)}} - W_{t_{j-1}^{(n)}}]\}.$$

From property (ii) we have that

$$E\{[W_{t_i^{(n)}} - W_{t_{i-1}^{(n)}}]\,[W_{t_j^{(n)}} - W_{t_{j-1}^{(n)}}]\}$$

$$= \begin{cases} \frac{N_0}{2}[t_i^{(n)} - t_{i-1}^{(n)}] & \text{if } i = j \\ 0 & \text{if } i \neq j\,. \end{cases} \quad \text{(VI.D.3)}$$

Substituting (VI.D.3) into (VI.D.2), we have

$$E\{\int_0^T f_t\,dW_t \int_0^T g_t\,dW_t\} = \lim_{\Delta_n \to 0} \sum_{i=1}^{n} f_{\xi_i^{(n)}} g_{\xi_i^{(n)}} [t_i^{(n)} - t_{i-1}^{(n)}]$$

$$\text{(VI.D.4)}$$

$$= \int_0^t f_t\, g_t\, dt\,,$$

which is the desired result.

□

The properties given in Proposition VI.D.1 allow for a useful representation of the Wiener process, which we give in Proposition VI.D.2 after the following definition.

A set $\{\phi_k\}_{k=1}^{\infty}$ of orthonormal functions on $[0, T]$ is said to be *complete* if every square-integrable function f on $[0, T]$ has the representation

$$f_t = \sum_{k=1}^{\infty} \hat{f}_k \phi_k(t), \quad 0 \leqslant t \leqslant T, \qquad \text{(VI.D.5)}$$

where $\hat{f}_k \triangleq \int_0^T f_t \phi_k(t) dt$, $k = 1, 2, \ldots$, and where the convergence in (VI.D.5) is in the integrated-square sense,

$$\lim_{n \to \infty} \int_0^T \left| f_t - \sum_{k=1}^{n} \hat{f}_k \phi_k(t) \right|^2 dt = 0. \qquad \text{(VI.D.6)}$$

Proposition VI.D.2: Representation of the Wiener Process

Suppose that $\{\phi_k\}_{k=1}^{\infty}$ is a complete set of orthonormal functions on $[0, T]$. Then $\{W_t; 0 \leqslant t \leqslant T\}$ has the representation

$$W_t = \sum_{k=1}^{\infty} \hat{W}_k \int_0^t \phi_k(u) du, \quad 0 \leqslant t \leqslant T, \qquad \text{(VI.D.7)}$$

where $\hat{W}_k \triangleq \int_0^T \phi_k(t) dW_t$, $k = 1, 2, \ldots$, and where the sum is a mean-square sum.

Proof: Choose $t \in [0, T]$. We have

$$E\{[W_t - \sum_{k=1}^{n} \hat{W}_k \int_0^t \phi_k(u) du]^2\}$$

$$= E\{W_t^2\} - 2 \sum_{k=1}^{n} E\{W_t \hat{W}_k\} \int_0^t \phi_k(u) du \qquad \text{(VI.D.8)}$$

$$+ \sum_{k=1}^{n} \sum_{l=1}^{n} E\{\hat{W}_k \hat{W}_l\} \int_0^t \phi_k(u) du \int_0^t \phi_l(u) du.$$

Note that $E\{W_t^2\} = (N_0/2)t$. Also, from Proposition VI.D.1 we have

$$E\{W_t \hat{W}_k\} = \frac{N_0}{2} \int_0^t \phi_k(u)du$$

and

$$E\{\hat{W}_k \hat{W}_l\} = \frac{N_0}{2} \int_0^T \phi_k(t)\phi_l(t)dt .$$

Applying orthonormality of $\{\phi_k\}_{k=1}^{\infty}$, (VI.D.8) thus becomes

$$\frac{N_0}{2}\left[t - \sum_{k=1}^{n}\left(\int_0^t \phi_k(u)du\right)^2\right].$$

Define the function $1^{(t)}$ by

$$1_u^{(t)} = \begin{cases} 1 \text{ if } 0 \leqslant u \leqslant t \\ 0 \text{ if } t < u \leqslant T , \end{cases} \tag{VI.D.9}$$

and define $\hat{1}_k^{(t)} \triangleq \int_0^T 1_u^{(t)} \phi_k(u)du \equiv \int_0^t \phi_k(u)du$, $k=1, 2, \ldots$. Since $\int_0^T [1_u^{(t)}]^2 du = t < \infty$, the completeness of $\{\phi_k\}$ implies that

$$\lim_{n\to\infty} \int_0^T \left[1_u^{(t)} - \sum_{k=1}^{n} \hat{1}_k^{(t)} \phi_k(u)\right]^2 du = 0. \tag{VI.D.10}$$

We have straightforwardly that

$$\begin{aligned} \int_0^T \left[1_u^{(t)} - \sum_{k=1}^{n} \hat{1}_k^{(t)} \phi_k(u)\right]^2 du &= t - \sum_{k=1}^{n} [\hat{1}_k^{(t)}]^2 \\ &\equiv t - \sum_{k=1}^{n} \left(\int_0^t \phi_k(u)du\right)^2 . \end{aligned} \tag{VI.D.11}$$

Comparing (VI.D.9) with (VI.D.11) and applying (VI.D.10), we have $\sum_{k=1}^{n}\hat{W}_k\int_0^t\phi_k(u)du \rightarrow W_t$ (m.s.), as was to be shown.

□

Proposition VI.D.2 gives a representation for the Wiener process in terms of any complete set of orthonormal functions. This result will be used below in deriving the likelihood ratio for detecting Gaussian signals in additive white noise. However, we first give the following interesting result which is a consequence of Proposition VI.D.2.

Proposition VI.D.3: Conversion of Gaussian Processes to the Wiener Process

Suppose that $\{Y_t; 0\leqslant t\leqslant T\}$ is a second-order zero-mean Gaussian random process with eigenvalues $\{\lambda_k\}_{k=1}^{\infty}$ and corresponding orthonormal eigenfunctions $\{\psi_k\}_{k=1}^{\infty}$. Assume that $\{\psi_k\}_{k=1}^{\infty}$ is complete and $\lambda_k>0, k=1,2,\cdots$. Then the process

$$Z_t \triangleq \sum_{k=1}^{\infty}(\hat{Y}_k/\lambda_k^{1/2})\int_0^t\psi_k(u)du, \quad 0\leqslant t\leqslant T, \qquad \text{(VI.D.12)}$$

with $\hat{Y}_k=\int_0^T Y_t\psi_k(t)dt$, $k=1,2,\ldots$, is a unit (N_0=2) Wiener process.

Proof: The sequence of coefficients $\{\hat{W}_k\}_{k=1}^{\infty}$ defined in Proposition VI.D.2 is easily seen to be a sequence of i.i.d. $N(0, N_0/2)$ random variables. As we have shown in Section VI.B, $\{\hat{Y}_k\}_{k=1}^{\infty}$ is a sequence of independent $N(0,\lambda_k)$ random variables. This implies that $\{\hat{Y}_k/\lambda_k^{1/2}\}_{k=1}^{\infty}$ is a sequence of i.i.d. $N(0,1)$ random variables. Thus, comparing (VI.D.12) with (VI.D.7), we see that the sum of (VI.D.12) must be statistically identical to that of (VI.D.7) with $N_0/2$=1; i.e., $\{Z_t; 0\leqslant t\leqslant T\}$ of (VI.D.12) must be a unit Wiener process.

□

Comments: Proposition VI.D.3 is a "whitening" result; i.e., it shows how to obtain a Wiener process from an arbitrary Gaussian process satisfying the two given conditions. This Wiener process is in fact equivalent to the original process $\{Y_t ; 0 \leqslant t \leqslant T\}$ since $\hat{Y}_k$ can be obtained as $\hat{Y}_k = \lambda_k^{1/2} \int_0^T \phi_k(t) dZ_t$, $k=1,2,\ldots$, which then gives Y_t via its Karhunen-Loéve representation. The required conditions, that $\{\psi_k\}_{k=1}^{\infty}$ be complete and $\lambda_k > 0$, $k=1,2,\ldots$, essentially assume that the process $\{Y_t ; 0 \leqslant t \leqslant T\}$ is rich enough to "fill up" the space of square-integrable functions. Note that $\{Z_t ; 0 \leqslant t \leqslant T\}$ is obtained linearly from $\{Y_t ; 0 \leqslant t \leqslant T\}$. Thus the results of the preceding section, which have been derived for signals in additive Wiener processes, can in principle be applied to signals in any additive Gaussian noise provided that the transformation of Proposition VI.D.3 is allowable.

VI.D.2 The Detection of Gaussian Signals in White Noise

We now turn to the problem of detecting a Gaussian signal $\{S_t ; 0 \leqslant t \leqslant T\}$ in the model (VI.D.1). We assume throughout that $\{S_t ; 0 \leqslant t \leqslant T\}$ is a mean-square continuous random process, independent of $\{W_t ; 0 \leqslant t \leqslant T\}$, whose autocovariance function C_S has eigenvalues $\{\lambda_k\}_{k=1}^{\infty}$ and corresponding orthonormal eigenfunctions $\{\psi_k\}_{k=1}^{\infty}$. We assume also that $\{\psi_k\}_{k=1}^{\infty}$ is complete. This assumption is made for simplicity in the analysis, and all of the results derived in the following paragraphs apply even if $\{\psi_k\}_{k=1}^{\infty}$ is not complete.

Consider the mean-square sum

$$\sum_{k=1}^{\infty} \hat{X}_k \int_0^t \psi_k(u) du , \tag{VI.D.13}$$

where $\hat{X}_k \triangleq \int_0^T \psi_k(t) dX_t$, $k=1,2,\ldots$. Under H_0, where $X_t = W_t$, Proposition VI.D.2 implies that this series equals X_t. Under H_1, we note that

$$\hat{X}_k = \int_0^T \psi_k(t)d\int_0^t S_u\,du + \int_0^T \psi_k(t)dW_t$$

$$= \int_0^T \psi_k(t)S_t\,dt + \int_0^t \psi_k(t)dW_t \qquad \text{(VI.D.14)}$$

$$= \hat{S}_k + \hat{W}_k,$$

where $\hat{W}_k$ is as in Proposition VI.D.2 and where $\hat{S}_k$ is the coefficient of ψ_k in the Karhunen-Loéve expansion of $\{S_t; 0 \leqslant t \leqslant T\}$. Since the convergence in the Karhunen-Loéve expansion is uniform in t we can write

$$\sum_{k=1}^{\infty} \hat{s}_k \int_0^t \phi_k(u)du = \int_0^t \sum_{k=1}^{\infty} \hat{S}_k \psi_k(u)du = \int_0^t S_u\,du. \qquad \text{(VI.D.15)}$$

Thus (VI.D.13) is a valid representation for X_t under H_1 as well.

Equation (VI.D.13) reduces the continuous-time observation $\{X_t; 0 \leqslant t \leqslant T\}$ to the equivalent discrete-time observation $\{\hat{X}_k\}_{k=1}^{\infty}$. It is straightforward to show that this sequence is a sequence of independent Gaussian random variables under either hypothesis with $\hat{X}_k \sim N(0, N_0/2)$ under H_0 and $\hat{X}_k \sim N(0, \lambda_k + N_0/2)$ under H_1. Thus (VI.D.1) is reduced for Gaussian signals to the equivalent hypothesis pair

$$H_0: \hat{X}_k \sim N\left(0, \frac{N_0}{2}\right), \quad k = 1, 2, \ldots$$

versus (VI.D.16)

$$H_1: \hat{X}_k \sim N\left(0, \lambda_k + \frac{N_0}{2}\right), \quad k = 1, 2, \ldots,$$

to which we can apply Grenander's theorem to investigate the likelihood ratio and singularity issues.

The following proposition results.

Proposition VI.D.4: The Likelihood Ratio for Gaussian Signals in White Noise

Consider the hypothesis testing problem of (VI.C.1) with $\{S_t\,; 0 \leqslant t \leqslant T\}$ Gaussian. Then $P_0 \equiv P_1$ and

$$\frac{dP_1}{dP_0}(X_0^T) = \left[\prod_{k=1}^{\infty} \frac{N_0/2}{\lambda_k + N_0/2}\right] \times \exp\left[\frac{1}{N_0}\sum_{k=1}^{\infty} \lambda_k \hat{X}_k^2 / \left(\lambda_k + \frac{N_0}{2}\right)\right], \tag{VI.D.17}$$

where the sum is a mean-square sum.

Proof: This Proposition follows from Grenander's theorem (Proposition IV.B.2). We note that the problem of testing (VI.D.16) on the basis of the first n observables, $\hat{X}_1, \ldots, \hat{X}_n$, is a particular case of the problem of testing for a Gaussian signal vector in an additive noise vector discussed in Chapter III [see (III.B.85)]. From this, the logarithm of the likelihood ratio for (VI.D.16) based on $\hat{X}_1, \hat{X}_2, \ldots, \hat{X}_n$ is given straightforwardly by

$$\log f_n(X_0^T) = -\sum_{k=1}^{n} \log(1 + 2\lambda_k / N_0) + \frac{1}{N_0}\sum_{k=1}^{n} \lambda_k \hat{X}_k^2 / \left(\lambda_k + \frac{N_0}{2}\right). \tag{VI.D.18}$$

Using the inequality $0 \leqslant \log(1+x) \leqslant x$ for $x \geqslant 0$ and the fact that $\lambda_k \geqslant 0$ for all k, we have

$$0 \leqslant \sum_{k=1}^{\infty} \log(1 + 2\lambda_k / N_0) \leqslant \frac{2}{N_0}\sum_{k=1}^{\infty} \lambda_k < \infty. \tag{VI.D.19}$$

Thus to investigate the convergence in probability of $f_n(X_0^T)$, we need only consider the second term of (VI.D.18). By Cauchy's criterion [see (VI.C.17)], this term converges in mean-square under H_1 if

$$\lim_{n\to\infty} \sup_{m \geqslant n} E_1\left[\left(\sum_{k=n+1}^{m} \lambda_k \hat{X}_k^2 \Big/ \left(\lambda_k + \frac{N_0}{2}\right)\right)^2\right] = 0. \quad \text{(VI.D.20)}$$

We have

$$E_1\left[\left(\sum_{k=n+1}^{m} \lambda_k \hat{X}_k^2 \Big/ \left(\lambda_k + \frac{N_0}{2}\right)\right)^2\right] = Var_1\left(\sum_{k=n+1}^{m} \lambda_k \hat{X}_k^2 \Big/ \left(\lambda_k + \frac{N_0}{2}\right)\right)$$

$$+ \left[E_1\left(\sum_{k=n+1}^{m} \lambda_k \hat{X}_k^2 \Big/ \left(\lambda_k + \frac{N_0}{2}\right)\right)\right]^2$$

$$= \sum_{k=n+1}^{m} \lambda_k^2 Var_1(\hat{X}_k^2) \Big/ \left(\lambda_k + \frac{N_0}{2}\right)^2$$

$$+ \left(\sum_{k=n+1}^{m} \lambda_k E\{\hat{X}_k^2\} \Big/ \left(\lambda_k + \frac{N_0}{2}\right)\right)^2, \quad \text{(VI.D.21)}$$

where we have used the independence of $\hat{X}_{n+1}, \ldots, \hat{X}_m$ in the second equality. Note that under $H_1, X_k \sim N(0, \lambda_k + N_0/2)$, which implies that $Var_1(\hat{X}_k^2) = 2(\lambda_k + N_0/2)^2$ and $E_1\{\hat{X}_k^2\} = \lambda_k + N_0/2$. The quantity on the left-hand side of (VI.D.20) then becomes

$$\lim_{n\to\infty} \sup_{m\geq n} \left[2 \sum_{k=n+1}^{m} \lambda_k^2 + \left(\sum_{k=n+1}^{m} \lambda_k \right)^2 \right]$$

$$= \lim_{n\to\infty} \left[2 \sum_{k=n+1+1}^{\infty} \lambda_k^2 \left(\sum_{k=n+1}^{\infty} \lambda_k \right)^2 \right]. \tag{VI.D.22}$$

The fact that $\sum_{k=1}^{\infty}\lambda_k < \infty$ implies that $\sum_{k=1}^{\infty}\lambda_k^2 < \infty$, and these together imply that the limit in (VI.D.22) is zero.

Thus $\log f_n(X_0^T)$ converges in mean-square under H_1, from which it follows that $\log f_n(X_0^T)$, and hence $f_n(X_0^T)$, converges in probability under H_1. This with Grenander's theorem implies that $P_1 << P_0$ and that $dP_1/dP_0(X_0^T)$ is given by (VI.D.17). To complete the proof we must show that $P_0 << P_1$. This can be accomplished by reversing the role of H_0 and H_1 in Grenander's theorem and repeating the above.

□

The observation-dependent term in the likelihood ratio (VI.D.17) is

$$\sum_{k=1}^{\infty} \lambda_k (\hat{X}_k)^2 / \left(\lambda_k + \frac{N_0}{2} \right) \triangleq T_Q . \tag{VI.D.23}$$

Note that since $\hat{X}_k = \int_0^T \psi_k(t) dX_t$, we can rewrite T_Q as

$$T_Q = \int_0^T \int_0^T Q(t,u) dX_t \, dX_u , \tag{VI.D.24}$$

where

$$Q(t,u) \triangleq \sum_{k=1}^{\infty} \left[\frac{\lambda_k}{\lambda_k + \frac{N_0}{2}} \right] \psi_k(t)\psi_k(u), \quad 0 \leq t, u \leq T . \tag{VI.D.25}$$

Returning to the heuristic model "$dX_t = Y_t dt$" with Y_t

representing signal plus white noise, the test statistic T_Q becomes

$$T_Q = \int_0^T \int_0^T Q(t,u) Y_t Y_u \, dt du, \tag{VI.D.26}$$

which is a quadratic form in Y_0^T. Thus, as in the case of vector observations [see (III.B.86)], the optimum detector structure for a Gaussian signal in additive white noise is quadratic in the observations.

The analogy between the vector and waveform cases is further seen by noting that the function Q of (VI.D.25) is the solution to the integral equation

$$C_S(t,u) = \int_0^T Q(t,s) C_S(s,u) ds + \frac{N_0}{2} Q(t,u), \quad 0 \leq t, u \leq T, \tag{VI.D.27}$$

while the matrix $\mathbf{Q}$ of the quadratic statistic $(\underline{y}^T \mathbf{Q} \underline{y})$ for vector observations solves $\mathbf{\Sigma}_S = \mathbf{Q} \mathbf{\Sigma}_S + \sigma^2 \mathbf{Q}$, where σ^2 is the variance of the white-noise vector in (III.B.85). Writing this matrix equation explicitly in terms of the elements of $\mathbf{\Sigma}_S$ and $\mathbf{Q}$ gives

$$(\mathbf{\Sigma}_S)_{k,l} = \sum_{j=1}^{m} \mathbf{Q}_{k,j} (\mathbf{\Sigma}_S)_{j,l} + \sigma^2 \mathbf{Q}_{k,l}, \quad 1 \leq k, l \leq n, \tag{VI.D.28}$$

which is the discrete-time version of (VI.D.27).

It is interesting to note that (VI.D.27) is a continuous-time version of the Wiener-Hopf equation (V.C.18) for estimating S_t from Y_0^T. In particular, $\tilde{s}_t \triangleq \int_0^T Q(t,u) Y_u du$ [or, more properly, $\int_0^T Q(t,u) dX_u$] can be shown to be the minimum-mean-squared-error linear estimate of S_t from the observation under the model of H_1.[†] Thus T_Q can be interpreted as an *estimator-correlator*,

$$T_Q = \int_0^T \tilde{S}_t Y_t \, dt \,, \qquad \text{(VI.D.29)}$$

which correlates the observed waveform with an estimated version of the signal. Note that $\tilde{S}_t$ is not a causal estimator, so that (VI.D.29) is not a preferred implementation of the detector. However, as we shall see in the following subsection, the likelihood ratio (VI.D.17) also has the representation [‡]

$$\frac{dP_1}{dP_0}(X_0^T) = \exp\left[\frac{2}{N_0}\int_0^T \hat{S}_t \, dX_t - \frac{1}{N_0}\int_0^T (\hat{S}_t)^2 dt\right], \qquad \text{(VI.D.30)}$$

where $\hat{S}_t$ is the causal MMSE estimator of S_t from signal plus noise. Equation (VI.D.30) is quite interesting in that it is exactly the form of the likelihood ratio for detecting known signals in white noise given by the Cameron-Martin formula (VI.C.54) with the known signal $\{S_t\,; 0 \leqslant t \leqslant T\}$ replaced by the causal estimator $\{\hat{S}_t\,; 0 \leqslant t \leqslant T\}$ of the random signal $\{S_t\,; 0 \leqslant t \leqslant T\}$.

Error probability expressions for likelihood ratio tests in the model of (VI.D.1) can be computed only for some very special cases. This is not surprising in view of the situation for the analogous discrete-time problem. Performance must thus be evaluated by using bounds or approximations. [See, e.g., Mazo and Salz (1965)]. Performance of quadratic detectors is sometimes assessed in terms of the so-called *generalized signal-to-noise ratio* or *deflection* criterion, defined by

$$\frac{[E_1\{T_Q(X_0^T)\} - E_0\{T_Q(X_0^T)\}]^2}{Var_0[T_Q(X_0^T)]}. \qquad \text{(VI.D.31)}$$

---

† This estimation problem will be discussed in Chapter VII.

‡ The validity of the representation (VI.D.30) requires correct interpretation of the integral $\int_0^T \hat{S}_t \, dX_t$, which we have yet to define. This point is discussed in section VI.D.3.

Although this quantity is not directly related to error probability, it is used often because of its tractability to measure the effectiveness of the quadratic statistic in separating the two hypotheses [see, e.g., Baker (1969)].

It is interesting to comment on the question of singularity in the problem of detecting Gaussian signals in Gaussian noise. We have seen that for the deterministic signal problem we either have the relationship of equivalence or that of singularity between P_0 and P_1, with no intermediate alternatives. We also have seen that for detecting a mean-square continuous signal in white noise, we always have equivalence between P_0 and P_1. It is no accident that we only arrive at $P_0 \equiv P_1$ or $P_0 \perp P_1$ in these problems. In particular it happens that if P_0 and P_1 are any two *Gaussian measures* (i.e., measures under which the observations are Gaussian), we always have $P_0 \equiv P_1$ or $P_0 \perp P_1$, with no other possibility. This dichotomy was first published independently by Feldman (1958) and Hajek (1958), and there are many interesting results giving conditions under which one of the two alternatives is assured. A discussion of these can be found in Grenander (1981).

One interesting example of such a result deals with observation processes that have rational spectra. A continuous-time random process is said to have a *rational spectrum* if its autocovariance function can be written as

$$C(t,s) = \frac{1}{2\pi} \int_{-\infty}^{\infty} e^{i\omega(t-s)} \phi(\omega) d\omega, \qquad \text{(VI.D.32)}$$

where $\phi(\omega)$ is a function of the form

$$\phi(\omega) = \frac{\sum_{k=1}^{m} a_k \omega^{2k}}{\sum_{k=1}^{n} b_k \omega^{2k}}, \quad -\infty < \omega < \infty, \qquad \text{(VI.D.33)}$$

where the a_k 's and b_k 's are constants. A dichotomy result for such processes is the following result due in part to Slepian (1958), Feldman (1958), and others. If Y_0^T is a Gaussian process with rational spectra ϕ_0 and ϕ_1 under H_0 and H_1, respectively, then $P_0 \equiv P_1$ if and only if

$$\lim_{\omega \to \infty} [\phi_1(\omega)/\phi_0(\omega)] = 1. \qquad \text{(VI.D.34)}$$

Suppose, for example, that we have the model

$$\begin{aligned} &H_0: Y_t = N_t, \quad 0 \leqslant t \leqslant T \\ \text{versus} \\ &H_1: Y_t = S_t + N_t, \quad 0 \leqslant t \leqslant T \end{aligned} \qquad \text{(VI.D.35)}$$

where $\{S_t; 0 \leqslant t \leqslant T\}$ and $\{N_t; 0 \leqslant t \leqslant T\}$ are independent Gaussian signal and noise processes with rational spectra ϕ_S and ϕ_N, respectively. This fits the situation above with $\phi_0 = \phi_N$ and $\phi_1 = \phi_N + \phi_S$, so that the necessary and sufficient condition for equivalence is

$$\lim_{\omega \to \infty} \phi_S(\omega)/\phi_N(\omega) = 0. \qquad \text{(VI.D.36)}$$

A heuristic way of interpreting condition (VI.D.36) is that the class of signal processes for which (VI.D.35) is a nonsingular problem is the class of those processes that cannot change as quickly as the noise process. That such signals provide a nonsingular detection problem is not surprising. It is also not surprising that signal processes that can change *faster* than the noise can [in which case $\phi_S(\omega)/\phi_N(\omega) \to \infty$] would provide a singular problem. However, what is perhaps surprising is that a signal having the same spectral shape as that of the noise [in which case $0 < \lim_{\omega \to \infty} \phi_S(\omega)/\phi_N(\omega) < \infty$] can be detected perfectly (i.e., $P_0 \perp P_1$) on a finite time interval.

A further interesting example of a Gaussian dichotomy result is due to Shepp (1966), which deals with the equivalence and singularity of measures relative to Wiener measures. In the context a signal detection, this result treats the following problem:

$$H_0: X_t = W_t, \quad 0 \leqslant t \leqslant T$$

$$H_1: X_t = \int_0^t S_u \, du + W_t, \quad 0 \leqslant t \leqslant T,$$

where $\{W_t; 0 \leqslant t \leqslant T\}$ is a Wiener process and $\{S_t; 0 \leqslant t \leqslant T\}$ is a Gaussian signal with mean $E\{S_t\}$ and covariance $C_S(t, u)$. Shepp's result asserts that $P_0 \equiv P_1$ if and only if $\int_0^T (E\{S_t\})^2 dt < \infty$ and $\int_0^T \int_0^T C_S^2(t, u) dt du < \infty$; otherwise, $P_0 \perp P_1$. When $P_0 \equiv P_1$, the likelihood ratio is

$$\frac{dP_1}{dP_0}(X_0^T) = \left[\prod_{k=1}^{\infty} (1+2\lambda_k / N_0)^{-1}\right] \exp\{\frac{2}{N_0} g(X_0^T)\},$$

where

$$g(X_0^T) \triangleq \int_0^T E\{S_t\} dX_t - \tfrac{1}{2} \int_0^T [E\{S_t\}]^2 dt$$

$$+ \tfrac{1}{2} \int_0^T \int_0^T Q(t, u) dX_t dX_u,$$

with Q the solution to (VI.D.27). Note that this expression reduces to the Cameron-Martin formula when $C_S \equiv 0$, and the formula of Proposition VI.D.4 when $E\{S_t\} \equiv 0$. Note further that $\int_0^T \int_0^T C_S^2(t, u) dt du$ is always finite for C_S continuous and $\int_0^T E^2\{S_t\} dt < \infty$ for $E\{S_t\}$ of bounded variation, so Shepp's result generalizes each of the results derived previously for detection in Gaussian white noise.

VI.D.3 The Estimator-Correlator Representation of the Likelihood Ratio for Stochastic Signals

In Section VI.D.2 we derived the likelihood ratio for the hypothesis pair

$$H_0: X_t = W_t, \quad 0 \leqslant t \leqslant T$$

versus

$$H_1: X_t = W_t + \int_0^t S_u\, du, \quad 0 \leqslant t \leqslant T, \tag{VI.D.37}$$

where $\{W_t; 0 \leqslant t \leqslant T\}$ is a Wiener noise process and $\{S_t; 0 \leqslant t \leqslant T\}$ is a zero-mean mean-square continuous Gaussian signal. This likelihood ratio is given by Proposition VI.D.4 as

$$\log \frac{dP_1}{dP_0}(X_0^T) = \sum_{k=1}^{\infty} \log \frac{N_0/2}{\lambda_k + N_0/2} + \frac{1}{N_0} \sum_{k=1}^{\infty} \lambda_k \hat{X}_k^2 / (\lambda_k + N_0/2), \tag{VI.D.38}$$

where $\hat{X}_k \triangleq \int_0^T \psi_k(t) dX_t$ and where $\{\lambda_k\}$ and $\{\psi_k\}$ are the eigenvalues and orthonormal eigenfunctions of the signal autocovariance function. We noted [see (VI.D.30)] that dP_1/dP_0 for (VI.D.37) also has the interesting representation

$$\frac{dP_1}{dP_0}(X_0^T) = \frac{2}{N_0} \int_0^T \hat{S}_t\, dX_t - \frac{1}{N_0} \int_0^T (\hat{S}_t)^2 dt, \tag{VI.D.39}$$

where $\hat{S}_t$ is the MMSE estimate of S_t from X_0^t under H_1. In this section we derive (VI.D.39) and discuss its generalization to the case of non-Gaussian signals. For simplicity, we take $T = N_0/2 = 1$, the more general case being a simple modification of this case.

Before deriving (VI.D.39), it is necessary to be somewhat more precise about what this formula means. Note that the estimator $\hat{S}_t$ in (VI.D.39) will be a function of the observation X_0^t for each t, so $\{\hat{S}_t; 0 \leqslant t \leqslant 1\}$ is a random process. Although we have defined integrals such as $\int_0^1 h_t\, dX_t$ for deterministic integrands h_t, we have not done so for situations in which the integrand h_t is itself a random process. Thus, to derive (VI.D.39), we must first define what we mean by integrals such as $\int_0^1 \hat{S}_t\, dX_t$.

There are, in fact, several ways of defining $\int_0^1 \hat{S}_t\, dX_t$, which unfortunately are not equivalent. To illustrate the difficulty in defining the integral of one random process with respect to another, we consider the problem of defining $\int_0^1 W_t\, dW_t$, where $\{W_t\,; 0 \leqslant t \leqslant 1\}$ is a (unit) Wiener process. Suppose that we try to use the usual definition

$$\int_0^1 W_t\, dW_t \overset{(m.s.)}{=} \lim_{\Delta_n \to 0} \sum_{i=1}^{n} W_{\xi_1^{(n)}}(W_{t_i^{(n)}} - W_{t_{i-1}^{(n)}}), \tag{VI.D.40}$$

where $0 = t_0^{(n)} < t_1^{(n)} < \ldots < t_n^{(n)} = 1$, $\xi_i^{(n)} \in [t_{i-1}^{(n)}, t_i^{(n)}]$, $i = 1, \ldots, n$ and $\Delta_n = \max_{1 \leqslant i \leqslant n} |t_i^{(n)} - t_{i-1}^{(n)}|$. If this integral exists (i.e., if the limit is the same for all choices of t_i's and ξ_i's), then we should have, for example, that

$$E\{\int_0^1 W_t\, dW_t\} = \lim_{\Delta_n \to 0} \sum_{i=1}^{n} E\{W_{\xi_i^{(n)}}(W_{t_i^{(n)}} - W_{t_{i-1}^{(n)}})\} \tag{VI.D.41}$$

$$= \lim_{\Delta_n \to 0} \sum_{i=1}^{n} (\xi_i^{(n)} - t_{i-1}^{(n)}).$$

It is easily seen that the rightmost quantity in (VI.D.41) can be any number between 0 and 1, depending on how the $\xi_i^{(n)}$'s are chosen within the intervals $[t_{i-1}^{(n)}, t_i^{(n)}]$. For example, choosing $\xi_i^{(n)} = t_{i-1}^{(n)}$ for $i = 1, \ldots, n$ gives $E\{\int_0^1 W_t\, dW_t] = 0$ and choosing $\xi_i^{(n)} = t_i^{(n)}$ for $i = 1, \ldots, n$ gives $E\{\int_0^1 W_t\, dW_t\} = 1$. Thus (VI.D.40) cannot exist in the sense that we would like for it to.

The basic difficulty here is that the integrand and the integrator are (statistically) dependent on one another. For this particular case in which both are the same Wiener process, the degree of dependence of $W_{\xi_i^{(n)}}$ on the increment $(W_{t_i^{(n)}} - W_{t_{i-1}^{(n)}})$ depends critically on the placement of $\xi_i^{(n)}$ within the interval $[t_{i-1}^{(n)}, t_i^{(n)}]$. Thus any useful definition of the integral of one random process with respect to another must be more specific than our previous definition of the integral of a deterministic function with respect to a random process. In particular, we must specify where in the intervals

$[t_{i-1}^{(n)}, t_i^{(n)}]$ the $\xi_i^{(n)}$ are to be placed. How this is done is more or less a matter of convention, and there are advantages and disadvantages associated with any of the possible ways of doing this. However, the most common definition uses the choice $\xi_i^{(n)} = t_{i-1}^{(n)}$. This choice is particularly useful for formulations of signal detection and estimation problems, so we will adopt it here. Thus we will define the integral of a random process $\{Y_t ; 0 \leq t \leq T\}$ with respect to another random process $\{X_t ; 0 \leq t \leq T\}$ as

$$\int_0^T Y_t \, dX_t \overset{(m.s.)}{=} \lim_{\Delta_n \to 0} \sum_{i=1}^{n} Y_{t_{i-1}^{(n)}} (X_{t_i^{(n)}} - X_{t_{i-1}^{(n)}}), \quad \text{(VI.D.42)}$$

where the $t_i^{(n)}$'s and Δ_n are as before. As is usual $\int_0^1 Y_t \, dX_t$ exists only when the limit in (VI.D.42) exists and is unique (with probability 1) for all choices of partitions.

The integral of (VI.D.42) is usually known as the *Ito stochastic integral*, and the definition can be generalized somewhat from (VI.D.42) to enlarge the class of integrable processes [see, e.g., Wong and Hajek (1985)]. However, the definition of (VI.D.42) is sufficient for our purposes. Note that this definition interprets the "increment" dX_t as a *forward increment*, i.e., $dX_t = X_{t+dt} - X_t$.†

The stochastic integral (VI.D.42) obeys the usual linearity property of integrals; i.e., $\int_0^T [\alpha Y_t + \beta Z_t] dX_t = \alpha \int_0^T Y_t \, dX_t + \beta \int_0^T Z_t \, dX_t$ for scalars α and β. However, this integral does not satisfy all the rules of ordinary calculus. This is illustrated by the following result.

Proposition VI.D.5: The Ito Correction Term

Suppose that $\{W_t ; 0 \leq t \leq 1\}$ is a unit Wiener process and the process $\{X_t ; 0 \leq t \leq 1\}$ is defined by

---

† An alternative definition of the stochastic integral is the Stratonovich stochastic integral, which interprets dX_t as a backward increment (i.e., takes $\xi_i^{(n)} = t_i^{(n)}$). The properties of this integral are somewhat different from those of the Ito integral, and each of the two definitions has advantages and disadvantages.

$$X_t = X_0 + \int_0^t \phi_u \, dW_u, \quad 0 \leqslant t \leqslant 1, \qquad \text{(VI.D.43)}$$

where ϕ is a square-integrable function on [0, 1] and X_0 is a random variable. Then $\int_0^1 X_t \, dX_t$ exists and is given by

$$\int_0^1 X_t \, dX_t = ½ (X_1^2 - X_0^2) - ½ \int_0^1 \phi_u^2 du. \qquad \text{(VI.D.44)}$$

Proof: For a partition $t_0^{(n)}, ..., t_n^{(n)}$ of [0, 1] we can write [since $a(b-a) = ½(b^2 - a^2) - ½(b-a)^2$]

$$\sum_{i=0}^{n-1} X_{t_i^{(n)}}(X_{t_{i+1}^{(n)}} - X_{t_i^{(n)}}) = ½ \sum_{i=0}^{n-1} X_{t_{i+1}^{(n)}}^2$$

$$- ½ \sum_{i=0}^{n-1} X_{t_i^{(n)}}^2 - ½ \sum_{i=0}^{n-1} (X_{t_{i+1}^{(n)}} - X_{t_i^{(n)}})^2 \qquad \text{(VI.D.45)}$$

$$= ½ (X_1^2 - X_0^2) - ½ \sum_{i=0}^{n-1} \left[\int_{t_i^{(n)}}^{t_{i+1}^{(n)}} \phi_u \, dW_u \right]^2.$$

Note that the variables $Y_i = \int_{t_i^{(n)}}^{t_{i+1}^{(n)}} \phi_u \, dW_u$, $i = 0, 1, ..., n-1$ are independent $N(0, \int_{t_i^{(n)}}^{t_{i+1}^{(n)}} \phi_u^2 du)$ random variables.

Consider the sequence of random variables

$$Z_n = \sum_{i=0}^{n-1} \left[\int_{t_i^{(n)}}^{t_{i+1}^{(n)}} \phi_u \, dW_u \right]^2, \; n = 1, 2, ... \,. \qquad \text{(VI.D.46)}$$

We have $E\{Z_n\} = \sum_{i=0}^{n-1} \int_{t_i^{(n)}}^{t_{i+1}^{(n)}} \phi_u^2 du = \int_0^1 \phi_u^2 du$, and

$$E[(Z_n - \int_0^1 \phi_u^2 du)^2] = Var(Z_n)$$

(VI.D.47)

$$= \sum_{i=0}^{n-1} Var[(\int_{t_i^{(n)}}^{t_{i+1}^{(n)}} \phi_u \, dW_u)^2].$$

It is straightforward to show that with $X \sim N(0, \sigma^2)$, we have $Var(X^2) = 2\sigma^4$. Thus

$$E[(Z_n - \int_0^1 \phi_u^2 du)^2] = 2\sum_{i=0}^{n-1} (\int_{t_i^{(n)}}^{t_{i+1}^{(n)}} \phi_u^2 du)^2$$

$$\leqslant 2\delta_n \sum_{i=0}^{n-1} \int_{t_i^{(n)}}^{t_{i+1}^{(n)}} \phi_u^2 du = 2\delta_n \int_0^1 \phi_u^2 du,$$

where

$$\delta_n \triangleq \max_{0 \leqslant i \leqslant n-1} \int_{t_i^{(n)}}^{t_{i+1}^{(n)}} \phi_u^2 du. \qquad \text{(VI.D.48)}$$

Combining (VI.D.45) to (VI.D.48), we have

$$\sum_{i=0}^{n-1} X_{t_i^{(n)}}(X_{t_{i+1}^{(n)}} - X_{t_i^{(n)}}) = \tfrac{1}{2}(X_1^2 - X_0^2) - \tfrac{1}{2}\int_0^1 \phi_u^2 du + \epsilon_n,$$

where $E\{\epsilon_n^2\} \leqslant 2\delta_n$. Note that $\Delta_n \to 0$ implies $\delta_n \to 0$. Thus, as $\Delta_n \to 0$, $\sum_{i=0}^{n-1} X_{t_i^{(n)}}(X_{t_{i+1}}^{(n)} - X_{t_i}^{(n)})$ converges in the mean-square sense and its limit is given by (VI.D.44).

□

Note that the formula (VI.D.44) is different from the corresponding formula from ordinary calculus. In particular, if f is a continuous function of bounded variation on [0, 1], we have the usual formula

$$\int_0^1 f_t\, df_t = \tfrac{1}{2}(f_1^2 - f_0^2). \qquad \text{(VI.D.49)}$$

The "correction term" in (VI.D.44) involves the so-called *quadratic variation* of $\{X_t\,; 0 \leqslant t \leqslant 1\}$ given by

$$\int_0^1 (dX_t)^2 = \lim_{\Delta_n \to 0} \sum_{i=0}^{n-1} (X_{t_{i+1}^{(n)}} - X_{t_i^{(n)}})^2 = \int_0^1 \phi_t^2 dt\,. \qquad \text{(VI.D.50)}$$

For a continuous function of bounded variation, the quadratic variation is zero since $(df_t)^2 \sim (dt)^2$, whereas for a random process of the form (VI.D.43), $(dX_t)^2$ behaves like $\phi_t^2 dt$. Further discussion of this phenomenon is contained in Chapter VII. Meanwhile, the result of Proposition VI.D.5 contains the property of stochastic integrals needed for our immediate purposes.

We now return to the likelihood ratio of (VI.D.38). Note from Proposition VI.D.1 that the Radon-Nikodym derivative dP_1/dP_0 is defined in terms of its behavior in integrals with respect to P_0. Thus, to show that (VI.D.39) is a representation for the likelihood ratio, it is only necessary to show that (VI.D.38) and (VI.D.39) are equal with P_0-probability 1. (This will also imply that they are equal with P_1-probability 1 since $P_1 \ll P_0$.)

The variables $\hat{X}_k$ are given by $\hat{X}_k = \int_0^1 \psi_k(t)dX_t$ and under P_0, $\{X_t\,; 0 \leqslant t \leqslant 1\}$ is a Wiener process. Thus, applying Proposition VI.D.5, we can write

$$\hat{X}_k^2 = 2\int_0^1 \Big[\int_0^t \psi_k(u)dX_u\Big] d\Big[\int_0^t \psi_k(u)dX_u\Big] + \int_0^1 \psi_k^2(t)dt$$

$$= 2\int_0^1 \Big[\int_0^t \psi_k(u)dX_u\Big]\psi_k(t)dX_t + \int_0^1 \psi_k^2(t)dt \qquad \text{(VI.D.51)}$$

with P_0-probability 1, since $d\int_0^t \psi_k(u)dW_u = \psi_k(t)dW_t$. We can use (VI.D.51) and the uniformity of the convergence of $\sum_{k=1}^{\infty} [\lambda_k/(\lambda_k+1)]\psi_k(t)\psi_k(u)$ [this follows from Mercer's theorem and the fact that $0 \leqslant \lambda_k/(\lambda_k+1) \leqslant \lambda_k$] to write

$$\log \frac{dP_1}{dP_0}(X_0^1) = \sum_{k=1}^{\infty} \log[1/(1+\lambda_k)]$$

$$+ \int_0^1 [\int_0^t Q(t,u) dX_u] dX_t \qquad \text{(VI.D.52)}$$

$$+ \tfrac{1}{2} \int_0^1 Q(t,t) dt$$

with P_0-probability 1, where, as before in (VI.D.22), Q is defined as

$$Q(t,u) = \sum_{k=1}^{\infty} \frac{\lambda_k}{1+\lambda_k} \psi_k(t) \psi_k(u), \quad 0 \leqslant t, u \leqslant 1.$$

We noted in Section VI.C that Q satisfies the integral equation

$$C_S(t,u) = \int_0^1 Q(t,s) C_S(s,u) ds$$

$$\text{(VI.D.53)}$$

$$+ Q(t,u), \quad 0 \leqslant t, u \leqslant 1,$$

which, as we shall see in Chapter VII, is the Wiener-Hopf equation for estimating S_t from $\int_0^u S_v dv + W_u$, $0 \leqslant u \leqslant 1$, via $\tilde{S}_t = \int_0^1 Q(t,u) dX_u$. This function Q is known as the *Fredholm resolvent* associated with C_S. In general, the estimator $\tilde{S}_t$ is noncausal. The MMSE causal estimator of S_t from $\int_0^u S_v dv + W_u$, $0 \leqslant u \leqslant T$, is given *by*†

$$\hat{S}_t = \int_0^t h(t,u) dX_u, \; 0 \leqslant t \leqslant 1, \qquad \text{(VI.D.54)}$$

† These results are derived in Chapter VII.

where h is the solution to the casual Wiener-Hopf equation

$$C_S(t,u) = \int_0^t h(t,s)C_S(s,u)ds$$

(VI.D.55)

$$+ h(t,u), \quad 0 \leqslant u \leqslant t \leqslant 1.$$

(This equation can be shown to have a unique continuous solution.) As with the Wiener-Kolmogorov problem in Chapter V, h is not simply a truncated version of Q. However, Q can be obtained from h via the relationship (see Exercise 10)

$$Q(t,u) = h(t,u) + h(u,t)$$

(VI.D.56)

$$- \int_0^1 h(s,t)h(s,u)ds, \quad 0 \leqslant t, u \leqslant 1,$$

where we take $h(x,y)$ to be zero if $y > x$.

Substituting (VI.D.56) into (VI.D.52), we have

$$\log \frac{dP_1}{dP_0}(X_0^1) = \sum_{k=1}^{\infty} \log [1/(1+\lambda_k)]$$

$$+ \int_0^1 \int_0^t h(t,u)dX_u]dX_t$$

(VII.D.57)

$$- \int_0^1 [\int_0^t \{\int_0^1 h(s,t)h(s,u)ds\}dX_u]dX_t$$

$$+ \int_0^1 h(t,t)dt - \tfrac{1}{2} \int_0^1 \int_0^1 h^2(s,t)dsdt .$$

Consider the middle term on the right-hand side of (VI.D.57). On interchanging order of integration, using the fact that

$h(s,t)=0$ for $t>s$, and applying Proposition VI.D.1, this term becomes

$$\int_0^1 [\int_0^s \{\int_0^T h(s,u)dX_u\} h(s,t)dX_t]ds =$$

$$\tfrac{1}{2}\int_0^1 [\int_0^s h(s,u)dX_u]^2 ds - \tfrac{1}{2}\int_0^1\int_0^s h^2(s,u)duds. \qquad \text{(VI.D.58)}$$

Combining (VI.D.54), (VI.D.57), and (VI.D.58), we have

$$\log \frac{dP_1}{dP_0}(X_0^1) = \int_0^1 \hat{S}_t\, dX_t - \tfrac{1}{2}\int_0^1 [\hat{S}_t]^2 dt$$

$$+ \left[\int_0^1 h(t,t) - \sum_{k=1}^{\infty} \log(1+\lambda_k)\right]. \qquad \text{(VI.D.59)}$$

The quantity $d(\lambda) \triangleq \Pi_{k=1}^{\infty}(1+\lambda^{-1}\lambda_k)$ is known as the *Fredholm determinant* of C_S, and it can be shown [see, e.g., Kailath (1969)] that

$$\log d(\lambda) = \int_0^1 h_\lambda(t,t)dt, \qquad \text{(VI.D.60)}$$

where h_λ is the solution to the equation

$$C_S(t,u) = \int_0^1 h_\lambda(t,s)C_S(s,u)ds$$

$$+ \lambda^{-1}h_\lambda(t,u), \quad 0 \leqslant t \leqslant 1. \qquad \text{(VI.D.61)}$$

Putting $\lambda=1$ in (VI.D.61), we have that $\Pi_{k=1}^{\infty}(1+\lambda_k)=\int_0^1 h(t,t)dt$, and the likelihood ratio of

(VI.D.59) thus becomes

$$\log \frac{dP_1}{dP_0}(X_0^1) = \int_0^1 \hat{S}_t \, dX_t - 1/2 \int_0^1 [\hat{S}_t]^2 dt \, , \quad \text{(VI.D.62)}$$

as was to be shown.

The likelihood ratio formula (VI.D.62) is interesting in that it says that the likelihood ratio for detecting the Gaussian signal $\{S_t; 0 \leqslant t \leqslant 1\}$ in white noise is the same as that for detecting a known signal in white noise, when the known signal is replaced by the estimate $\{\hat{S}_t; 0 \leqslant t \leqslant 1\}$. The derivation above does not provide much intuition as to why this should be so. However, the following result sheds some light on this.

Proposition VI.D.6: The Innovations Theorem for Gaussian Signals

Consider the observation model under H_1, i.e.,

$$X_t = \int_0^t S_u \, du + W_t \, , \quad 0 \leqslant t \leqslant 1, \quad \text{(VI.D.63)}$$

where $\{S_u; 0 \leqslant u \leqslant 1\}$ is a mean-square continuous Gaussian signal independent of $\{W_t; 0 \leqslant t \leqslant 1\}$. We can write

$$X_t = \int_0^t \hat{S}_u \, du + I_t \, , \, 0 \leqslant t \leqslant 1, \quad \text{(VI.D.64)}$$

where $\{I_t; 0 \leqslant t \leqslant 1\}$ is a (unit) Wiener process with $(I_t - I_s)$ independent of $\{X_u; 0 \leqslant u \leqslant s\}$ for all $1 \geqslant t \geqslant s \geqslant 0$.

Proof: We first note that $I_t = W_t + \int_0^t (S_u - \hat{S}_u) du$. Since both $\{S_t; 0 \leqslant t \leqslant 1\}$ and $\{W_t; 0 \leqslant t \leqslant 1\}$ are zero-mean Gaussian processes, $\{\hat{S}_t; 0 \leqslant t \leqslant 1\}$ is also a zero-mean and Gaussian (it is a linear transformation of $S_0^1 + W_0^1$). Thus $\{I_t; 0 \leqslant t \leqslant 1\}$ is a zero-mean Gaussian process, and we need only show that $E\{I_t I_s\} = \min\{t, s\}$ to show that $\{I_t; 0 \leqslant t \leqslant 1\}$ is a Wiener process.

Consider $E\{(I_t - I_s)^2\}$ for $1 \geqslant t \geqslant s \geqslant 0$. We have

$$E\{(I_t - I_s)^2\} = E\{(W_t - W_s)^2\}$$

$$+ 2\int_s^t E\{(W_t - W_s)(S_u - \hat{S}_u)\}du \qquad \text{(VI.D.65)}$$

$$+ \int_s^t \int_s^t E\{(S_u - \hat{S}_u)(S_v - \hat{S}_v)\}dudv .$$

Since $\{W_t ; 0 \leqslant t \leqslant 1\}$ is a Wiener process, the first term on the right-hand side of (VI.D.65) equals $(t - s)$. Consider the expectation in the center integral on the right-hand side of (VI.D.65). We can write

$$E\{(W_t - W_s)(S_u - \hat{S}_u)\} = E\{(W_t - W_u)(S_u - \hat{S}_u)\}$$

$$+ E\{(W_u - W_s)(S_u - \hat{S}_u)\}$$

$$= E\{(W_t - W_u)\}E\{(S_u - \hat{S}_u)\} \qquad \text{(VI.D.66)}$$

$$+ E\{(W_u - W_s)(S_u - \hat{S}_u)\}$$

$$= E\{(W_u - W_s)(S_u - \hat{S}_u)\},$$

where the second equality follows from the fact that $(W_t - W_u)$ is independent of S_u (by assumption) and of $\hat{S}_u$ [by the fact that $\hat{S}_u$ depends only on S_0^u and W_0^u, both of which are independent of $(W_t - W_u)$], and where the third equality follow from the fact that $E\{W_t - W_u\} = E\{S_u - \hat{S}_u\} = 0$. Note that the integrand in the third term on the right-hand side of (VI.D.65) is symmetric in (u, v), and using this together with the above, we can write

$$+\,2\int_0^t E\{(S_u-\hat{S}_u)(W_u-W_s+\int_s^u (S_v-\hat{S}_v)dv\,]\}du$$

(VI.D.67)

$$=(t-s)+2\int_0^t E\{(S_u-\hat{S}_u)(I_u-I_s)\}du\,.$$

Note that $(I_u - I_s)$ is a linear function of X_0^u for all $u \geqslant s$; thus

$$E\{(S_u-\hat{S}_u)(I_u-I_s)\}=0$$

for all $u \geqslant s$ by the orthogonality principle, and $E\{(I_t-I_s)^2\}=t-s$ for $1 \geqslant t \geqslant s \geqslant 0$.

Now since $(I_t - I_s)=\frac{1}{2}\{(I_t-I_0)^2+(I_s-I_0)^2-(I_t-I_s)^2\}$, we have

$$E\{I_t I_s\}=\tfrac{1}{2}\,[E\{(I_t-I_0)^2\}+E\{(I_s-I_0)^2\}-E\{(I_t-I_s)^2\}]$$

$$=\tfrac{1}{2}\,[t+s-|t-s|]=\min\{t\,,s\},\quad 0\leqslant t\,,s\leqslant 1.$$

Thus $\{I_t\,; 0\leqslant t \leqslant 1\}$ is a Wiener process.

To see that $(I_t - I_s)$ is independent of X_0^s for all $1 \geqslant t \geqslant s \geqslant 0$, we first note that for $1 \geqslant t \geqslant s \geqslant u \geqslant 0$,

$$E\{(I_t-I_s)X_u\}=E\{(W_t-W_s)X_u\}+\int_s^t E\{(S_v-\hat{S}_v)X_u\}dv=0,$$

where we have used the facts that $(W_t - W_s)$ is independent of X_u and that $(S_v - \hat{S}_v)$ is orthogonal to X_u for all $s \leqslant v \leqslant t$. Thus $(I_t - I_s)$ is orthogonal to X_u for all $u \leqslant s$, and so $(I_t - I_s)$ is independent of X_0^s since both processes are Gaussian.

□

The process $\{I_t\,;0\leqslant t\leqslant 1\}$ defined in the proposition above is the *innovations process* for $\{X_t\,;0\leqslant t\leqslant 1\}$. At each time, as with its discrete-time counterpart discussed in Chapter V, it represents that part of the observations that cannot be predicted from the past. To see this heuristically, we note that the change in the observation over an interval $(t\,,t+dt\,)$ is given by

$$(X_{t+dt}-X_t\,)=\int_t^{t+dt}\hat{S}_u\,du\;+(I_{t+dt}-I_t\,).$$

Since $h\,(t\,,u\,)$ is continuous, the process $\{\hat{S}_t\,;0\leqslant t\leqslant 1\}$ has continuous sample paths. Thus as $dt\rightarrow 0$, we can write

$$(X_{t+dt}-X_t\,)\sim\hat{S}_t\,dt\;+(I_{t+dt}-I_t\,). \qquad \text{(VI.D.68)}$$

That is, the change in the observation over $(t\,,t+dt\,)$ consists of a part, $\hat{S}_t\,dt$, completely dependent on the past (X_0^t) and the *innovation*, $(I_{t+dt}-I_t\,)$, that is completely independent of the past.

This representation allows us to give a heuristic, but illuminating derivation of the estimator-correlator form of the likelihood ratio (VI.D.39). In particular, suppose that for $0\leqslant t\leqslant 1$ we let $L_t\,(X_0^1\,)$ denote the likelihood ratio based on observations up to time t only; i.e., $L_t\,(X_0^1\,)$ depends only on X_0^t. Then, for each $dt>0$ with $1/dt$ an integer, we can write

$$\log L_1(X_0^1\,)=\sum_{k=1}^{1/dt}[\,\log L_{kdt}\,(X_0)-\log L_{(k-1)dt}\,(X_0^1\,)], \qquad \text{(VI.D.69)}$$

where we have simply added and subtracted identical terms from $\log L_1(X_0^1\,)$. Now, for infinitesimal dt, having already observed X_0^t, observing X_t^{t+dt} is like observing the process $\{Y_u\,;t\leqslant u\leqslant t+dt\,\}$ satisfying the model

$$H_0: Y_u = (W_{t+u} - W_t), \quad t \leqslant u \leqslant t+dt$$

versus (VI.D.70)

$$H_1: Y_u = \hat{S}_t du + (I_{t+u} - I_t), \quad t \leqslant u \leqslant t+dt,$$

where we have used (VI.D.68). Since, given X_0^t, $\hat{S}_t$ is known and $(W_{t+u} - W_t)$ and $(I_{t+u} - I_t)$ are identical Wiener processes on $t \leqslant u \leqslant t+dt$, the incremental change in log-likelihood ratio obtained by observing X_t^{t+dt} is given by the known-signal-in-white-noise (Cameron-Martin) formula as

$$\log L_{t+dt}(X_0^1) - \log L_t(X_0^1)$$

$$\sim \int_t^{t+dt} \hat{S}_t dX_u - \tfrac{1}{2} \int_t^{t+dt} (\hat{S}_t)^2 du \tag{VI.D.71}$$

$$= \hat{S}_t (X_{t+dt} - X_t) - \tfrac{1}{2} (\hat{S}_t)^2 dt.$$

Inserting this into (VI.D.69), we thus have that

$$\log L_1(X_0^1) \sim \sum_{k=1}^{1/dt} [\hat{S}_{(k-1)dt} [X_{kdt} - X_{(k-1)dt}] - \tfrac{1}{2} \hat{S}^2_{(k-1)dt} dt]$$

(VI.D.72)

$$\underset{(dt \to 0)}{\rightarrow} \int_0^1 \hat{S}_t dX_t - \tfrac{1}{2} \int_0^1 (\hat{S}_t)^2 dt.$$

Although heuristic, the derivation above gives an intuitive feeling for why the estimator-correlator likelihood-ratio formula is valid. Moreover, the reason why the Ito (forward-increment) integral appears in this formula should also be clear from this derivation. In particular, since $\hat{S}_t$ is continuous, the approximation (VI.D.68) can just as well be replaced by

$$(X_{t+dt} - X_t) \sim \hat{S}_{t_o} dt + (I_{t+dt} - I_t) \tag{VI.D.73}$$

for any t_o between t and $t+dt$. However, only by choosing

$t_o = t$ do we have $\hat{S}_{t_o}$ depending completely on X_0^t, and so the formula (VI.D.71) is valid only in this one case. This choice of $t = t_o$ leads to the interpretation of $\int_0^1 \hat{S}_t \, dX_t$ as $\int_0^1 \hat{S}_t (X_{t+dt} - X_t)$, the Ito integral.

The heuristic proof of the estimator-correlator formula given above can be made rigorous, and this method of proof is perhaps more illuminating than the earlier proof, in that it points to the essential reason for the validity of the estimator-correlator formula, namely the Wiener statistics of the innovations process $I_t \triangleq X_t - \int_0^t \hat{S}_u \, du$, $0 \leqslant t \leqslant 1$. Note that this derivation would still be valid regardless of the nature of the signal provided that $\hat{S}_u$ has some continuity property and that, for each s, $\{(I_t - I_s); s \leqslant t \leqslant 1\}$ is a Wiener process independent of X_0^s. Surprisingly, these properties of the innovations are valid for a much broader class of signal models than the Gaussian model treated here.

In particular, consider the hypothesis-testing problem

$$H_0: X_t = W_t, \quad 0 \leqslant t \leqslant 1$$

versus

$$H_1: X_t = \int_0^t S_u \, du + W_t, \quad 0 \leqslant t \leqslant 1, \tag{VI.D.74}$$

where $\{S_t; 0 \leqslant t \leqslant 1\}$ is a signal process (not necessarily Gaussian) satisfying the conditions

$$E\{\int_0^1 |S_t| dt\} < \infty \tag{VI.D.75a}$$

and

$$\int_0^1 S_t^2 dt < \infty \;\; w.p.\, 1, \tag{VI.D.75b}$$

and $\{W_t; 0 \leqslant t \leqslant 1\}$ is a (unit) Wiener process such that for each $s \in [0, 1]$, the process $\{W_t - W_s; s \leqslant t \leqslant 1\}$ is independent of $\{X_0^s, W_0^s\}$. The latter condition is automatically satisfied if the signal is independent of the noise. Under these assumptions, it can be shown that the innovations process

$$I_t \triangleq X_t - \int_0^t \hat{S}_u \, du , \qquad \text{(VI.D.76)}$$

where $\hat{S}_t \triangleq E_1\{S_t | X_0^t\}, 0 \leqslant t \leqslant 1$, is a unit Wiener process whose forward increments $\{(I_t - I_s); s \leqslant t \leqslant 1\}$ are independent of X_0^s for each $s \in [0, 1]$. Moreover, it can also be shown that these assumptions guarantee that $P_1 << P_0$ and that †

$$\log \frac{dP_1}{dP_0}(X_0^1) = \int_0^1 \hat{S}_t \, dX_t - \tfrac{1}{2} \int_0^1 (\hat{S}_t)^2 dt . \qquad \text{(VI.D.77)}$$

Note that we have not assumed that the signal and noise are independent under H_1, but merely that future increments of the noise are independent of the signal. This allows for situations in which S_t depends on the past of the observations, X_0^t, a phenomenon that arises in situations in which S_t may be generated in part from the observations by a feedback mechanism.

As noted above, the representation of (VI.D.77) follows basically from the Wiener property of the innovations process. Thus the innovations result is the most surprising one. The proof given for the Gaussian case with independent signal and noise (Proposition VI.D.6) can easily be modified to show that under the assumptions made on (VI.D.74), $\{I_t ; 0 \leqslant t \leqslant 1\}$ satisfies

$$E\{I_t - I_s | X_0^s\} = 0, \quad 0 \leqslant s \leqslant t \leqslant 1 \qquad \text{(VI.D.78a)}$$

and

$$E\{(I_t - I_s)^2 | X_0^s\} = t - s , \quad 0 \leqslant s \leqslant t \leqslant 1. \qquad \text{(VI.D.78b)}$$

† To obtain this generality it is necessary to generalize the definition of the Ito integral slightly to use a Lebesgue-Stieltjes (rather than Riemann-Stieltjes) integral and to relax the mean-square convergence in its definition to convergence in probability. This issue is discussed, for example, in Wong and Hajek (1985).

A Wiener process has continuous sample paths with probability 1; also, conditions (VI.D.75) assure that $\int_0^t S_u\,du$ and $\int_0^t \hat{S}_u\,du$ have continuous sample paths with probability 1. Thus the innovations process, which is $W_t + \int_0^t S_u\,du - \int_0^t \hat{S}_u\,du$, has continuous sample paths with probability 1. Any process satisfying (VI.D.78a) is said to be a *martingale* with respect to $\{X_0^s, 0 \leqslant s \leqslant 1\}$. It can be shown that any martingale with continuous sample paths that satisfies (VI.D.78b) is a Wiener process with increments $\{(I_t - I_s); s \leqslant t \leqslant 1\}$ independent of X_0^s for each $s \in [0, 1]$. Thus the innovations properly follow from this result.†

Since $E_1\{S_t | X_0^t\}$ is the condition-mean (MMSE) estimator of S_t from X_0^t under H_1, the likelihood ratio formula of (VI.D.77) has the same estimator-correlator interpretation that it has for the Gaussian case. Since the conditions on the signal for which (VI.D.77) is valid are quite weak, this is a very general representation for the likelihood ratio for signals in white noise. This formula was developed under various sets of assumptions on the signal by several researchers, including Duncan (1968, 1970), Kailath (1969, 1971), and Stratonovich and Sosulin (1965). The formula developed by Stratonovich and Sosulin uses the Stratonovich definition of the stochastic integral and was derived for the case of a Gaussian signal by Schweppe (1965). The particular result quoted here is found in Kailath (1969).

This formula can be used to derive likelihood ratios for signals in other than white noise processes provided that there is a white-noise floor. For example, suppose that we have the hypothesis pair

$$\begin{aligned} &H_0: Y_t = N_t, \quad 0 \leqslant t \leqslant 1 \\ &\text{versus} \\ &H_1: Y_t = N_t + S_t, \quad 0 \leqslant t \leqslant 1, \end{aligned} \qquad \text{(VI.D.79)}$$

where $\{N_t; 0 \leqslant t \leqslant 1\}$ is a noise process of the form

† This martingale result is known as the Lévy-Doob theorem [see, e.g., Doob (1953)]. Actually, the validity of the estimator-correlator formula also implies the Wiener property of the innovations under some weak conditions. This is a special case of a result known as Girsanov's theorem [see, e.g., Lipster and Shiryayev (1977)]. Thus the estimator-correlator formula and the properties of the innovations are essentially equivalent statements of the same property.

$$N_t = N_t^w + N_t^c,\ 0 \leqslant t \leqslant 1, \tag{VI.D.80}$$

where $\{N_t^w, 0 \leqslant t \leqslant 1\}$ is a unit white noise and $\{N_t^c; 0 \leqslant t \leqslant 1\}$ is a noise process satisfying the assumptions (VI.D.75). Then we can introduce a third hypothesis, $H: Y_t = N_t^w,\ 0 \leqslant t \leqslant 1$, and use the probability measure, P, corresponding to this hypothesis as a "catalyst" to get dP_1/dP_0 via

$$\frac{dP_1}{dP_0}(X_0^1) = \frac{dP_1}{dP}(X_0^1) / \frac{dP_0}{dP}(X_0^1)$$

$$= \frac{\exp\{\int_0^1 (\hat{S}_t + \hat{N}_t^1) dX_t - \tfrac{1}{2} \int_0^1 (\hat{S}_t + \hat{N}_t^1)^2 dt\}}{\exp\{\int_0^1 \hat{N}_t^0 dX_t - \tfrac{1}{2} \int_0^1 (\hat{N}_t^0)^2 dt\}} \tag{VI.D.81}$$

where $\hat{N}_t^j = E_j\{N_t^c | X_0^t\}$ and $\hat{S}_t = E_1\{S_t | X_0^t]$ with, as usual, $X_t = \int_0^t Y_u\, du,\ 0 \leqslant t \leqslant 1$.

As a final comment, we note that the estimator-correlator form for non-Gaussian signals should be thought of primarily as a *representation* for the likelihood ratio, since it give dP_1/dP_0 in terms of $\{\hat{S}_t, 0 \leqslant t \leqslant 1\}$, which is generally quite difficult to determine for non-Gaussian signals. In other words, this formula converts the detection problem into an estimation problem which is at least as difficult to solve. (Recall that for the Gaussian case, in which the estimator $\hat{S}_t$ is relatively easy to find because of its linearity, the likelihood ratio was easily derivable from the Karhunen-Loéve expansion.) In the following chapter we turn to this problem of estimation of continuous-time signals.

VI.E Exercises

1. Consider the detection problem

$$H_0: Y_t = N_t \quad , \quad 0 \leqslant t \leqslant 1$$

versus

$$H_1: Y_t = N_t \Theta s_t \ , \quad 0 \leqslant t \leqslant 1,$$

where $\{N_t ; 0 \leqslant t \leqslant T\}$ is a zero-mean, mean-square continuous Gaussian process and where $\{s_t ; 0 \leqslant t \leqslant T\}$ is a known signal. The parameter Θ equals +1 and -1 with equal probabilities and independently of the noise. Find the likelihood ratio and an expression for the ROC's.

2. Consider the following hypothesis pair:

$$H_0: Y_t = N_t + kJ_t + E_0 \sin(\omega_0 t), \quad 0 \leqslant t \leqslant T$$

versus

$$H_1: Y_t = N_t + kJ_t + E_0 \cos(\omega_0 t), \quad 0 \leqslant t \leqslant T$$

where E_0 and ω_0 are known positive numbers; $T = 2\pi / \omega_0$; $\{N_t , 0 \leqslant t \leqslant T\}$ is a zero-mean white Gaussian process with spectral height $N_0/2$; and $\{J_t , 0 \leqslant t \leqslant T\}$ is a zero-mean Gaussian random process, independent of $\{N_t , 0 \leqslant t \leqslant T\}$, with autocovariance function

$$C_J(t, s) = \cos(\omega_0 |t - s|), \quad 0 \leqslant t , s \leqslant T .$$

Assume that H_0 and H_1 are equally likely.
(a) Assuming that the parameter k is known, derive the minimum-probability-of-error detector and find the corresponding probability or error.
(b) Suppose that a detector is designed as in part (a) under the assumption that $k=0$, but that the actual value of $k = k_0 > 0$. Find the probability of error resulting from using the $k=0$ detector when $k = k_0 > 0$.

3. Consider the hypothesis testing problem

$$H_0: Y_t = N_t, 0 \leqslant t \leqslant T$$

versus

$$H_1: Y_t = N_t + \sum_{i=0}^{M-1} a_i p(t - iT_c) \cos(\omega_0 t + \Theta), 0 \leqslant t \leqslant T$$

where $\{N_t, 0 \leqslant t \leqslant T\}$ is a white Gaussian noise process with spectral height $N_0/2$; the sequence $a_0, a_1, \ldots, a_{M-1}$ is a known signal sequence; the phase Θ is independent of the noise and is uniformly distributed on $[0, 2\pi]$; the function $p(t)$ is given by

$$p(t) = \begin{cases} 1, & 0 \leqslant t \leqslant T_c \\ 0, & \text{otherwise.} \end{cases}$$

Assume that T_c is an integral number of periods of the sinusoid and that $T = mT_c$.
(a) Find the likelihood ratio for testing between H_0 and H_1. Draw a block diagram of the receiver structure.
(b) Find the threshold for false-alarm probability α.

4. Prove the validity of Eq. (VI.B.19). (This result is known as the *Lebesgue decomposition*).

5. Consider the hypothesis testing problem

$$H_0: Y_t = W_t \quad , \quad 0 \leqslant t \leqslant T$$

versus

$$H_1: Y_t = \mu t + W_t \, , \quad 0 \leqslant t \leqslant T ,$$

where $\{W_t, 0 \leqslant t \leqslant T\}$ is a standard (i.e., unit) Wiener process. Find the likelihood ratio.

6. Consider the detection problem

$$H_0: Y_t = N_t \quad , \quad 0 \leq t \leq 1$$

versus

$$H_1: Y_t = N_t + t^2, \quad 0 \leq t \leq 1$$

where $\{N_t, 0 \leq t \leq 1\}$ is a zero-mean Gaussian random process with autocovariance function

$$C_N(t, u) = \begin{cases} t^2, & 0 \leq t \leq u \leq 1 \\ u^2, & 0 \leq u \leq t \leq 1. \end{cases}$$

Show that $P_0 \equiv P_1$ and find dP_1/dP_0.

7. Consider the detection problem

$$H_0: Y_t = N_t \quad , \quad 0 \leq t \leq T$$

versus

$$H_1: Y_t = N_t + m_0, \quad 0 \leq t \leq T$$

where m_0 is a positive constant and $\{N_t, 0 \leq t \leq T\}$ is zero-mean and Gaussian.
(a) Suppose $C_N(t, u) = N_0 e^{-\alpha|t-u|}, 0 \leq t, u \leq T$, where $\alpha > 0$ is fixed. Show that

$$\log \frac{dP_1}{dP_0}(Y_0^T) = \frac{m_0}{2N_0}[Y_0 + Y_T - m_0 + \alpha \int_0^T (Y_t - \frac{m_0}{2})dt].$$

(b) Suppose

$$C_N(t, u) = \begin{cases} N_0(1 - |t-u|), & |t-u| \leq 1 \\ 0 \quad , & |t-u| > 1. \end{cases}$$

Assuming $T < 1$, show that

$$\log \frac{dP_1}{dP_0}(Y_0^T) = \frac{m_0}{N_0(2-T)}(Y_0 + Y_T - m_0).$$

8. Consider the model of (IV.C.47). Suppose we design a Neyman-Pearson detector based on a signal model $m_t = \int_0^t s_u \, du$, $0 \leqslant t \leqslant T$, but that the actual signal is given by $\int_0^t r_u \, du$, $0 \leqslant t \leqslant T$, where $\{r_u; 0 \leqslant u \leqslant T\}$ is a deterministic signal.
   (a) Find P_F and P_D of the designed detector under the actual signal conditions.
   (b) Assuming $\int_0^T r_t^2 dt = \int_0^T s_t^2 dt = P > 0$, sketch the detection probability from (a) as a function of $\int_0^T r_t s_t \, dt$.

9. Consider the model of (VI.C.1) with white Gaussian noise. Find optimum signals under the constraints $\max\{\int_0^T [s_t^0]^2 dt, \int_0^T [s_t^1]^2\} \leqslant P$, where $P > 0$ is fixed.

10. Show that Eq. (VI.D.56) is valid.

VII SIGNAL ESTIMATION IN CONTINUOUS TIME

VII.A Introduction

In Chapter VI we treated the problem of signal detection with continuous-time observations. In this chapter we consider the problem of signal estimation in continuous time. We treat three basic problems: *parameter estimation* for signals of known form (up to a set of unknown parameters) observed in additive Gaussian noise; *linear/Gaussian estimation* in which either we assume that the signals and noise of interest are Gaussian processes or we restrict attention to linear estimators; and *nonlinear filtering*, in which we derive estimators for non-Gaussian random signals generated by nonlinear differential equations when observed in additive Gaussian noise. In all cases, we consider primarily the case of white Gaussian noise, although as we have seen in Chapter VI, other Gaussian noise models can be transformed to this model, so that these results are more general.

VII.B Estimation of Signal Parameters

We consider first the problem of estimating the parameters of continuous-time signals embedded in noise. As in the discrete-time problem, we adopt an observation model

$$Y_t = s_t(\theta) + N_t, \quad 0 \leqslant t \leqslant T, \qquad \text{(VII.B.1)}$$

where the dependence of the waveform $\{s_t(\theta); 0 \leqslant t \leqslant T\}$ on θ is known, and where $\{N_t : 0 \leqslant t \leqslant T\}$ represents white Gaussian noise with spectral height $N_0/2$. As usual, we integrate Y_t to obtain an appropriate mathematical model:

$$X_t = \int_0^t s_u(\theta)du + W_t, \quad 0 \leqslant t \leqslant T, \qquad \text{(VII.B.2)}$$

where $\{W_t; 0 \leqslant t \leqslant T\}$ is the relevant Wiener process.

To apply the estimation techniques of Chapter IV to estimate the parameter θ in (VII.B.2), it is necessary to specify a family $\{p_\theta; \theta \in \Lambda\}$ of densities on the observation space (Γ, G). To do this we need a single measure μ on (Γ, G) that dominates all of the measures $\{P_\theta; \theta \in \Lambda\}$ generated by the model (VII.B.2). We know from Example VI.C.1 that, assuming that $s_t(\theta)$ is is square-integrable on $[0, T]$, the measure P_θ is dominated by the probability measure described by the signal-absent condition $X_t = W_t, 0 \leqslant t \leqslant T$. (This measure is called *Wiener measure.*) If we denote this measure by μ, then the density p_θ is $dP_\theta/d\mu$, which is simply the likelihood ratio of (VI.C.65), i.e.,

$$p_\theta(X_0^T) = \exp\{\frac{2}{N_0}\int_0^T s_t(\theta)dX_t - \frac{1}{N_0}\int_0^T s_t^2(\theta)dt\}. \qquad \text{(VII.B.3)}$$

Thus we assume throughout that $\{s_t(\theta); 0 \leqslant t \leqslant T\}$ is square-integrable for each $\theta \in \Lambda$, and (VII.B.3) thus gives a density class for X_0^T.

Having determined the family $\{p_\theta; \theta \in \Lambda\}$, we may now apply the techniques of Chapter IV to the estimation of θ. For example, if θ is actually drawn from a random variable Θ with *a priori* probability density function w, Bayesian estimates are computed from the *a posteriori* density

$$w(\theta | X_0^T) = \frac{p_\theta(X_0^T)w(\theta)}{\int_\Lambda p_\theta(X_0^T)w(\theta)d\theta}. \qquad \text{(VII.B.4)}$$

Similarly, the non-Bayesian approaches of minimum-variance unbiased estimation and maximum-likelihood estimation apply directly to the model $\{p_\theta; \theta \in \Lambda\}$.

It is of interest to consider the asymptotics of maximum-likelihood estimates in the model of (VII.B.2) in some detail. To do so, we assume that the parameter set Λ is an open interval of $\mathbb{R}$. Extension to vector parameters can be made as in the discrete-time case.

We first note that Fisher's information for this model is given by

$$I_\theta \triangleq E_\theta\{[\frac{\partial}{\partial\theta} \log p_\theta(X_0^T)]^2\}$$

$$= E_\theta\{[\frac{2}{N_0}\frac{\partial}{\partial\theta}[\int_0^T s_t(\theta)dX_t - \tfrac{1}{2}\int_0^T s_t^2(\theta)dt\;]]^2\}$$

$$= \frac{4}{N_0^2}E_\theta\{[\int_0^T s_t'(\theta)(dX_t - s_t(\theta)dt\;)]^2\} \qquad \text{(VII.B.5)}$$

$$= \frac{4}{N_0^2}E_\theta\{[\int_0^T s_t'(\theta)dW_t\,]^2\} = \frac{2}{N_0}\int_0^T [s_t'(\theta)]^2 dt\;,$$

where in the third equality we have made the assumption that partial differentiation with respect to θ can be taken inside the integrals [here $s_t'(\theta) = \partial s_t(\theta)/\partial\theta$], in the fourth equality we have used (VII.B.2), and in the final equality we have used Proposition VI.D.1. Dividing (VII.B.5) by T yields the *rate* at which X_0^T generates information about θ, and we will assume for the moment that, as $T \to \infty$, this rate approaches a nonzero, finite number i_θ; i.e., we assume that

$$0 < i_\theta \triangleq \lim_{T\to\infty} \frac{2}{N_0 T}\int_0^T [s_t'(\theta)]^2 dt < \infty, \qquad \text{(VII.B.6)}$$

for all $\theta \epsilon \Lambda$. This assumption means that our observations generate information about the parameter at a rate that is asymptotically linear. Note that for the i.i.d. discrete-time problem treated in Section IV.D, the observations generate information about the parameter at exactly a linear rate since I_θ based on $Y_1, \ldots, Y_n$ is of the form $n i_\theta$ in that case.

The likelihood equation for the family of densities (VII.B.3) is given by

$$\frac{\partial}{\partial\theta}\left[\frac{2}{N_0}\int_0^T s_t(\theta)dX_t - \frac{1}{N_0}\int_0^T s_t^2(\theta)dt\right]\Bigg|_{\theta=\hat{\theta}_{ML}(X_0^T)} = 0. \quad \text{(VII.B.7)}$$

Assuming again that the partial differentiation with respect to θ can be taken inside the integrals, (VII.B.7) is equivalent to

$$J_\theta(X_0^T)\Big|_{\theta=\hat{\theta}_{ML}(X_0^T)} = 0. \quad \text{(VII.B.8)}$$

with

$$J_\theta(X_0^T) \triangleq \frac{1}{T}\int_0^T s_t'(\theta)[dX_t - s_t(\theta)dt], \quad \text{(VII.B.9)}$$

where $s_t'(\theta)$ is as before and where we have divided by $2T/N_0$ for convenience. Denoting the true parameter value by θ_0, and using (VII.B.2) we can write

$$\begin{aligned} J_\theta(X_0^T) &= \frac{1}{T}\int_0^T s_t'(\theta)[dW_t + [s_t(\theta_0) - s_t(\theta)]dt] \\ &= \frac{1}{T}\int_0^T s_t'(\theta)dW_t + \frac{1}{T}\int_0^T s_t'(\theta)[s_t(\theta_0) - s_t(\theta)]dt. \end{aligned} \quad \text{(VII.B.10)}$$

From Proposition VI.D.1, the first term on the right-hand side of (VII.B.10) has a $N(0, I_\theta/T^2)$ distribution, where I_θ is from (VII.B.5). Since $I_\theta/T \to i_\theta < \infty$, the variance of this first term approaches zero as $T \to \infty$, so this term converges in mean-square (and hence in probability) to zero as $T \to \infty$. Thus, on assuming that the second term in (VII.B.10) converges, we see that under P_{θ_0}

$$J_\theta(X_0^T) \overset{i.p.}{\rightarrow} \lim_{T\to\infty} \frac{1}{T}\int_0^T s_t'(\theta)[s_t(\theta_0) - s_t(\theta)]dt$$

(VII.B.11)

$$\triangleq J(\theta;\theta_0).$$

The function $J(\theta;\theta_0)$ has a root at $\theta=\theta_0$. So the asymptotic situation here is quite similar to that in the discrete-time case, in that we would expect $\hat{\theta}_{ML}(X_0^T)$ to converge in some sense to the correct parameter value if $J(\theta;\theta_0)$ has a unique root at $\theta=\theta_0$. In fact, the following result, analogous to Proposition IV.D.3, holds.

Proposition VII.B.1: Consistency

Suppose that θ_0 is the true value of the parameter, $J(\theta;\theta_0)$ has an isolated root at $\theta=\theta_0$ at which it changes sign, and $J_\theta(X_0^T)$ is continuous as a function of θ in a neighborhood of θ_0 for each T. Then there is a (continuous-time) sequence of solutions to $J_\theta(X_0^T)=0$ converging in probability to θ_0 as $T\to\infty$.

Remarks: The proof of this result is almost identical to that of Proposition IV.D.3 and is left as an exercise. Note that a corollary to Proposition VII.B.1 is that if $J_\theta(X_0^T)$ has a unique root for each T, then this sequence of roots is consistent.

Asymptotic normality of solutions to the likelihood equation can also be established under fairly mild conditions. In particular, the following can be established using techniques virtually identical to those used in proving Proposition IV.D.4.

Proposition VII.B.2: Asymptotic Normality

Suppose that $\{\hat{\theta}_T(X_0^T); T\geq 0\}$ is a consistent sequence of roots to the equation $J_\theta(X_0^T)=0$, and suppose further that $\partial^2 s_t(\theta)/\partial\theta^2$ and $\partial^3 s_t(\theta)/\partial\theta^3$ exist and are uniformly bounded in $T\geq 0$ and $\theta\in\Lambda$. Then

$$\sqrt{T}\,[\hat{\theta}_T(X_0^T) - \theta_0] \rightarrow N(0, 1/i_{\theta_0})$$

where θ_0 is the true parameter value and i_θ is as defined in (VII.B.6).

Remark: Note that this proposition implies that MLEs in the model of (VII.B.2) are asymptotically efficient in the same sense as those for discrete-time data.

Although the discussion above has considered the asymptotic properties of MLEs as $T \rightarrow \infty$ with $0 < i_\theta < \infty$, we are often interested in other asymptotics in estimation problems of this type. In particular, it is of interest to consider asymptotics as the amount of information about θ, I_θ, increases without bound. The analysis above is a particular case of this more general situation since there we assumed that $I_\theta \sim i_\theta T$. To study this more general case, let us drop the assumption $0 < i_\theta < \infty$ and rewrite the likelihood equation (VII.B.9) as

$$\left. \frac{\int_0^T s_t'(\theta)[dX_t - s_t(\theta)dt]}{\int_0^T [s_t'(\theta)]^2 dt} \right|_{\theta = \hat{\theta}_{ML}(X_0^T)} = 0. \quad \text{(VII.B.12)}$$

Expanding as in (VII.B.10), the left-hand side of (VII.B.12) becomes

$$\frac{\int_0^T s_t'(\theta)dW_t}{\int_0^T [s_t'(\theta)]^2 dt} + \frac{\int_0^T s_t'(\theta)[s_t(\theta_0) - s_t(\theta)]dt}{\int_0^T [s_t'(\theta)]^2 dt}. \quad \text{(VII.B.13)}$$

The leftmost quantity in (VII.B.13) is easily seen to have a $N(0, 1/I_\theta)$ distribution. Thus if $I_\theta \rightarrow \infty$, this term converges in the mean-square sense to zero, and asymptotically as $I_\theta \rightarrow \infty$, the left-hand side of the likelihood equation behaves as a function of θ like

$$\lim_{I_\theta \to \infty} \left[\frac{\frac{1}{N_0} \int_0^T s_t'(\theta)[s_t(\theta_0) - s_t(\theta)]dt}{I_\theta} \right] \triangleq \tilde{J}(\theta; \theta_0). \quad \text{(VII.B.14)}$$

Consistency of the solutions to the likelihood equation as $I_\theta \to \infty$ then follows within the appropriate assumption on $\tilde{J}$, similarly to the result Proposition VII.B.1.

Asymptotic normality of consistent solutions to the likelihood equation when $I_{\theta_0} \to \infty$ follows from a straightforward modification of the proof of Proposition VII.B.2. In particular, under the same regularity conditions as in Proposition VII.B.2, we have

$$I_{\theta_0}^{1/2} [\hat{\theta}_T(X_0^T) - \theta_0] \to N(0, 1) \quad \text{(VII.B.15)}$$

in distribution as $I_{\theta_0} \to \infty$ for any consistent sequence of solutions to the likelihood equation.

Note that I_{θ_0} can increase without bound under any of the conditions $T \to \infty$, $N_0 \to 0$, $\int_0^T [s_t'(\theta_0)]^2 dt \to \infty$, or combinations thereof. Aside from the case $T \to \infty$, of particular interest is the case $N_0 \to 0$ with other parameters fixed, in which we can examine the high signal-to-noise ratio behavior of estimates. Note that the solution to the likelihood equation does not depend on N_0, so as $N_0 \to 0$ with other parameters fixed, we are examining a *fixed* estimate whose *distribution* is changing with N_0. The following example illustrates the application of this type of analysis.

Example VII.B.1: Estimation of Arrival Time

In a number of applications, one is interested in the time at which a signal of known form arrives at a receiver. The canonical problem of this type is the *radar ranging* problem, in which one transmits a signal of known form which reflects from a target and returns to the receiving antenna. By determining the arrival time of the return signal, the travel time to and from the target, and hence the distance (or *range*) to the target, can be found.

A parametric signal model for estimation of arrival time is

$$s_t(\theta) = s(t - \theta), \quad 0 \leqslant t \leqslant T \qquad \text{(VII.B.16)}$$

where $\{s(t); t \in \mathbb{R}\}$ is a known waveform which is zero for $t < 0$, and $\theta \geqslant 0$ is the arrival time of the signal. Fisher's information in this case is

$$I_\theta = \frac{2}{N_0} \int_0^T [s'(t - \theta)]^2 dt \qquad \text{(VII.D.17)}$$

with $s'(t) = ds(t)/dt$. For the signals used in applications of arrival-time estimation, we typically have $\int_0^\infty [s'(t - \theta)]^2 dt < \infty$, so that the asymptotic (in T) information rate i_θ is zero. This means that we cannot use the $T \to \infty$ asymptotics to analyze the MLE here. However, we are often interested in the high signal-to-noise ratio performance of arrival time estimators, so we consider the $N_0 \to 0$ asymptotics.

As $N_0 \to 0$, consistent solutions to the likelihood equation will have asymptotic error variance I_θ^{-1}. If we let $T \to \infty$, we see that

$$I_\theta = \frac{2}{N_0} \int_0^\infty [s'(t)]^2 dt$$
$$= \frac{1}{\pi N_0} \int_{-\infty}^\infty \omega^2 |S(\omega)|^2 d\omega, \qquad \text{(VII.B.18)}$$

where $S(\omega)$ is the Fourier transform of $s(t)$. [The second equality in (VII.C.18) is a straightforward application of Parceval's relationship.] Thus the asymptotic error variance is

$$Var_\theta(\hat{\theta}_\infty) \sim \left[\frac{2E_s}{N_0} \beta^2 \right]^{-1}, \qquad \text{(VII.B.19)}$$

where $E_s \triangleq \int_0^\infty s^2(t)dt \equiv 1/2\pi\int_{-\infty}^{\infty}|S(\omega)|^2 d\omega$ is the signal energy, and β is the *root-mean-square signal bandwidth* defined by

$$\beta \triangleq \left[\frac{1}{2\pi}\int_{-\infty}^{\infty}\omega^2|S(\omega)|^2 d\omega \Big/ \frac{1}{2\pi}\int_{-\infty}^{\infty}|S(\omega)|^2 d\omega\right]^{1/2}. \qquad \text{(VII.B.20)}$$

Thus we see that in the limiting case, the accuracy of estimating the arrival time is improved either by increasing the signal-to-noise ratio, $2E_s/N_0$, or by increasing the bandwidth of the signal. The latter phenomenon agrees with intuition since a wide bandwidth corresponds to a sharply edged signal, and one would expect to estimate the arrival time best with a sharp pulse.

It is interesting to note that the log-likelihood function is given here by

$$\log p_\theta(X_0^T) = \qquad \text{(VII.B.21)}$$

$$\frac{2}{N_0}\int_0^T s(t-\theta)dX_t - \frac{1}{N_0}\int_0^T s^2(t-\theta)dt.$$

If we assume that the signal has a finite duration $D < T$ and that θ is known to lie within $[0, T-D]$, then the second term on the right-hand side of this expression does not depend on θ. (Usually, $T >> D$ and these conditions are reasonable.) In this case, the MLE chooses a value of θ that maximizes $\int_0^\infty s(t-\theta)dX_t$. In terms of implementation, we return to the original observed waveform Y_t, which can be thought of heuristically as the derivative of X_t. So, to implement the MLE we would compute $\int_0^\infty s(t-\theta)Y_t dt$ for each θ, and choose as our estimate that θ yielding the largest value of this integral. At first glance this appears to be difficult computationally; note, however, that $\int_0^\infty s(t-\theta)Y_t dt$ is the output at time θ of the linear time-invariant filter with impulse response $h(t)=s(-t)$, due to the input $\{Y_t; 0 \leqslant \leqslant T\}$. Thus we apply Y_t to the filter $h(t)$, and the MLE of θ is the time at which the output peaks.

VII.C Linear/Gaussian Estimation

VII.C.1: Estimation in White Noise

We now turn to the problem of estimating a mean-square continuous Gaussian signal in independent additive white Gaussian noise. That is, we consider the observational model

$$Y_t = S_t + N_t, \quad 0 \leqslant t \leqslant T, \tag{VII.C.1}$$

where $\{S_t, 0 \leqslant t \leqslant T\}$ is a zero-mean Gaussian signal with continuous autocovariance function $C_S(t, u)$ and where $\{N_t; 0 \leqslant t \leqslant T\}$ is white Gaussian noise (independent of the signal) with spectral height $N_0/2$. As usual, to deal with this problem analytically, we integrate $\{Y_t; 0 \leqslant t \leqslant T\}$ to yield the model

$$X_t = \int_0^t S_u \, du + W_t, \quad 0 \leqslant t \leqslant T, \tag{VII.C.2}$$

where $\{W_t, 0 \leqslant t \leqslant T\}$ is a Wiener process.

For fixed $t \in [0, T]$ we are interested in finding the best estimator of S_t given observations $X_0^s = \{X_u; 0 \leqslant u \leqslant s\}$, for some $s \in [0, T]$ where optimality is defined in the minimum-mean-squared-error (MMSE) sense. To do so, we first find the best *linear* estimator of S_t, given X_0^s, and then argue that this is in fact the best estimator globally.

To derive the best linear estimator, fix $t \in [0, T]$ and consider linear estimators of the form

$$\hat{S}_t = \int_0^s h(t, v) dX_v, \tag{VII.C.3}$$

where $h(t, v)$ is a continuous function of v. For such an estimator to be the best linear estimator of S_t given X_0^s, it is necessary and sufficient (see Section V.C) that the error $(\hat{S}_t - S_t)$ be orthogonal to the observations, i.e., that

$$E\{(\hat{S}_t - S_t)X_u\} = 0, \; 0 \leqslant u \leqslant s. \qquad \text{(VII.C.4)}$$

Inserting (VII.C.2) and (VII.C.3) into (VII.C.4) and expanding, we see that (VII.C.4) is equivalent to

$$E\{W_u \int_0^s h(t,v)dW_v\} + E\{\int_0^u S_q dq \int_0^s h(t,v)dW_v\}$$

$$+ E\{W_u \int_0^s h(t,v)S_v dv\}$$

$$+ E\{\int_0^u S_q dq \int_0^s h(t,v)S_v dv\} \qquad \text{(VII.C.5)}$$

$$= E\{S_t W_u\} + E\{S_t \int_0^u S_q dq\},$$

$$0 \leqslant u \leqslant s.$$

Using the fact that signal and noise are independent and property (vi) of Proposition VI.D.1, (VII.C.5) reduces to†

$$\frac{N_0}{2} \int_0^u h(t,v)dv + \int_0^u \int_0^s h(t,v)C_S(v,q)dvdq$$

$$\qquad \text{(VII.C.6)}$$

$$= \int_0^u C_S(t,q)dq, \quad 0 \leqslant u \leqslant s.$$

Taking the partial derivatives with respect to u of both sides of (VII.C.6), we see that $\hat{S}_t$ of (VII.C.3) is optimum if and only if h satisfies the equation

---

† Recall that order of mean-square integration and expectation can be interchanged.

$$\frac{N_0}{2}h(t,u) + \int_0^s h(t,v)C_S(v,u)dv$$
$$= C_S(t,u), \quad 0 \leq u \leq s. \tag{VII.C.7}$$

With t fixed, (VII.C.7) is a Fredholm integral equation of the second kind. It can be shown that by virtue of the assumed continuity of C_S, (VII.C.7) has a unique solution $h(t,u)$, continuous in u [see, e.g., Lovitt (1950) for a discussion of Fredholm equations].

We conclude that (VII.C.3) with h given as the solution to (VII.C.7) is the MMSE linear estimator of S_t given X_0^s. Note that (VII.C.7) is a continuous time Wiener-Hopf equation, and we have already mentioned it for the two particular cases $s=t$ and $s=T$ in connection with the problem of detecting Gaussian signals in white Gaussian noise.

We now would like to show that $\hat{S}_t$ of (VII.C.3) is not only the best *linear* estimator given X_0^s, but that it is also the best estimator of S_t given X_0^s among all possible estimators. To do so we note that the most general class of estimators of interest is the set of random variables that are functions of X_0^s and that have finite second moments. Let us denote this class by G_0^s. The following result is easily proved.

Proposition VII.C.1: The Orthogonality Principle

$$E\{(Y - S_t)^2\} = \min_{Z \in G_0^s} E\{(Z - S_t)^2\} \tag{VII.C.8}$$

if and only if

$$E\{(Y - S_t)Z\} = 0 \tag{VII.C.9}$$

for all $Z \in G_0^s$. Moreover, if Y_1 and Y_2 satisfy (VII.C.9) for all $Z \in G_0^s$, then $Y_1 = Y_2$ with probability 1.

From Chapter IV we know that $Y = E\{S_t | X_0^s\}$ is the solution to (VII.C.8), and thus the orthogonality condition

(VII.C.9) is simply a characterization of the conditional mean of S_t given X_0^s.

Consider $\hat{S}_t$ of (VII.C.3). We know that $(\hat{S}_t - S_t)$ is uncorrelated with X_0^s. Moreover, $(\hat{S}_t - S_t)$ is jointly Gaussian with X_0^s since it is linearly derived from the Gaussian pair (S_t, X_0^s). Thus $(\hat{S}_t - S_t)$ is in fact *independent* of X_0^s which implies that for and $Z \in G_0^s$

$$E\{(\hat{S}_t - S_t)Z\} = E\{(\hat{S}_t - S_t)\}E\{Z\}.$$

Since $E\{\hat{S}_t\} = E\{S_t\} = 0$, (VII.C.9) follows. So we see that for the Gaussian model the best linear estimate $\hat{S}_t$ of (VII.C.3) is also the globally best estimator.

Alternatively, it should be noted that the Gaussian assumption was not used in showing that $\hat{S}_t$ is the best linear estimator in the model of (VII.D.2) with $E\{W_t W_u\} = (N_0/2)\min\{t, u\}$, $Cov(S_t, S_u) = C_S(t, u)$, $Cov(S_t W_u) = 0$, and $E\{S_t\} = E\{W_t\} = 0$, $0 \leqslant u, t \leqslant T$. Thus (VII.D.3) is, in fact, the best linear estimator for any signal in noise satisfying these second-order properties, whether they are Gaussian or not.

All of the results above can be extended straightforwardly to the case in which signal and noise are correlated, with the assumption of joint Gaussianity of signal and noise being required to show that the best linear estimator is also the globally optimum estimator in this case. Similarly, the situation in which the noise consists of the sum of a mean-square continuous component and a white component can be handled straightforwardly.

VII.C.2 The Linear Innovations Process

Consider again the observation model

$$X_t = \int_0^t S_u \, du + W_t, \quad 0 \leqslant t \leqslant T, \qquad \text{(VII.C.10)}$$

where $E\{W_t W_u\} = \min\{t, u\}$, $C_S(t, u)$ is continuous, $E\{W_t S_u\} = 0$, and $E\{S_t\} = E\{W_t\} = 0$ for all $0 \leqslant t, u \leqslant T$. For each $t \in [0, T]$, let $\hat{S}_t$ denote the best linear estimator of S_t from X_0^t [i.e., (VII.C.3) with $s = t$ and $N_0/2 = 1$]. From the results in Section VI.D, it follows that the process

$$I_t = X_t - \int_0^t \hat{S}_u \, du, \quad 0 \leq t \leq T \qquad \text{(VII.C.11)}$$

is of zero mean with covariance $C_I(t,u) = \min\{t,u\}$ and that $(I_t - I_u)$ is uncorrelated with X_0^u for all $0 \leq u \leq t \leq T$. Furthermore, if the processes in (VII.C.10) are Gaussian, then I_0^T is also Gaussian (so it is a Wiener process) and the uncorrelatedness of $(I_t - I_u)$ and X_0^u becomes independence. We shall term I_t the *linear innovations process* to distinguish it from the already defined innovations process $\{X_t - \int_0^t E\{S_u | X_0^u\} du, 0 \leq t \leq T\}$. Of course, these two processes coincide for the Gaussian case.

Note that the formation of the linear innovations process is basically a linear whitening operation on X_0^T, since I_0^T has the covariance of integrated white noise. Note also that unlike the prewhitening discussed in Section VI.C, I_0^T is obtained causally from X_0^T. However, we do not know whether or not the linear innovations process is in any way equivalent to the original observations since we have not shown that X_0^T can be obtained back from I_0^T. This, in fact, can be done, as we show in the following paragraphs.

Note that X_0^T is related to I_0^T via the relationship

$$X_t = I_t + \int_0^t \hat{S}_u \, du, \quad 0 \leq t \leq T. \qquad \text{(VII.C.12)}$$

Thus if we could somehow obtain $\hat{S}_u$ from I_0^u for each $u \in [0,T]$, then X_0^T could be reconstructed from I_0^T via (VII.C.12). For each $t \in [0,T]$, let us denote by $\tilde{S}_t$ the MMSE linear estimate of S_t given I_0^t. Then, for example, if it were true that $\hat{S}_t = \tilde{S}_t$ for all $t \in [0,T]$, X_0^T could be obtained in this way.

It can be shown using the orthogonality principle and the definition of I_0^T that $\tilde{S}_t$ is given by

$$\tilde{S}_t = \int_0^t g(t, u) dI_u, \quad 0 \leqslant t \leqslant T, \qquad \text{(VII.C.13)}$$

where g is defined by

$$g(t, u) = C_S(t, u) - \int_0^u C_S(t, u) h(u, v) dv, \qquad \text{(VII.C.14)}$$

$$0 \leqslant t \leqslant T,$$

and where $h(u, v)$ is the impulse response yielding $\hat{S}_u$ [i.e., $\hat{S}_u = \int_0^u h(u, v) dX_v$]. Replacing I_u with $X_u - \int_0^u \hat{S}_v dv$ we have

$$\hat{S}_t = \int_0^t g(t, u) dX_u - \int_0^t g(t, u) \hat{S}_u du$$

$$= \int_0^t g(t, u) dX_u - \int_0^t g(t, u) \int_0^u h(u, v) dX_v du \qquad \text{(VII.C.15)}$$

$$= \int_0^t g(t, u) dX_u - \int_0^t \int_v^t g(t, u) h(u, v) du dX_v$$

$$= \int_0^t [g(t, u) - \int_u^t g(t, v) h(v, u) dv] dX_u,$$

where the third equality is obtained by interchanging the order of integration in u and v and the fourth equality is obtained by reversing the roles of the variables of integration in the second integral. Inserting (VII.C.14) into (VII.C.15), we arrive after some rearranging at the following result:

$$\tilde{S}_t = \int_0^t f(t, u) dX_u, \qquad \text{(VII.C.16)}$$

where

$$f(t,u) \triangleq C_S(t,u) - \int_0^t Q_t(u,v)C_S(t,v)dv \qquad \text{(VII.C.17)}$$

and

$$Q_t(u,v) \triangleq h(u,v) + h(v,u)$$

$$- \int_0^t h(\alpha,u)h(\alpha,v)d\alpha, \quad 0 \leqslant u, v \leqslant t, \qquad \text{(VII.C.18)}$$

with $h(x,y)$ taken to be zero for $t \geqslant x > y \geqslant 0$.

The function $Q_t(u,v), 0 \leqslant u, v \leqslant t$, is the *resolvent kernel* (or *Fredholm resolvent*) associated with $C_S(u,v), 0 \leqslant u, v \leqslant t$. As we have seen in Chapter VI, Q_t solves the integral equation

$$C_S(u,v) = \int_0^t Q_t(u,\alpha)C_S(\alpha,v)d\alpha$$

$$+ Q_t(u,v), \quad 0 \leqslant u, v \leqslant t.$$

Q_t is equivalently known as the *resolvent* of the integral equation

$$y(u) = \int_0^t C_S(u,v)x(v)dv + x(u),$$

$$0 \leqslant u \leqslant t, \qquad \text{(VII.C.19)}$$

because the unique continuous solution to this equation for continuous y is given by

$$x(u) = y(u) - \int_0^t Q_t(u,v)y(v)dv,$$

(VII.C.20)

$$0 \leqslant u \leqslant t.$$

From (VII.C.17), (VII.C.18) and (VII.C.20) we see that $f(t, u)$ satisfies the equation

$$C_S(t, u) = \int_0^t C_S(u, v)f(t, v)dv + f(t, u), \quad 0 \leqslant u \leqslant t,$$

which is the Wiener-Hopf equation [(VII.C.7) with $s = t$ and $N_0/2 = 1$] determining $h(t\ u)$. By the uniqueness of the solution to (VII.C.7), we have $f(t, u) = h(t, u)$, $0 \leqslant u \leqslant t$, and thus

$$\tilde{S}_t = \int_0^t h(t, u)dX_u = \hat{S}_t.$$

So we have shown that $\hat{S}_t$ can be determined from I_0^t and hence that X_0^t can be obtained from I_0^t. To be more specific, the linear innovations I_0^T is obtained from a causal linear transformation of X_0^T, and this transformation has a causal linear inverse transformation via which X_0^T can be obtained from I_0^T. This pair of transformations can be written as

$$I_t = \int_0^t [1 - \int_u^t h(u, v)dv]dX_u, \quad 0 \leqslant t \leqslant T \qquad \text{(VII.C.21)}$$

and

$$X_t = \int_0^t [1 + \int_u^t g(u, v)dv]dI_u, \quad 0 \leqslant t \leqslant T, \qquad \text{(VII.C.22)}$$

where h and g are as defined in (VII.C.14). This implies that I_0^T is an equivalent observation to X_0^T, and particularly that best linear estimates based on I_0^T are also the best linear estimates based on X_0^T.

The properties of the linear innovations process extend straightforwardly to the vector case. In particular, suppose that we have n-dimensional vector observations given by

$$\underline{X}_t = \int_0^t \underline{S}_u \, du + \underline{W}_t, \; 0 \leqslant t \leqslant T, \qquad \text{(VII.C.23)}$$

where $\{\underline{S}_t ; 0 \leqslant t \leqslant T\}$ is a zero-mean vector signal process with continuous autocovariance†

$$\mathbf{C}_S(t, u) \triangleq E\{\underline{S}_t \underline{S}_u'\}, \quad 0 \leqslant u, t \leqslant T,$$

and $\{\underline{W}_t ; 0 \leqslant t \leqslant T\}$ is a vector process with zero mean and autocovariance

$$E\{\underline{W}_t \underline{W}_u'\} = \mathbf{I} \min\{t, u\}, \quad 0 \leqslant u, t \leqslant T,$$

where $\mathbf{I}$ is the $n \times n$ identity matrix. Note that $\underline{W}_0^T$ is a vector of uncorrelated scalar processes each with autocovariance $\min\{t, u\}$. If $\underline{W}_0^T$ is also Gaussian, it is a vector Wiener process and can be thought of as the integral of a vector of independent Gaussian white noises.

Suppose that further that $\underline{S}_u$ and $\underline{W}_t$ are uncorrelated for all t and u in $[0, T]$. The best linear estimate of $\underline{S}_t$ given $\underline{X}_0^T$ is given by‡

---

† In this section, the superscript T denoting transposition is replaced with a prime in order to avoid confusion with the observation endpoint T.

‡ The vector m.s. integral $\int_0^t \mathbf{h}(t, u) d\underline{X}_u$ can be defined simply as the random vector whose j^{th} component is

$$\sum_{k=1}^{n} \int_0^t h_{jk}(t, u) dX_u^{(k)},$$

where h_{jk} is the jk^{th} element of $\mathbf{h}$ and $X^{(k)}$ is the kth component of $\underline{X}$.

$$\underline{\hat{S}}_t = \int_0^t \mathbf{h}(t, u) d\underline{X}_u, \qquad \text{(VII.C.24)}$$

where the $n \times n$ matrix function $\mathbf{h}(t, u)$ is the solution to the matrix integral equation

$$\mathbf{C}_S(t, u) = \int_0^t \mathbf{h}(t, v) \mathbf{C}_S(v, u) dv + \mathbf{h}(t, u), \qquad \text{(VII.C.25)}$$

$$0 \leqslant u \leqslant t.$$

This result follows immediately from the necessary and sufficient orthogonality condition

$$E\{(\underline{S}_t - \underline{\hat{S}}_t)\underline{X}_u'\} = \mathbf{O}, \quad 0 \leqslant u \leqslant t, \qquad \text{(VII.C.26)}$$

where $\mathbf{O}$ is the $n \times n$ matrix of all zeros, and the fact that

$$E\{(\int_0^T \mathbf{A}_t d\underline{W}_t)(\int_0^T \mathbf{B}_t d\underline{W}_t)'\} \qquad \text{(VII.C.27)}$$

$$= \int_0^T \mathbf{A}_t \mathbf{B}_t' dt,$$

where $\mathbf{A}_t$ and $\mathbf{B}_t$ are matrix functions with square integrable elements and $\underline{W}_0^T$ has autocovariance function $\mathbf{I} \min\{t, u\}$.

Analogously to the scalar case, the innovations process,

$$\underline{I}_t \triangleq \underline{X}_t - \int_0^t \underline{\hat{S}}_u du, \quad 0 \leqslant t \leqslant T, \qquad \text{(VII.C.28)}$$

for this model satisfies

$$E\{\underline{I}_t \underline{I}_u'\} = \mathbf{I}\ \min\{t, u\}, \quad 0 \leqslant u, t \leqslant T \qquad \text{(VII.C.29)}$$

and

$$E\{(\underline{I}_t - \underline{I}_s)\underline{X}_u'\} = \mathbf{O}, \quad 0 \leqslant u \leqslant s \leqslant t \leqslant T. \qquad \text{(VII.C.30)}$$

Furthermore, $\underline{X}_0^T$ can be recovered from $\underline{I}_0^T$ via

$$\underline{X}_t = \int_0^t [\mathbf{I} + \int_u^t \mathbf{g}(u, v)dv]d\underline{I}_u, \quad 0 \leqslant t \leqslant T, \qquad \text{(VII.C.31)}$$

where

$$\mathbf{g}(t, u) = \mathbf{C}_S(t, u) - \int_0^u \mathbf{C}_S(t, v)\mathbf{h}'(u, v)dv,$$

$$0 \leqslant u \leqslant t \leqslant T.$$

Of course, if signal and noise processes are Gaussian (and jointly Gaussian), then $\hat{\underline{S}}_t$ of (VII.C.24) is the global MMSE estimator and $\underline{I}_0^T$ becomes a vector Wiener process with increments $(\underline{I}_t - \underline{I}_s)$ independent of $\underline{X}_0^s$ for all $t \geqslant s \geqslant 0$.

VII.C.3 The Continuous-Time Kalman-Bucy Filter

An interesting application of the innovation representation for an observation of signal in white noise is in deriving the continuous time version of Kalman-Bucy filter discussed in Chapter V. In the following paragraphs we will accomplish this derivation briefly, referring the reader to Chapter V for motivation for the Kalman-Bucy problem.

The behavior of many physical systems in continuous-time can be modeled over a time interval, say $[0, T]$, by specifying at each time $t \in [0, T]$ a vector of *states*, say $\underline{X}_t$, which evolves according to a linear differential equation of the form

$$\dot{\underline{X}}_t = \mathbf{A}_t \underline{X}_t + \mathbf{B}_t \underline{U}_t , \quad 0 \leqslant t \leqslant T \qquad \text{(VII.C.32a)}$$

driven by a random process $\{\underline{U}_t ; t \in [0, T]\}$. Here, for each t, the state vector $\underline{X}_t$ is n-dimensional, the random vector $\underline{U}_t$ is m-dimensional, and $\mathbf{A}_t$ and $\mathbf{B}_t$ are matrices conformal with these dimensions. The states of such systems can often be observed via a noisy linear observation process

$$\underline{Y}_t = \mathbf{C}_t \underline{X}_t + \underline{N}_t , \quad 0 \leqslant t \leqslant T , \qquad \text{(VII.C.32b)}$$

where the *observation* $\underline{Y}_t$ and the *observation noise* $\underline{N}_t$ are r-dimensional, and $\mathbf{C}_t$ is of dimension $r \times n$.

Given such a model, we are often interested in *state estimation*; that is, the production of estimates of $\underline{X}_t$ from observations of $\underline{Y}_u$ over some interval of times $u \in [0, s]$, with $t < s$ representing *smoothing*, $t = s$ representing *filtering*, and $t > s$ representing *prediction*. As in Chapter V and the preceding sections of this chapter, we can derive optimum linear estimates by specifying a model for the second-order statistics of all processes involved, and these estimates are globally optimum whenever the processes are Gaussian. Here we adopt the former approach; however, the second interpretation of the resulting filter interpretation should be kept in mind.

We model the state-driving process (called the *process noise*) $\{\underline{U}_t ; 0 \leqslant t \leqslant T\}$ and the observation noise $\{\underline{N}_t ; 0 \leqslant t \leqslant T\}$ as zero-mean uncorrelated unit vector white noises. That is, we assume that

$$Cov\ (\underline{U}_t , \underline{U}_s) = \mathbf{I}_m \delta(t - s), 0 \leqslant t , s \leqslant T , \qquad \text{(VII.C.33a)}$$

$$Cov\ (\underline{N}_t , \underline{N}_s) = \mathbf{I}_r \delta(t - s), 0 \leqslant t , s \leqslant T , \qquad \text{(VII.C.33b)}$$

and

$$Cov\ (\underline{U}_t , \underline{N}_s) = \mathbf{O} , \quad 0 \leqslant t , s \leqslant T , \qquad \text{(VII.C.33c)}$$

where $\mathbf{I}_m$ and $\mathbf{I}_r$ are the $m \times m$ and $r \times r$ identity matrices, respectively, and where $\mathbf{O}$ is the matrix of all zeros. Thus we are assuming that the noises are componentwise and timewise uncorrelated with themselves and each other for all times t and s with $t \neq s$ and that they are componentwise uncorrelated with themselves and with each other for all times $t = s$. (The relaxation of these assumptions is discussed below.) We also assume that the matrix functions $\{\mathbf{A}_t ; 0 \leqslant t \leqslant T\}$, $\{\mathbf{B}_t ; 0 \leqslant t \leqslant T\}$, and $\{\mathbf{C}_t ; 0 \leqslant t \leqslant T\}$ (known as the *state*, *input*, and *observation* matrices, respectively) are piecewise-continuous functions of time. This assumption assures the validity of various operations made below and will not be mentioned explicitly in this context.

Given an initial state, we can rewrite the state equation (VII.C.31) as

$$\underline{X}_t = \underline{X}_0 + \int_0^t \mathbf{A}_u \underline{X}_u \, du + \int_0^t \mathbf{B}_u \, d\underline{V}_u , \qquad 0 \leqslant t \leqslant T, \tag{VII.C.34}$$

where $\underline{V}_t \triangleq \int_0^t \underline{U}_s \, ds$ is a process with autocovariance $\mathbf{C}(t, u) = \mathbf{I}_m \min\{t, u\}$. Similarly, we can integrate the observation equation to give

$$\underline{Z}_t \triangleq \int_0^t \underline{Y}_u \, du = \int_0^t \mathbf{C}_u \underline{X}_u \, du + \underline{W}_t , \qquad 0 \leqslant t \leqslant T, \tag{VII.C.35}$$

where $\mathbf{C}_W(t, u) = \mathbf{I}_r \min\{t, u\}$. Thus the model of (VII.C.34) and (VII.C.35) is a well-behaved version of the model (VII.C.31) and (VII.C.32) when the state and observation noises are white. A more compact notation for writing (VII.C.34) and (VII.C.35) is

$$d\underline{X}_t = \mathbf{A}_t \underline{X}_t \, dt + \mathbf{B}_t \, d\underline{V}_t , \quad 0 \leqslant t \leqslant T \tag{VII.C.36}$$

and

$$d\underline{Z}_t = \mathbf{C}_t \underline{X}_t dt + d\underline{W}_t, 0 \leqslant t \leqslant T. \qquad \text{(VII.C.37)}$$

To complete the statistical model, we assume that the initial state vector, $\underline{X}_0$, is a random vector with zero mean and covariance matrix $\mathbf{P}_0$, and that it is uncorrelated with the noise processes $\underline{V}_0^T$ and $\underline{W}_0^T$.

We will now derive a recursive estimator for $\underline{X}_t$ from the observations $\underline{Z}_0^t$ based on the model above. To do so we first note that the observation equation (VII.C.35) is in the form of (VII.C.23) with $\underline{S}_u = \mathbf{C}_u \underline{X}_u$. If we denote by $\hat{\underline{S}}_t$ and $\underline{X}_t$ the best linear estimates of $\underline{S}_t$ and $\underline{X}_t$, respectively, based on $\underline{Z}_0^t$, it is easy to see that $\hat{\underline{S}}_t = \mathbf{C}_t \hat{\underline{X}}_t$. Thus in view of our previous discussion, the linear innovations process is

$$\underline{I}_t = \underline{Z}_t - \int_0^t \mathbf{C}_u \hat{\underline{X}}_u du, \quad 0 \leqslant t \leqslant T, \qquad \text{(VII.C.38)}$$

and this process has the same autocovariance as $\{\underline{W}_t ; 0 \leqslant t \leqslant T\}$ with increments $(\underline{I}_t - \underline{I}_s)$ that are uncorrelated with past observations $\underline{Z}_0^s$ for all $0 \leqslant s \leqslant t \leqslant T$. Moreover, the linear innovations are equivalent to the observations for linear estimation, so we should be able to write $\underline{X}_t$ in the form

$$\hat{\underline{X}}_t = \int_0^t \mathbf{g}(t, u) d\underline{I}_u \qquad \text{(VII.C.39)}$$

for some matrix impulse response $\mathbf{g}$.

To find $\mathbf{g}$ we can use the orthogonality principle, from which we have

$$E\{(\underline{X}_t - \hat{\underline{X}}_t)\underline{I}_u'\} = \mathbf{O}, \quad 0 \leqslant u \leqslant t. \qquad \text{(VII.C.40)}$$

Combining (VII.C.39) and (VII.C.40), we see that $\mathbf{g}$ must satisfy

$$E\{\underline{X}_t \underline{I}_u'\} = E\{(\int_0^t \mathbf{g}(t, v) d\underline{I}_v) \underline{I}_u'\},$$

or, equivalently,

$$\mathbf{C}_{XI}(t, u) = \int_0^u \mathbf{g}(t, v) dv, \quad 0 \leqslant u \leqslant t.$$

On differentiating, we then have that

$$\mathbf{g}(t, u) = \frac{\partial}{\partial u} \mathbf{C}_{XI}(t, u), \quad 0 \leqslant u \leqslant t. \qquad \text{(VII.C.41)}$$

[Note that (VII.C.41) is a general formula having no dependence on the state model (VII.C.34).] Noting that

$$\underline{I}_u = \int_0^u \mathbf{C}_v (\underline{X}_v - \hat{\underline{X}}_v) dv + \underline{W}_u, \qquad \text{(VII.C.42)}$$

we can write $\mathbf{C}_{XI}(t, u)$ as

$$\mathbf{C}_{XI}(t, u) \triangleq E\{\underline{X}_t \underline{I}_u'\}$$

$$= \int_0^u E\{\underline{X}_t (\underline{X}_v - \hat{\underline{X}}_v)'\} \mathbf{C}_v' dv + E\{\underline{X}_t \underline{W}_u'\}. \qquad \text{(VII.C.43)}$$

Since $\underline{X}_t$ is produced linearly from $\underline{X}_0$ and $\underline{V}_0^t$, both of which are uncorrelated with W_0^T, we have $E\{\underline{X}_t \underline{W}_u'\} = \mathbf{O}$. Thus on combining (VII.C.41) and (VII.C.43), we have

$$\mathbf{g}(t, u) = E\{\underline{X}_t (\underline{X}_u - \hat{\underline{X}}_u)'\} \mathbf{C}_u', \quad 0 \leqslant u \leqslant t \leqslant T. \qquad \text{(VII.C.44)}$$

Using (VII.C.44), we can write a recursion for $\hat{\underline{X}}_t$. To see this, we rewrite $\mathbf{g}$ as

$$\mathbf{g}(t,u) = E\{(\underline{X}_t - \underline{X}_u)(\underline{X}_u - \hat{\underline{X}}_u)'\}\mathbf{C}_u'$$

$$+ E\{\underline{X}_u(\underline{X}_u - \hat{\underline{X}}_u)'\}\mathbf{C}_u' .$$

From the state equation (VII.C.34), we have

$$\underline{X}_t - \underline{X}_u = \int_u^t \mathbf{A}_v \underline{X}_v \, dv + \int_u^t \mathbf{B}_v \, d\underline{V}_v ,$$

from which

$$\mathbf{g}(t,u) = \int_u^t \mathbf{A}_v E\{\underline{X}_v(\underline{X}_u - \hat{\underline{X}}_u)'\}\mathbf{C}_u' \, dv$$

$$+ E\{(\int_u^t \mathbf{B}_v \, d\underline{V}_v)(\underline{X}_u - \hat{\underline{X}}_u)'\} \qquad \text{(VII.C.45)}$$

$$+ E\{\underline{X}_u(\underline{X}_u - \hat{\underline{X}}_u)'\}\mathbf{C}_u' .$$

From (VII.C.44), the first term on the right-hand side of (VII.C.45) is $\int_u^t \mathbf{A}_v \mathbf{g}(v,u)dv$. The second term on the right-hand side of (VII.C.45) is zero since $\underline{X}_u$ and $\hat{\underline{X}}_u$ are generated linearly from $\underline{X}_0, \underline{V}_0^u$, and $\underline{W}_0^u$, all of which are uncorrelated with the increments $\{\underline{V}_v - \underline{V}_u ; u \leqslant v \leqslant T\}$ from which $\int_u^t \mathbf{B}_v \, d\underline{V}_v$ is generated. The third term on the right-hand side of (VII.C.45) depends only on u and we will denote it by $\mathbf{K}_u$. Thus $\mathbf{g}(t,u)$ can be written as

$$\mathbf{g}(t,u) =$$

$$\int_u^t \mathbf{A}_v \mathbf{g}(v,u)dv + \mathbf{K}_u, \quad 0 \leqslant u \leqslant t \leqslant T . \qquad \text{(VII.C.46)}$$

Using (VII.C.46) we have

$$\underline{\hat{X}}_t = \int_0^t g(t,u)d\underline{I}_u = \int_0^t \int_u^t \mathbf{A}_v g(v,u) dv d\underline{I}_u + \int_0^t \mathbf{K}_u d\underline{I}_u$$

$$= \int_0^t \mathbf{A}_v \int_0^v g(v,u)d\underline{I}_u dv + \int_0^t \mathbf{K}_u d\underline{I}_u \qquad \text{(VII.C.47)}$$

$$= \int_0^t \mathbf{A}_v \underline{\hat{X}}_v dv + \int_0^t \mathbf{K}_u d\underline{I}_u, \quad 0 \leqslant t \leqslant T,$$

or, more compactly,

$$d\underline{\hat{X}}_t = \mathbf{A}_t \underline{\hat{X}}_t + \mathbf{K}_t d\underline{I}_t, 0 \leqslant t \leqslant T \qquad \text{(VII.C.48)}$$

with initial condition $\underline{\hat{X}}_0 = \underline{0}$.

Comparing (VII.C.36) and (VII.C.48) we see that the optimum state estimator is generated by a state equation with the same dynamics (i.e., the same *state matrix* or *plant*) as the original state, but with input matrix given by $\mathbf{K}_t$. This equation is driven by the innovations process which behaves statistically like the integral of white noise. Thus the equation generating the state estimator is quite similar to the original state equation, and the analogy with the discrete-time case is apparent.

In order to specify the state estimator $\underline{\hat{X}}_t$ completely via (VII.C.48), we need an expression for the gain matrix $\mathbf{K}_t$. We have, by definition, that $\mathbf{K}_t = E\{\underline{X}_t \ (\underline{X}_t - \underline{\hat{X}}_t)'\}\mathbf{C}_t'$. Note that $\underline{\hat{X}}_t$ is generated linearly from $\underline{I}_0^t$, so it is uncorrelated with the estimation error, $\underline{X}_t - \underline{\hat{X}}_t$. Thus $\mathbf{K}_t$ can be rewritten as

$$\mathbf{K}_t = E\{(\underline{\hat{X}}_t - \underline{X}_t)(\underline{\hat{X}}_t - \underline{X}_t)'\}\mathbf{C}_t' = \mathbf{P}_t \mathbf{C}_t', \qquad \text{(VII.C.49)}$$

where

$$\mathbf{P}_t \triangleq E\{(\underline{X}_t - \underline{\hat{X}}_t)(\underline{X}_t - \underline{\hat{X}}_t)'\}$$

is the covariance matrix of the estimation error at time t. So in order to find the gain $\mathbf{K}_t$, we must find a way of determining $\mathbf{P}_t$.

Consider the error process $\underline{e}_t \triangleq \underline{X}_t - \hat{\underline{X}}_t$, and let $\mathbf{C}_e(t, u)$ denotes its autocovariance function. Note that, by virtue of (VII.C.34) and (VII.C.47) $\{\underline{e}_t; t \in [0, T]\}$ satisfies the following equation

$$\underline{e}_t = \underline{e}_0 + \int_0^t \mathbf{A}_u \underline{e}_u \, du + \int_0^t \mathbf{B}_u \, d\underline{V}_u - \int_0^t \mathbf{K}_u \, d\underline{I}_u, \quad 0 \leqslant t \leqslant T, \tag{VII.C.50}$$

where $\underline{e}_0 = \underline{X}_0 - \hat{\underline{X}}_0 = \underline{X}_0$. This relationship can be used to derive a differential equation from which $\mathbf{P}_t \triangleq \mathbf{C}_e(t, t)$ can be determined. To see this we assume initially that $\underline{e}_0 \equiv \underline{0}$, (or equivalently that $\mathbf{P}_0 = \mathbf{O}$). This assumption simplifies the derivation. With this assumption and using (VII.C.50), we can write

$$\begin{aligned}
\mathbf{P}_t \triangleq E\{\underline{e}_t \underline{e}'_t\} &= \int_0^t \mathbf{A}_u E\{\underline{e}_u \underline{e}'_t\} du \\
&\quad + E\{(\int_0^t \mathbf{B}_u \, d\underline{V}_u - \int_0^t \mathbf{K}_u \, d\underline{I}_u)\underline{e}'_t\} \\
&= \int_0^t \mathbf{A}_u \mathbf{C}_e(u, t) du \\
&\quad + E\{(\int_0^t \mathbf{B}_u \, d\underline{V}_u)\underline{e}'_t\},
\end{aligned} \tag{VII.C.51}$$

where we have used the fact that $\underline{e}_t$ is orthogonal to $\underline{I}_u$ for all $u \in [0, t]$. Again using (VII.C.50), (VII.C.51) becomes

$$\mathbf{P}_t = \int_0^t\int_0^t \mathbf{A}_u \mathbf{C}_e(u,v)\mathbf{A}_v' \, du dv$$

$$+ \int_0^t \mathbf{A}_u E\{\underline{e}_u (\int_0^t \mathbf{B}_v d\underline{V}_v - \int_0^t \mathbf{K}_v d\underline{I}_v)'\} du$$

$$+ E\{\int_0^t \mathbf{B}_u d\underline{V}_u (\int_0^t \mathbf{A}_v \underline{e}_v dv)'\}$$

$$+ E\{\int_0^t \mathbf{B}_u d\underline{V}_u (\int_0^t \mathbf{B}_v d\underline{V}_v - \int_0^t \mathbf{K}_v d\underline{I}_v)'\}$$

$$= \int_0^t\int_0^t \mathbf{A}_u \mathbf{C}_e(u,v)\mathbf{A}_v' \, du dv \qquad \text{(VII.C.52)}$$

$$+ \int_0^t \mathbf{A}_u E\{\underline{e}_u (\int_0^t \mathbf{B}_v d\underline{V}_v)'\} du$$

$$+ \int_0^t E\{(\int_0^t \mathbf{B}_u d\underline{V}_u)\underline{e}'_v\}\mathbf{A}_v' \, dv$$

$$+ E\{(\int_0^t \mathbf{B}_u d\underline{V}_u)(\int_0^t \mathbf{B}_v d\underline{V}_v)'\}$$

$$- E\{(\int_0^t \mathbf{A}_u \underline{e}_u du + \int_0^t \mathbf{B}_u d\underline{V}_u)(\int_0^t \mathbf{K}_v d\underline{I}_v)'\},$$

where the second equality results from combining terms involving $\int_0^t \mathbf{K}_v d\underline{I}_v$ into the final term on the right.

Consider this final term. From (VII.C.50) we have that

$$E\{(\int_0^t \mathbf{A}_u \underline{e}_u\, du + \int_0^t \mathbf{B}_u\, d\underline{V}_u)(\int_0^t \mathbf{K}_v\, d\underline{I}_v)'\}$$

$$= E\{(\underline{e}_t + \int_0^t \mathbf{K}_u\, d\underline{I}_u)(\int_0^t \mathbf{K}_v\, d\underline{I}_v)'\}$$

$$= E\{(\int_0^t \mathbf{K}_u\, d\underline{I}_u)(\int_0^t \mathbf{K}_v\, d\underline{I}_v)'\} \tag{VII.C.53}$$

$$= \int_0^t \mathbf{K}_u \mathbf{K}_u'\, du ,$$

where the second equality follows from the orthogonality of $\underline{e}_t$ and $\underline{I}_0^t$ and the third equality follows from the fact that $\mathbf{C}_I(u,v)=\mathbf{I}_r \min\{u,v\}$. Also, since $\underline{e}_u$ is produced linearly from $\underline{e}_0$, $\underline{V}_0^u$, and $\underline{W}_0^u$, it is orthogonal to $(\underline{V}_s - \underline{V}_u)$ for all $s \geqslant u$. Thus the second term on the right-hand side of (VII.C.52) becomes

$$\int_0^t \mathbf{A}_u E\{\underline{e}_u (\int_0^t \mathbf{B}_v\, d\underline{V}_v)'\} du$$

$$= \int_0^t \mathbf{A}_u E\{\underline{e}_u (\int_0^u \mathbf{B}_v\, d\underline{V}_v)'\} du , \tag{VII.C.54}$$

and a similar relationship holds for the third term. Inserting (VII.C.53) and (VII.C.54) into (VII.C.52), and noting that $\mathbf{C}_V(u,v)=\mathbf{I}_m \min\{u,v\}$, we have

$$\mathbf{P}_t = \int_0^t \int_0^t \mathbf{A}_u \mathbf{C}_e(u, v) \mathbf{A}_v' \, du dv$$

$$+ \int_0^t \mathbf{A}_u E\{\underline{e}_u (\int_0^u \mathbf{B}_v d\underline{V}_v)'\} du$$

$$+ \int_0^t E\{(\int_0^v \mathbf{B}_u d\underline{V}_u) \underline{e}_v'\} \mathbf{A}_v' \, dv + \int_0^t \mathbf{B}_u \mathbf{B}_u' \, du \qquad \text{(VII.C.55)}$$

$$- \int_0^t \mathbf{K}_u \mathbf{K}_u' \, du , \quad 0 \leqslant t \leqslant T.$$

We can differentiate both sides of (VII.C.55), which yields

$$\dot{\mathbf{P}}_t = \int_0^t \mathbf{A}_t \mathbf{C}_e(t, v) \mathbf{A}_v' \, dv$$

$$+ \int_0^t \mathbf{A}_u \mathbf{C}_e(u, t) \mathbf{A}_t' \, du + \mathbf{A}_t E\{\underline{e}_t (\int_0^t \mathbf{B}_v d\underline{V}_v)'\}$$

$$+ E\{(\int_0^t \mathbf{B}_u d\underline{V}_u) \underline{e}'_t\} \mathbf{A}_t' + \mathbf{B}_t \mathbf{B}_t' - \mathbf{K}_t \mathbf{K}_t' \qquad \text{(VII.C.56)}$$

$$= \mathbf{A}_t [\int_0^t \mathbf{C}_e(t, v) \mathbf{A}_v' \, dv + E\{\underline{e}_t (\int_0^t \mathbf{B}_v d\underline{V}_v)'\}]$$

$$+ [\int_0^t \mathbf{A}_u \mathbf{C}_e(u, t) du + E\{(\int_0^t \mathbf{B}_u d\underline{V}_u) \underline{e}_t'\}] \mathbf{A}_t'$$

$$+ \mathbf{B}_t \mathbf{B}_t' - \mathbf{K}_t \mathbf{K}_t'$$

$$= \mathbf{A}_t \mathbf{P}_t' + \mathbf{P}_t \mathbf{A}_t' + \mathbf{B}_t \mathbf{B}_t' - \mathbf{K}_t \mathbf{K}_t',$$

where the third equality follows from (VII.C.51). Since $\mathbf{P}_t' = \mathbf{P}_t$ and $\mathbf{K}_t = \mathbf{P}_t \mathbf{C}_t'$, (VII.C.56) becomes

$$\dot{\mathbf{P}}_t = \mathbf{A}_t \mathbf{P}_t' + \mathbf{P}_t \mathbf{A}_t' + \mathbf{B}_t \mathbf{B}_t' - \mathbf{P}_t \mathbf{C}_t' \mathbf{C}_t \mathbf{P}_t, \qquad 0 \leqslant t \leqslant T, \tag{VII.C.57}$$

which gives a differential equation for $\mathbf{P}_t$. As assumed, the initial condition for this equation is $\mathbf{P}_0 = \mathbf{O}$. However, it is straightforward to show that (VII.C.57) holds for arbitrary initial state covariance $\mathbf{P}_0$ using the above technique. This derivation is left as an exercise.

Before summarizing the above results for the continuous-time Kalman-Bucy filter, we introduce two modifications into the assumptions. The first of these is to relax (VII.C.33a) and (VII.C.33b) to

$$Cov\ (\underline{U}_t, \underline{U}_s) = \mathbf{Q}_t \delta(t-s), \quad 0 \leqslant t, s \leqslant T \tag{VII.C.58a}$$

and

$$Cov\ (\underline{N}_t, \underline{N}_s) = \mathbf{R}_t \delta(t-s), \quad 0 \leqslant t, s \leqslant T, \tag{VII.C.58b}$$

where for each $t \in [0, T]$, $\mathbf{Q}_t$ and $\mathbf{R}_t$ are covariance matrices with $\mathbf{R}_t > 0$. Assuming that these matrix functions are piecewise-continuous functions of t, the filter derived above can easily be modified to account for this change. In particular, as discussed in Chapter III, we can write $\mathbf{Q}_t$ and $\mathbf{R}_t$ as

$$\mathbf{Q}_t = \mathbf{Q}_t^{1/2} \mathbf{Q}_t^{1/2} \text{ and } \mathbf{R}_t = \mathbf{R}_t^{1/2} \mathbf{R}_t^{1/2},$$

where $\mathbf{Q}_t^{1/2}$ and $\mathbf{R}_t^{1/2}$ are symmetric and $\mathbf{R}_t^{1/2} > 0$. Thus $\underline{U}_t$ is $\mathbf{Q}_t^{1/2}$ times a white noise with covariance $\mathbf{I}_m \delta(t-s)$ and $\underline{N}_t$ is $\mathbf{R}_t^{1/2}$ times a white noise with covariance $\mathbf{I}_r \delta(t-s)$. From this and the invertibility of $\mathbf{R}_t^{1/2}$, it follows straightforwardly that the change in the filter equations caused by this modification in assumptions is to replace $\mathbf{B}_t$ by $\mathbf{B}_t\mathbf{Q}_t^{1/2}$, $\mathbf{K}_t$ by $\mathbf{P}_t\mathbf{C}_t^T\mathbf{R}_t^{-1}$, and $\mathbf{C}_t$ by $\mathbf{R}_t^{1/2}\mathbf{C}_t$.

The second modification in the assumptions is to change $E\{\underline{X}_0\}=\underline{0}$ to $E\{\underline{X}_0\}=\underline{m}_0$. This change merely changes the means of the processes $\{\underline{X}_t\,;0\leqslant t\leqslant T\}$ and $\{\underline{Y}_t\,;0\leqslant t\leqslant T\}$, and as discussed in Chapter V, the MMSE linear estimator for nonzero means is a simple modification of that for the zero-mean case. In the Kalman-Bucy model, it is straightforward to show that this change in assumption merely changes the initial condition of the estimator equation to $\hat{\underline{X}}_0=\underline{m}_0$.

We now summarize the continuous-time Kalman-Bucy model and optimal estimator in the following proposition.

Proposition VII.C.1: The Continuous Time Kalman-Bucy Filter

Suppose that $\{\underline{X}_t\,;0\leqslant t\leqslant T\}$ and $\{\underline{Z}_t\,;0\leqslant t\leqslant T\}$ are generated by the model

$$d\underline{X}_t=\mathbf{A}_t\underline{X}_t\,dt+\mathbf{B}_t\,d\underline{V}_t\,,\quad 0\leqslant t\leqslant T \quad \text{(VII.C.59a)}$$

with initial condition $\underline{X}_0$, and

$$d\underline{Z}_t=\mathbf{C}_t\underline{X}_t\,dt+d\underline{W}_t\,,\, 0\leqslant t\leqslant T\,, \quad \text{(VII.C.59b)}$$

where $E\{\underline{V}_t\}=E\{\underline{W}_t\}=\underline{0}$, $\mathbf{C}_V(t,s)=\int_0^{\min\{t,s\}}\mathbf{Q}_u\,du$, $\mathbf{C}_W(t,s)=\int_0^{\min\{t,s\}}\mathbf{R}_u\,du$, and $\mathbf{C}_{VW}(t,s)=\mathbf{O}$, for all $0\leqslant t,s\leqslant T$, and where $\mathbf{A}_t,\mathbf{B}_t,\mathbf{C}_t,\mathbf{Q}_t$, and $\mathbf{R}_t$, are all piecewise continuous functions of t with $\mathbf{R}_t>0$ for each $t\in[0,T]$. Suppose that further that the initial condition $\underline{X}_0$ has mean $\underline{m}_0$, covariance $\mathbf{P}_0$, and is uncorrelated with $\underline{V}_0^T$ and $\underline{W}_0^T$. Then the MMSE linear estimate of $\underline{X}_t$ given $\underline{Z}_0^t$ is generated by the equation

$$d\hat{\underline{X}}_t=\mathbf{A}_t\hat{\underline{X}}_t\,dt+\mathbf{K}_t\,d\underline{I}_t\,,\,\hat{\underline{X}}_0=\underline{m}_0, \quad \text{(VII.C.60)}$$

where $d\underline{I}_t=d\underline{Z}_t-\mathbf{C}_t\hat{\underline{X}}_t\,dt$,

$$\mathbf{K}_t = \mathbf{P}_t \mathbf{C}_t' \mathbf{R}_t^{-1}, \tag{VII.C.61}$$

and $\mathbf{P}_t \triangleq E\{(\underline{X}_t - \hat{\underline{X}}_t)(\underline{X}_t - \hat{\underline{X}}_t)'\}$ satisfies the *Riccati differential equation*

$$\dot{\mathbf{P}}_t = \mathbf{A}_t \mathbf{P}_t + \mathbf{P}_t \mathbf{A}_t' + \mathbf{B}_t \mathbf{Q}_t \mathbf{B}_t' - \mathbf{P}_t \mathbf{C}_t' \mathbf{R}_t^{-1} \mathbf{C}_t \mathbf{P}_t \tag{VII.C.62}$$

with initial condition $\mathbf{P}_0$.

As in the discrete-time case the optimum filter here can be thought of as a replica of the original system dynamics driven by the innovations through the gain $\mathbf{K}_t$, or as a system with state matrix $\{(\mathbf{A}_t - \mathbf{K}_t \mathbf{C}_t), 0 \leqslant t \leqslant T\}$ driven by the observations $d\underline{Z}^t$ through the gain $\mathbf{K}_t$. Interpreting $d\underline{Z}_t$ as $\underline{Y}_t\, dt$ where $\underline{Y}_t$ is the original white-noise-corrupted observation, these two interpretations of the filter are illustrated in Fig. VII.C.1. Further aspects of the continuous-time Kalman-Bucy filtering problem are developed in the exercises. For additional discussion of this problem the reader is

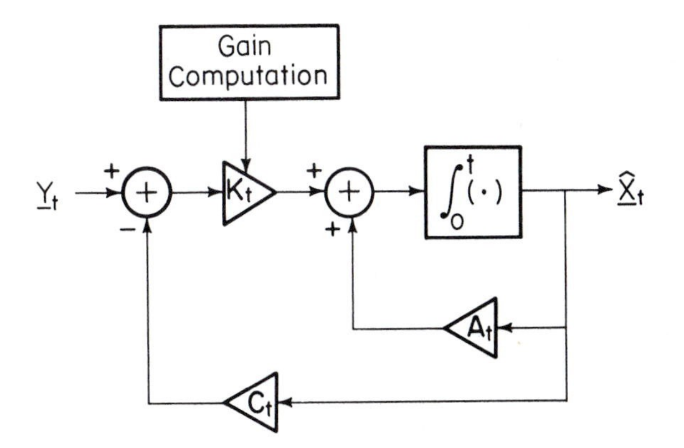

Fig. VII.C.1: The Kalman-Bucy filter in continuous time.

referred to the books by VanTrees (1968) and Kailath (1981).

VII.C.4 Further Aspects of the Linear/Gaussian Problem

The linear/Gaussian estimation results of the preceding sections can be fit within a more general context similar to that for the discrete-time case discussed in Section VC. In the following paragraphs, we outline some of these ideas. In particular, consider two second-order random processes $\{X_t ; t \in \mathbb{R}\}$ and $\{Y_t ; t \in \mathbb{R}\}$ which, for the purposes of discussion, we will assume to have zero means. Suppose that we observe $\{Y_t ; t \in \mathbb{R}\}$ over some time interval $[a, b]$ and we wish to estimate X_t from this observation for some fixed t.

Linear estimates of X_t lie in the set H_a^b of all random variables of the form

$$\sum_{i=1}^{n} c_i Y_{t_i}$$

for $n \geqslant 1, c_i \in \mathbb{R}$ and $t_i \in [a, b], i=1, ..., n$, and mean-square limits of all such sums. Thus the linear MMSE estimation problem here is

$$\min_{Z \in H_a^b} E\{(Z - X_t)^2\}, \qquad \text{(VII.C.63)}$$

which yields the necessary and sufficient orthogonality conditions for an optimum estimator $\hat{X}_t$:

$$E\{(X_t - \hat{X}_t)Z\} = 0, \quad Z \in H_a^b, \qquad \text{(VII.C.64)}$$

or equivalently,

$$E\{(\hat{X}_t - X_t)Y_s\} = 0, \quad a \leqslant s \leqslant b. \qquad \text{(VII.C.65)}$$

An interesting subset of H_a^b is the set of linear filtrations of $\{Y_t ; a \leqslant t \leqslant b\}$; i.e., the set of all estimates of the form

$$\int_a^b h(t,\tau)Y_\tau d\tau, \qquad \text{(VII.C.66)}$$

with $\int_a^b\int_a^b h(t,\tau)h(t,\sigma)C_Y(\tau,\sigma)d\tau d\sigma<\infty$. (Here, as usual, C_Y denotes the autocovariance function of $\{Y_t;t\in\mathbb{R}\}$.) Applying (VII.C.65), and the interchangeability of mean-square limits and expectations, it is straightforward to see that an estimate of the form (VII.C.66) is optimum if and only if h satisfies the integral equation

$$C_{XY}(t,s)=\int_a^b h(t,\tau)C_Y(\tau,s)d\tau, \quad a\leqslant s\leqslant b, \qquad \text{(VII.C.67)}$$

where C_{XY} is the cross-covariance function of $\{X_t;t\in\mathbb{R}\}$ and $\{Y_t;t\in\mathbb{R}\}$.

Equation (VII.C.67) is the continuous-time version of the *Wiener-Hopf equation* introduced in (V.C.18). This equation is a Fredholm equation of the first kind, and as mentioned earlier, solutions to this type of equation often contain singularities. This is a manifestation of the fact that unlike the analogous discrete-time quantity, (VII.C.66) does not model a sufficiently large class of estimates to solve the linear MMSE estimation problems of interest in practice unless one allows singularities in $h(t,\tau)$. For example, depending on the model, one might consider estimates that contain not only filtered versions of $\{Y_t;a\leqslant t\leqslant b\}$ but also weighted samples of $\{Y_t;a\leqslant t\leqslant b\}$ itself and of derivatives of $\{Y_t;a\leqslant t\leqslant b\}$.

Two types of problem for which the class of estimates described by (VII.C.66) is large enough to contain most solutions of interest are problems in which the observation interval is infinite (i.e., $a=-\infty$ and/or $b=+\infty$) and problems in which $\{Y_t;t\in\mathbb{R}\}$ has a white component uncorrelated with $\{X_t;t\in\mathbb{R}\}$.

The infinite-observation interval problem of most interest is the continuous-time versions of the Wiener-Kolmogorov problem. In particular by assuming that $\{X_t;t\in\mathbb{R}\}$ and $\{Y_t;t\in\mathbb{R}\}$ are jointly and individually wide-sense stationary (i.e., that C_{XY} and C_Y are functions only of the difference of their arguments) and taking $a=-\infty$, we have the causal and noncausal versions of this problem by taking

$b=t$ and $b=+\infty$, respectively. It is straightforward to show that, in either case, attention can be restricted to time-invariant filters; i.e., that $h(t,\tau)$ can be taken to be $h(t-\tau)$ for some function $h:\mathbb{R}\to\mathbb{R}$. Moreover, the Wiener-Hopf equation reduces to

$$C_{XY}(\tau)=\int_{-\infty}^{\infty} h(\alpha)C_Y(\tau-\alpha)d\alpha,\ -\infty<\tau<\infty \qquad \text{(VII.C.68)}$$

in the noncausal case, and to

$$C_{XY}(\tau)=\int_{0}^{\infty} h(\alpha)C_Y(\tau-\alpha)d\alpha,\quad 0\leqslant\tau<\infty \qquad \text{(VII.C.69)}$$

in the causal case.

Analogously to the discrete-time case, both (VII.C.68) and (VII.C.69) can be solved by using Fourier transforms. To see this, assume that C_{XY}, h, and C_Y have Fourier transforms ϕ_{XY}, H, and ϕ_Y, respectively. (Note that ϕ_{XY} is the cross-spectrum of $\{X_t;t\in\mathbb{R}\}$ and $\{Y_t;t\in\mathbb{R}\}$, and ϕ_Y is the spectrum of $\{Y_t;t\in\mathbb{R}\}$.) Since (VII.C.68) can be written as $C_{XY}(\tau)=(h*C_Y)(\tau)$, where $*$ denotes the convolution operation, (VII.C.68) is equivalent to

$$\phi_{XY}(\omega)=H(\omega)\phi_Y(\omega),\ -\infty<\omega<\infty. \qquad \text{(VII.C.70)}$$

Thus, as in the discrete-time case, we see that the transfer function of the optimum noncausal filter is given simply by

$$H_0(\omega)=\frac{\phi_{XY}(\omega)}{\phi_Y(\omega)}. \qquad \text{(VII.C.71)}$$

The solution of (VII.C.69) is somewhat more difficult because of the causality constraint on h. As with the discrete-time case, this solution involves the factorization of the observation spectrum ϕ_Y into the product of causal and anticausal parts ϕ_Y^+ and ϕ_Y^- $[\equiv(\phi_Y^+)^*]$, respectively. This can always be done provided ϕ_Y satisfies the continuous-time

Paley-Wiener condition (see, e.g., Wong (1983)):

$$\int_{-\infty}^{\infty} |\log \phi_Y(\omega)|/(1+\omega^2) d\omega < \infty. \qquad \text{(VII.C.72)}$$

Having thus factored ϕ_Y, the transfer function of the solution to (VII.C.70) can be shown to be given by

$$H_0^+(\omega) = \frac{1}{\phi_Y^+(\omega)} \left[\frac{\phi_{XY}(\omega)}{\phi_Y^-(\omega)} \right]_+, \qquad \text{(VII.C.73)}$$

where the operation $[H(\omega)]_+$ denotes taking the causal part in an additive spectral decomposition; i.e., $[H(\omega)]_+ = F\{h^+(t)\}$ where

$$h^+(t) = \begin{cases} h(t) & \text{if } t \geqslant 0 \\ 0 & \text{if } t < 0. \end{cases} \qquad \text{(VII.C.74)}$$

That (VII.C.73) solves (VII.C.69) can be shown, for example, by direct substitution. This solution has the same intuitive interpretation as the analogous discrete-time filter; namely as a causal prewhitener, $1/\phi_Y^+$, followed by a MMSE causal estimator of X_t from the prewhitened observation.

As noted above, a second interesting class of estimation problems for which solutions can be derived from the Wiener-Hopf equation (VII.C.67) is that in which the observation $\{Y_t; t \in \mathbb{R}\}$ contains a components of white noise uncorrelated with $\{X_t; t \in \mathbb{R}\}$. (Of course, this type of model does not fit the second-order model and the usual interpretation of white noise should be kept in mind.) In particular if we assume that $C_Y(t,s) = C_C(t,s) + (N_0/2)\delta(t-s)$ then the Wiener-Hopf equation becomes

$$C_{XY}(t,s) = \int_a^b h(t,\tau)C_C(\tau,s)d\tau + \frac{N_0}{2}h(t,s), \tag{VII.C.75}$$

$$a \leqslant s \leqslant b.$$

Equation (VII.C.75) is a Fredholm equation of the second kind, and assuming C_{XY} and C_C are continuous and $N_0/2>0$, it has a unique continuous solution.

The principal example of interest is that of signal estimation in additive uncorrelated white noise, for which $X_t = S_t$ and

$$Y_t = S_t + N_t, \quad a \leqslant t \leqslant b, \tag{VII.C.76}$$

with $\{N_t; -\infty < t < \infty\}$ white, in which case, $C_{XY} = C_C = C_X$. The Wiener-Hopf equation thus reduces to (VII.C.7). Note that the Kalman-Bucy problem is a special case of (VII.C.76) (the extension of the above to the vector case is straightforward), and the Kalman-Bucy filter can be derived directly from the Wiener-Hopf equation [see, e.g., VanTrees (1968)]. Our derivation used this approach in part since the equation above (VII.C.41) is in fact the Wiener-Hopf equation for estimating the state from the innovation.

VII.D Nonlinear Filtering

Solution techniques for the linear/Gaussian signal estimation problems discussed in Section VII.C are relatively well developed. Unfortunately, when it is necessary to abandon the linear constraint and the Gaussian model, the problem of optimum signal estimation in continuous time becomes quite difficult. The basic difficulty is the same as that in the continuous-time detection problem for non-Gaussian models, namely, the construction of suitable densities for non-Gaussian measures on function spaces. Nevertheless, some progress in this problem can be made, particularly for models in which signals and observations are generated by nonlinear dynamical models, i.e., for nonlinear versions on the Kalman-Bucy problem. Since this class of models also provides good approximations to the behavior of a large number of physical phenomena, optimum filtering procedures for nonlinear dynamical models are of interest in a number of applications. In this section we consider this nonlinear filtering problem. As a complete and rigorous treatment of this subject can easily fill a book of its own, our discussion will be limited to a summary of key ideas in this area. The reader interested in a more detailed treatment of this subject is referred to the book by Lipster and Shiryayev (1977). Also of interest are the review articles by Beneš (1987) and Marcus (1984).

The behavior of a large number of physical phenomena can be modeled by nonlinear stochastic differential equations of the form

$$dX_t = m(X_t, t)dt + \sigma(X_t, t)dW_t, \; t \geqslant 0, \quad \text{(VII.D.1)}$$

with an initial condition X_0, where m and σ are real-valued functions on $\mathbb{R} \times (0, \infty)$ with $\sigma > 0$, and where $\{W_t ; t \geqslant 0\}$ is a standard Wiener process independent of X_0. As before, the notation in (VII.D.1) is short hand for the integral equation

$$X_t = X_0 + \int_0^t m(X_s, s)ds + \int_0^t \sigma(X_s, s)dW_s, \quad \text{(VII.D.2)}$$

where, for our purposes, we interpret the second integral as an Ito stochastic integral. Note that a special case of (VII.D.1) is the linear case in which $m(X_t, t) = A_t X_t$ and $\sigma(X_t, t) = B_t$ for some deterministic functions A_t and B_t. Thus (VII.D.1) represents a generalization of the linear stochastic system model arising in the Kalman-Bucy problem. For the purposes of exposition, we will restrict our discussion of such models to the scalar case, the situation for the vector case being similar.

Equation (VII.D.1) or (VII.D.2) is not a meaningful model unless a random process satisfying this equation exists. Fortunately, existence and uniqueness of well-behaved solutions to (VII.D.2) are assured within simple smoothness and growth conditions on the functions m and σ. Assuming that $E\{X_0^2\} < \infty$, a sufficient condition for existence of such a process is the *linear growth condition*

$$|m(x,t)| + |\sigma(x,t)| \leqslant K[1+x^2]^{1/2} \qquad \text{(VII.D.3)}$$

for some $K>0$, and a sufficient condition for uniqueness is the *uniform Lipschitz condition*

$$|m(x,t)-m(y,t)| + |\sigma(x,t)-\sigma(y,t)| \leqslant K|x-y|, \qquad \text{(VII.D.4)}$$

for some $K>0$ and all $(x,y)\in \mathbb{R}^2$.

Henceforth, we assume that (VII.D.2) is a meaningful equation defining a unique random process with continuous sample paths. In such processes, the term $m(X_t,t)dt$ is usually known as the *drift* term and $\sigma(X_t,t)dW_t$ as the *variance* term.

Processes of the form (VII.D.1) are known as *diffusions*. Such processes are *Markov processes*; that is, they have the property that for each $s>0$ and conditioned on X_s, $\{X_t;t>s\}$ and $\{X_t;0\leqslant t<s\}$ are independent. Intuitively, this property is easy to see since for any $t\geqslant s\geqslant 0$, we can write

$$X_t = X_s + \int_s^t m(X_u,u)du + \int_s^t \sigma(X_u,u)dW_u. \qquad \text{(VII.D.5)}$$

Equation (VII.D.5) indicates that $\{X_t;t>s\}$ can be constructed completely from X_s and $\{W_t-W_s;t>s\}$. Thus, with X_s fixed, $\{X_t;t>s\}$ is generated independently of $\{X_t;t<s\}$ since $\{W_t-W_s;t>s\}$ is independent of all the past.

The class of all random processes satisfying the Markov property comprises a large fraction of the stochastic models used in practice and is much broader than the class of diffusions. However, it turns out that all continuous Markov processes that satisfy certain regularity conditions are diffusions representable by (VII.D.1) with drift given by

$$m(x,t) \triangleq \lim_{\Delta \to 0^+} \frac{E\{X_{t+\Delta} | X_t = x\} - x}{\Delta}, \qquad \text{(VII.D.6)}$$

and variance given by the positive square root of

$$\sigma^2(x,t) \triangleq \lim_{\Delta \to 0^+} \frac{Var\,(X_{t+\Delta} | X_t = x)}{\Delta}. \qquad \text{(VII.D.7)}$$

Thus the class of diffusion processes (and their vector counterparts) covers a fairly large subset of models arising in practice. [Note that $m(x,t)$ and $\sigma^2(x,t)$ from (VII.D.6) and (VII.D.7) are, respectively, the time derivatives at t of the conditional mean and variance of the process given $X_t = x$.]

Because of the Markov property, the statistics of diffusions are much simpler to characterize than those of general non-Gaussian random processes. In particular, for any positive integer n and a set of n times $0 \leq t_1 < t_2 < \ldots < t_{n-1} < t_n < \infty$, the joint density of $X_{t_1}, \ldots, X_{t_n}$ is given simply by [see, e.g., (III.B.99)]

$$p_{X_{t_1} \cdots X_{t_n}}(x_1, \ldots, x_n) = p_{X_{t_1}}(x_1) \prod_{k=2}^{n} p_{X_{t_k} | X_{t_{k-1}}}(x_k | x_{k-1}). \qquad \text{(VII.D.8)}$$

Thus all finite-dimensional distributions of the process can be obtained from the density of the initial condition, p_{X_0}, and the *transition densities* $p_{X_t | X_s}(x | y), t > s \geq 0$. Moreover, within regularity,† these transition densities satisfy the partial differential equation

$$\frac{\partial}{\partial t} p_{X_t | X_s}(x | y) = \mathbf{A}_t^* p_{X_t | X_s}(x | y), \; t > s, \qquad \text{(VII.D.9)}$$

where for a function $f(x)$, $\mathbf{A}_t^* f(x)$ denotes the operation

† A sufficient set of conditions is that $\sigma(x,t)$ be bounded away from zero and that $\sigma(x,t), m(x,t), \partial\sigma(x,t)/\partial x, \partial m(x,t)/\partial x$, and $\partial^2\sigma(x,t), \partial x^2$ all satisfy linear growth and uniform Lipschitz conditions.

$$\mathbf{A}_t^* f(x) = \tfrac{1}{2}\frac{\partial^2}{\partial x^2}[\sigma^2(x,t)f(x)] - \frac{\partial}{\partial x}[m(x,t)f(x)]. \tag{VII.D.10}$$

Equation (VII.D.9) is known as the *Fokker-Planck equation* and also as *Kolmogorov's forward equation of diffusion.*

One consequence of the Fokker-Planck equation is that the first-order density of the process satisfies the partial differential equation

$$\frac{\partial}{\partial t}p_{X_t}(x) = \mathbf{A}_t^* p_{X_t}(x), \quad t > 0,\ x \in \mathbb{R}, \tag{VII.D.11}$$

with boundary condition at $t=0$ given by the initial density $p_{X_0}(x)$. From (VII.D.11) we can get differential equations for expectations of random variables of the form $f(X_t)$, where f is a real-valued function on $\mathbb{R}$. In particular, note that

$$E\{f(X_t)\} = \int_{-\infty}^{\infty} f(x)p_{X_t}(x)dx. \tag{VII.D.12}$$

Thus, assuming order of integration and differentiation can be interchanged, we have

$$\frac{d}{dt}E\{f(X_t)\} = \int_{-\infty}^{\infty} f(x)\frac{\partial}{\partial t}p_{X_t}(x)dx = \int_{-\infty}^{\infty} f(x)\mathbf{A}_t^* p_{X_t}(x)dx. \tag{VII.D.13}$$

Assuming sufficient regularity[†] of f, integration by parts reduces (VII.D.13) to

$$\frac{d}{dt}E\{f(X_t)\} = E\{\mathbf{A}_t f(X_t)\}, t \geqslant 0, \qquad \text{(VII.D.14)}$$

with initial condition $E\{f(X_0)\}$, where the operation $\mathbf{A}_t$ is defined by

$$\mathbf{A}_t f(x) \triangleq m(x,t)\frac{d}{dx}f(x) + \tfrac{1}{2}\sigma^2(x,t)\frac{d^2}{dx^2}f(x).$$

For example, the mean of X_t, $\mu_t = E\{X_t\}$, corresponds to the case $f(x)=x$. From (VII.D.15) we thus have that

$$\dot{\mu}_t = E\{m(X_t, t)\}, t \geqslant 0, \text{ with } \mu_0 = E\{X_0\}, \qquad \text{(VII.D.15)}$$

where the dot over the variable denotes time differentiation. Similarly, the variance of X_t, $v_t = E\{X_t^2\} - \mu_t^2$, satisfies the differential equation

$$\dot{v}_t = \frac{d}{dt}E\{X_t^2\} - 2\mu_t \dot{\mu}_t$$

$$\text{(VII.D.16)}$$

$$= 2\, Cov\,(X_t, m(X_t, t)) + E\{\sigma^2(X_t, t)\}, t \geqslant 0,$$

with $v_0 = Var\,(X_0)$. Note that neither of these equations are easily solved in general due to the presence of expectations involving $m(X_t, t)$ and $\sigma^2(X_t, t)$. These quantities also satisfy differential equations of the form (VII.D.15); however, these equations involve expectations of further nonlinear

[†] Sufficient conditions are that $\partial^2 f(x)/\partial x^2$ exists, is continuous, and satisfies

$$\lim_{|x|\to\infty} m(x,t)f(x)p_{X_t}(x) = \lim_{|x|\to\infty}\left[\frac{\partial}{\partial x}f(x)\right]\left[\frac{\partial}{\partial x}\sigma(x,t)p_{X_t}(x)\right]$$

$$= \lim_{|x|\to\infty}\left[\frac{\partial^2}{\partial x^2}f(x)\right][\sigma(x,t)p_{X_t}(x)] = 0.$$

functions of X_t, which lead to further equations, etc. As we shall see, a similar difficulty arises in the nonlinear filtering problem.

An exception to this difficulty is the linear case, $m(X_t, t) = A_t X_t$ and $\sigma(X_t, t) = B_t$ where A_t and B_t are known functions, for which (VII.D.15) and (VII.D.16) reduce to

$$\dot{\mu}_t = A_t \mu_t, \; t \geqslant 0, \; \mu_0 = E\{X_0\} \qquad \text{(VII.D.17)}$$

and

$$\dot{v}_t = 2A_t + B_t^2, \; t \geqslant 0, \; v_0 = Var\ (X_0). \qquad \text{(VII.D.18)}$$

Equations (VII.D.17) and (VII.D.18) are simple first-order linear differential equations with no coupling to other unknown quantities. Note that (VII.D.17) and (VII.D.18) are the mean and variance equations, respectively, in the Kalman filtering formulation with no observations ($C_t = 0, t \geqslant 0$).

It should be noted that the differential equation (VII.D.14) describing the evolution of $E\{f(X_t)\}$ can be derived directly from the diffusion equation (VII.D.1) without the intermediate step of deriving the Fokker-Planck equation. In particular, assuming that f has a continuous first and second derivatives f' and f'', respectively, we can use Taylor's theorem and (VII.D.1) to write for $\Delta > 0$,

$$\frac{f(X_{t+\Delta}) - f(X_t)}{\Delta}$$

$$\sim \frac{(X_{t+\Delta} - X_t)}{\Delta} f'(X_t) + \tfrac{1}{2} \frac{(X_{t+\Delta} - X_t)^2}{\Delta} f''(X_t)$$

$$\sim \left[m(X_t, t) + \sigma(X_t, t) \frac{(W_{t+\Delta} - W_t)}{\Delta} \right] f'(X_t)$$

$$+ \tfrac{1}{2} \frac{[m(X_t, t)\Delta + \sigma(X_t, t)(W_{t+\Delta} - W_t)]^2}{\Delta} f''(X_t)$$

$$= m(X_t, t)f'(X_t) + \sigma(X_t, t)f'(X_t)\frac{(W_{t+\Delta} - W_t)}{\Delta}$$

$$+ \frac{\Delta}{2}m^2(X_t, t)f''(X_t)$$

$$\text{(VII.D.19)}$$

$$+ m(X_t, t)\sigma(X_t, t)f''(X_t)(W_{t+\Delta} - W_t)$$

$$+ \tfrac{1}{2}\sigma^2(X_t, t)f''(X_t)\frac{(W_{t+\Delta} - W_t)^2}{\Delta}.$$

On noting that $(W_{t+\Delta} - W_t)$ is $N(0, \Delta)$ and is independent of X_t, taking expectations on both sides of (VII.D.19) leads to

$$\frac{E\{f(X_{t+\Delta})] - E\{f(X_t)\}}{\Delta} \sim E\{m(X_t, t)f'(X_t)\}$$

$$+ \frac{\Delta}{2}E\{m^2(X_t, t)f''(X_t)\} + \tfrac{1}{2}E\{\sigma^2(X_t, t)f''(X_t)\}.$$

Upon assuming that $E\{m^2(X_t, t)f''(X_t)\} < \infty$ and allowing $\Delta \to 0$, this expression becomes

$$\frac{d}{dt}E\{f(X_t)\} = E\{m(X_t, t)f'(X_t) + \tfrac{1}{2}\sigma^2(X_t, t)f''(X_t)\}$$

$$= E\{\mathbf{A}_t f(X_t)\},$$

which is (VII.D.14).

Note that, since $E\{f(X_t)\} = \int_{-\infty}^{\infty} f(x)p_{X_t}(x)dx$, we can think of $p_{X_t}(x)$ for a particular x as $E\{\delta(x - X_t)\}$, where δ is the Dirac delta function. By reversing the operations by which (VII.D.13) leads to (VII.D.14), the moment equation (VII.D.14) gives

$$\frac{d}{dt}\int_{-\infty}^{\infty} f(x)p_{X_t}(x)dx = \int_{-\infty}^{\infty} f(x)A_t^* p_{X_t}(x)dx,$$

which, upon allowing f to be the required impulse yields the equation $\partial p_{X_t}/\partial t = A_t^* p_{X_t}$, for the evolution of the density. (The Fokker-Planck equation itself can be derived in this way by conditioning the expectations above on $X_s = y$.) Although this final step is not rigorous since f was assumed above to have a continuous second derivative, it can be made rigorous for sufficiently regular m and σ by representing the Dirac delta function as a limit of smooth functions. Thus we see that the equation for the evolution of expectations and that for the evolution of the density contain basically the same information about the diffusion. We return to this point again when deriving the basic equations for nonlinear filtering.

VII.D.1 Basic Equations of Nonlinear Filtering

We now consider the situation in which the diffusion $\{X_t ; t \geq 0\}$ generated by (VII.D.1) is observed through a noisy, nonlinear observation

$$Y_t = h(X_t, t) + N_t, \; t \geq 0, \qquad \text{(VII.D.20)}$$

where h is real-valued function on $\mathbb{R} \times (0, \infty)$ and, as in the Kalman filtering formulating, $\{N_t ; t \geq 0\}$ is a zero-mean Gaussian process, independent of $\{W_t ; t \geq 0\}$ and X_0, with $E\{N_t N_s\} = R_t \delta(t-s)$, where $R_t > 0$ for each $t \geq 0$. To obtain a more rigorous model, we integrate (VII.D.20) to yield observations $\{Z_t ; t \geq 0\}$ given by

$$dZ_t = h(X_t, t)dt + R_t^{1/2} dV_t, \; t \geq 0, \qquad \text{(VII.D.21)}$$

where $\{V_t ; t \geq 0\}$ is a standard Wiener process independent of X_0 and $\{W_t ; t \geq 0\}$.

Our objective is to estimate at each $t \geq 0$ some given function of X_t, say $f(X_t)$, from observations up to time t. As we know from Chapter IV, the minimum-mean-squared-error estimator is the conditional mean

$$\widehat{f(X_t)} = E\{f(X_t)|Z_0^t\}. \qquad \text{(VII.D.22)}$$

Note that (VII.D.22) can be computed by

$$\widehat{f(X_t)} = \int_{-\infty}^{\infty} f(x) q_{X_t}(x) dx , \qquad \text{(VII.D.23)}$$

where $q_{X_t}(x)$ denotes the conditional density of X_t given Z_0^t. [The functional dependence of $q_{X_t}(x)$ on Z_0^t should be kept in mind, although we suppress it notationally for the sake of simplicity.] Note that in the absence of observations [i.e., $h(x,t) \equiv 0$], $q_{X_t}(x)$ is simply $p_{X_t}(x)$ and $\widehat{f(X_t)}$ satisfies the differential equation (VII.D.15). Thus if we could find an equation for the conditional density $q_{X_t}(x)$ similar to that for $p_{X_t}(x)$ arising from the Fokker-Planck equation, a similar differential equation for the conditional mean (VII.D.22) could be found. Note, however, that $\{q_{X_t}(x); t \geqslant 0\}$ is a random process for each $x \in \mathbb{R}$, and so such an equation would necessarily be stochastic.

It turns out that within regularity, such an equation is indeed valid. In particular, $q_{X_t}(x)$ satisfies the following stochastic partial differential equation.

$$dq_{X_t}(x) = \mathbf{A}_t^* q_{X_t}(x) dt$$

$$+ q_{X_t}(x)[h(x,t) - \widehat{h(X_t,t)}]R_t^{-1} dI_t , \qquad \text{(VII.D.24)}$$

$$t \geqslant 0,$$

where, here and in what follows, the carat over a function of X_t denotes the conditional mean of that function conditioned on Z_0^t. The process $\{I_t; t \geqslant 0\}$ in (VII.D.24) is the (nonlinear) innovations process

$$I_t \triangleq Z_t - \int_0^t \widehat{h(X_s, s)} ds, \; t \geq 0, \qquad \text{(VII.D.25)}$$

and the initial condition is $q_{X_0}(x) = p_{X_0}(x)$. Equation (VII.D.24) is sometimes known as *Kushner's equation*. Note that this equation reduces to the Fokker-Planck equation (VII.D.12) for p_{X_t} when there are no observations (i.e., $h \equiv 0$). Otherwise, the evolution of the conditional density is dictated by a combination of the system dynamics (through the operator $\mathbf{A}_t^*$) and the observations. Note that the set of equations (VII.D.24) for $x \in \mathbb{R}$ gives a recursion for the density $q_{X_t}(x)$ since the right-hand side involves only $q_{X_t}(x)$, its derivatives, and $\widehat{h(X_t, t)}$, which can be computed from q_{X_t}. We will not derive (VII.D.24) at this point. An outline of its derivation will be given below.

We can now combine (VII.D.23) and (VII.D.24) to obtain a stochastic differential equation for the evolution of $\widehat{f(X_t)}$. In particular, let us first rewrite (VII.D.24) in its integral form

$$q_{X_t}(x) = p_{X_0}(x) + \int_0^t q_{X_s}(x) ds$$
$$\text{(VII.D.26)}$$
$$+ \int_0^t \mathbf{A}_s^* q_{X_s}(x)[h(x, s) - \widehat{h(X_s, s)}] R_s^{-1} dI_s .$$

Assuming interchanges of orders of integration are permissible, we then have

$$\widehat{f(X_t)} = \int_{-\infty}^{\infty} f(x)q_{X_t}(x)dx$$

$$= \int_{-\infty}^{\infty} f(x)p_{X_0}(x)dx + \int_0^t \int_{-\infty}^{\infty} f(x)\mathbf{A}_s^* q_{X_s}(x)dx$$

$$+ \int_0^t \int_{-\infty}^{\infty} f(x)q_{X_s}(x)[h(x,s) - \widehat{h(X_s,s)}]dxdI_s$$

$$= E\{f(X_0)\} + \int_0^t \int_{-\infty}^{\infty} q_{X_s}(x)\mathbf{A}_s f(x)dxds \tag{VII.D.27}$$

$$+ \int_0^t [\int_{-\infty}^{\infty} f(x)h(x,s)q_{X_s}(x)dx$$

$$- \int_{-\infty}^{\infty} f(x)q_{X_s}(x)dx\widehat{h(X_s,s)}]R_s^{-1}dI_s$$

$$= E\{f(X_0)\} + \int_0^t \widehat{\mathbf{A}_s f(X_s)}ds$$

$$+ \int_0^t [\widehat{f(X_s)h(X_s,s)} - \widehat{f(X_s)}\widehat{h(X_s,s)}]R_s^{-1}dI_s,$$

where, in the second equality, we have made use of integration by parts as in (VII.D.15). Equation (VII.D.27) can be rewritten as

$$d\widehat{f(X_t)} = \widehat{\mathbf{A}_t f(X_t)}dt$$

$$+ [\widehat{f(X_t)h(X_t,t)} - \widehat{f(X_t)}\widehat{h(X_t,t)}]R_t^{-1}dI_t, \tag{VII.D.28}$$

$$t > 0, \widehat{f(X_0)} = E\{f(X_0)\}.$$

Note that, the term $\widehat{f(X_t)h(X_t,t)} - \widehat{f(X_t)}\widehat{h(X_t,t)}$ appearing in this equation is the conditional covariance of $f(X_t)$ and $h(X_t,t)$ conditioned on Z_0^t.

An equation similar to (VII.D.28) can be derived for the case in which $f(X_t)$ also depends explicitly on t; i.e., we can consider $f(X_t,t)$. The only modification is that the right-hand side of (VII.D.28) contains an extra term, $\widehat{\dot{f}(X_t,t)}dt$, where $\dot{f} = \partial f / \partial t$. Note that as in the unobserved case ($h \equiv 0$), the equation (VII.D.24) for the evolution of the conditional density $q_{X_t}(x)$ follows from (VII.D.28) by letting $f(X_t) \to \delta(x - X_t)$. Thus the two equations (VII.D.24) and (VII.D.28) are essentially equivalent.

Although (VII.D.28) describes the evolution of the filter for estimating $f(X_t)$ from Z_0^t, this equation does not generally provide an implementation of that filter. This is because implementation of (VII.D.28) also requires implementation of filters for $\widehat{\mathbf{A}_t^* f}$, $\widehat{fh}$, and $\hat{h}$, which in turn require implementation of further nonlinear filters, and so on. Unless the set of quantities involved in these equations close after a finite number of applications of (VII.D.28), the filter for $\widehat{f(X_t)}$ derived in this way will be infinite dimensional. In contrast, as noted above, the equation for the conditional density is a strict recursion. However, it is a recursion for an infinite-dimensional parameter since the conditional density is, of course, function-valued. Unfortunately, this difficulty is alleviated only in some special cases.

These cases and other aspects of the filtering equation (VII.D.28) are discussed following the examples below. Despite the general unsuitability of the optimum filtering equation from the viewpoint of filter implementation, it is useful in suggesting approximations to the optimum filters, which in fact do lead to filters that are useful for practical applications. A brief discussion of some practical filters arising as approximations to (VII.D.28) also follows the examples.

Example VII.D.1: Direct State Estimation - Evolution of the Conditional Mean

Commonly, one is interested in direct estimation of the diffusion X_t itself. This corresponds to the case $f(x)=x$, and the filtering equation (VII.D.28) becomes

$$d\hat{X}_t = m(X_t, t)dt \\ + [\widehat{X_t h(X_t, t)} - \hat{X}_t \widehat{h(X_t, t)}]R_t^{-1}dI_t . \tag{VII.D.29}$$

Note that in the absence of observations ($h \equiv 0$), (VII.D.29) reduces to (VII.D.15).

For linear observations [i.e., $h(X_t, t)=C_t X_t$], (VII.D.29) can be written as

$$\begin{aligned} d\hat{X}_t &= \widehat{m(X_t, t)}dt + C_t[\widehat{X_t^2} - \hat{X}_t^2]R_t^{-1}dI_t \\ &= \widehat{m(X_t, t)}dt + C_t P_t R_t^{-1}dI_t , \end{aligned} \tag{VII.D.30}$$

where $P_t \triangleq \widehat{X_t^2} - \hat{X}_t^2$ is the conditional variance of X_t given Z_0^t; i.e., P_t is the conditional MSE. Note that the innovation term, $C_t P_t R_t^{-1}dI_t$, is identical in form to the corresponding term in the Kalman filter [see (VII.C.61)] Moreover, if we also assume a linear drift term, $m(X_t, t)=A_t X_t$, the filtering equation (VII.D.31) reduces further to

$$d\hat{X}_t = A_t \hat{X}_t dt + C_t P_t R_t^{-1}dI_t . \tag{VII.D.31}$$

Equation (VII.D.31) is identical in form to the filtering equation arising in the Kalman-Bucy model. However, there is a basic difference in that P_t in (VII.D.31) may possibly depend on Z_0^t. We consider this issue in the following example.

Example VII.D.2: Direct State Estimation - Evolution of the Conditional Variance

The performance of the estimator $\hat{X}_t$ of Example VII.D.1 can be measured in terms of the conditional MSE, $P_t = \widehat{X_t^2} - \hat{X}_t^2$, introduced above. To consider the behavior of this quantity, we can seek an expression for $dP_t = d\widehat{X_t^2} - d(\hat{X}_t^2)$. The first term, $d\widehat{X_t^2}$, is easy since it corresponds to the case $f(x) = x^2$, and direct application of (VII.D.28) yields

$$d\widehat{X_t^2} = 2\widehat{X_t m(X_t, t)}dt + \widehat{\sigma^2(X_t, t)}dt$$
$$+ [\widehat{X_t^2 h(X_t, t)} - \widehat{X_t^2}\widehat{h(X_t, t)}]R_t^{-1}dI_t. \tag{VII.D.32}$$

To treat the second term, $d\hat{X}_t^2$, one may be tempted to write $d\hat{X}_t^2 = 2\hat{X}_t d\hat{X}_t$ and then use the evolution equation of $\hat{X}_t$ to eliminate $d\hat{X}_t$. However, note from (VII.D.30) that $\hat{X}_t$ contains an Ito stochastic integral with respect to the innovation process $\{I_s; 0 \leqslant s \leqslant t\}$. We have already seen in Section VI.D (see Proposition VI.D.5) that processes of this type do not generally obey the rule $\int_0^t X_s dX_s = \frac{1}{2}(X_t^2 - X_0^2)$, so a more careful consideration of this term is needed.

A correct expression for $d\hat{X}_t^2$ can be obtained from the so-called *Ito differentiation rule* or *Ito formula*, which we now state.

Proposition VII.D.1: The Ito Differentiation Rule

Suppose that F_t is a random process of the form

$$dF_t = G_t dt + H_t dW_t, \quad 0 \leqslant t \leqslant T \tag{VII.D.33}$$

where G_0^T and H_0^T are random processes satisfying $P(\int_0^T |G_t| dt < \infty) = P(\int_0^T H_t^2 dt < \infty) = 1$, and where W_0^T is a Wiener process such that, for each $s \in (0, T)$, $\{W_t - W_s; s \leqslant t \leqslant T\}$ is independent of (G_0^s, H_0^s, W_0^s). Suppose further that g is a real-valued function on $\mathbb{R}$ with continuous first and second derivatives g' and

g'', respectively. Define a process $J_t = g(F_t)$. Then

$$dJ_t = g'(F_t)dF_t + \tfrac{1}{2}g''(F_t)H_t^2, \quad 0 \leqslant t \leqslant T. \qquad \text{(VII.D.34)}$$

Remarks

(1) The formula in ordinary calculus corresponding to (VII.D.33) is $dJ_t = g'(F_t)dF_t$. The appearance of the *correction term*, $\tfrac{1}{2}g''(F_t)H_t^2dt$, is due to the fact that $(dW_t)^2$ for a Wiener process behaves basically like dt rather than $(dt)^2$. This translates into the behavior $(dF_t)^2 \sim H_t^2dt$ for processes of the form (VII.D.33). In particular, as a plausibility argument for (VII.D.34) we can use a Taylor series expansion for $dg(F_t)$ to get

$$\begin{aligned} dg(F_t) &\sim g(F_{t+dt}) - g(F_t) \\ &\sim g'(F_t)(F_{t+dt} - F_t) \\ &\quad + \tfrac{1}{2}g''(F_t)(F_{t+dt} - F_t)^2 \\ &\sim g'(F_t)dF_t + \tfrac{1}{2}g''(F_t)(dF_t)^2. \end{aligned} \qquad \text{(VII.D.35)}$$

Now, from (VII.D.33), we have the expression

$$(dF_t)^2 = G_t^2(dt)^2 + 2G_tH_t(dW_t)dt + H_t^2(dW_t)^2,$$

of which the first term is $O((dt)^2)$, the second term is $O((dt)^{3/2})$, and the third term is $\sim H_t^2dt$. Thus, as noted above, $(dF_t)^2 \sim H_t^2dt$, and (VII.D.34) follows. A rigorous proof of this result is found, for example, in Lipster and Shiryayev (1977).

(2) The particular case of (VII.D.35) in which $g(x) = x^2$, $G_t = 0$, and H_t is a deterministic function (say ϕ_t) is the same as (VI.D.44). Since all g satisfying the assumptions of the proposition are locally quadratic, the quadratic case is really the quintessential case.

(3) On combining (VII.D.33) and (VII.D.34), we see that

$$dJ_t = [g'(F_t)G_t + \tfrac{1}{2}g''(F_t)H_t^2]dt + g'(F_t)H_t\,dW_t. \qquad \text{(VII.D.36)}$$

Thus J_t is also a process of the form (VII.D.33), where G_t is replaced by $[g'(F_t)G_t + \tfrac{1}{2}g''(F_t)H_t^2]$ and H_t by $g'(F_t)H_t$. So we see that the class of processes of the form (VII.D.34) is closed under smooth (memoryless) nonlinear transformations. This class of processes (which includes the diffusions) is sometimes known as the class of *Ito processes*.

Now let us return to the problem at hand, namely, the evolution of the conditional variance P_t. Before we can apply Ito's formula to $(\hat{X}_t)^2$ we must first put $d\hat{X}_t$ in the form (VII.D.33). Noting that

$$dI_t = dZ_t - \widehat{h(X_t,t)}dt = [h(X_t,t) - \widehat{h(X_t,t)}]dt + R_t^{1/2}dV_t,$$

the filtering equation for $\hat{X}_t$ becomes

$$d\hat{X}_t = \{\widehat{m(X_t,t)} + [\widehat{X_t h(X_t,t)} - \hat{X}_t\widehat{h(X_t,t)}] \times R_t^{-1}[h(X_t,t) - \widehat{h(X_t,t)}]\}dt + [\widehat{X_t h(X_t,t)} - \hat{X}_t\widehat{h(X_t,t)}]R_t^{-1/2}dV_t.$$

Since V_t is a Wiener process with $\{V_t - V_s; t \geqslant s\}$ independent of all past quantities, Ito's formula can be applied to yield

$$d\hat{X}_t^2 = 2\hat{X}_t d\hat{X}_t + [\widehat{X_t h(X_t,t)} - \hat{X}_t\widehat{h(X_t,t)}]^2 R_t^{-1}dt. \qquad \text{(VII.D.37)}$$

On combining (VII.D.30), (VII.D.32), and (VII.D.37) and rearranging terms, we obtain the following stochastic differential equation for the conditional variance of X_t:

$$dP_t = [2\widehat{(X_t - \hat{X}_t)m(X_t, t)} + \widehat{\sigma^2(X_t, t)}$$

$$- [\widehat{X_t h(X_t, t)} - \hat{X}_t \widehat{h(X_t, t)}]^2 R_t^{-1}]dt \qquad \text{(VII.D.38)}$$

$$+ [\widehat{(X_t - \hat{X}_t)^2 h(X_t, t)} - P_t \widehat{h(X_t, t)}] R_t^{-1} dI_t, \; t > 0,$$

with $P_0 = Var(X_0)$.

In general, this variance equation is quite complicated. For linear observations $(h(X_t, t) = C_t X_t)$, this equation simplifies somewhat to

$$dP_t = [2\widehat{(X_t - \hat{X}_t)m(X_t, t)} + \widehat{\sigma^2(X_t, t)} + C_t^2 P_t^2 R_t^{-1}]dt$$
(VII.D.39)
$$+ \widehat{(X_t - \hat{X}_t)^3} R_t^{-1} dI_t.$$

Note the quadratic term $C_t^2 P_t^2 R_t^{-1} dt$ on the right-hand side of (VII.D.39). This term also appears in the Riccati equation (VII.C.62) for the error variance in the Kalman-Bucy model. If we assume that the drift term is also linear $[m(X_t, t) = A_t X_t]$, then (VII.D.39) reduces further to

$$dP_t = [2A_t P_t + \widehat{\sigma^2(X_t, t)} + C_t^2 P_t^2 R_t^{-1}]dt$$
(VII.D.40)
$$+ \widehat{(X_t - \hat{X}_t)^3} R_t^{-1} dI_t.$$

[The term $2A_t P_t$ in (VII.D.40) also appears in the Kalman-Bucy variance equation]. Note that although the filter equation (VII.D.31) under the assumption of linear observations and drift is identical in form to the Kalman-Bucy filter equation, (VII.D.40) points to an important difference in these two situations; namely, the variance P_t in (VII.D.31) satisfies (VII.D.40), so it generally depends on the observations.

Equation (VII.D.40) can be simplified further by assuming that $\sigma(x,t)$ does not depend on x, say $\sigma(x,t)=B_t$. Then (VII.D.40) reduces to

$$dP_t = [2A_t P_t + B_t^2 + C_t^2 P_t^2 R_t^{-1}]dt + \widehat{(X_t - \hat{X}_t)^3} R_t^{-1} dI_t . \qquad \text{(VII.D.41)}$$

Thus, we see that even within a completely linear model, the conditional variance P_t is still observation dependent due to the innovation term containing the conditional third central moment $\widehat{(X_t - \hat{X}_t)^3}$. However, if we make a final assumption that the initial condition X_0 is Gaussian, then all processes in the model are jointly Gaussian and X_t becomes conditionally Gaussian given Z_0^t. Since the Gaussian distribution is symmetric about its mean, all of its odd central moments are zero. This implies in particular that $\widehat{(X_t - \hat{X}_t)^3}=0$ and

$$dP_t = [2A_t P_t + B_t^2 + C_t^2 P_t^2 R_t^{-1}]dt , \qquad \text{(VII.D.42)}$$

which is the Kalman-Bucy variance equation.

VII.D.2 A Derivation of the Nonlinear Filtering Equations

We now present a brief derivation of the equation (VII.D.28):

$$d\widehat{f(X_t)} = \widehat{\mathbf{A}_t f(X_t)}dt + [\widehat{f(X_t) h(X_t,t)} - \widehat{f(X_t)}\widehat{h(X_t,t)}] R_t^{-1} dI_t ,$$

which describes the evolution of the MMSE estimator of $f(X_t)$ from Z_0^t. As noted above, particularizing this equation to $f(X_t)=\delta(x-X_t)$ yields the equation (VII.D.24) describing the evolution of the conditional density. In the process of deriving (VII.D.28), we will also find an equation that yields the same information as the density equation (VII.D.24) but with a simpler form. For the sake of simplicity, we assume in the following that $R_t = 1$ for all $t \geqslant 0$. The modification for general R_t is straightforward.

Our objective is to find an expression for the conditional expectation $E\{f(X_t)|Z_0^t\}$. To do this we begin by considering the conditional statistics of Z_0^t given X_0^t and then use Bayes formula to convert to the desired conditioning. Note that if we know the sample path of the diffusion up to time t, say $X_0^t = x_0^t$, then conditioned on this knowledge Z_0^t is simply a known waveform in an additive Wiener process; i.e.,

$$dZ_s = h(x_s, s)ds + dV_s, \quad 0 \leqslant s \leqslant t. \quad \text{(VII.D.43)}$$

Assuming that $\int_0^t h^2(x_s, s)ds < \infty$ (which will always be true if h is bounded and will be true w.p.1 if h is continuous in both variables), we can write a probability density for Z_0^t with respect to Wiener measure. In particular, this density will be the likelihood ratio between the hypothesis that Z_0^t is a standard Wiener process versus the alternative that Z_0^t is given by (VII.D.43). This likelihood ratio, which we will denote by Λ_t, is given by the Cameron-Martin formula as

$$\Lambda_t(x_0^t; Z_0^t) = \exp\{\int_0^t h(x_s, s)dZ_s - \tfrac{1}{2}\int_0^t h^2(x_s, s)ds\}.$$

Suppose that we let $P_{X_0^t}$ denote the probability measure[†] associated with X_0^t. Then, by Bayes formula, it follows that

† Note: X_0^t and Z_0^t are random elements in the space of continuous functions on [0, t], which has a natural σ-algebra [see Dunford and Schwartz (1958)]. The measure $P_{X_0^t}$ and the measures involved in defining Λ_t are measures on this space.

$$\widehat{f(X_t)} = \frac{\int f(X_t)\Lambda_t(x_0^t; Z_0^t)P_{X_0^t}(dx_0^t)}{\int \Lambda_t(x_0^t; Z_0^t)P_{X_0^t}(dx_0^t)}$$

(VII.D.44)

$$\triangleq \frac{N_t(Z_0^t)}{D_t(Z_0^t)},$$

where $N_t(Z_0^t)$ and $D_t(Z_0^t)$ denote the numerator and denominator, respectively, of the middle term of this equation.

We will derive an equation for the evolution of $f(X_t)$ by first finding equations for $N_t(Z_0^t)$ and $D_t(Z_0^t)$, and then combining them. Toward this end, we give the following generalization of the Ito differentiation formula to vector processes.

Proposition VII.D.2: The Vector Ito Formula

Suppose that $\{\underline{W}_t; t \geqslant 0\}$ is a k-dimensional vector of independent standard Wiener processes and that $\{\underline{F}_t; t \geqslant 0\}$ is an m-dimensional vector of random processes generated by the equation

$$d\underline{F}_t = \underline{G}_t\,dt + \mathbf{H}_t\,d\underline{W}_t,\ 0 \leqslant t \leqslant T, \qquad \text{(VII.D.45)}$$

where $\underline{G}_t$ is an m-vector of random processes and $\mathbf{H}_t$ is an $m \times n$ matrix of random processes such that $\{\underline{W}_t - \underline{W}_s; t \geqslant s\}$ is independent of $(\underline{G}_0^s, \mathbf{H}_0^s, \underline{W}_0^s)$ for all $s \geqslant 0$, and that each component of $\underline{G}_t$ and each element of $\mathbf{H}_t$ is absolutely integrable on $[0, T]$ with probability one. [The meaning of (VII.D.45) is that the $j^{\underline{th}}$ component of $\underline{F}_t$ is given by $F_t^{(j)} = F_0^{(j)} + \int_0^t G_s^{(j)} ds + \sum_{k=1}^m \int_0^t H_s^{(j,k)} dW_s^{(k)}$, $j = 1, 2, ..., m$]. Suppose further that g is a real-valued function on $\mathbb{R}^m$ with continuous second partials. Then

$$dg(\underline{F}_t) = \sum_{j=1}^{m} g_{x_j}(\underline{F}_t) dF_t^{(j)}$$

$$+ \tfrac{1}{2} \sum_{i=1}^{m} \sum_{j=1}^{m} g_{x_i, x_j}(\underline{F}_t) d < F^{(i)}, F^{(j)} >_t, \qquad \text{(VII.D.46)}$$

where $g_{x_i}(\underline{x}) = \partial g(\underline{x}) / \partial x_i$, $g_{x_i, x_j}(\underline{x}) = \partial^2 g(\underline{x}) / \partial x_i \partial x_j$, and

$$d < F^{(i)}, F^{(j)} >_t = (\mathbf{H}_t \mathbf{H}_t^T)_{i,j} dt \qquad \text{(VII.D.47)}$$

with $< F^{(i)}, F^{(j)} >_0 = 0$.

Remarks

(1) The proof of this result is similar to that of Proposition VII.D.1, after one expands $dg(\underline{F}_t) \sim g(\underline{F}_{t+dt}) - g(\underline{F}_t)$ in a vector Taylor series. This is perhaps more obvious if we rewrite (VII.D.46) as

$$dg(\underline{F}_t) = [\nabla g(\underline{F}_t)]^T d\underline{F}_t + \tfrac{1}{2} tr\{\mathbf{D}g(\underline{F}_t) \mathbf{H}_t \mathbf{H}_t^T\} dt$$

$$= [\nabla g(\underline{F}_t)]^T d\underline{F}_t + \tfrac{1}{2} tr\{\mathbf{H}_t^T \mathbf{D}g(\underline{F}_t) \mathbf{H}_t^T\} dt, \qquad \text{(VII.D.48)}$$

where ∇g is the gradient of g, $\mathbf{D}g$ is the matrix of second partials of g, and $tr\{\cdot\}$ denotes the trace operation (i.e., $tr\,\mathbf{A} = \sum_{i=1}^{n} A_{i,i}$).

If we expand

$$dg(\underline{F}_t) \sim g(\underline{F}_{t+dt}) - g(\underline{F}_t) \sim [\nabla g(\underline{F}_t)]^T (\underline{F}_{t+dt} - \underline{F}_t)$$

$$+ \tfrac{1}{2}(\underline{F}_{t+dt} - \underline{F}_t)^T \mathbf{D}g(\underline{F}_t)(\underline{F}_{t+dt} - \underline{F}_t)$$

$$\sim [\nabla g(\underline{F}_t)]^T d\underline{F}_t + \tfrac{1}{2} d\underline{F}_t^T \mathbf{D}g(\underline{F}_t) d\underline{F}_t$$

and use (VII.D.45), we have

$$dg(\underline{F}_t) \sim [\nabla g(\underline{F}_t)]^T d\underline{F}_t + \tfrac{1}{2}\underline{G}_t^T \mathbf{D}g(\underline{F}_t)\underline{G}_t (dt)^2$$

$$+ \underline{G}_t^T \mathbf{D}g(\underline{F}_t)\mathbf{H}_t (d\underline{W}_t) dt \qquad \text{(VII.D.49)}$$

$$+ \tfrac{1}{2}(d\underline{W}_t)^T \mathbf{H}_t^T \mathbf{D}g(\underline{F}_t)\mathbf{H}_t d\underline{W}_t .$$

Note that as dt vanishes, the only significant terms in (VII.D.49) are those of $O(dt)$. Thus the second and third terms on the right-hand side vanish for infinitesimal dt. The fourth term on the right-hand side is $\tfrac{1}{2}\sum_{i=1}^m \sum_{j=1}^m (\mathbf{M}_t)_{ij} dW_t^{(i)} dW_t^{(j)}$, where $\mathbf{M}_t = \mathbf{H}_t^T \mathbf{D}g(\underline{F}_t)\mathbf{H}_t$. We know that $(dW_t^{(i)})^2 \sim dt$. Since the components of W_t are independent it can be shown that $dW_t^{(i)} dW_t^{(j)} \sim o(dt)$. Thus this fourth term is asymptotically $\tfrac{1}{2}\sum_{i=1}^m (\mathbf{M}_t)_{ii} dt = \tfrac{1}{2} tr(\mathbf{H}_t^T \mathbf{D}g(\underline{F}_t)\mathbf{H}_t) dt$, and (VII.D.48) follows.

(2) The process $<F^{(i)}, F^{(j)}>_t$ defined in (VII.D.47) is known as the *quadratic variation* process of $F_t^{(i)}$ and $F_t^{(j)}$. For a single Ito process $dF_t = G_t dt + H_t dW_t$, the quadratic variation is given simply by $d<F, F>_t = H_t^2 dt$. For the vector case, if the components are uncoupled in the sense that H_t is a diagonal matrix, then $d<F^{(i)}, F^{(i)}>_t = (H_{ii})^2 dt$ and $d<F^{(i)}, F^{(j)}>_t = 0$ for $i \neq j$.

Let us now return to the problem of finding $d\widehat{f(X_t)}$. We begin by considering the process $M_t \triangleq f(X_t)\Lambda_t(X_0^t; Z_0^t)$. We can write this process as $f(X_t)e^{F_t}$, where

$$dF_t = h(X_t, t)dZ_t - \tfrac{1}{2}h^2(X_t, t)dt$$
$$= \tfrac{1}{2}h^2(X_t, t)dt - h(X_t, t)dV_t . \qquad \text{(VII.D.50)}$$

Also, recall that X_t satisfies the diffusion equation

$$dX_t = m(X_t, t)dt + \sigma(X_t, t)dW_t . \qquad \text{(VII.D.51)}$$

Thus since V_t and W_t are independent Wiener processes, $\begin{pmatrix} F_t \\ X_t \end{pmatrix}$

is a vector Ito process of the form (VII.D.45) with $m = k = 2$,

$$\underline{G}_t = \begin{bmatrix} m(X_t, t) \\ \tfrac{1}{2}h^2(X_t, t) \end{bmatrix}, \text{ and } \mathbf{H}_t = \begin{bmatrix} \sigma(X_t, t) & 0 \\ 0 & -h(X_t, t) \end{bmatrix}.$$

We can write $M_t = g(X_t, F_t)$, where $g(x_1, x_2) = f(x_1)e^{x_2}$, so we can apply the vector Ito differential rule to M_t. Noting that

$$\nabla g(x) = \begin{pmatrix} f'(x_1)e^{x_2} \\ f(x_1)e^{x_2} \end{pmatrix} \text{ and } \mathbf{Dg}(x) = \begin{pmatrix} f''(x_1)e^{x_2} & f'(x_1)e^{x_2} \\ f'(x_1)e^{x_2} & f(x_1)e^{x_2} \end{pmatrix}$$

and that

$$d\langle X, F\rangle_t = d\langle F, X\rangle_t = 0,$$

$$d\langle X, X\rangle_t = \sigma^2(X_t, t)dt,$$

$$d\langle F, F\rangle_t = h^2(X_t, t)dt,$$

we then have (assuming that f'' is continuous)

$$\begin{aligned} dM_t &= f'(X_t)e^{F_t}dX_t + f(X_t)e^{F_t}dF_t \\ &+ \tfrac{1}{2}f''(X_t)e^{F_t}\sigma^2(X_t, t)dt \qquad \text{(VII.D.52)} \\ &+ \tfrac{1}{2}f(X_t)e^{F_t}h^2(X_t, t)dt. \end{aligned}$$

Equation (VII.D.52) can be rewritten using

$$\Lambda_t = e^{F_t},$$

$$M_t = f(X_t)\Lambda_t,$$

and

$$dF_t = h(X_t, t)dZ_t - \tfrac{1}{2}h^2(X_t, t)dt$$

to yield an explicit expression in terms of X_0^t and Z_0^t:

$$d[f(X_t)\Lambda_t] = f'(X_t)\Lambda_t dX_t + f(X_t)h(X_t, t)\Lambda_t dZ_t + \tfrac{1}{2}f''(X_t)\sigma^2(X_t, t)\Lambda_t dt. \quad \text{(VII.D.53)}$$

For a functional $\alpha(x_0^t)$, let $\overline{\alpha(X_0^t)}$ denote the integral $\int \alpha(x_0^t)\Lambda_t(x_0^t; Z_0^t)P_{X_0^t}(dx_0^t)$. [Note that $\overline{\alpha(X_0^t)}$ is a function of Z_0^t.] Thus, for example, $N_t(Z_0^t) = \overline{f(X_t)}$ and $D_t(Z_0^t) = \overline{1}$. Applying this linear operation to both sides of (VII.D.53), we obtain an equation for $dN_t(Z_0^t)$, namely,

$$dN_t(Z_0^t) = \overline{f'(X_t)dX_t} + [\overline{f(X_t)h(X_t, t)}]dZ_t + \tfrac{1}{2}[\overline{f''(X_t)\sigma^2(X_t, t)}]dt. \quad \text{(VII.D.54)}$$

Consider the first term, $\overline{f'(X_t)dX_t}$. This term is the expectation under the unconditional distribution of X_0^t of $f'(X_t)\Lambda_t(X_0^t; Z_0^t)dX_t$ with Z_0^t fixed (not *conditioned on* Z_0^t). Since $dX_t = m(X_t, t)dt + \sigma(X_t, t)dW_t$ and dW_t is independent of all the past, we see that $\overline{f'(X_t)dX_t} = \overline{f'(X_t)m(X_t, t)}dt$. Thus, (VII.D.54) becomes

$$dN_t(Z_0^t) = \overline{[f'(X_t)m(X_t,t) + ½f''(X_t)\sigma^2(X_t,t)]}dt$$

$$+ \overline{f(X_t)h(X_t,t)}dZ_t$$

$$= \overline{\mathbf{A}_t f(X_t)}dt + \overline{f(X_t)h(X_t,t)}dZ_t \quad \text{(VII.D.55)}$$

$$= [\overline{\mathbf{A}_t f(X_t)} + \overline{f(X_t)h(X_t,t)}h(X_t,t)]dt$$

$$+ \overline{f(X_t)h(X_t,t)}dV_t.$$

Since $D_t(Z_0^t)$ is the special case of $N_t(Z_0^t)$ with $f(x)=1$, it follows immediately that

$$dD_t(Z_0^t) = \overline{h(X_t,t)}h(X_t,t)dt + \overline{h(X_t,t)}dV_t. \quad \text{(VII.D.56)}$$

Since V_t is a Wiener process, we see from (VII.D.55) and (VII.D.56) that (N_t/D_t) forms a vector Ito process with $m=2, k=1$,

$$\underline{G}_t = \begin{pmatrix} \overline{\mathbf{A}_t f(X_t)} + \overline{f(X_t)h(X_t,t)}h(X_t,t) \\ \overline{h(X_t,t)}h(X_t,t) \end{pmatrix},$$

and

$$\mathbf{H}_t = \begin{pmatrix} \overline{f(X_t)h(X_t,t)} \\ \overline{h(X_t,t)} \end{pmatrix}.$$

From (VII.D.44), we have that $\widehat{f(X_t)}=N_t(Z_0^t)/D_t(Z_0^t)= g(N_t,D_t)$, where $g(x_1,x_2)=x_1/x_2$. On taking partials we have

$$\nabla g(\underline{x}) = \begin{pmatrix} 1/x_2 \\ \\ -x_1/x_2^2 \end{pmatrix} \text{ and } \mathbf{D}g(\underline{x}) = \begin{pmatrix} 0 & -1/x_2^2 \\ -1/x_2^2 & 2x_1/x_2^3 \end{pmatrix}.$$

Also,

$$d <N, D>_t = d <D, N>_t = [\overline{f(X_t)h(X_t, t)}]\ [\overline{h(X_t, t)}]dt ,$$

$$d <N, N>_t = [\overline{f(X_t)\, h(X_t, t)}]^2 dt ,$$

and

$$d <D, D>_t = [\overline{h(X_t, t)}]^2 dt .$$

Combining the above to apply the vector Ito formula† yields

$$d\widehat{f(X_t)} = \frac{1}{D_t} dN_t - \frac{N_t}{D_t^2} dD_t$$

$$- \frac{1}{D_t^2}[\overline{f(X_t)h(X_t, t)}][\overline{h(X_t, t)}]dt \qquad \text{(VII.D.57)}$$

$$+ \frac{N_t}{D_t^3}[\overline{h(X_t, t)}]^2 dt$$

Now using (VII.D.55) and (VII.D.56), we have

† Note that ∇g and $\mathbf{D}g$ are discontinuous at $x_2=0$. However, D_t is positive with probability one (for reasons that will be discussed below), so this discontinuity can effectively be ignored.

$$d\widehat{f(X_t)} = \left[\frac{\overline{\mathrm{A}_t f(X_t)}}{D_t} - \frac{[\overline{f(X_t)h(X_t,t)}]}{D_t}\frac{[\overline{h(X_t,t)}]}{D_t} + \frac{N_t}{D_t}\left[\frac{\overline{h(X_t,t)}}{D_t}\right]^2 \right] dt \qquad \text{(VII.D.58)}$$

$$+ \left[\frac{\overline{f(X_t)h(X_t,t)}}{D_t} - \frac{N_t}{D_t}\frac{\overline{h(X_t,t)}}{D_t} \right] dZ_t .$$

We now note that for any random variable of the form $g(X_t)$, we have $\widehat{g(X_t)} = \overline{g(X_t)}/D_t$. Using this, the fact that $\widehat{f(X_t)} = N_t/D_t$, and the definition of the innovation ($dI_t = dZ_t - \hat{h}dt$), (VII.D.58) reduces to

$$d\widehat{f(X_t)} = \widehat{\mathrm{A}_t f(X_t)}dt \qquad \text{(VII.D.59)}$$

$$+ [\widehat{f(X_t)h(X_t,t)} - \widehat{f(X_t)}\widehat{h(X_t,t)}]dI_t ,$$

which is the filtering equation (VII.D.28) with $R_t \equiv 1$. The initial condition $f(X_0)$ can be obtained from (VII.D.44) by noting that $\Lambda_0 \equiv 1$.

Remarks

(1) The Unnormalized Conditional Density. From (VII.D.55) the quantity $\overline{f(X_t)}$ satisfies the evolution equation

$$d\overline{f(X_t)} = \overline{\mathrm{A}_t f(X_t)}dt + \overline{f(X_t)h(X_t,t)}dZ_t , \qquad \text{(VII.D.60)}$$

which is somewhat simpler than the evolution equation for $\widehat{f(X_t)}$. Since the equation for $\widehat{f(X_t)}$ leads to the equation for the evolution of $q_{X_t}(x)$, this suggests that we might use (VII.D.60) to find another "density" related to $q_{X_t}(x)$ that obeys a simpler equation than that for $q_{X_t}(x)$. In particular, suppose that we define a function

$$\rho_{X_t}(x) = \overline{\delta(x - X_t)}$$

$$\equiv \int \delta(x - x_t)\Lambda_t(x_0^t; Z_0^t)P_{X_0^t}(dx_0^t), \tag{VII.D.61}$$

where δ is the Dirac delta function. [Note that $\rho_{X_t}(x)$ is a function of Z_0^t]. Then we have that

$$q_{X_t}(x) = \widehat{\delta(x - X_t)} = \frac{\overline{\delta(x - X_t)}}{N_t} = \frac{\rho_{X_t}(x)}{N_t}.$$

Moreover, it is easy to see that

$$\overline{f(X_t)} = \int_{-\infty}^{\infty} f(x)\rho_{X_t}(x)dx, \tag{VII.D.62}$$

so that $N_t \triangleq \overline{1} = \int_{-\infty}^{\infty} \rho_{X_t}(x)dx$ and

$$q_{X_t}(x) = \frac{\rho_{X_t}(x)}{\int_{-\infty}^{\infty} \rho_{X_t}(x)dx}. \tag{VII.D.63}$$

From (VII.D.63) we see that for each t, ρ_{X_t} is an *unnormalized conditional density* for X_t given Z_0^t, in the sense that it has the same shape as q_{X_t} but perhaps does not integrate to unity. Since q_{X_t} can be obtained from ρ_{X_t}, the latter quantity is an equally useful quantity to compute.

Combining (VII.D.60) and (VII.D.62), we have

$$d \int_{-\infty}^{\infty} f(y)\rho_{X_t}(y)dy$$

$$= [\int_{-\infty}^{\infty} \mathbf{A}_t f(y)\rho_{X_t}(y)dy]\, dt$$

$$+ [\int_{-\infty}^{\infty} f(y)h(y,t)\rho_{X_t}(y)dy]\, dZ_t \qquad \text{(VII.D.64)}$$

$$= \int_{-\infty}^{\infty} f(y)\mathbf{A}_t^* \rho_{X_t}(y)\, dt$$

$$+ [\int_{-\infty}^{\infty} f(y)h(y,t)\rho_{X_t}(y)dy] dZ_t .$$

Letting $f(y)$ approach $\delta(x-y)$, we obtain the *Zakai equation*:

$$d\,\rho_{X_t}(x) = \mathbf{A}_t^* \rho_{X_t}(x)dt + h(x,t)\rho_{X_t}(x)dZ_t . \qquad \text{(VII.D.65)}$$

For comparison, recall that $q_{X_t}(x)$ evolves according to (for $R_t \equiv 1$)

$$dq_{X_t}(x) = \mathbf{A}_t^* q_{X_t}(x)dt$$

$$+ q_{X_t}(x)[h(x,t) - \widehat{h(X_t,t)}][dZ_t - \widehat{h(X_t,t)}dt]. \qquad \text{(VII.D.66)}$$

Comparing the evolution equations for ρ_{X_t} and q_{X_t}, we see that the former equation is the much simpler of the two. In particular, since $\widehat{h(X_t,t)} = \int_{-\infty}^{\infty} h(x,t)q_{X_t}(x)dx$, it follows that (VII.D.66) is nonlinear in q_{X_t} through the terms $q\hat{h}dZ$, $qh\hat{h}dt$, and $q(\hat{h})^2dt$. Equation (VII.D.65), on the other hand, is linear in ρ_{X_t} and thus is much simpler analytically. [Note that (VII.D.65) is not a linear differential equation since the product of ρ_{X_t} and dZ_t appears on the right-hand side.

However, this is equation is *bilinear*; i.e., it is linear in each of ρ_{X_t} and dZ_t with the other fixed.] Moreover, (VII.D.65) is much simpler computationally. In order to propagate ρ_{X_t} numerically at a given x one needs only to know ρ_{X_t}, ρ'_{X_t}, and ρ''_{X_t} at x. These are all *local* properties that can be computed from knowledge of ρ_{X_t} in a neighborhood of x. In contrast, numerical propagation of q_{X_t} at a particular x requires computation of the integral $\int_{-\infty}^{\infty} h(x,t) q_{X_t}(x) dx$, which is a global quantity, requiring knowledge of q_{X_t} for all x.

The unnormalized density ρ_{X_t} has a further role in nonlinear filtering in that the issue of finite dimensionality of optimum filters can be explored through the evolution equation for ρ_{X_t}. This point is discussed below.

(2) The Estimator-Correlator Formula for Signals Generated by Diffusions. Consider the signal-detection problem described by the hypothesis pair

$$H_0: dZ_t = dV_t, \quad 0 \leqslant t \leqslant 1$$

versus (VII.D.67)

$$H_1: dZ_t = S_t dt + dV_t, \quad 0 \leqslant t \leqslant 1,$$

where the signal S_t is a function $h(X_t, t)$ of a diffusion process $\{X_t; 0 \leqslant t \leqslant 1\}$, and $\{V_t; 0 \leqslant t \leqslant 1\}$ is a Wiener process independent of $\{X_t; 0 \leqslant t \leqslant 1\}$. Assume that $\int_0^1 S_t^2 dt < \infty$ with probability one.

Suppose that we take observations up to some time $t \in (0, 1)$. Conditioned on $X_0^t = x_0^t$, the likelihood ratio for (VII.D.67) is given from the Cameron-Martin formula as

$$\exp\{\int_0^t h(x_u, u) dZ_u - 1/2 \int_0^t h^2(x_u, u) du\},$$

which is $\Lambda_t(x_0^t; Z_0^t)$. Thus the unconditioned likelihood ratio based on observations up to time t, say L_t, is found by averaging $\Lambda_t(x_0^t; Z_0^t)$ over the distribution of X_0^t; i.e.,

$$L_t = \int \Lambda_t(x_0^t; Z_0^t) P_{X_0^t}(dx_0^t) = \overline{1} = D_t(Z_0^t),$$

where D_t is the denominator term from (VII.D.44). So, using (VII.D.56), L_t satisfies the stochastic differential equation

$$dL_t = \overline{h(X_t, t)} dZ_t = \overline{S}_t dZ_t = \hat{S}_t L_t dZ_t, \quad \text{(VII.D.68)}$$

where we have used the fact that $\hat{S}_t = \overline{S}_t / D_t$.

Now consider $\log L_1$. Writing

$$dL_t = \hat{S}_t L_t S_t dt + \hat{S}_t L_t dV_t$$

and using Ito's formula with $g(x) = \log(x)$, we have

$$\begin{aligned} d \log L_t &= \frac{1}{L_t} dL_t + \tfrac{1}{2} \left(\frac{-1}{L_t^2} \right) (\hat{S}_t L_t)^2 dt \\ &= \hat{S}_t dZ_t - \tfrac{1}{2} (\hat{S}_t)^2 dt . \end{aligned} \quad \text{(VII.D.69)}$$

Converting (VII.D.69) to integral form yields

$$\log L_t = \log L_0 + \int_0^t \hat{S}_u dZ_u - \tfrac{1}{2} \int_0^t [\hat{S}_u]^2 du. \quad \text{(VII.D.70)}$$

Since $L_0 = 1$ and $L_1 = dP_1 / dP_0 (Z_0^1)$, (VII.D.70) implies that

$$\frac{dP_1}{dP_0}(Z_0^1) = \exp\left\{ \int_0^1 \hat{S}_t dZ_t - \tfrac{1}{2} \int_0^1 (\hat{S}_t)^2 dt \right\}, \quad \text{(VII.D.71)}$$

which is the estimator-correlator formula for the likelihood ratio given in Section VI.D.

Note that we have derived (VII.D.71) without exploiting directly any properties of the innovations process $dI_t = dZ_t - \hat{S}_t dt$. As noted in Section VI.D, the desirable properties of the innovations in fact follow essentially from the likelihood ratio formula (VII.D.71). Of course, (VII.D.71) follows for more general signal processes then those generated by diffusions; however, the validity of this detection formula was first recognized for the diffusion case [see, e.g., Duncan (1968) or Stratonovich and Sosulin (1965)].

(3) Finite-Dimensionality in Nonlinear Filtering. As we have noted before, the equations for the evolution of the conditional density (normalized or unnormalized) are infinite dimensional recursions. The possibility of obtaining finite-dimensional optimum estimation filters in the nonlinear model rests essentially on the possibility of obtaining a finite-dimensional set of quantities which can be propagated independently of other quantities and which form a sufficient statistic for the conditional density. This in fact is what happens in the Gaussian Kalman-Bucy model, in which all conditional densities are Gaussian and are thus determined by their means and variances. These in turn can be propagated independently of other quantities for the Kalman-Bucy model, as we have seen in (VII.D.31) and (VII.D.42).

The possibility of a finite-dimensional sufficient statistics in nonlinear filtering models can be examined through the unnormalized density equation $d\rho = A_t^* \rho dt + h \rho dZ_t$. It turns out that because this equation is bilinear, the question of finite-dimensional realizability of this equation can be answered through the finite-dimensionality of an associated *Lie algebra* of vector fields. We will not elaborate on this theory here; however, we mention two interesting cases that have been treated in this context. One is the *cubic sensor* problem

$$dX_t = dW_t$$

$$dZ_t = X_t^3 dt + dV_t,$$

which can be shown to inherently infinite dimensional despite its simplicity [see Hazewinkel and Marcus (1982)]; and the so-called *Beneš class* of problems

$$dX_t = m(X_t)dt + dW_t$$

$$dZ_t = X_t dt + dV_t .$$

with $m'(x)+m^2(x)=ax^2+bx+c$ for constants a, b, and c (with $a \geqslant -1$), which turns out to admit a 10-dimensional sufficient statistic [see Beneš (1981)].

The reader interested in details of this aspect of nonlinear filtering is referred to Brockett and Clark (1980) or Marcus (1984).

VII.D.3 Practical Approximations to Optimum Nonlinear Filters

Since the optimum nonlinear filtering equations admit a finite-dimensional implementation only in some special circumstances, it is of interest to consider finite-dimensional approximations to these equations.

Consider, for example, the equation (VII.D.30) for the evolution of the direct state estimator:

$$d\hat{X}_t = \widehat{m(X_t, t)}dt + [\widehat{X_t h(X_t, t)} - \hat{X}_t \widehat{h(X_t, t)}]R_t^{-1}dI_t$$

$$= \widehat{m(X_t, t)}dt + [\widehat{X_t h(X_t, t)} - \hat{X}_t \widehat{h(X_t, t)}]R_t^{-1}dZ_t$$

$$- [\widehat{X_t h(X_t, t)} - \hat{X}_t \widehat{h(X_t, t)}]R_t^{-1}\widehat{h(X_t, t)}dt .$$

A basic difficulty with this equation is the appearance on the right-hand side of the quantities $\hat{m}$, $\widehat{Xh}$, and $\hat{h}$. Suppose, however, that the estimation error, $(X_t - \hat{X}_t)$, can be assumed to be small most of the time. Then, assuming that $m(x, t)$ and $h(x, t)$ are smooth functions of x, we can approximate $m(X_t, t)$ and $h(X_t, t)$ with

$$m(X_t, t) \cong m(\hat{X}_t, t) + (X_t - \hat{X}_t) m'(\hat{X}_t, t) \qquad \text{(VII.D.72)}$$

and

$$h(X_t, t) \cong h(\hat{X}_t, t) + (X_t - \hat{X}_t) h'(\hat{X}_t, t), \qquad \text{(VII.D.73)}$$

where $m'(x,t) = \partial m(x,t)/\partial x$ and $h'(x,t) = \partial h(x,t)/\partial x$. Now, taking conditional expectations in (VII.D.72) and (VII.D.73), and noting that $E\{X_t - \hat{X}_t | Z_0^t\} = 0$, we have that

$$\widehat{m(X_t, t)} \cong m(\hat{X}_t, t) \qquad \text{(VII.D.74)}$$

and

$$\widehat{h(X_t, t)} \cong h(\hat{X}_t, t). \qquad \text{(VII.D.75)}$$

Also, similarly considering $Xh - \hat{X}\hat{h}$, we arrive at

$$\widehat{X_t h(X_t, t)} - \hat{X}_t \widehat{h(X_t, t)} = \overline{(X_t - \hat{X}_t) h(X_t, t)}$$
$$\text{(VII.D.76)}$$
$$\cong \widehat{(X_t - \hat{X}_t)^2} h'(\hat{X}_t, t) = P_t h'(\hat{X}_t, t),$$

where P_t is the conditional variance of X_t given Z_0^t. On combining (VII.D.74) through (VII.D.76), we arrive at an approximate expression for the evolution of the conditional mean; namely,

$$d\hat{X}_t \cong m(\hat{X}_t, t)dt + P_t h'(\hat{X}_t, t) R_t^{-1}$$
$$\text{(VII.D.77)}$$
$$\times [dZ_t - h(\hat{X}_t, t)dt].$$

Note that the right-hand side of (VII.D.77) involves only dZ_t, $\hat{X}_t$, and P_t. Thus if we could compute P_t, this approximate filter could be implemented. Consider the propagation equation (VII.D.38) for the variance:

$$dP_t = [2\widehat{(X_t - \hat{X}_t)m(X_t, t)} + \widehat{\sigma^2(X_t, t)}$$

$$- [\widehat{X_t h(X_t, t)} - \hat{X}_t \widehat{h(X_t, t)}]^2 R_t^{-1}]dt \qquad \text{(VII.D.78)}$$

$$+ [\widehat{(X_t - \hat{X}_t)^2 h(X_t, t)} - P_t \widehat{h(X_t, t)}]R_t^{-1} dI_t .$$

If we apply the approximations above to this equation and additionally approximate $\sigma^2(x, t)$ by a first-order Taylor series, we get

$$dP_t \cong [2P_t m'(\hat{X}_t, t) + \sigma^2(\hat{X}_t, t) - P_t^2[h'(\hat{X}_t, t)]^2 R_t^{-1}]dt$$

$$\text{(VII.D.79)}$$

$$+ \widehat{(X_t - \hat{X}_t)^3} h'(\hat{X}_t, t) R_t^{-1}[dZ_t - h(\hat{X}_t, t)dt].$$

Unfortunately, the right-hand side of (VII.D.79) involves not only dZ_t, $\hat{X}_t$, and P_t, but also involves the conditional third central moment, $\widehat{(X_t - \hat{X}_t)^3}$. If we repeat the approximation procedure above to obtain an equation for this quantity, we would find it depending on the conditional fourth central moment, and so on. Thus, even with these approximations we would still get an infinite set of equations. However, if we make the further assumption that the error $(X_t - \hat{X}_t)$ is symmetrically distributed about its mean value of zero (which is true, for example, if the errors are Gaussian), then $\widehat{(X_t - \hat{X}_t)^3} = 0$ and (VII.D.77) and (VII.D.79) become a closed set of equations:

$$d\hat{X}_t = m(\hat{X}_t, t)dt + P_t h'(\hat{X}_t, t) R_t^{-1}$$

$$\text{(VII.D.80)}$$

$$\times [dZ_t - h(\hat{X}_t, t)]dt$$

and

$$\dot{P}_t = 2P_t m'(\hat{X}_t, t) + \sigma^2(\hat{X}_t, t)$$
$$- P_t^2[h'(\hat{X}_t, t)]^2 R_t^{-1}. \qquad \text{(VII.D.81)}$$

The filter described by (VII.D.80) and (VII.D.81) is known as the *extended Kalman filter*, and it is a very commonly used approximation to the optimum nonlinear filter. This filter is depicted in Fig. VII.D.1. Note that unlike the Kalman filter, this filter requires feeding of the state estimates into the gain computation filter. The following example illustrates the use of these equations.

Example VII.D.3: Phase Tracking

One inherently nonlinear estimation problem that arises frequently in applications is that of phase tracking, in which we have noisy observations of a sinusoid from which we wish to estimate the sinusoid's phase. In particular, consider the observation model

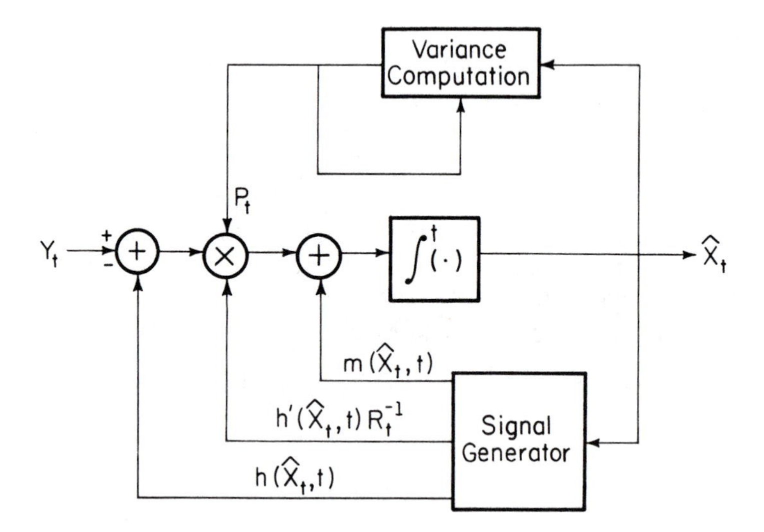

Fig. VII.D.1: The extended Kalman filter.

$$dZ_t = C \sin(\omega_0 t + X_t)dt + \left[\frac{N_0}{2}\right]^{1/2} dV_t, \quad \text{(VII.D.82)}$$

where C and ω_0 are a known constant amplitude and frequency, respectively, and the noise spectral height $N_0/2$ is constant. This model corresponds to $h(x,t) = A \sin(\omega_0 t + x)$, and the extended Kalman filtering equations for this case thus become

$$d\hat{X}_t = m(\hat{X}_t, t)dt + (2CP_t/N_0)\cos(\omega_0 t + \hat{X}_t) \times [dZ_t - C \sin(\omega_0 t + \hat{X}_t)dt] \quad \text{(VII.D.83)}$$

and

$$\dot{P}_t = 2P_t m'(\hat{X}_t, t) + \sigma^2(\hat{X}_t, t) - 2C^2 P_t^2 \cos^2(\omega_0 t + \hat{X}_t)/N_0. \quad \text{(VII.D.84)}$$

The phase diffusion X_t is sometimes assumed to arise from first-order time-invariant linear model,

$$dX_t = -aX_t dt + dW_t, \quad \text{(VII.D.85)}$$

where $a > 0$. In this case the extended Kalman filtering equations reduce to

$$d\hat{X}_t = -a\hat{X}_t + (2CP_t/N_0)\cos(\omega_0 t + \hat{X}_t) \times [dZ_t - C \sin(\omega_0 t + \hat{X}_t)dt], \quad \text{(VII.D.86)}$$

and

$$\dot{P}_t = -2aP_t + 1$$
$$- 2C^2 P_t^2 \cos^2(\omega_0 t + \hat{X}_t)/N_0. \qquad \text{(VII.D.87)}$$

Equation (VII.D.86) is essentially a low-pass filter ($d\hat{X}_t = -a\hat{X}_t\,dt$) with 3-dB bandwidth 2a driven by the nonlinear feedback

$$(2CP_t/N_0)\cos(\omega_0 t + \hat{X}_t)[dZ_t - \sin(\omega_0 t + \hat{X}_t)dt].$$

Typically, the phase bandwidth (which is also 2a) is much smaller than the carrier frequency ω_0. Note that in this case, the term $\cos(\omega_0 t + \hat{X}_t)\sin(\omega_0 t + \hat{X}_t) = \frac{1}{2}\sin(2\omega_0 t + 2\hat{X}_t)$ represents a high-frequency signal that will not pass the estimation filter. Thus the estimator equation becomes approximately,

$$d\hat{X}_t = -a\hat{X}_t + (2CP_t/N_0)\cos(\omega_0 t + \hat{X}_t)dZ_t. \qquad \text{(VII.D.88)}$$

Similarly, the term $2C^2P_t^2\cos^2(\omega_0 t + \hat{X}_t)/N_0$ in the variance equation (VII.D.87) equals $C^2P_t^2[1+\cos(2\omega_0 t + 2\hat{X}_t)]/N_0$, the second term of which will not pass the filter $\dot{P}_t = -2aP_t$ if $a << \omega_0$. Thus the variance equation reduces to

$$\dot{P}_t = -2aP_t + 1 - C^2P_t^2/N_0, \qquad \text{(VII.D.89)}$$

which, incidentally, is the Riccati equation for Kalman-Bucy filtering in this model with $h(x,t)$ replaced by Cx. Assuming this variance achieves a steady-state value, P_∞, the steady-state phase tracker becomes

$$d\hat{X}_t = -a\hat{X}_t + K_\infty\cos(\omega_0 t + \hat{X}_t)dZ_t, \qquad \text{(VII.D.90)}$$

where $K_\infty = 2CP_\infty/N_0$.

The nonlinear filter of (VII.D.90) (with the usual interpretation of "$dZ_t = Y_t\,dt$") is depicted in Fig. VII.D.2. This filter is known as a *phase-locked loop* and similar filters

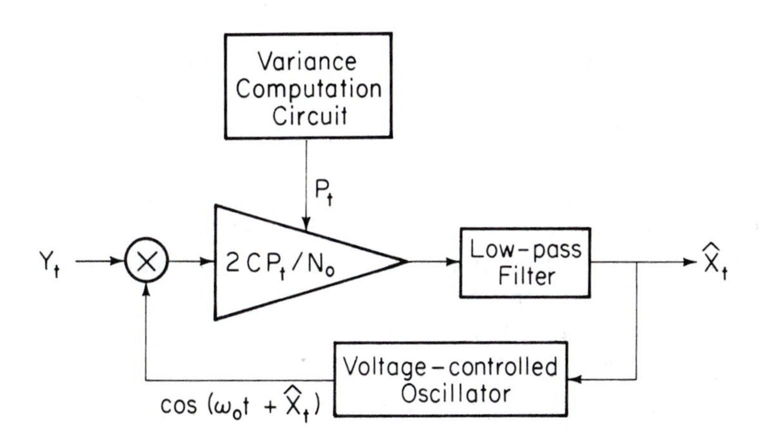

Fig. VII.D.2: Phase-locked Loop: An approximation to an optimum phase tracker.

are widely used in practice. Note that the steady-state performance of this loop is approximately P_∞, which is the positive root of the equation

$$0 = -2aP_\infty + 1 - C^2P_\infty^2/N_0;$$

i.e.,

$$P_\infty = \frac{aN_0}{C^2}\left(\left[1 + \frac{C^2}{aN_0}\right]^{1/2} - 1\right).$$

The quantity C^2/aN_0 is a measure of signal-to-noise ratio (SNR) in the loop since the average power of $C\sin(\omega_0 t + X_t)$ is proportional to C^2, and the white noise of spectral height $N_0/2$ when passed through a filter with 3-dB bandwidth 2a will have an approximate average output power of aN_0. As C^2/aN_0 increases, we have $P_\infty \sim [aN_0/C^2]^{1/2}$; i.e., the steady-state tracking accuracy decreases as the inverse of the square root of the SNR. Note that the high-SNR case is the one in which the performance of the extended Kalman filter should be closest to that of the MMSE nonlinear filter since $(X_t - \hat{X}_t)$ should be small in this case.

The reader interested in learning more about phase-locked loops and their behavior is referred to the book by Viterbi (1968).

The extended Kalman filter is a useful approximation to the optimum nonlinear filter for many applications. However, if the estimation error is not small enough on the average, this filter gives a poor approximation to the optimum one. Better approximations can be obtained by using higher-order Taylor series expansions in approximating the nonlinearities.

The next step up from the extended Kalman filter is the *second-order filter*, which (as its name implies) uses second-order approximations for these nonlinearities. In particular, on expanding m as

$$m(X_t, t) \cong m(\hat{X}_t, t) + (X_t - \hat{X}_t)m'(\hat{X}_t, t) + \tfrac{1}{2}(X_t - \hat{X}_t)^2 m''(\hat{X}_t, t),$$

and h similarly, we obtain the following second-order approximations to quantities in the filtering equation:

$$\widehat{m(X_t, t)} \cong m(\hat{X}_t, t) + \tfrac{1}{2}P_t m''(\hat{X}_t, t),$$

$$\widehat{h(X_t, t)} \cong h(\hat{X}_t, t) + \tfrac{1}{2}P_t h''(\hat{X}_t, t),$$

and

$$\widehat{(X_t - \hat{X}_t)h(X_t, t)} \cong [P_t h'(\hat{X}_t, t) + \tfrac{1}{2}\widehat{(X_t - \hat{X}_t)^3} h''(\hat{X}_t, t)].$$

On repeating the assumption that $\widehat{(X_t - \hat{X}_t)^3} = 0$, the evolution equation for $\hat{X}_t$ becomes (to second-order)

$$d\hat{X}_t \cong [m(\hat{X}_t, t) + \tfrac{1}{2}P_t m''(\hat{X}_t, t)]dt$$

$$+ P_t h'(\hat{X}_t, t)R_t^{-1}[dZ_t - [h(\hat{X}_t, t) - \tfrac{1}{2}P_t h''(\hat{X}_t, t)]dt]. \quad \text{(VII.D.91)}$$

We again need an equation for P_t to complete the picture. Applying the procedure above [including setting $\widehat{(X_t - \hat{X}_t)^3} = 0$] to the evolution equation for P_t and using

$$\widehat{(X_t - \hat{X}_t)^2 h(X_t, t)} - P_t \widehat{h(X_t, t)}$$

$$= \widehat{(X_t - \hat{X}_t)^2 [h(X_t, t) - \widehat{h(X_t, t)}]}$$

$$\cong \tfrac{1}{2}\widehat{(X_t - \hat{X}_t)^4} h'(\hat{X}_t, t),$$

we obtain

$$dP_t \cong \{2P_t m'(\hat{X}_t, t) + \sigma^2(\hat{X}_t, t)$$

$$+ P_t[\sigma'(\hat{X}_t, t)]^2 + P_t \sigma(\hat{X}_t, t)\sigma''(\hat{X}_t, t)$$

$$- P_t^2[h'(\hat{X}_t, t)]^2 R_t^{-1}\}dt \quad \text{(VII.D.92)}$$

$$+ \tfrac{1}{2}\widehat{(X_t - \hat{X}_t)^4} h''(\hat{X}_t, t)R_t^{-1}$$

$$\times [dZ_t - (h(\hat{X}_t, t) + \tfrac{1}{2}P_t h''(\hat{X}_t, t))dt].$$

We now must contend with the conditional fourth central moment $\widehat{(X_t - \hat{X}_t)^4}$. If we try to derive a second-order approximate evolution equation for this quantity, it will involve the conditional fifth and sixth central moments. The symmetric-error assumption gets rid of the fifth moment, but computation of the sixth moment will involve the eighth moment, and so on. Thus we must make some additional assumption to break this coupling to higher-order moments. Note that if the error $(X_t - \hat{X}_t)$ were Gaussian, this coupling would be broken since the fourth and second central moments of a Gaussian random variable are related by

$$\widehat{(X_t - \hat{X}_t)^4} = 3P_t^2. \tag{VII.D.93}$$

In the absence of any more realistic assumption, we arbitrarily assume that (VII.D.93) holds approximately and the equation for P_t thus reduces to

$$\begin{aligned} dP_t \cong \{ & P_t[2m'(\hat{X}_t, t) + \sigma(\hat{X}_t, t)\sigma''(\hat{X}_t, t) + (\sigma'(X_t, t))^2] \\ & + \sigma^2(\hat{X}_t, t) - P_t^2[h'(\hat{X}_t, t)]^2 R_t^{-1}\} dt \\ & + \frac{3}{2} P_t^2 h''(\hat{X}_t, t) R_t^{-1} \\ & \times \{dZ_t - (h(\hat{X}_t, t) + \tfrac{1}{2} P_t h''(\hat{X}_t, t))dt\}. \end{aligned} \tag{VII.D.94}$$

Equations (VII.D.92) and (VII.D.94) are now a closed set of equations representing a second-order approximation to the nonlinear filtering equation, with the additional assumptions that $\widehat{(X_t - \hat{X}_t)^3} \cong 0$ and $\widehat{(X_t - \hat{X}_t)^4} \cong 3P_t^2$. This filter is illustrated in Fig. VII.D.3. Note that this filter feeds not only the state estimate $\hat{X}_t$ into the gain computation but also the direct observation, Y_t.

As an example, consider the phase-tracking problem of Example VII.D.3, with phase model $dX_t = -aX_t + dW_t$. Assuming that $0 < a \ll \omega_0$, the filter equation in this case is of the same approximate form as before; i.e.,

$$d\hat{X}_t \cong -a\hat{X}_t dt + (2CP_t / N_0)\cos(\omega_0 t + \hat{X}_t) dZ_t. \tag{VII.D.95}$$

However, the variance equation now becomes

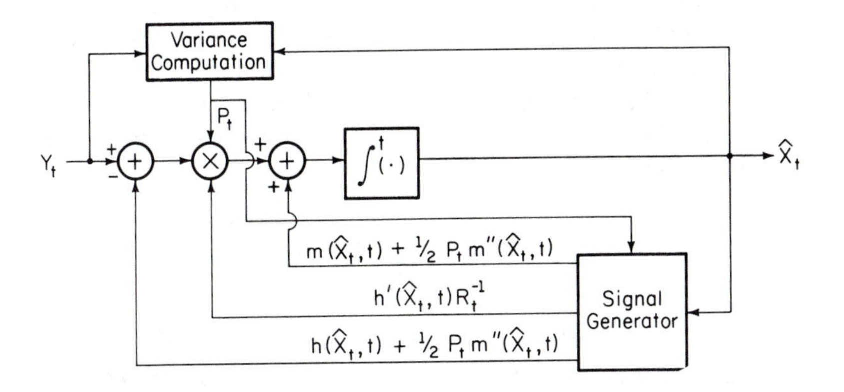

Fig. VII.D.3: The second-order filter.

$$dP_t \cong \left[2aP_t + 1 - \frac{C^2P_t^2}{2N_0}(5 + 3/2P_t)\right]dt$$

$$- \frac{3P_t^2}{N_0}C \sin(\omega_0 t + \hat{X}_t)dZ_t . \qquad \text{(VII.D.96)}$$

Thus the second-order version of the phase estimator in this case has the same form as the extended Kalman filter except that the gain computation in the second-order filter is data-dependent.

Before leaving the subject of approximate nonlinear filters, we mention the fact that the linear Kalman-Bucy filter can also be though of as an approximation to an optimum non-linear filter in some situations. To see this, suppose that we have a deterministic function $x_t^{(o)}, t \geq 0$, that satisfies a non-linear differential equation

$$dx_t^{(o)} = m(x_t^{(o)}, t)dt , \quad t \geq 0. \qquad \text{(VII.D.97)}$$

For example, $\{x_t^{(o)}; t \geq 0\}$ might represent the nominal motion or trajectory in one dimension of some vehicle such as an aircraft, and the function $m(x_t^{(o)}, t)$ might represent the dynamics of the vehicle as well as the effects of deterministic controls applied to the vehicle. Suppose that the actual trajectory $\{X_t; t \geq 0\}$ is perturbed from the nominal trajectory $\{x_t^{(o)}; t \geq 0\}$ because of random disturbances (such as turbulence in the case of an aircraft). We can model its behavior by a diffusion

$$dX_t = m(X_t, t)dt + \sigma_t dW_t, \qquad \text{(VII.D.98)}$$

where the term $\sigma_t dW_t$ represents the random disturbances.

Assume access to the usual noisy observations process

$$dZ_t = h(X_t, t)dt + R_t^{1/2} dV_t, \qquad \text{(VII.D.99)}$$

where, as before, $\{W_t\}$ and $\{V_t\}$ are assumed to be independent standard Wiener processes. Estimating X_t from Z_0^t is equivalent to estimating δX_t from $\tilde{Z}_0^t$, where δX_t is the *deviation* of X_t from the nominal trajectory, i.e.,

$$\delta X_t \triangleq X_t - x_t^{(o)}, t \geq 0,$$

and $\tilde{Z}_t$ is the deviation of the observations from $h(x_t^{(o)}, t)$;

$$d\tilde{Z}_t = dZ_t - h(x_t^{(o)}, t)dt.$$

Using the evolution equation for $\hat{X}_t$ and the fact that $x_t^{(o)}$ is deterministic, it is easy to see that

$$d\widehat{(\delta X_t)} = [\widehat{m(X_t,t)} - m(x_t^{(o)}t)]dt$$

$$+ Cov(\delta X_t, h(X_t,t) - h(x_t^{(o)},t) | Z_0^t) R_t^{-1} \quad \text{(VII.D.100)}$$

$$\times [d\tilde{Z}_t - [h(X_t,t) - h(x_t^{(o)},t)]dt].$$

Now suppose we assume that the deviations in the actual trajectory from the nominal trajectory are small. Then we can use the approximations

$$m(X_t, t) - m(x_t^{(o)}, t) \cong m'(x_t^{(o)}, t)\delta X_t$$

and

$$h(X_t, t) - h(x_t^{(o)}, t) \cong h'(x_t^{(o)}, t)\delta X_t,$$

to yield the approximate estimator equation

$$d\widehat{(\delta X_t)} \cong m'(x_t^{(o)}, t)\widehat{\delta X_t}$$

$$+ P_t h'(x_t^{(o)}, t) R_t^{-1} \quad \text{(VII.D.101)}$$

$$\times [d\tilde{Z}_t - h'(x_t^{(o)}, t)\widehat{\delta X_t} dt].$$

Note that implementation of (VII.D.101) requires only the conditional variance $P_t = Var(X_t | Z_0^t) \equiv Var(\delta X_t | Z_o^t)$ and the observation $d\tilde{Z}_t$. Consider the evolution equation for P_t. Applying the above approximation to (VII.D.38) yields the approximate variance equation

$$dP_t \cong [2P_t m'(x_t^{(o)}, t) + \sigma_t^2 - [P_t h'(x_t^{(o)}, t)]^2 R_t^{-1}]dt$$

$$+ \widehat{(\delta X_t - \widehat{\delta X_t})^3} R_t^{-1}[d\tilde{Z}_t - h'(x_t^{(o)}, t)\widehat{\delta X_t} dt]. \qquad \text{(VII.D.102)}$$

Again invoking the assumption that the error is symmetrically distributed about its mean, we set $\widehat{(\delta X_t - \widehat{\delta X_t})^3} = 0$ and (VII.D.102) becomes

$$\dot{P}_t \cong 2P_t m'(x_t^{(o)}, t) + \sigma_t^2$$

$$- [P_t h'(x_t^{(o)}, t)]^2 R_t^{-1}. \qquad \text{(VII.D.103)}$$

Note that (VII.D.101) and (VII.D.103) are the equations for optimal filtering in the Kalman-Bucy model with $A_t = m'(x_t^{(o)}, t)$, $B_t = \sigma_t$, $C_t = h'(x_t^{(o)}, t)$, and R_t as is. Thus, the approximate filter is the optimum filter for the model

$$d(\delta X_t) = m'(x_t^{(o)}, t)(\delta X_t)dt + \sigma_t dW_t$$

$$d\tilde{Z}_t = h'(x_t^{(o)}, t)(\delta X_t)dt + R_t^{1/2} dV_t, \qquad \text{(VII.D.104)}$$

which is simply a linearization of (VII.D.98) and (VII.D.99) about the nominal trajectory $x_t^{(o)}$. This type of linearization is in fact the way in which the linear state space model often arises in practice, since the purpose of state estimators derived from such models is frequently for use in regulators that control the process of interest to stay on a nominal state trajectory.

VII.E Exercises

1. Consider the model $Y_t = N_t + \Theta s_t, 0 \leqslant t \leqslant T$, where $\{N_t, 0 \leqslant t \leqslant T\}$ is a zero-mean mean-square continuous Gaussian noise process; $\{s_t, 0 \leqslant t \leqslant T\}$ is a known continuous signal; and Θ is a $N(\mu, \sigma^2)$ random variable independent of the noise.

   (a) Assuming the detection problem

$$H_0: Y_t = N_t, \quad 0 \leqslant t \leqslant T$$

versus

$$H_1: Y_t = N_t + s_t, \quad 0 \leqslant t \leqslant T$$

is nonsingular, find the MMSE estimate of Θ given $\{Y_t, 0 \leqslant t \leqslant T\}$. Find the corresponding MMSE.
(b) Consider in detail the particular case of (a) in which

$$C_N(t,u) = \frac{N_0}{2}\delta(t-u) + \lambda \alpha_t \alpha_u,$$

$$0 \leqslant t, u \leqslant T,$$

where λ and $\{\alpha_t; 0 \leqslant t \leqslant T\}$ are known and satisfy $\int_0^T \alpha_t^2 dt = 1$ and $\lambda > -N_0/2$.
(c) Assuming now that $\{N_t, 0 \leqslant t \leqslant T\}$ is a Wiener process with autocovariance

$$C_N(t,u) = \frac{N_0}{2} \min\{t, u\}$$

and $s_t = -E_0 \sin(\omega_0 t)/\omega_0$ where $E_0 > 0$ and $\omega_0 = n\pi/T$ for integer n, discuss the quantities of (a) as σ^2 varies in relation to $E_0^2 T/N_0$.

2. (a) Consider the observation model

$$Y_t = A\sin(2\pi t + \Phi) + N_t, \quad 0 \leqslant t \leqslant 1,$$

where $\{N_t, 0 \leqslant t \leqslant 1\}$ is a zero-mean Gaussian random process with continuous autocovariance function $C_N(t,u)$, and A is a $N(\mu, \sigma^2)$ random variable independent of $\{N_t, 0 \leqslant t \leqslant 1\}$. Assuming Φ is known, find the MMSE estimate of A from observations $\{Y_t, 0 \leqslant t \leqslant 1\}$.
(b) Repeat (a) assuming instead that $\{N_t, 0 \leqslant t \leqslant 1\}$ is a white Gaussian noise with unit spectral height.
(c) Repeat (b) assuming further that Φ is a random

variable, independent of A and $\{N_t, 0 \leqslant t \leqslant 1\}$, and uniformly distributed on $[0, 2\pi]$.

3. Verify Eq. (VII.C.13).

4. Show that (VII.C.20) solves the integral equation (VII.C.19).

5. Suppose a random process X_t is described by the differential equation

$$\frac{dX_t}{dt} = -aX_t + U_t, \quad t \geqslant 0$$

where a is a positive constant; X_0 is a zero-mean random variable with variance $\sigma_0^2 > 0$; and $\{U_t; t \geqslant 0\}$ is a zero-mean white noise which is independent of X_0 and has autocovariance

$$C_U(t, \tau) = \sigma_U^2 \delta(t - \tau).$$

Suppose further that we observe a random process given by

$$Y_t = X_t + N_t, \quad t \geqslant 0$$

where $\{N_t; t \geqslant 0\}$ is a zero-mean white noise process, independent of X_0 and $\{U_t; t \geqslant 0\}$, with autocovariance

$$C_N(t, \tau) = \sigma_N^2 \delta(t - \tau)$$

(a) Identify the elements of the Kalman filtering problem for the estimation of X_t from Y_0^t.
(b) Derive expressions for the mean squared estimation error and Kalman gain as functions of the parameters a, σ_0^2, σ_U^2, σ_N^2, and the time t. Draw a block diagram of the Kalman-Bucy filter.
(c) Discuss the steady-state case ($t \to \infty$). Consider the

cases: $a^2 >> (\sigma_U^2/\sigma_N^2)$; $a^2 = (\sigma_U^2/\sigma_N^2)$; and $a^2 << (\sigma_U^2/\sigma_N^2)$.

6. Consider the situation in part (b) of Exercise 1 with $\lambda=0$. Let $\hat{\theta}_T$ corresponding estimate of Θ. Show that, as T progresses, $\hat{\theta}_T$ can be computed recursively using the Kalman-Bucy filter. Find the steady-state ($T \to \infty$) Kalman gain and MMSE.

7. Suppose $\{S_t, t \geq 0\}$ and $\{X_t, t \geq 0\}$ are two random processes generated by the differential equations:

$$\frac{d}{dt} S_t = X_t, \quad t \geq 0$$

$$\frac{d}{dt} X_t = 0, \quad t \geq 0$$

with initial condition S_0 and X_0 which are independent $N(\mu, \sigma^2)$ random variables. Suppose further that we observe a process

$$Y_t = S_t + N_t, \quad t \geq 0$$

where $\{N_t; t \geq 0\}$ is a zero-mean Gaussian white noise with spectral height $N_0/2$, independent of S_0 and X_0.
(a) Solve the appropriate Riccati differential equation to find the covariance matrix of the error vector

$$\begin{pmatrix} S_t - \hat{S}_t \\ X_t - \hat{X}_t \end{pmatrix}$$

where $\hat{S}_t = E\{S_t | Y_0^t\}$ and $\hat{X}_t = E\{X_t | Y_0^t\}$.
(b) Suppose that our initial information about S_0 and X_0 is very poor; i.e., suppose that $\sigma^2 \to \infty$. Specify the estimates $\hat{S}_t$ and $\hat{X}_t$.

8. Consider the model

$$\dot{X}_t = -AX_t + U_t, \ t \geqslant 0$$

$$Y_t = C_t X_t + N_t, \ t \geqslant 0$$

where $\{N_t\,; t \geqslant 0\}$ and $\{U_t\,; t \geqslant 0\}$ are independent white Gaussian noises with spectral heights q and r, respectively, and where X_0 is Gaussian and independent of $\{N_t\,; t \geqslant 0\}$ and $\{U_t\,; t \geqslant 0\}$.

(a) Suppose $C_t = 1$ for all t, and A is a random variable taking positive values $a_1, a_2, \ldots, a_m$ with probabilities $p_1, p_2, \ldots, p_m$, respectively. Find the steady-state MMSE estimate of X_t given $\{Y_s\,; 0 \leqslant s \leqslant t\}$.
(b) Suppose $C_t = \sin(\omega_0 t)$ and A is fixed and positive. Find the steady-state MMSE estimate of X_t given $\{Y_s\,; 0 \leqslant s \leqslant t\}$. (Note that this is not a time-invariant filter.)

9. Show that the filter of (VII.C.73) solves the causal Wiener-Hopf equation (VII.C.69).

10. Prove (VII.D.44)

REFERENCES

Anderson, B. D. O., and J. B. Moore (1979), *Optimal Filtering* (Prentice-Hall: Englewood Cliffs, NJ).

Apostol, A. (1974), *Mathematical Analysis, 2nd Ed.* (Addison-Wesley: Reading, MA).

Ash, R. B., and M. F. Gardner (1975), *Topics in Stochastic Processes* (Academic: New York).

Baker, C. R. (1969), "On the Deflection of a Quadratic Linear Test Statistic," *IEEE Trans. Inform. Theory*, vol. IT-15, pp. 16-21.

Beneš, V. E. (1981), "Exact Finite Dimensional Filters for Certain Diffusions with Nonlinear Drift," *Stochastics*, vol. 5, pp. 65-92.

Beneš, V. E. (1987), "Nonlinear Filtering: Problems, Examples, Applications," Chapter 1 in *Advances in Statistical Signal Processing - Volume 1: Estimation*, H. V. Poor, Ed. (JAI Press: Greenwich, CT).

Bierman, G. J. (1977), *Factorization Methods for Discrete Sequential Estimation* (Academic: New York).

Billingsley, P. (1979), *Probability and Measure* (Wiley: New York).

Boekee, D. E., and J. C. Ruitenbeck (1981), "A Class of Lower Bounds on the Bayesian Probability of Error," *Inform. Sciences*, vol. 25, pp. 21-25.

Breiman, L. (1968), *Probability* (Addison-Wesley: Reading, MA).

Brockett, R., and J. M. C. Clark (1980), "The Geometry of the Conditional Density Equation," in *Analysis and Optimization of Stochastic Systems*, O. L. R. Jacobs, Ed. (Academic: New York).

Carlyle, J. W. (1968), "Nonparametric Methods in Detection Theory," Chapter 8 in *Communication Theory*, A. V. Balakrishnan, Ed. (McGraw-Hill: New York).

DeBruijn, N. G. (1961), *Asymptotic Methods in Analysis, 2nd Ed.* (North-Holland: Amsterdam).

Desoer, C. R., (1970), *Notes for a Second Course on Linear Systems* (D. Van Nostrand: Princeton, NJ).

Doob, J. L. (1953), *Stochastic Processes* (Wiley: New York).

Duncan, T. E. (1968), "Evaluation of Likelihood Functions," *Inform. Control*, vol. 13, pp. 62-74.

Duncan, T. E. (1970), "Likelihood Functions for Stochastic Signals in White Noise," *Inform. Control*, vol. 16, pp. 303-310.

Dunford, N., and J. T. Schwartz (1958), *Linear Operators-Part I* (Wiley: New York).

Feldman, J. (1958), "Equivalence and Perpendicularity of Gaussian Process," *Pacific J. Math.*, vol. 8, pp. 699-708.

Feldman, J. (1960), "Some Classes of Equivalent Gaussian Process on an Interval," *Pacific J. Math.*, vol. 10, pp. 1211-1220.

Ferguson, T. S. (1967), *Mathematical Statistics: A Decision Theoretic Approach* (Academic: New York).

Girsanov, J. V. (1960), "On Transforming a Certain Class of Stochastic Process by Absolutely Continuous Substitution of Measures," *Theory Prob. Appl.*, vol. 5, pp. 285-301.

Goodwin G. C., and K. S. Sin (1984), *Adaptive Filtering, Prediction and Control* (Prentice-Hall: Englewood Cliffs, NJ).

Grenander, U. (1981), *Abstract Inference* (Wiley: New York).

Hajek, J. (1958), "On a Property of Normal Distributions of Any Stochastic Process," *Czech. Math. J.*, vol. 8, pp. 610-617.

Hajek, J., and Z. Sidak (1967), *Theory of Rank Tests* (Academic: New York).

Hampel, F. R., *et al.* (1986), *Robust Statistics: The Approach Based on Influence Functions* (Wiley: New York).

Hazewinkel, M., and S. Marcus (1982), "On Lie Algebras and Finite Dimensional Filtering," *Stochastics*, vol. 5, pp. 29-62.

Honig, M. L., and D. G. Messerschmidt (1984), *Adaptive Filters: Structures, Algorithms, and Applications* (Kluwer: Boston).

Huber, P. J. (1965), "A Robust Version of the Probability Ratio Test," *Ann. Math. Stat.*, vol. 36, pp. 1753-1758.

Huber, P. J. (1981), *Robust Statistics* (Wiley: New York).

Huber, P. J., and V. Strassen (1973), "Minimax Tests and the Neyman-Pearson Lemma for Capacities," *Ann. Statist.*, vol. 1, pp. 251-263.

Kailath, T. (1966), "Some Integral Equations with 'Nonrational' Kernels," *IEEE Trans. Inform. Theory*, vol. IT-12, pp. 442-447.

Kailath, T. (1969), "A General Likelihood Ratio Formula for Random Signals in Gaussian Noise," *IEEE Trans. Inform. Theory.*, vol. IT-15, pp. 350-361.

Kailath, T. (1971), "The Structure of Radon Nikodym Derivatives with Respect to Wiener and Related Measures," *Ann. Math. Stat.*, vol. 42, pp. 1054-1067.

Kailath, T. (1981), *Lectures on Wiener and Kalman Filtering* (Springer-Verlag: New York).

Kassam, S. A., and H. V. Poor (1985), "Robust Techniques for Signal Processing: A Survey," *Proc. IEEE*, vol. 73, pp. 433-481.

Kassam, S. A., and J. B. Thomas (1980), *Nonparametric Detection: Theory and Applications* (Dowden, Hutchinson & Ross: Stroudsburg, PA).

Kendall, M. G. (1948), *Rank Correlation Methods* (Griffin: London).

Kendall, M. G., and A. Stuart (1961), *The Advanced Theory of Statistics-Vol. 2* (Hafner: New York).

Kobayashi, H., and J. B. Thomas (1967), "Distance Measures and Related Criteria," *Proc. 5th Ann. Allerton Conf. Circuit and System Theory*, Monticello, IL, pp. 491-500.

Kullback, S. (1959), *Information Theory and Statistics* (Wiley: New York).

Lehmann, E. L. (1983), *Theory of Point Estimation* (Wiley: New York).

Lehmann, E. L. (1986), *Testing Statistical Hypotheses* (Wiley: New York).

Lipster, R. S., and A. N. Shiryayev (1977), *Statistics of Random Processes I: General Theory* (Springer-Verlag: New York).

Ljung, L., and T. Soderstrom (1982), *Theory and Practice of Recursive Identification* (MIT Press: Cambridge, MA).

Lovitt, W. V. (1950), *Linear Integral Equations* (Dover: New York).

Lugannani, R., and S. Rice (1980), "Saddle Point Approximation for the Distribution of the Sum of Independent Random Variables," *Adv. Appl. Prob.*, vol. 12, pp. 475-490.

Lukacs, E. (1960), *Characteristic Functions* (Hafner: New York).

Marcus, S. I. (1984), "Algebraic and Geometric Methods in Nonlinear Filtering," *SIAM J. Control Optimization*, vol. 22, pp. 817-844.

Martin, R. D., and S. C. Schwartz (1971), "Robust Detection of a Known Signal in Nearly Gaussian Noise," *IEEE Trans. Inform. Theory*, vol. IT-17, pp. 50-56.

Mazo, J. E., and J. Salz (1965), "Probability of Error for Quadratic Detectors," *Bell Syst. Tech. J.*, vol. 44, pp. 2165-2186.

Nevel'son, M. B., and R. Z. Has'minskii (1973), *Stochastic Approximation and Recursive Estimation* (American Mathematical Society: Providence, RI).

Noether, G. E. (1955), "On a Theorem of Pitman," *Ann. Math. Stat.*, vol. 26, pp. 64-68.

Oppenheim, A. V., and R. W. Schafer (1975), *Digital Signal Processing* (Prentice-Hall: Englewood Cliffs, NJ).

Papoulis, A. (1986), *Probability, Random Variables and Stochastic Processes* (McGraw-Hill: New York).

Parzen, E. (1962), *Stochastic Processes* (Holden-Day: San Francisco).

Schweppe, F. C. (1965), "Evaluation of Likelihood Functions for Gaussian Signals," *IEEE Trans. Inform. Theory*, vol. IT-11, pp. 61-70.

Shepp, L. A. (1966), "Radon-Nikodym Derivatives of Gaussian Measures," *Ann. Math. Stat.*, vol. 37, pp. 321-353.

Skorohod, A. V. (1974), *Integration in Hilbert Space* (Springer-Verlag: New York).

Slepian, D. (1958), "Some Comments on the Detection of Gaussian Signals in Gaussian Noise," *IRE Trans. Inform. Theory*, vol. IT-4, pp. 65-68.

Stranovich, R. L., and Yu. G. Sosulin (1965), "Optimal Detection of a Diffusion Process in White Noise," *Radio Eng. Electron. Phys.*, vol. 10, pp. 704-713.

Tantaratana, S. (1986), "Sequential Detection of a Positive Signal," Chapter 7 in *Communications and Networks: A Survey of Recent Advances*, I. F. Blake and H. V. Poor, Eds. (Springer-Verlag: New York).

Thomas, J. B. (1971), *An Introduction to Applied Probability and Random Processes* (Wiley: New York).

Thomas, J. B. (1986), *An Introduction to Applied Probability* (Springer-Verlag; New York).

Trench, W. F. (1964), "An Algorithm for the Inversion of Finite Toeplitz Matrices," *J. SIAM*, vol. 12, pp. 515-522.

Van Trees, H. L. (1968), *Detection, Estimation and Modulation Theory-Part I* (Wiley: New York).

Viterbi, A. (1968), *Principles of Coherent Communications* (Wiley: New York).

Wong, E. (1983), *Introduction to Random Processes* (Springer-Verlag: New York).

Wong, E., and B. Hajek (1985), *Stochastic Process in Engineering Systems* (Springer-Verlag: New York).

Yao, K., and R. M. Tobin (1976), "Moment Space Upper and Lower Bounds for Digital Systems with Intersymbol Interference," *IEEE Trans. Inform. Theory*, vol. IT-22, pp. 65-74.

INDEX